RF Electronics for Electronic Warfare

For a complete listing of titles in the
Artech House Electronic Warfare Library,
turn to the back of this book.

RF Electronics for Electronic Warfare

Richard A. Poisel

ARTECH
HOUSE

BOSTON | LONDON
artechhouse.com

Library of Congress Cataloging-in-Publication Data
A catalog record for this book is available from the U.S. Library of Congress.

British Library Cataloguing in Publication Data
A catalogue record for this book is available from the British Library.

Cover design by John Gomes

ISBN 13: 978-1-63081-705-3

685 Canton Street
Norwood, MA 02062

10 9 8 7 6 5 4 3 2 1

This book is dedicated to

Joshua and Loren Poisel, and their children
Harrison and Alexandra.

Brandon Poisel, and his son Jacob..

Table of Contents

Preface

This book is about the techniques and methods for generating RF signals for electronic warfare (EW) applications, and, in particular, for anti-communications purposes. This includes forming the signals from scratch with oscillators of various kinds, modulating those signals with disruptive information (usually noise), and amplifying those signals to levels adequate for transmitting to a target receiver with enough energy to disrupt the successful reception of the target signal by the intended receiver.

Almost all applications of EW technologies used to be military in nature, or at least of interest only for government uses. That has changed somewhat in recent years as the application of wireless communications has exploded into the private sector with cellular communications. The intent in all cases, however, is the same: to prevent the successful transfer of information from a sender (transmitter) to one or more receivers. For military applications this intent is reasonably clear. That is, to disrupt command and control (C2) links to prevent the successful C2 of forces or the transfer of targeting data, for example, from a radar to a missile system. For commercial applications the uses of EW techniques are more dubious, and, for the most part, are probably illegal. We will not delve into these latter applications here—our attention will apply to the former—military counter-C2 applications.

EW is also applicable to other forms of RF spectrum uses, such as radars. While many of the methods and techniques presented herein are applicable to such applications, we will not discuss any of them here, again, focusing on the counter C2 uses of EW.

This book begins, in the introduction chapter, with thoughts that are intended to establish the motivation for discussing EW in the first place. Also contained in the first chapter is a flow diagram of signals within EW systems. This information is to put into perspective the taxonomy of an EW system, and, in particular, where the RF signal generation is performed.

The second chapter discusses the fundamentals of modulation and the modulators used in the modulation process. In general, the carrier (target) signals originate in RF oscillators. While these signals are occasionally useful in and of themselves, normally they must be modulated with some sort of signal that is more effective at counter-C2 that the unmodulated continuous wave (CW) signal. By modulated we mean that some characteristic(s) of the oscillator signal are modified—the characteristics available are the amplitude (power level), frequency, and phase. All of these topics are covered in Chapter 2.

Properties of the aforementioned oscillators, also known in EW parlance as exciters, are presented in Chapter 3. There are various types of oscillators that can be used, both analog and digital. It is safe to say that most modern oscillators, and in particular those used in EW systems, are both tunable (variable frequency) and are digitally tuned.

The signals from oscillators are generally fairly weak and must be boosted in power. This is the function of the power amplifier, although not all uses for amplifiers require high power. Chapter 4 provides an introduction to RF amplifiers where the characteristics discussed are common to both power and non-power amplifiers. The active devices discussed most herein, which provide the amplification of the signals are solid-state. That is, two forms of transistors: bipolar junction transistors (BJT) and metal-oxide-semiconductor field-effect transistors (MOSFET) and their derivatives. Both of these technologies are available in high-power forms over broad frequency ranges.

Characteristics of these two forms of transistors are further explored in Chapter 5. The underlying parameters of the transistor families are discussed and compared. The architectures of amplifiers are presented and comparisons of modern solid-state technologies are provided.

The next two chapters drill down further into BJTs (Chapter 6) and MOSFETs (Chapter 7). Models of the devices are presented and we show how these models are used in analysis and design of amplifiers. The salient properties are discussed to include the various circuit configurations, and when and why a particular architecture is better in some applications than others.

Biasing is often not discussed in depth when active devices are considered. It is not clear if this topic is too mundane to enter into the discussion—the RF performance is more interesting. Or it may be assumed that the topic has been covered elsewhere. In any case, the fact is that biasing should be just as important as any other part of the design process. Therefore, we discuss biasing the active devices in Chapter 8.

RF power amplifiers are considered in Chapter 9. Multistage architectures are the mainstay, and as such, interstage matching is a major consideration, to include matching the output stage to an antenna via transmission lines. The second half of Chapter 9 presents digital configurations of power amplifiers. These architectures are the most modern configurations and present the most efficient topologies available, and efficiency is one of the most important parameters when considering amplifiers for EW applications.

Chapter 10 delves into the stability of power amplifiers. Stability is another principal concern for power amplifier design and implementation. We show how to determine an amplifier's stability and to influence the design to ensure an amplifier is stable.

Another major consideration is the linearity of power amplifiers. Linearity is extremely important in EW applications since there are extensive communication links in most cases. Many of these links are friendly. If an EW amplifier is non-linear, it spills power into unintended channels close by to the targeted channel, many of which are not to be interfered with. Therefore linearity is an important concern. We discuss linearity in depth in Chapter 11.

The energy that is not radiated from an EW system by the intended route—that is, the antenna, is dissipated as heat. The origins of this heat are the active devices within the system. Chapter 12 considerations this heat and its dissipation under the topic of thermal management, and how to deal with it.

The last chapter, Chapter 13, considers the design of power supplies for EW systems, and in particular, power amplifiers.

Chapter 1

Electronic Warfare

1.1 Introduction

Most (all?) modern militaries rely on radio frequency (RF) communications for *command and control* (C2) of tactical forces. Electronic warfare (EW) is a means for exploiting and denying access to such RF communications, while simultaneously protecting the RF spectrum for use by friendly forces.

EW has been a constituent of warfare essentially since RF communications have been used—since late in the 19th century. In fact, if ship-to-ship communications with flags is included (and it can be) then it is older than the usage of RF communications.

RF communications interconnect information systems, the latter of which may be people or machines, or both.

This chapter is structured as follows. We begin with a general overview of electronic warfare, what it is, and where it fits in the operational scheme of warfighting, primarily from the viewpoint of land-based forces. These schemes are extremely oversimplified and are intended simply to put EW in context. We then cover the frequencies normally of concern to military forces and how those frequencies are normally used. Last, a general-purpose model in the form of a flow diagram of an EW system is introduced and the principal components of the model involved with RF signal generation are pointed out.

1.2 Overview of Electronic Warfare

The taxonomy of EW consists of three mutually supportive domains: electronic support, electronic attack, and electronic protect. An overview of EW showing these three components is illustrated in Figure 1.1.

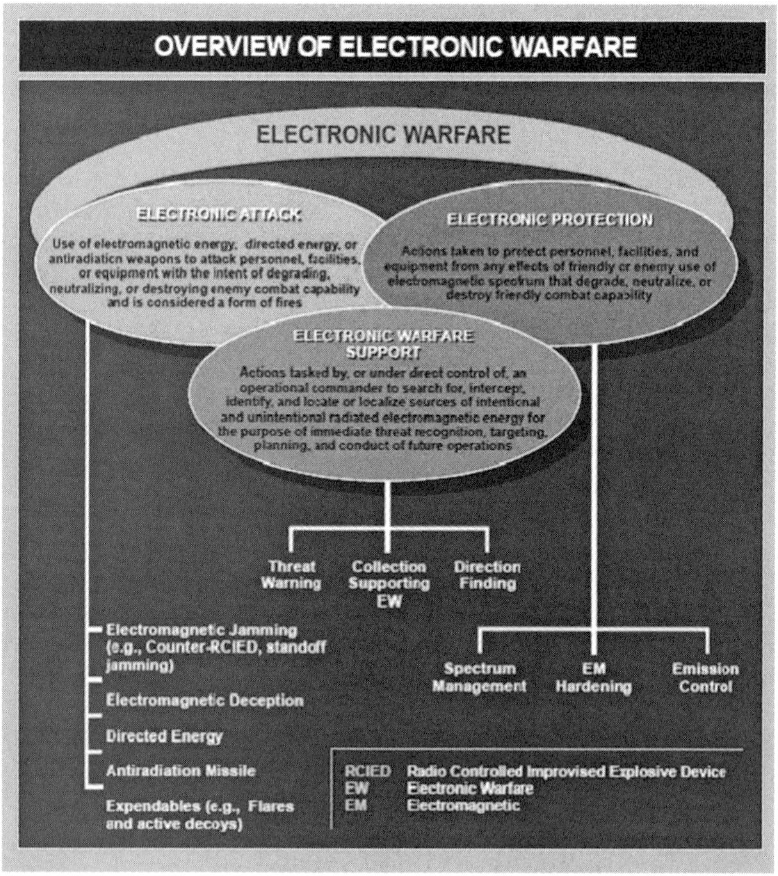

Figure 1.1 Overview of EW. (Source: [1].)

1.2.1 Electronic Support

Electronic support (ES) consists of those operations and actions of sensing the *electromagnetic* (EM) environment for the purpose of detecting and identifying targets of interest, primarily in support of *electronic attack* (EA), but also useful for establishing and maintaining the *electronic order of battle*[1] (EOB).

1.2.2 Electronic Attack

Electronic attack (EA) can consist of many possible actions, from the physical destruction of a piece of equipment or capability so that it is no longer usable without repair or replacement to simply delaying the exchange of information between two points, possibly by jamming the receiver of an RF network.

The possible constituents of EA can be thought of as follows:

- *Destruction*—as mentioned.
- *Divert and distract*—diversion of a target's attention and resources; hide other attacks or delay their discovery; script kiddie attacks while others are working much more subtly.
- *Distort*—Target's content versus conduit; perception management; psychological operations.
- *Control*—feeding misinformation (e.g. mimicking a known signal so well that the receiver cannot distinguish phony signal from real signal); use of various techniques to mask identity of penetrating party into a network or system.
- *Delay*—as mentioned.

1.2.3 Electronic Protect

Electronic protect (EP) consists of those actions and activities that preclude and/or minimize an adversary from conducting EA against friendly forces. Examples of EP are communication encryption and spread spectrum communications.

1.2.4 EW Effects

The following is extracted from [1];

> EW is waged to secure and maintain the use of the EMS for friendly forces and to deny the enemy an actual or perceived advantage in the same. EW can be applied

---

[1] The EOB is the database maintained, and frequently plotted on a map, of the electronic entities being faced in the battlespace.

from the air, land, sea, and space by manned and unmanned systems; and is employed to achieve desired effects of: control, detection, denial, deception, disruption and degradation, destruction, and/or protection.

All EW effects will involve various levels of control; while controlling use of the EMS through proper application of EW is advantageous, when not properly coordinated and integrated, it may adversely affect forces through fratricide or eliminate targets of high intelligence value. For example, an ill-timed jamming package may highlight an otherwise unobserved reconnaissance team or interfere with that team's use of the EMS. EW actions enable commanders to achieve the following effects as defined in joint publication (JP) 3-13.1.

(1) *Control.* Control of the EMS is achieved by effective management/ coordination of friendly systems while countering adversary systems. EA limits adversary use of the EMS; EP secures use of the EMS for friendly forces; and ES enables the commander's accurate estimate of the situation in the operational area. All three must be carefully integrated to be effective.

Additionally, commanders should ensure maximum integration among EW, communications, ISR, and other IO capabilities.

(2) *Detection.* Detection is the active and passive monitoring of the battlespace for EM threats such as RF, EO, laser, IR and ultraviolet threats. This definition acknowledges that the actual threat may be the adversary perpetrating the EM interference, but it is through the use of the EMS that enables friendly forces to detect the adversary. It is the essential first step in EW for effective exploitation, targeting defensive planning, and force protection. Friendly forces must have the capability to detect and characterize interference as hostile (jamming) or nonhostile (intentional or not intentional) EM interference.

(3) *Denial.* Denial is controlling the information an adversary receives via the EMS and preventing the acquisition of accurate information about friendly forces. Denial can be done by traditional jamming techniques, expendable countermeasures, or destructive measures. This definition does not exclude the possibility that future Modular Forces may also have to address other denial capabilities, such as nontraditional jamming and denial of service.

(4) *Deception.* Deception is confusing or misleading an adversary by using some combination of human produced, mechanical, or electronic means. Through the use of the EMS, EW manipulates the adversary's decision loop, making it difficult to establish an accurate perception of objective reality.

(5) *Disruption and Degradation.* Disruption and degradation techniques interfere with the enemy's use of the EMS (such as counter-C2) to limit their combat capabilities. This is achieved with electronic jamming, electronic deception and electronic intrusion. These enhance attacks on hostile forces and act as force multiplier by increasing adversary uncertainty, while reducing uncertainty for friendly forces. Advanced EA techniques offer the opportunity to nondestructively disrupt or degrade adversary infrastructure.

(6) *Destruction.* When used in the EW context, destruction is the elimination of targeted adversary systems. Sensors and C2 nodes are lucrative targets because their destruction seriously hampers the enemy's perceptions and ability to

coordinate actions. ES supports destruction by providing target location or targeting information and combat assessment.

Adversary systems that use the EMS can be destroyed by a variety of weapons and techniques, ranging from conventional munitions to DE weapons. While destruction of adversary equipment is an effective means to deny the adversary use of the EMS, the duration of denial will depend on the adversary's ability to reconstitute.

(7) *Protection.* Protection is the use of physical properties, operational tactics, techniques, and procedures (TTP), as well as planning and employment processes to protect our use of the EMS. This includes ensuring that joint offensive EW activities do not electronically kill, destroy, or degrade our intelligence sensors and information systems. Protection is achieved by component hardening, EMCON, frequency management/deconfliction, and employing other IO and kinetic means to counterattack and defeat adversary attempts to control the EMS.

Frequency management and deconfliction includes the capabilities to detect, characterize, geolocate, and mitigate EM interference that can affect operations. Additionally, organizational structures like a joint commander's electronic warfare staff or EWCC enhance EP through deconfliction of EW efforts. This definition does not attempt to provide an exhaustive list of all intelligence sensors and information systems such as unmanned platforms and C2 links. Even though these specific sensors and systems are not listed in this definition, they must also be protected from our use of the EMS.

We can see from the above that EA permeates throughout the effects desired from EW. Herein we focus specifically on the ability to conduct EA against communication C2 targets, recognizing that according to the above, EW involves considerably more than this limited set.

1.3 Operational Scenarios

An EW intercept operational scenario is shown in Figure 1.2. We will not delve much into this scenario here, as it has been covered extensively elsewhere [3 –12]. We include it here to remind the reader that there are two fundamental functions performed by EW systems, not just jamming but the associated function of intercept and analysis of target signals.

Active jamming operational scenario is shown in Figure 1.3. This is the scenario of primary concern here. A low-level RF signal is generated in the EW system and amplified to levels that are sufficient to transmit toward a target receiver in an attempt to preclude the target transmit signal from being accurately received at the target receiver.

The jamming signal may be, and usually is, modulated with some form of baseband signal. For reasons that make theoretical sense, that modulating signal is usually noise. It has been shown that under reasonable assumptions about the target

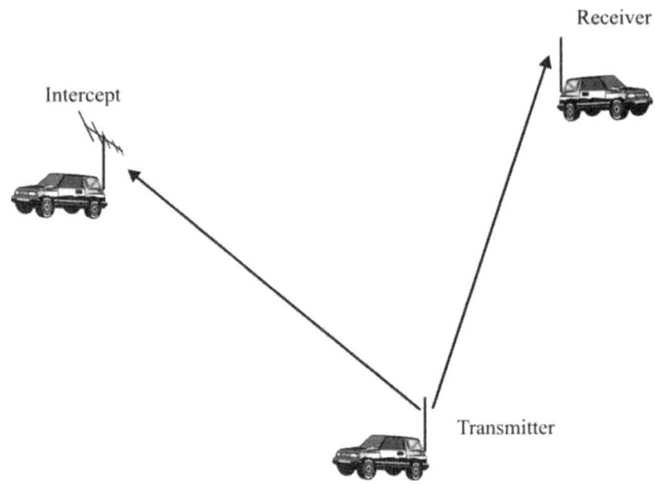

Figure 1.2 EW system intercept operational scenario. (Source: [4]. © 2008 Artech House. Reprinted with permission.)

signals, *additive white Gaussian noise* (AWGN) is the best waveform for jamming many types of target signals [2, 3].

The scenarios illustrated in Figures 1.2 [4] and 1.3 [5] are, of course, greatly simplified. The EA scenario is never stand-alone, whereas there may be

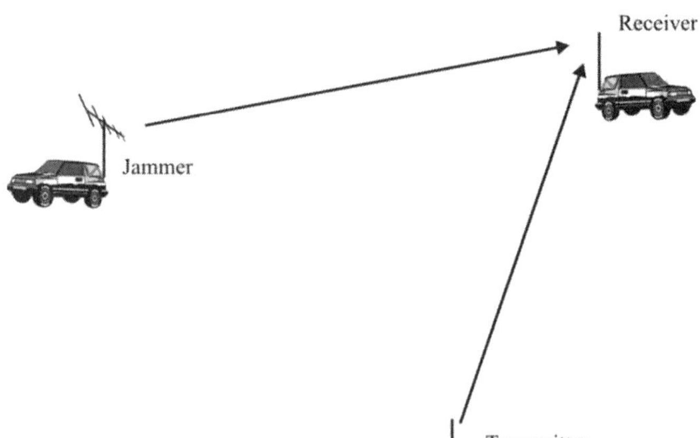

Figure 1.3 EW system jamming operational scenario. (Source: [5]. © 2008 Artech House. Reprinted with permission.)

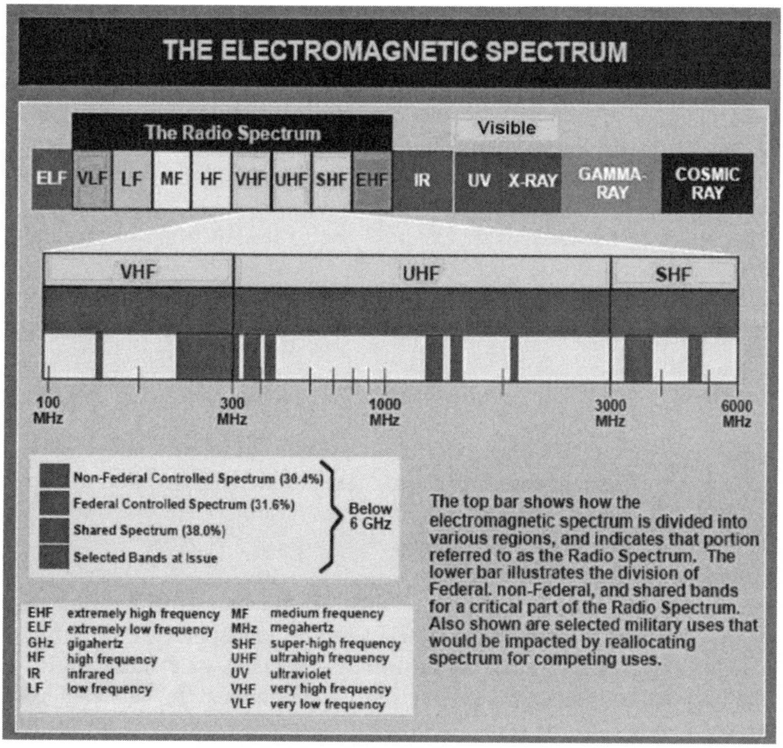

Figure 1.4 The electromagnetic spectrum. (Source: [1].)

circumstances where the ES scenario can be. They are, however, almost always part of a much larger operational scenario.

1.4 EW Target Frequency Ranges

The frequencies of interest to design and analysis of communication EW systems cover a considerable range. We consider the international ranges of interest to tactical ground-based forces here. The list is basically the same for other types of forces with some extensions (medium frequency, low frequency, and very low frequency for the Navy, for example).

A summary of the utilities that require use of the EMS and where their limits are located is illustrated in Figure 1.4.

1.4.1 High Frequency

Range: 3 MHz – 30 MHz

The high-frequency (HF) range of frequencies is used primarily for long-haul communications. Tactical forces can use these frequencies for *line-of-sight* (LOS) C2 *push-to-talk* (PTT) functions, however. Some teletype and facsimile transmissions occur in the HF range as well.

Near-vertical incidence skywave (NVIS) in this range is used for special-purpose short-range communications. NVIS HF signals are propagated essentially straight up toward zenith and are reflected by the ionosphere straight back down; due to scattering it comes back to Earth in a cone that with radius of 50–100 km from the transmitter. One such application is communication with helicopter nap-of-the-Earth flights.

Citizens-band (CB) communications is a cheap, effective, widely used short range communications capability that resides in the upper HF frequency range (around 27 MHz in the U.S.). They are readily applied to nonmilitary but adversarial force situations.

1.4.2 Very High Frequency

Range: 30 MHz – 300 MHz

The low *very high frequency* (VHF) (30–90 MHz) has long been used internationally for LOS C2 PTT communications. During the Cold War this was the principal range used by both Soviet and NATO forces in Europe and elsewhere.

Also contained in this region are C2 for air, both commercial and military—that is, air traffic control.

Commercial FM is located from 88 MHz to 108 MHz (U.S.) and the commercial broadcast TV region for channels 2 through 6 is located between 54 to 88 MHz.

There are several other services provided in the VHF frequency range. It should be noted that while there is some international cooperation on the use of these (and other) frequencies, individual countries generally cooperated in service assignments.

1.4.3 Ultrahigh Frequency

Range: 300 MHz – 3,000 MHz

Perhaps the best-known uses of the ultrahigh frequency (UHF) range are the cellular personal communication systems (in the U.S. 824 MHz–896 MHz, 1,850 MHz–1,990 MHz, 3G: 1,220 MHz–2,155 MHz, 4G: 2,490 MHz–2,690 MHz) and Global Positioning System (GPS):

L1: 1,575.42 MHz—Coarse acquisition (C/A) and encrypted precision [P()Y)] codes

L2: 1,227.60 MHz

There are several other services located in this frequency region.

1.4.4 Superhigh Frequency (SHF)

Range: 3 GHz – 30 GHz

Superhigh frequency (SHF) is used primarily for satellite communications.

1.4.5 Summary

Tactical communications that may be of interest for ES and EA are of several varieties and are used for many purposes. For all but those in the lower frequencies (HF and below) the jammer and target receiver must be able to see each other (radio line-of-sight).

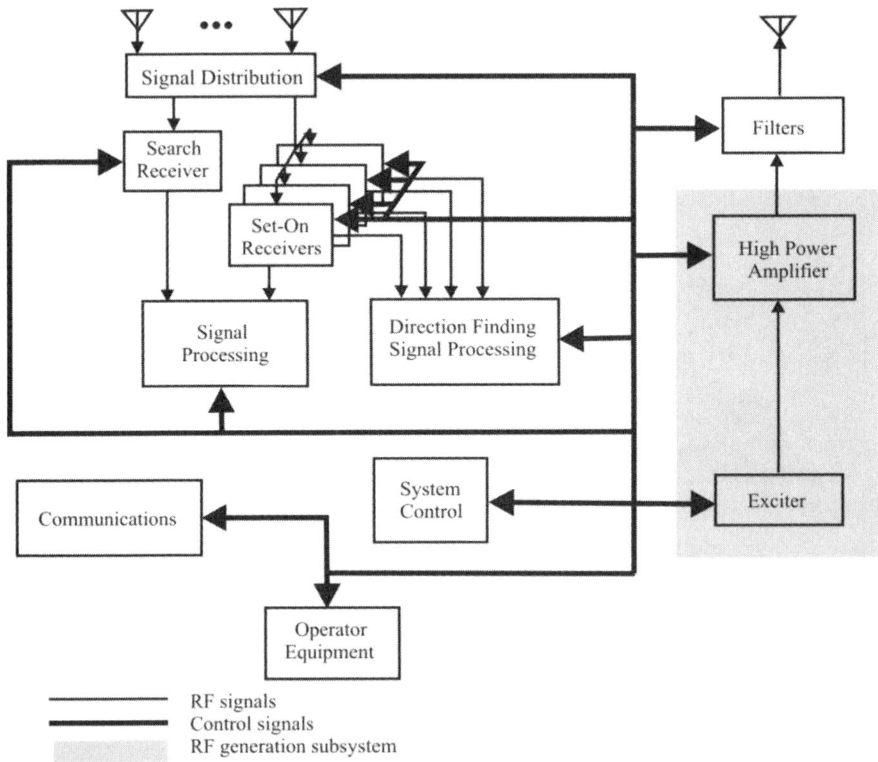

Figure 1.5 EW system flow diagram. (Source: [6]. © 2008 Artech House. Reprinted with permission.)

1.5 EW System Block Diagram

The generalized and simplified flow diagram of a communication EW system is depicted in Figure 1.5 [6]. All of the relevant subsystems are included, but the principal subsystems of interest here are those in the shading. Coverage of the other subsystems shown in Figure 1.5 are contained in [3–12].

1.5.1 RF Signal Generation

In particular, the exciter (sometimes called an oscillator or synthesizer) is where the RF signals originate. They are usually modulated with some form of jamming modulation (noise for example), amplified, usually several times in the *high-power amplifier* (HPA, or just PA), to form the EW RF output signal emanating from the EW system.

Filters are sometimes used at the output of the PA in order to reduce spectral leakage into adjacent communication channels. Aside from the necessity of handling the PA power levels and of usually the requirement to be rapidly tunable, there is nothing special about this filtering so we do not include a discussion of such filtering herein.

Generating RF signals in EW systems is required for a number of purposes, the main one being the jamming signal. That is the signal that is greatly amplified and is intended to be transmitted to a target receiver to interrupt communications.

1.6 Concluding Remarks

We cover several ways RF signals are generated in EW systems in the remainder of this book. This chapter is intended to put EW in context with other entities in a battlespace, and it provides an introduction to where RF signal generation resides in EW systems.

RF signals are formed at a low level in the exciter and are modulated with some form of jamming signal. That composite signal is amplified, usually several times, to form a signal powerful enough that when emitted by the antenna, it can be used to disrupt the target communication signals.

References

[1] TRADOC Pam 525-7-6: Concept Capability Plan for Army Electronic Warfare Operations, August 2007.

[2] Basar, T., "The Gaussian Test Channel with an Intelligent Jammer," *IEEE Transactions on Information Theory,* Vol. IT-29, No. 1, Jan. 1983, pp. 152–157.

[3] Poisel, R. A., *Modern Communication Jamming Principles and Techniques*, 2nd Ed., Norwood, MA: Artech House, 2011.

[4] Poisel, R. A., *Introduction to Communication Electronic Warfare Systems*, 2nd Ed., Norwood, MA: Artech, 2008, Figure 1.2.

[5] Poisel, R. A., *Introduction to Communication Electronic Warfare Systems,* 2nd Ed., Norwood, MA: Artech House, 2008, Figure 1.3.

[6] Poisel, R. A., *Introduction to Communication Electronic Warfare Systems*, 2nd Ed., Norwood, MA: Artech House, 2008, Figure 1.4.

[7] Poisel, R. A. *Antenna Systems and Electronic Warfare Applications,* Norwood, MA: Artech, 2012.

[8] Poisel, R. A. *Target Acquisition in Communication Electronic Warfare Systems*, Norwood, MA: Artech House, 2004.

[9] Poisel, R. A., *Electronic Warfare Target Location Methods*, 2nd Ed., Norwood, MA: Artech House, 2012.

[10] Poisel, R. A., *Introduction to Communications Electronic Warfare Systems*, 2nd Ed., Norwood, MA: Artech House, 2008.

[11] Poisel, R. A., *Foundations of Communications Electronic Warfare*, Norwood, MA: Artech House, 2008.
[12] Poisel, R. A., *Electronic Warfare Receivers and Receiving Systems*, Norwood, MA: Artech House, 2014.

Chapter 2

Modulation and Modulators

2.1 Introduction

Modulation is used in communication systems to impose an information-bearing signal onto a carrier. The carrier is typically much higher in frequency than the modulating signal because the modulating signal is normally not appropriate for efficient transmission. Signals at higher frequencies can be transmitted and received with equipment that is reasonably sized.

The parameters of the carrier that can be changed by imposing modulating signals are its amplitude, frequency, and phase, or combinations of these. Modification of the amplitude is referred to as *amplitude modulation* (AM), while modifying the frequency is referred to as *frequency modulation* (FM). Varying the phase is referred to as *phase modulation* (PM).

Modulators are used in EW systems to impose the modulating jamming signal onto the RF carrier for transmission. Depending on the type of modulation to be imposed on the carrier, the type of modulator used varies. There are two fundamental types of modulation: amplitude and phase. The latter can be further decomposed into frequency modulation, where the instantaneous frequency of the carrier is changed, and phase modulation, where the instantaneous phase of the carrier is changed. In all cases, unless an unmodified tone is the jamming signal, a modulating signal is used as the source of the jamming signal.

Representative modulators and demodulators for AM, FM, and PM will be discussed in this chapter, along with their noise characteristics, and other deviations from ideal behavior.

Modulation converts a baseband signal to a passband counterpart. Figure 2.1 illustrates this process in the frequency domain. From another point of view, modulation varies certain parameters of a carrier (e.g., a sinusoid) according to the baseband signal. These two definitions are equivalent because a passband signal can always be expressed as

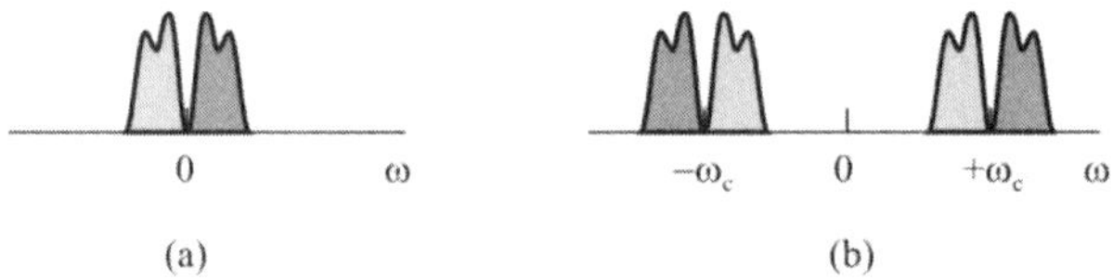

Figure 2.1 AM modulation converts a baseband signal (a) to a higher frequency ω_c illustrated in (b).

$$x(t) = a(t)\cos[\omega_c t + \theta(t)] \tag{2.1}$$

where $a(t)$ and $\theta(t)$ are generally functions of time. In other words, we begin with a periodic carrier signal $A\cos(\omega_c t + \theta)$ and vary its amplitude or phase to perform modulation.

Our primary intent in this chapter is to discuss signal modulators. However, because EW systems are intended to counter receiving systems, we cover the necessary points addressing the reception/detection of signals as well.

This chapter is structured as follows. We first present a discussion of AM modulation and modulators as AM is the oldest form of modulation. We then introduce PM and FM. The characteristics of these two forms of modulation are quite similar and it only makes sense to examine these characteristics together. Some PM and FM high-level reference designs are included. We conclude the chapter with a brief discussion of noise and noise generators as noise is probably the most common form of modulating waveform.

2.1.1 Additive White Gaussian Noise Channel

In order to compare the noise tolerance of different receivers, we assume the transmitted signals power is constant and the received signal is corrupted by AWGN having a psd equal to $N_0/2$ as illustrated in Figure 2.2. The AWGN assumption proves adequate in considering the inherent noise of the receiver – in particular the thermal noise associated with the components (we will see later that the source of noise in electronic circuits is due to the capacitance in the circuit assuming that the circuit is in thermal equilibrium). However, the noise introduced by adjacent

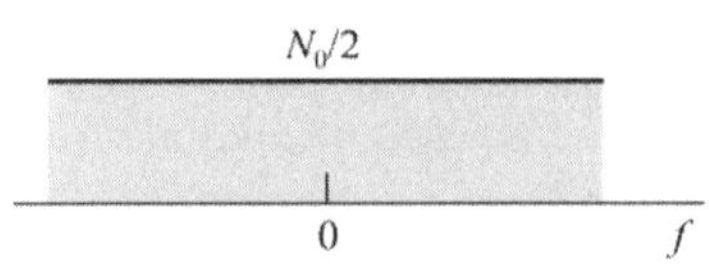

Figure 2.2 Noise power spectral density.

interferers is generally not white and must be considered separately. Furthermore, the noise in some demodulators (e.g., envelope detectors) is better modeled as multiplicative rather than additive.

2.2 Modulations

Modern RF communication signaling is accomplished in one of two ways: in analog form or in digital form. This is, however, misleading because straightforward digital signaling is not performed at RF. That form is limited to computer networking. If the signaling goes over the RF medium it is always done as mentioned above: by modifying one or more of the three parameters of the RF signal. Therefore, to say that the signaling is analog or digital is referring to the modulat*ing* signal rather than the modulat*ed* signal.

Performance is different, however, depending on the form of the baseband (modulating) signal. For example, analog signals can sometimes tolerate more noise than digital signals unless the latter employ error coding to combat the effects of noise.

We address the two forms of modulation first.

2.3 Analog Modulations

2.3.1 Amplitude Modulation

For a baseband signal $s_{BB}(t)$, an AM waveform can be constructed from (2.1) as

$$s_{AM}(t) = s_{BB}(t) A_c \cos \omega_c t, \tag{2.2}$$

Illustrated in Figure 2.3 is a method of generating an AM signal along with the corresponding waveforms and spectra: multiplication by $\cos\omega_c t$ simply translates the spectrum of $s_{BB}(t)$ to the band around $\pm\omega_c$. Mathematically this is given by the convolution (denoted by $\otimes$) of the spectra of $s_{BB}(t)$ with the spectra of $\cos\omega_c t$, the latter of which is a pair of impulse functions located at ω_c as illustrated in Figure 2.3(b). Thus, the bandwidth of $s_{AM}(t)$ is twice that of $s_{BB}(t)$. The signal $A_c \cos \omega_c t$ is generated by a *local oscillator* (LO).

Except for broadcast radios and the video in analog television, AM finds limited use in modern wireless systems. This is because carrying information in the amplitude both makes the signal susceptible to noise and requires a highly linear power amplifier in the transmitter.

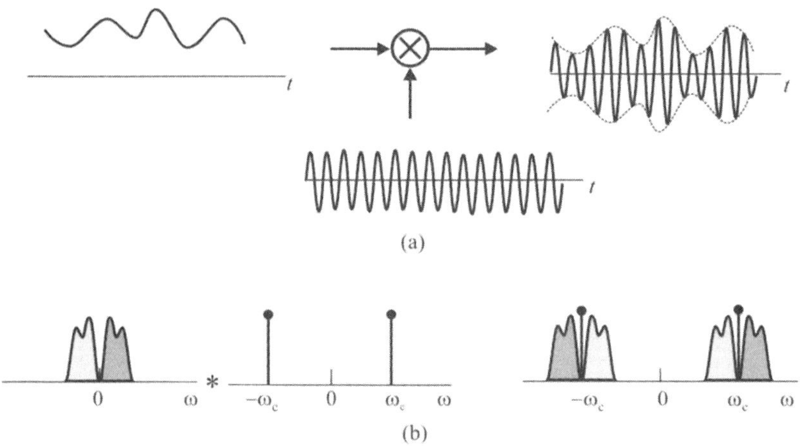

Figure 2.3 AM: (a) Time domain multiplication corresponds to (b) convolution in the frequency domain.

2.3.1.1 Baseband, Passband Signals, and Amplitude Modulation

A salient feature of analog information signals is that they are generally low frequency. Sometimes this is due to the nature of data itself such as the human voice which, has frequency components from 300 Hz to approximately 20 kHz.

Due to their low frequency content, the information signals have a spectrum such as that in the Figure 2.4. There are many low frequency components and the one-sided spectrum is located near the zero frequency. The frequency range of this signal extends from zero to a maximum frequency of f_m. We say that this signal has a bandwidth of f_m. In the time domain this analog baseband signal may look as shown in Figure 2.5.

Now let's modulate this signal, which means we are going to transfer it to a higher (usually much higher) frequency. Just as information signals are

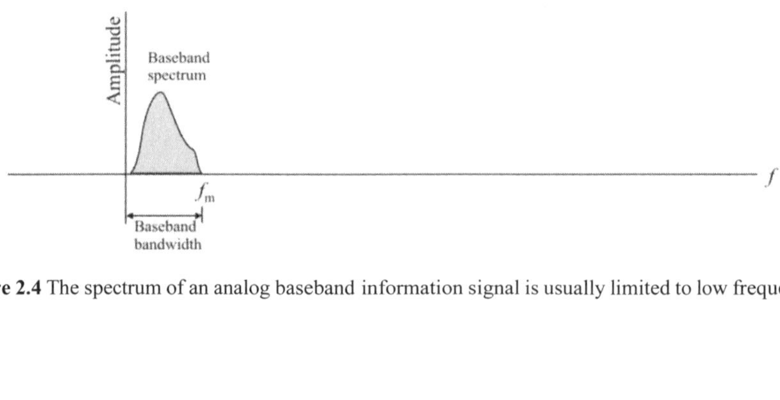

Figure 2.4 The spectrum of an analog baseband information signal is usually limited to low frequencies.

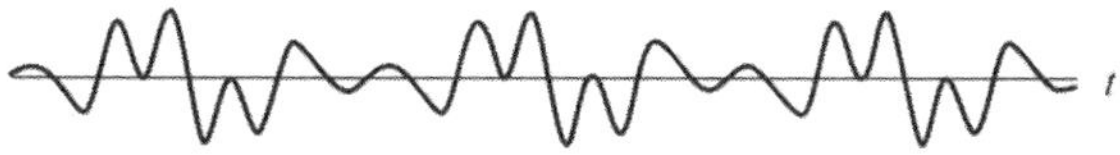

Figure 2.5 Analog signal in the time domain.

characterized by their low frequency, the transmission medium or carriers are characterized by their high frequency.

The simplest type of modulator for nearly all modulation schemes is called the product modulator consisting of a multiplier or a mixer [1] and a band-pass filter (Figure 2.6). Let's modulate the above signal using the product modulator, where $m(t)$ is the low frequency message signal and $c(t)$ is the high frequency carrier signal. The modulator takes these two signals and multiplies them.

$$s(t) = m(t) \times c(t)$$

The product modulator produces sums and differences of the frequencies of the two input signals in both the positive and negative frequency domains. Some of these sums and differences are desired responses and some may not be desired.

If the carrier frequency source in a product modulator is not perfectly stable, each deviation frequency will also produce its own sum and difference frequencies with the baseband signal. These are called *spurs* and are a characteristic of the mixing process. For this reason, simple mixer modulators and demodulators do not work well and further complexity is required in the form of phase lock loops or similar devices is introduced into the transmitter and receiver design.

The two-sided spectrum of the message signal is shown in Figure 2.7. After mixing, modulating, or heterodyning (all of these terms refer to the same thing), we get the spectrum shown in Figure 2.8. The spectrum is now shifted up (and down) to the carrier frequency, it is replicated on both sides of the y-axis, and we see that the spectrum is reversed left of the y-axis.

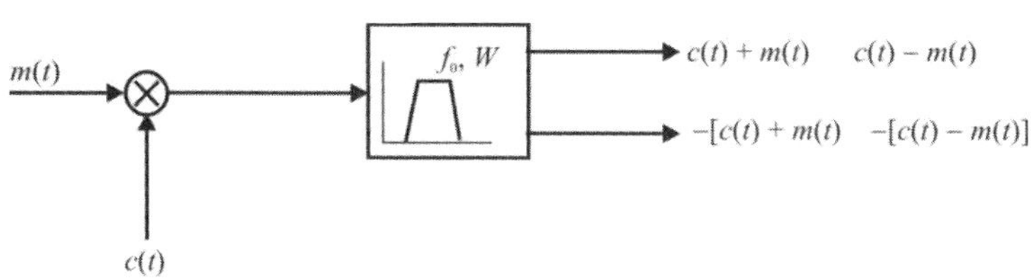

Figure 2.6 Product modulator.

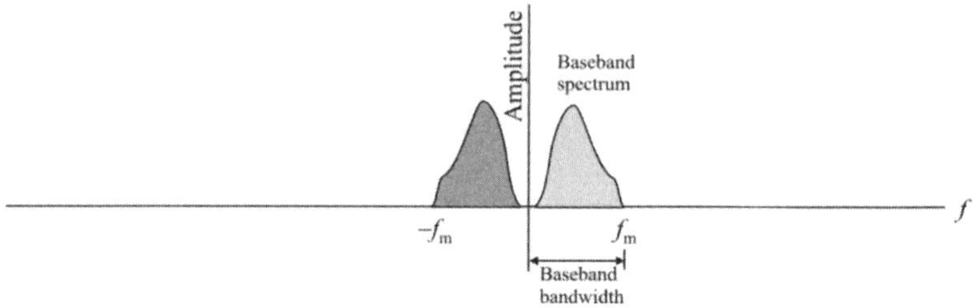

Figure 2.7 A baseband spectrum.

Another way to describe the process is that multiplication by a sinusoid shifts one copy of the spectrum up to f_c and another down to $-f_c$. The reason for this is explained by the Fourier transform of this signal which, is a product of two signals, one of them a sinusoid. Consider

$$s(t) = m(t) \times \cos \omega_c t$$

The Fourier transform of $s(t)$, denoted by $\mathcal{F}\{s(t)\}$, is given by the Fourier transform of the signal $m(t)$, half of it shifted up and half of it down. Thus

$$\mathcal{F}\{s(t)\} = \frac{1}{2}\left[S(f - f_c) + S(f + f_c)\right]$$

In the time domain, we see that this signal has much higher frequency—see Figure 2.9—but its envelope is still the original low frequency signal of Figure 2.5.

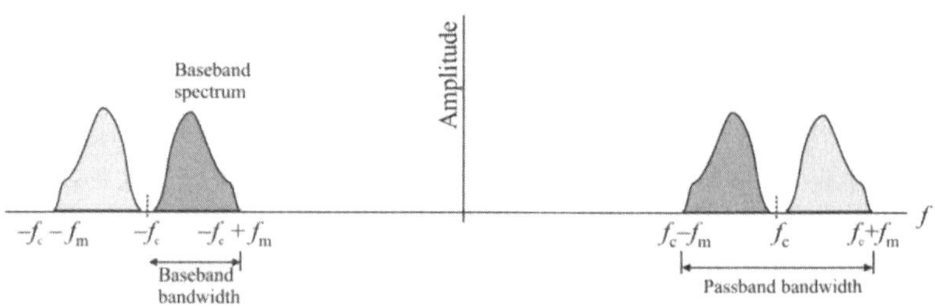

Figure 2.8 Passband spectrum of the same signal.

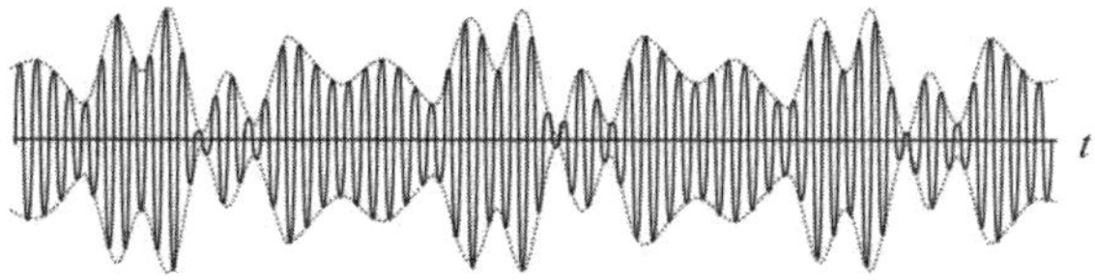

Figure 2.9 Output signal of a product modulator, the envelope of which is the information signal (see also Figure 2.5).

2.3.1.2 Baseband and Passband Signals

The modulating information signal is called the *baseband signal*. The bandwidth of this signal is denoted by f_m. In most cases of interest for communications EW the analog baseband signal will consist of low frequency components.

The multiplication of the baseband signal with a sinusoid carrier signal translates the baseband signal to $\pm f_c$. The translated signal is called the *passband signal*. This signal extends in range from $(-f_c - f_m)$ to $(f_c + f_m)$. The new signal has doubled in bandwidth. The passband signal bandwidth is double that of the baseband signal; see Figure 2.8.

2.3.1.3 Sidebands

Now note that in Figure 2.10, the passband spectrum has two parts (on each side of f_c) that are identical (except reversed in frequency). The upper part of the passband spectrum above the carrier is called the *upper sideband* and the one below is called the *lower sideband*. We notice that since the passband spectrum is symmetrical (not only about the *y*-axis but also about the carrier frequency) the upper sideband is the mirror image of the lower sideband. We do not need both to recover the signal.

This leads to the single and double sideband modulation techniques. In double sideband, we use the whole spectrum just as we show above. Both halves are used.

Figure 2.10 Filter passband to for the upper and lower sideband as separate signals. (Negative passband frequency range not shown.)

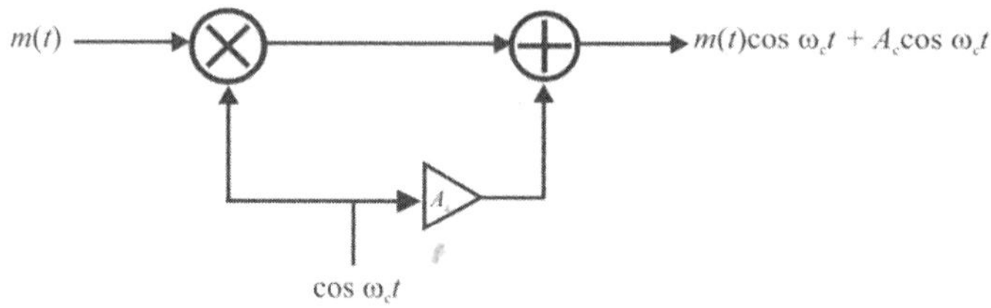

Figure 2.11 A basic AM modulator. When $A_c = 0$, DSB-SC is generated since there is no excess carrier in the transmitted signal. When $A_c > 0$, DSB-C is generated since some amount of carrier is added to the transmitted signal, determined by the amplitude A_c.

In single sideband modulation, we eliminate the lower or the upper band to separate out these signals as if they were two independent signals. Each half is enough to recover the signal.

Filter F_1 and Filter F_2 in Figure 2.10 are used to separate one sideband from the other. Using F_1 before transmitting, and only the lower sideband is transmitted. Use F_2 and only the upper side band is transmitted. We get two channels in place of one. SSB is used wherever there are bandwidth limitations. The most notable application is analog PSTN networks (which are rapidly disappearing). These signals have ideal characteristics for the use of SSB. There is very little signal content below 300 Hz so the SSB signal does not suffer much distortion. Also, telephone signals are bandwidth-limited, and SSB maximizes bandwidth usage. Analog HF communications is one area where the SSB modulation is used to this advantage.

2.3.1.4 Amplitude Modulation

We have already discussed that much of the building blocks of AM as SSB is a form of AM. The simplest form of AM is *double sideband* (DSB) modulation.

Double Sideband Modulation

Denote the modulating information signal by *m(t)*. A model of the AM modulation process is shown in Figure 2.11. The output of the mixer is given by

$$s(t) = m(t)\cos(\omega_c t) \tag{2.3}$$

and when $A_c = 0$, the transmitted signal is given by (2.3). The Fourier transform of this signal is given by

$$S(f) = \frac{1}{2}\left[\underbrace{M(f + f_c)}_{\text{Spectrum on the negative side}} + \underbrace{M(f - f_c)}_{\text{Spectrum on the positive side}}\right]$$

Figure 2.12 Double sideband modulation spectrum of received signal (Note only the positive side of the passband spectrum is shown.)

This is called DSB suppressed carrier (DSB-SC) since there is no extra carrier present in the transmitted signal.

When $A_c > 0$ in Figure 2.11, some of the carrier signal is added to this output of the mixer as shown. In that case we get

$$s_{\text{AMDSB}}(t) = m(t)\cos(\omega_c t) + \underbrace{A_c \cos(\omega_c t)}_{\text{Carrier}} \tag{2.4}$$

The Fourier transform of (2.4) is given by

$$S(f) = \frac{1}{2}\left[\underbrace{M(f - f_c) + A_c\delta(f - f_c)}_{\text{Spectrum on the positive side}} + \underbrace{M(f + f_c) + A_c\delta(f + f_c)}_{\text{Spectrum on the negative side}}\right]$$

The spectrum of this signal is shown in Figure 2.12.

We see the carrier signal pops up in the middle of the spectrum. We can put a filter around this signal and recover the carrier at the receiver. This is then fed to the demodulation circuitry later.

While this modulation is called double sideband modulation, it is also sometimes called AM DSB with carrier, DSB-C). It is the most basic form of AM modulation. From here on, we can do a variety of things such as suppress the carrier, use one band or the other, and so forth. All of these are variations of DSB AM.

An AM modulator that does not use a multiplier can be made with a nonlinear device and filtering. A block diagram of such a modulator using a squaring device as the nonlinear component is shown in Figure 2.13. The carrier and modulating signal are first summed then passed through nonlinearity, in this case a squaring device. The signals at the unwanted frequencies are filtered, leaving the AM modulated carrier at the output.

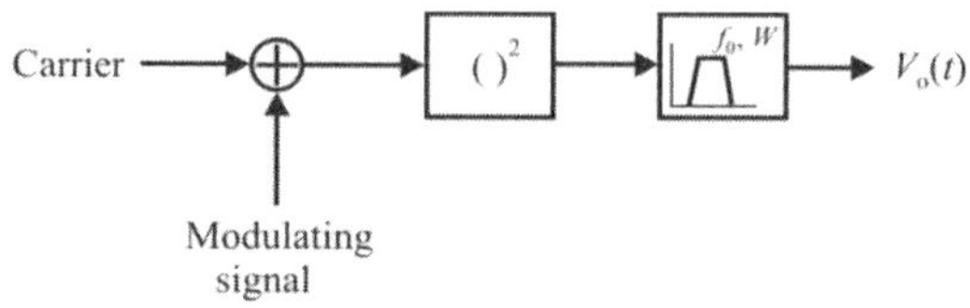

Figure 2.13 AM modulator with a nonlinearity (a squaring circuit).

An AM modulator implemented according to the block diagram shown in Figure 2.13 with a square law FET is illustrated in Figure 2.14 (bias circuitry is not shown). A square law n-channel FET will pass a drain-source current given by

$$I_{\mathrm{DS}} = k_1 (V_{\mathrm{G}} + V_{\mathrm{p}})^2 \tag{2.5}$$

where V_{p} is the FET's *pinch-off voltage* and V_{G} is the *gate voltage.* This expression is accurate provided that the gate voltage is in the correct range. In the circuit shown in Figure 2.14 we apply a gate voltage that is a combination of the output from a local oscillator, $A_{\mathrm{c}} \cos(\omega_{\mathrm{c}} t)$, and a modulation input, $m(t)$. Assuming that the input resistors are of equal value, the gate voltage is

$$V_{\mathrm{G}} = \frac{1}{2} m(t) + \frac{1}{2} A \cos(\omega_{\mathrm{c}} t) \tag{2.6}$$

This voltage produces a drain-source current given by

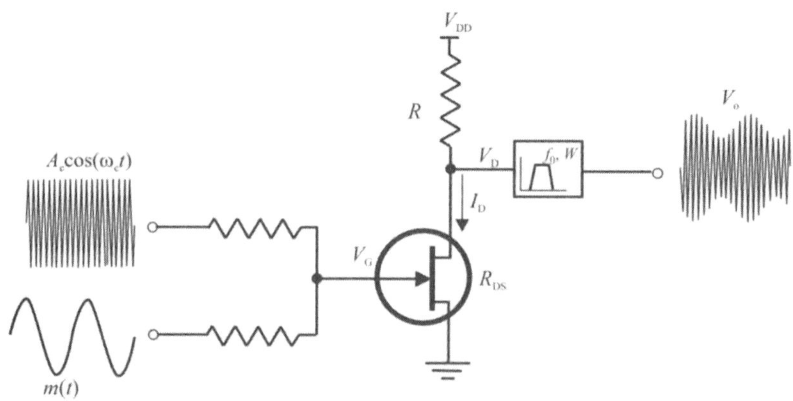

Figure 2.14 AM modulator using a JFET.

$$I_D = \frac{k_2}{4}[m(t) + A\cos(\omega_c t) + V_p]^2 \tag{2.7}$$

where k_2 and V_p are constants, the values of which depend on the particular FET. This can be expanded and re-written as

$$\begin{aligned} I_D(t) &= \frac{k_2}{4}[m^2(t) + V_p^2 + 2V_p m(t)] \\ &+ \frac{k_2}{4}[2m(t)A\cos(\omega_c t) + 2V_p \cos(\omega_c t)] \\ &+ \frac{k_2}{4}A^2\cos^2(\omega_c t) \end{aligned} \tag{2.8}$$

This produces a drain voltage of

$$V_D = V_{DD} - RI_{DS} \tag{2.9}$$

where R is the resistance between the FET's drain terminal and the bias voltage, V_{DD}. Since

$$\cos^2\theta = \frac{1}{2}[1 + \cos(2\theta)] \tag{2.10}$$

the last term in (2.8) is a combination of a DC current and a fluctuation at the frequency $2\omega_c$. In all cases of interest here the frequency content of $m(t)$ is much less than ω_c. This means that the first part of the expression consists of a steady current plus some fluctuations at frequencies well below ω_c. A bandpass filter can be used to eliminate the undesired components at frequencies away from ω_c so the output is

$$V_o(t) = \frac{-Rk_2}{4}[2m(t)A\cos(\omega_c t) + 2V_p A\cos(\omega_c t)] \tag{2.11}$$

which can be rewritten in the form

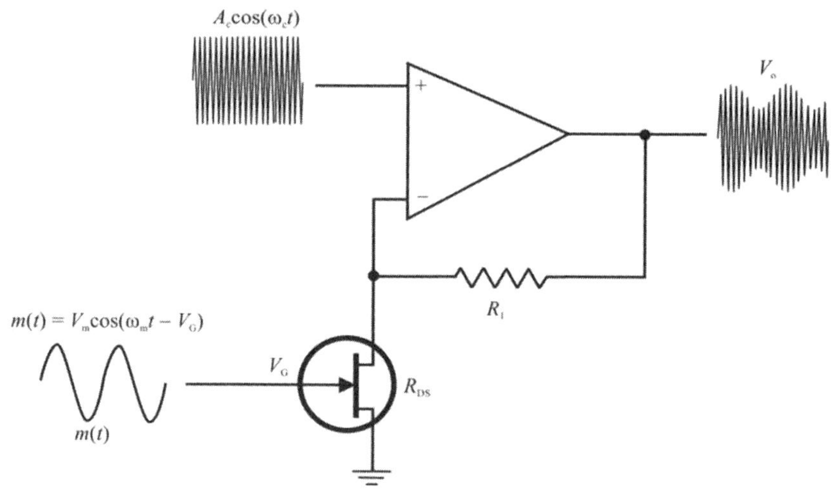

Figure 2.15 AM modulator with an op amp.

$$V_o(t) = A'_0[1 + m'(t)\cos(\omega_c t)] \tag{2.12}$$

where

$$m'(t) = \frac{m(t)}{V_p}; \qquad A'_0 = \frac{-Rk_2V_pA_c}{2} \tag{2.13}$$

which is the desired result. Thus, the output is a wave whose unmodulated amplitude is A'_0 and is amplitude modulated by an amount, $m'(t)$ proportional to the input modulating signal, $m(t)$. The circuit therefore behaves as an amplitude modulator.

AM Modulator with an Operational Amplifier

Another example of a simple AM modulator is shown in Figure 2.15, where a JFET is used in conjunction with an operational amplifier (again, bias circuitry not shown). It consists of a noninverting amplifier in which the carrier $v_c(t)$ with frequency ω_c, is applied to the noninverting input terminal. The gain of the amplifier

$$G = 1 + \frac{R_1}{R_{DS}} \tag{2.14}$$

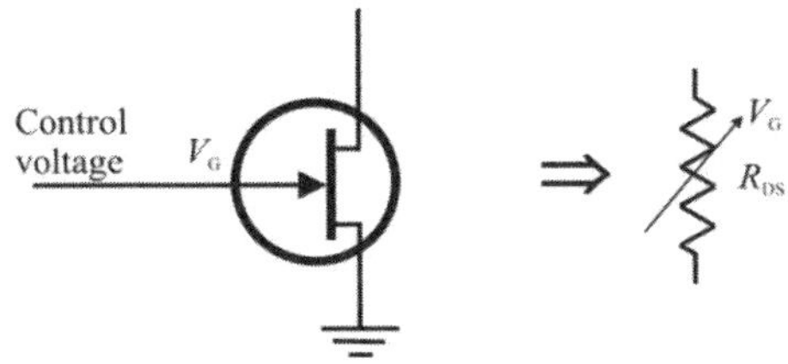

Figure 2.16 A JFET can be viewed as a voltage-controlled resistor.

can be varied through a voltage-controlled resistor R_{DS}. This resistor is implemented with a FET in this case, which serves as a variable resistor, as is shown in Figure 2.16. The voltage applied to the gate input terminal, V_G, determines the value of the resistor. The JFET is biased to operate in the linear region of the output characteristic as illustrated in Figure 2.17. The resistance ($R_{DS} = V_{DS}/I_D$) is proportional to the applied gate voltage. The gate terminal is connected to the modulating signal $m(t)$, with frequency ω_m. This signal will thus change the amplification of the amplifier in relation to the value of the modulating signal, as schematically indicated in Figure 2.16. This results in an AM modulated signal.

2.3.1.5 Double Sideband - Suppressed Carrier

In AM DSB modulation the carrier consumes a considerable amount of power and perhaps it would be better to not include it in the transmitted signal. But we want to somehow include the carrier information but without actually doing so. And we want to use the envelope detector as the receiver.

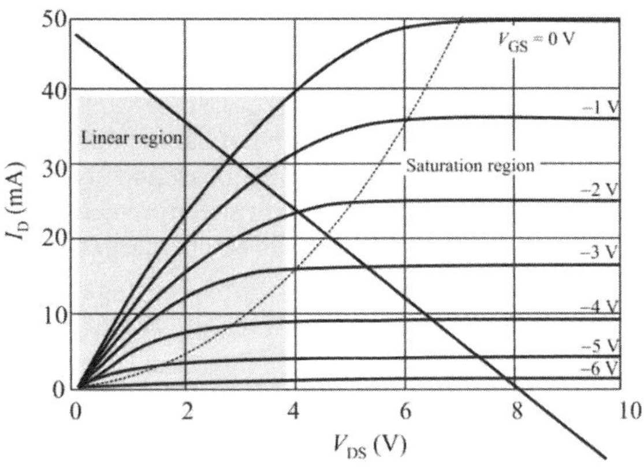

Figure 2.17 Output characteristic of JFET.

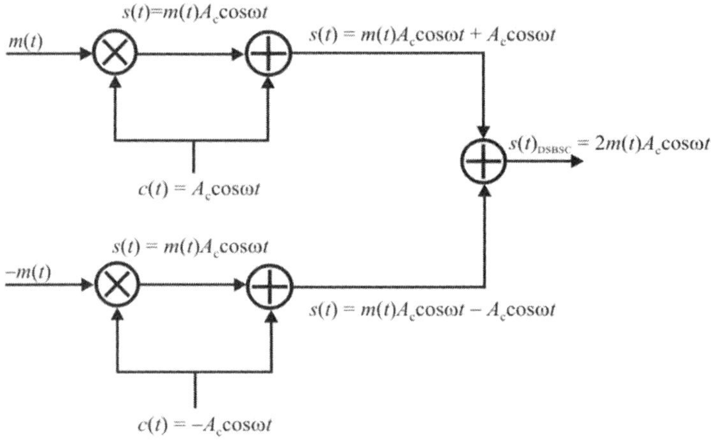

Figure 2.18 A balanced modulator results in suppression of the carrier.

Consider DSB-SC modulation. In this case no carrier in included. DSB-SC signals can be created by a modulator called the *balanced modulator*, the topology of one of which is shown in Figure 2.18.

This balanced modulator is basically two product modulators added together. The input to one is a negative information signal and a negative carrier. The product of this modulator when added to its positive counterpart results in canceling the carrier as we can see in the output.

An envelope detector cannot be used with DSB-SC signal because the envelope of the DSB-SC signal is not the same as the baseband signal. A more sophisticated demodulator is needed with this signal.

2.3.1.6 Generating Single Sideband Signals

In essence single sideband (SSB) transmission is a bandwidth conserving technique. The most notable point of SSB is that the SSB passband signal and the baseband signal occupy the same bandwidth, thus cutting spectrum needs in half.

There are two main ways that SSB signals can be generated:

- Filtering the unwanted sideband
- Phasing method

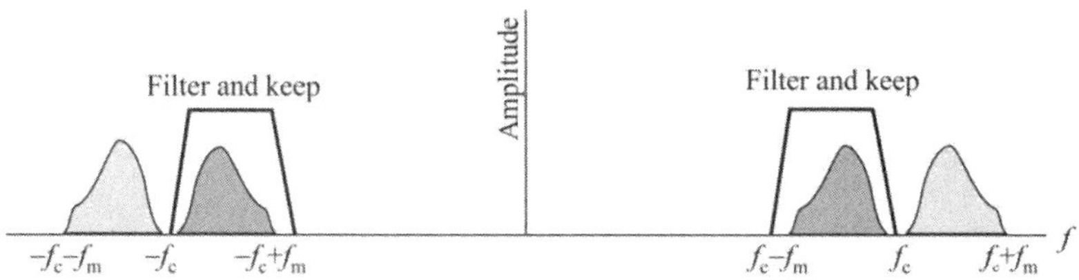

Figure 2.19 SSB filtering method where a passband filter after DSB-SC modulation results in eliminating all but one band.

Filtering Method of SSB Generation

The simplest solution is to take the DSB-SC signal and filter the unwanted band before transmission so that the unwanted side is not sent at all, as shown in Figure 2.19. By keeping only the part shown, the upper images are eliminated.

The most significant problem with this method is that it is difficult to construct practical filters with steep enough cutoffs at high frequencies. Such a filter ends up distorting the desired signal as well as including some of the unwanted sideband anyway. Assuming a 10MHz carrier modulated by a 50–5,000 Hz audio signal, a 13th order elliptic bandpass filter is required with crystal Qs as high as 768,019 for 0.1-dB passband ripple, and 60-dB rejection of the unwanted sideband.

Phasing Method of SSB Generation

The second method of generating SSB signals involves the use of Hilbert transform and the analytic signal. As a way of review, the Figure 2.20(a) shows the baseband spectrum of a signal while Figure 2.20(b) shows the Hilbert transform of the same signal. (Recall that the Hilbert transform rotates the positive frequency components.)

Modulating this signal generates

$$s_a(t) = m(t)\cos(\omega_c t) \tag{2.15}$$

The Hilbert transform of this signal, modulated by a sine wave, is

$$\hat{s}_a(t) = \hat{m}(t)\sin(\omega_c t) \tag{2.16}$$

Now we create a carrier which is the sum of (2.15) and (2.16) and get

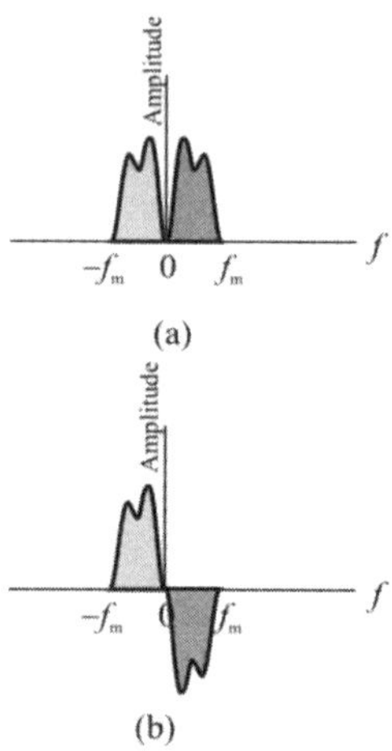

Figure 2.20 (a) Baseband spectrum (symmetric about the y-axis and (b) Hilbert transform of the same signal (antisymmetric about the y-axis)

$$s(t) = \underbrace{m(t)\cos(\omega_c t)}_{\text{Signal}} \pm \underbrace{\hat{m}(t)\sin(\omega_c t)}_{\text{It's Hilbert transform}} \tag{2.17}$$

$s(t)$ is the desired SSB signal, and is called the *analytic signal*. The resulting modulator is shown in Figure 2.21.

The SSB signal created in this way is essentially two signals in phase quadrature. The combination gives us the equation for the SSB signal. By changing the sign of the analytic signal, we can create either the upper sideband or the lower.

The Fourier transform of (2.17) is

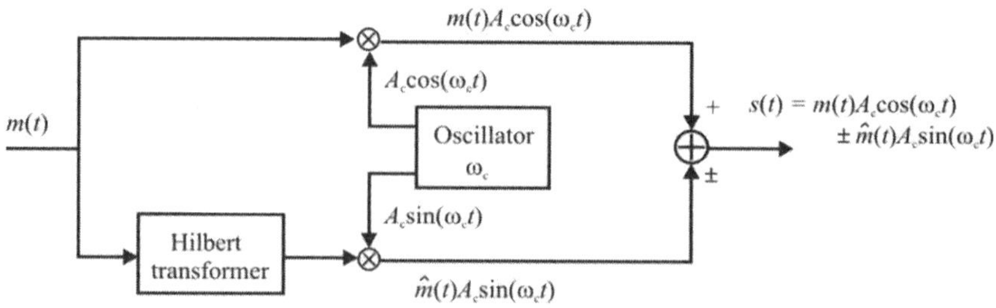

Figure 2.21 SSB modulator using the phasing method.

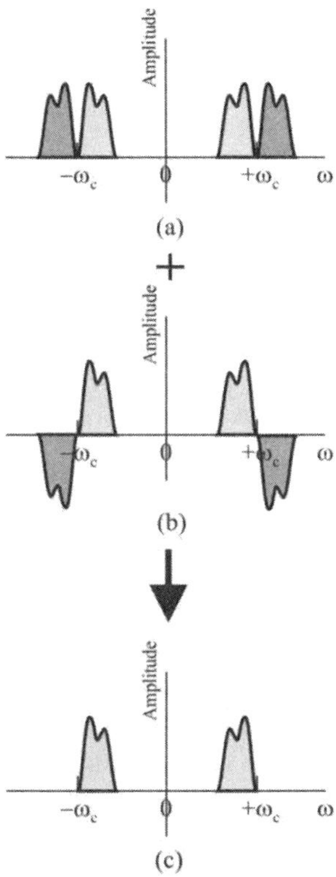

Figure 2.22 SSB spectrum of two parts: (a) spectrum of the first part, (b) spectrum of the second part, and (c) the sum of these two gives us the lower sideband; the difference would give the upper sideband.

$$S(\omega)=\frac{1}{2}\left[M(\omega-\omega_c)+M(\omega+\omega_c)\right]+\frac{1}{2j}\left[-jM(\omega-\omega_c)+jM(\omega+\omega_c)\right] \quad (2.18)$$

(The j multipliers simply shift the spectrum by $\pi/2$ radians.) Figure 2.22 shows the two parts of (2.18) and we see at once that adding these two representations give us a signal with only one sideband, upper or lower.

Another interesting point is that the sum of the two sidebands give us the DSB-SC waveform.

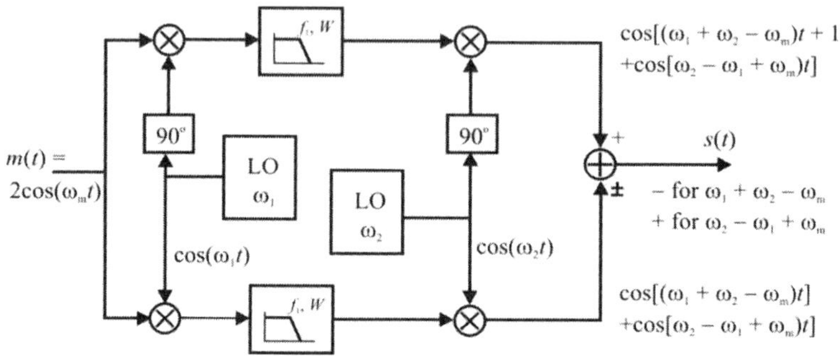

Figure 2.23 Weaver method of SSB generation.

Weaver Method of SSB Generation

The Weaver method of single sideband generation in Figure 2.23 is similar to the phase method except that the broadband 90 ° phase difference is implemented with quadrature mixing processing [2]. The frequency, ω_1, must be greater than the highest baseband frequency, ω_m, and the low-pass filter is required to block the sum frequency, $\omega_1 + \omega_m$. Although an accurate broadband 90° phase difference is not required, the two low-pass filters must closely track each other in amplitude and phase for adequate rejection of the unwanted sideband.

The configuration in Figure 2.24 shows a technique for producing a broadband 90° (quadrature) phase shift for the local oscillator (LO). A phase locked loop (PLL) with an integrating loop filter will maintain 90° phase accuracy. An integrating PLL will maintain a quadrature phase relationship when the loop is closed; however PLL circuits tend to be noisy. Sideband noise is troublesome in both SSB and FM systems, but SSB is less sensitive to phase noise problems in the LO.

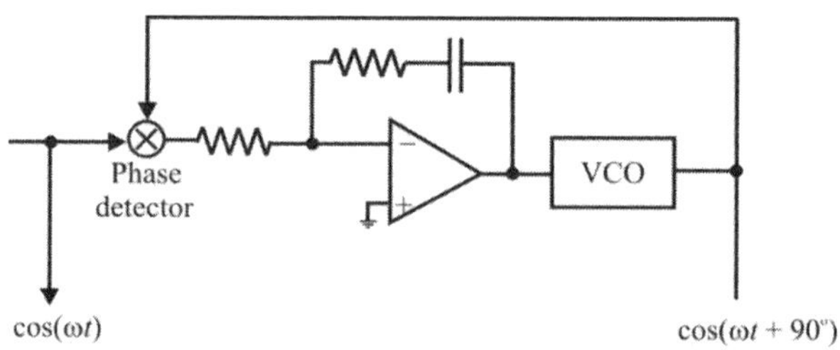

Figure 2.24 PLL quadrature synthesis.

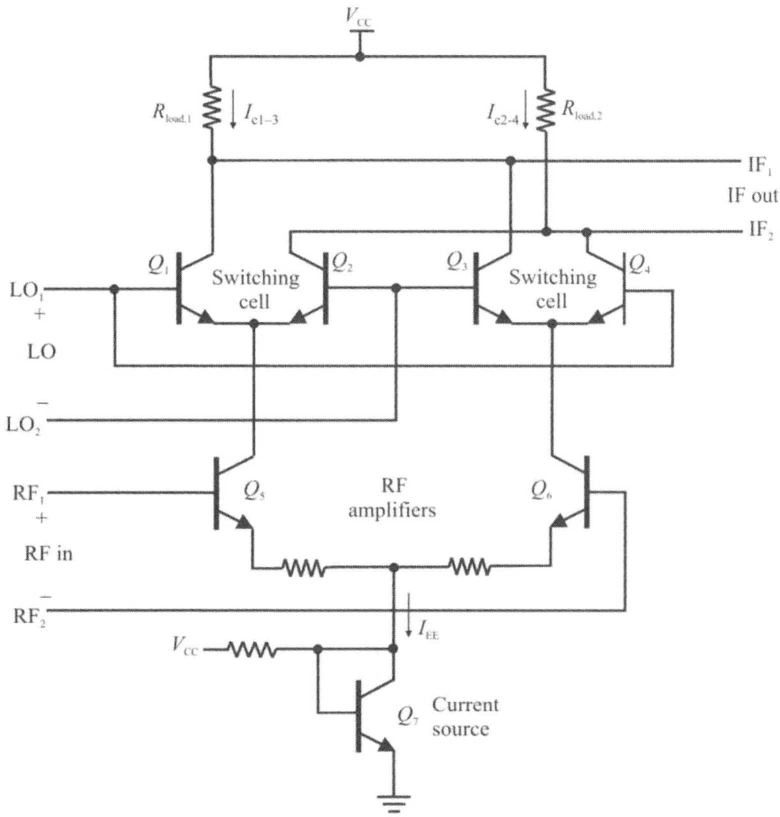

Figure 2.25 Gilbert cell mixer.

Gilbert Cells

A Gilbert cell is a cross-coupled differential amplifier, similar to the topology in Figure 2.25, where the gain is controlled by modulating the emitter bias current [3]. Gilbert cells are four-quadrant multipliers and have common applications as mixers, AGC amplifiers, amplitude modulators, DSB modulators, SSB modulators, AM detectors, SSB and DSB detectors, frequency doublers, squaring circuits, dividers, square-root circuits, and root-mean-square (RMS) measuring circuits. We'll briefly examine how the Gilbert cell works in this section.

The Gilbert cell uses a linear, time-varying circuit to achieve time-domain multiplication, and hence, frequency shifting.

The RF signal is input to a differential amplifier consisting of Q_5 and Q_6. The collectors of these transistors have a cross-connected set of four transistors, which are driven by a LO.

Consider what happens when V_{LO1} is large enough so that Q_1 and Q_4 turn on and V_{LO2} is small enough that Q_2 and Q_3 turn off. We see that Q_1 and Q_4 act as closed switches so that

- Q_5 is connected to $R_{load,1}$;
- Q_6 is connected to $R_{load,2}$.

as in a typical differential amplifier configuration with the output taken at what are here called the IF terminals.

Next, consider what happens when the opposite input occurs. Specifically, suppose V_{LO2} is large enough that Q_1 and Q_4 are off. Now we see that Q_2 and Q_3 act as closed switches so that

- Q_5 is connected to $R_{load,2}$;
- Q_6 is connected to $R_{load,1}$.

This also is a differential amplifier configuration, but with the outputs interchanged with respect to the previous case. In other words, the output (the IF) is almost the same as before, it has just been multiplied by –1.

Thus, the overall function of the Gilbert cell is to multiply in the time domain the input RF signal (at the RF frequency) by a square wave with value ±1 at the LO frequency.

The Gilbert cell is an active mixer in that the IF output signal is amplified because of the differential amplifier gain.

Mixer Mathematics

Let the RF input voltage be

$$V_{\text{RF}}(t) = A_{\text{RF}}\cos(\omega_{\text{RF}}t) \tag{2.19}$$

and define the mixing signal by the square wave shown in Figure 2.26. Since the mixing signal is a periodic waveform, we can expand it in a Fourier series as

$$V_{\text{LO}}(t) = \frac{4}{\pi}\left[\cos(\omega_{\text{LO}}t) - \frac{1}{3}\cos(3\omega_{\text{LO}}t) + \frac{1}{5}\cos(5\omega_{\text{LO}}t) - \cdots\right] \tag{2.20}$$

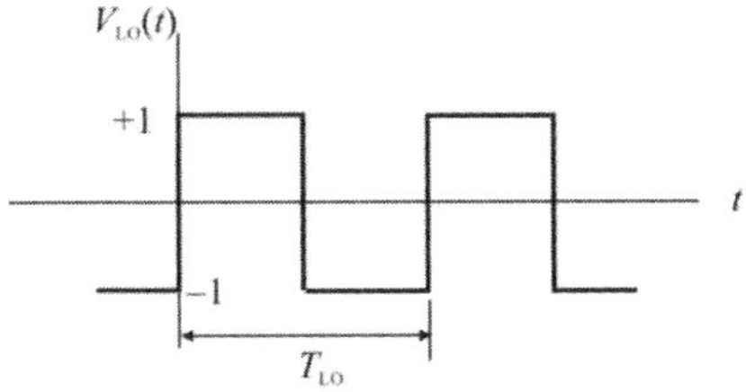

Figure 2.26 Square wave for the LO in Figure 2.25.

The Gilbert cell effectively multiplies both of the signals as specified by (2.19) and (2.20) in the time domain as

$$V(t) = V_{\mathrm{RF}}(t)V_{\mathrm{LO}}(t) \tag{2.21}$$

yielding

$$\begin{aligned} V(t) = {} & \frac{2A_{\mathrm{RF}}}{\pi}\left[\cos(\omega_{1\text{-}}t) - \frac{1}{3}\cos(3\omega_{3\text{-}}t) + \frac{1}{5}\cos(5\omega_{5\text{-}}t) - \cdots\right] \\ & + \frac{2A_{\mathrm{RF}}}{\pi}\left[\cos(\omega_{1+}t) - \frac{1}{3}\cos(3\omega_{3+}t) + \frac{1}{5}\cos(5\omega_{5+}t) - \cdots\right] \end{aligned} \tag{2.22}$$

where

$$\begin{aligned} \omega_{1+} &= \omega_{\mathrm{LO}} + \omega_{\mathrm{RF}} & \qquad \omega_{1-} &= |\omega_{\mathrm{LO}} - \omega_{\mathrm{RF}}| \\ \omega_{3+} &= 3\omega_{\mathrm{LO}} + \omega_{\mathrm{RF}} & \qquad \omega_{3-} &= |3\omega_{\mathrm{LO}} - \omega_{\mathrm{RF}}| \\ \omega_{5+} &= 5\omega_{\mathrm{LO}} + \omega_{\mathrm{RF}} & \qquad \omega_{5-} &= |5\omega_{\mathrm{LO}} - \omega_{\mathrm{RF}}| \end{aligned}$$

Note in (2.22) that we have the sum and difference signals present in the output (IF) voltage signal

$$V_{+}(t) = \frac{2A_{\mathrm{RF}}}{\pi}\cos(\omega_{+}t) \tag{2.23}$$

and

$$V_{-}(t) = \frac{2A_{\mathrm{RF}}}{\pi}\cos(\omega_{-}t) \tag{2.24}$$

as well as the third harmonic terms

$$V_{3+}(t) = \frac{2A_{\text{RF}}}{3\pi}\cos(\omega_{3+}t) \tag{2.25}$$

and

$$V_{3-}(t) = \frac{2A_{\text{RF}}}{3\pi}\cos(\omega_{3-}t) \tag{2.26}$$

and the fifth harmonic terms

$$V_{5+}(t) = \frac{2A_{\text{RF}}}{5\pi}\cos(\omega_{5+}t)$$

and

$$V_{5-}(t) = \frac{2A_{\text{RF}}}{5\pi}\cos(\omega_{5-}t)$$

and all odd higher-order harmonics.

Observe that the amplitudes of these harmonics are decreasing with increasing harmonic number. Also note that the RF signal, the LO signal, and even the mixer harmonics are not present in the output. This occurs because the Gilbert cell is a balanced mixer. However, in reality some (or all) of these signal components will be present in the output since we won't have a perfectly balanced mixer.

2.3.2 Phase and Frequency Analog Modulation

PM and FM are important concepts that are encountered not only within the context of modems but also in the analysis of such circuits as oscillators and frequency synthesizers.

In (2.1), if $a(t)$ is a constant, A_c, and the excess phase $\theta(t)$ is linearly proportional to the baseband signal, we say the carrier is *phase modulated*,

$$s_{\text{PM}}(t) = A_c \cos[\omega_c t + m s_{\text{BB}}(t)] \tag{2.27}$$

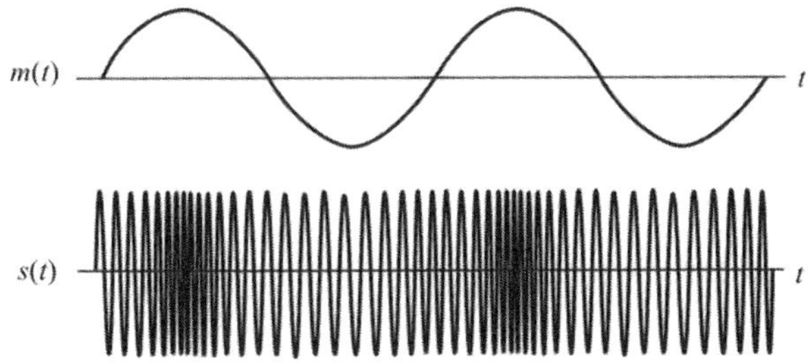

Figure 2.27 FM (analog).

where m is the phase modulation index. For example, if $s_{BB}(t) = \alpha t$, then $s_{PM}(t) = A_c \cos(\omega_c + m\alpha)t$, indicating that a ramp baseband waveform simply shifts the carrier frequency by a constant value.

Similarly, if the excess frequency, $d\theta/dt$, is linearly proportional to the baseband signal, we say the carrier is *frequency modulated*,

$$s_{FM}(t) = A_c \cos\left[\omega_c t + m_{FM} \int_{-\infty}^{t} s_{BB}(t)dt\right] \tag{2.28}$$

where m_{FM} is the frequency modulation index. For example, with a DC baseband signal, $s_{BB}(t) = A$, we have $s_{FM}(t) = A_c \cos(\omega_c + m_{FM}A)t$. Thus, a DC input to a frequency modulator shifts the carrier frequency by a constant value, a useful result in the context of VCOs.

Figure 2.27 shows a baseband signal and the corresponding frequency-modulated waveform. Note that in general it is not possible to determine whether a waveform is FM or PM if the baseband signal is unknown. Also, the maximum frequency deviation is equal to $m_d s_{BB}(t)/dt$ in PM and $m_{FM} s_{BB}(t)$ in FM.

The nonlinear dependence of $s_{PM}(t)$ and $s_{FM}(t)$ upon $s_{BB}(t)$ makes it difficult to calculate the bandwidth occupied by the modulated waveform. We first consider a simple case called *narrowband FM*, defined as a signal for which $m_{FM} \int s_{BB}(t)dt << 1\,\text{rad}$. The FM waveform can then be approximated as

$$s_{FM,NB}(t) \approx A_c \cos\omega_c t - A_c m_{FM} (\sin\omega_c t)\int s_{BB}(t)dt \tag{2.29}$$

Consequently, as shown in Figure 2.28, the spectrum consists of the carrier and that of $\int s_{BB}(t)dt$ translated to $\pm\omega_c$, indicating that the bandwidth of $s_{FM}(t)$ is somewhat

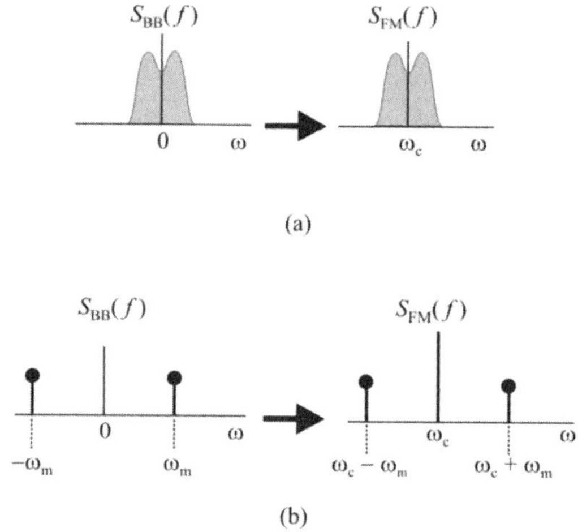

Figure 2.28 Narrowband FM: (a) continuous psd, only positive translation is shown and (b) monochromatic baseband signal, both positive and negative translations shown.

less than that of $s_{AM}(t)$ in Figure 2.3 because of the $1/s$ attenuation resulting from integration.

A special case of narrowband FM that proves quite useful is a sinusoidal baseband signal $s_{BB}(t) = A_m \cos \omega_m t$. In this case, we have

$$s_{FM,NB}(t) \approx A_c \cos \omega_c t - A_m A_c \frac{m_{FM}}{\omega_m} \sin \omega_c t \sin \omega_m t \tag{2.30}$$

$$= A_c \cos \omega_c t - \frac{A_m A_c m_{FM}}{2\omega_m} \cos(\omega_c - \omega_m)t + \frac{A_m A_c m_{FM}}{2\omega_m} \cos(\omega_c + \omega_m)t \tag{2.31}$$

The resulting spectrum, illustrated in Figure 2.28(b), consists of impulses at $\pm\omega_c$ and "sidebands" at $\pm(\omega_c \pm \omega_m)$.

Equations (2.30) and (2.31) yield three important points. First, for the narrowband FM assumption to hold, we must have $m_{FM} A_m / \omega_m << 1 \text{ rad}$. Second, as ω_m increases, the magnitude of the sidebands decreases; that is., higher modulating frequencies have less weighting in the FM signal. Third, if $s_{FM}(t) = A_c \cos[\omega_c t - (m_{FM} A_m / \omega_m) \sin \omega_m t]$, then the maximum frequency deviation is equal to $m_{FM} A_m$. Note that the spacing between the sidebands and the carrier is

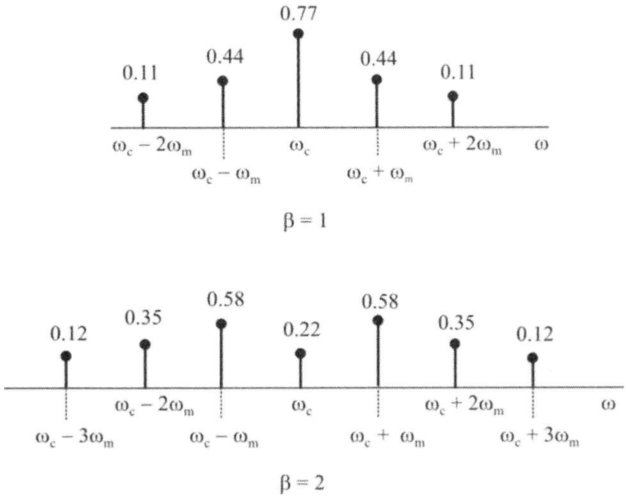

Figure 2.29 FM spectra for $\beta = 1$ and $\beta = 2$.

equal to ω_m and has no relationship with the maximum frequency deviation. While a useful concept in many communication systems, narrowband FM has limited application in modems, because of its low SNR, an issue described later.

As the next step toward a realistic case, we consider an FM signal with a sinusoidal baseband waveform, but without the restriction $m_{FM}A_m / \omega_m << 1\,\text{rad}$. If $s_{BB}(t) = A_m \cos\omega_m t$, then $s_{FM}(t) = A_c \cos[\omega\, ct - (m_{FM}A_m / \omega_m)\sin\omega_m t]$. This expression can be expanded to

$$s_{FM}(t) = A_c \sum_{n=-\infty}^{\infty} J_n(\beta)\cos(\omega_c + n\omega_m)t \tag{2.32}$$

where $J_n(.)$ denotes the nth order Bessel function of the first kind and $\beta = m_{FM}A_m / \omega_m$. If $\beta << 1$ rad (narrowband FM), then $J_0(\beta) \approx 1$, $J_{\pm 1}(\beta) \approx \pm\beta / 2$, and $J_n(\beta) \approx 0$ for $|n| > 1$. Thus,

$$s_{FM}(t) = A_c \cos\omega_c t + (A_c\beta / 2)\cos(\omega_c + \omega_m)t - (A_c\beta / 2)\cos(\omega_c - \omega_m)t \tag{2.33}$$

the same result as that in (2.31). As β approaches and exceeds unity, the coefficients $J_n(\beta)$ become significant for $|n| > 1$, introducing components at $\omega_c \pm n\omega_m$. Shown in Figure 2.29 are examples of the FM spectra for $\beta = 1$ and $\beta = 2$.

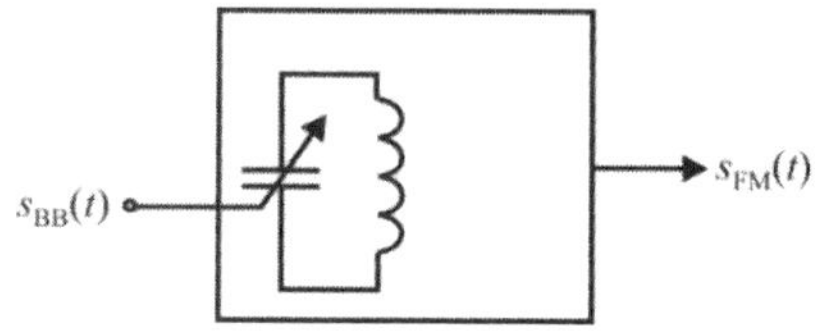

Figure 2.30 Voltage-controlled oscillator.

An important conclusion to be drawn from the above analysis is that the bandwidth of FM signals depends on m, A_m, and ω_m. In fact, the bandwidth W_{FM}, is defined as that containing 98% of the signal power, then

$$W_{FM} \approx 2(\beta + 1)W_{BB} \tag{2.34}$$

where W_{BB} is the bandwidth of the baseband signal. This is called *Carson's rule.*

The last attribute of FM systems to be considered is their power efficiency. Since an FM signal has a constant envelope and carries the information in the zero crossings, it can be processed by a nonlinear PA with no corruption of the baseband signal or increasing the required bandwidth. Thus, FM signals are well suited to high efficiency (inevitably nonlinear) PAs.

Analog FM is more widely used that analog PM, perhaps because it is easier to perform frequency modulation and demodulation. Figure 2.30 illustrates a simple frequency modulator, where the resonance frequency of an LC oscillator is varied in proportion to the amplitude of the baseband signal. The voltage dependent capacitor required here is usually a reverse biased varactor diode whose junction capacitance varies with $s_{BB}(t)$. This circuit is referred to as a voltage-controlled oscillator (VCO). We have much more to say about VCOs later.

2.3.2.1 Wideband FM Modulator

An example of an FM modulator suitable for modulation of wideband signals that uses a nonlinear device is illustrated in Figure 2.31. $s_1(t)$ is the product of $s_m(t)$ and $s_{LO}(t)$. This signal is applied to a nonlinear device generating $s_2(t)$, which has the form of (2.32). $s_2(t)$ is filtered with a band-pass filter according to the appropriate carrier frequency and the result, $s(t)$, is the desired WBFM modulated signal.

2.3.2.2 FM Modulation

The instantaneous frequency of a carrier wave is where the information is contained in FM. (In analog FM applications, the carrier frequency is varied in direct

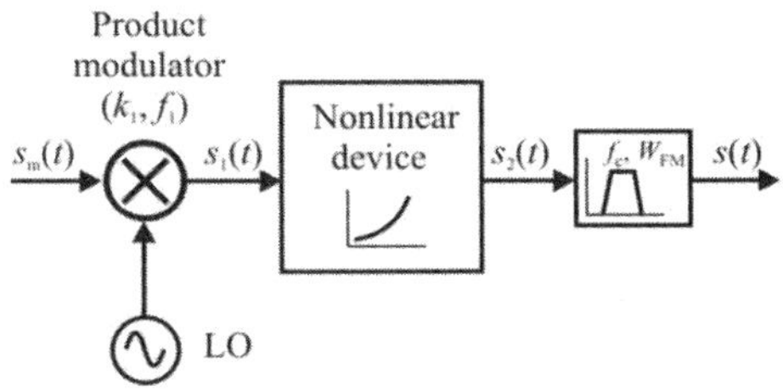

Figure 2.31 WBFM modulator.

proportion to changes in the amplitude of an input signal.) Digital data can be represented by shifting the carrier frequency among a set of discrete values, a technique known as frequency-shift keying. We cover the digital modulations later.

FM is commonly used at VHF radio frequencies for military *push-to-talk* (PTT) applications. It is also used for high-fidelity broadcasts of music and speech (FM broadcasting) in the same frequency range. Analog TV sound is also broadcast using FM. A narrowband form is used for voice communications in commercial and amateur radio settings, as well. The type of FM used in broadcast is generally called wideband FM, or WBFM. In two-way radio, narrowband FM (NBFM) is used to conserve bandwidth. In the military VHF frequency range, the spectrum is very crowded and it is imperative to use as little bandwidth for each radio network as possible.

Figure 2.27 illustrates an example of FM. Edwin Armstrong presented his paper, "A Method of Reducing Disturbances in Radio Signaling by a System of Frequency Modulation," which first described FM radio, before the New York section of the Institute of Radio Engineers on November 6, 1935.

FM requires a wider bandwidth than amplitude modulation for an equivalent modulating signal, but this also makes the signal more robust against noise and interference. FM is also more robust against simple signal amplitude fading phenomena since the information is conveyed in the instantaneous frequency and not the amplitude.

FM receivers inherently exhibit a phenomenon called *capture*, where the tuner is able to clearly receive the stronger of two stations being broadcast on the same frequency. Problematically, however, frequency drift or lack of selectivity may cause one station or signal to be suddenly overtaken by another on an adjacent channel. Frequency drift typically constituted a problem on very old or inexpensive receivers, while inadequate selectivity may plague any tuner.

The capture effect may be taken advantage of in EW settings. The EW signal need only be somewhat larger than the target signal at the target receiver to be effective at precluding successful communication. It has been demonstrated many times that a *jam-to-signal ratio* (JSR) of 0 dB is normally sufficient to accomplish this task.

2.3.2.3 Mathematical Basis of FM

If the information signal to be transmitted is given by $m(t)$, which is restricted in amplitude to be

$$|m(t)| \leq 1 \tag{2.35}$$

and the sinusoidal carrier is

$$x_{\mathrm{c}}(t) = A\cos(2\pi f_{\mathrm{c}} t) \tag{2.36}$$

where f_c is the carrier's base frequency in hertz and A is an arbitrary amplitude, the carrier will be modulated by the signal as in

$$x_{\mathrm{c}}(t) = A\cos\left(2\pi\int_0^t f(u)du\right) = A\cos\left(2\pi\int_0^t [f_{\mathrm{c}} + f_{\Delta} m(u)]du\right) \tag{2.37}$$

where, $f(t) = f_{\mathrm{c}} + f_{\Delta x_m}(t)$.

In this equation, $f(t)$ is the *instantaneous frequency* of the oscillator and f_Δ is the *frequency deviation*, which represents the maximum shift away from f_c in one direction, assuming $m(t)$ is limited to the range ±1.

For the simplified case when the modulating signal is a sine wave, the harmonic distribution of a sine wave carrier signal modulated by another sine wave modulating signal can be represented with Bessel functions—this provides a basis for a mathematical understanding of frequency modulation in the frequency domain.

2.3.2.4 Carson's Rule

Another equivalent expression for the *Carson's rule* given in (2.34), is that nearly all the power of an FM signal (98%) lies within a bandwidth of

$$W = 2(f_{\Delta} + f_{\mathrm{m}}) \tag{2.38}$$

where f_Δ is the peak deviation of the instantaneous frequency $f(t)$ from f_c (assuming $m(t)$ is in the range ±1) and f_m is the highest modulating frequency of $m(t)$.

2.3.2.5 Modulation Index

As with other modulation indices, in AM this quantity indicates by how much the modulated variable varies around its unmodulated level. For FM, it relates to the variations in the frequency of the carrier signal:

$$h = \frac{\Delta f}{f_\mathrm{m}} = \frac{f_\Delta |m(t)|}{f_\mathrm{m}} \tag{2.39}$$

With a tone-modulated FM wave, if the modulation frequency is held constant and the modulation index is increased, the (nonnegligible) bandwidth of the FM signal increases, but the spacing between spectra stays the same.

If the frequency deviation is held constant and the modulation index increased, the bandwidth stays roughly the same, but the spacing between spectra decreases.

2.3.3 PM Modulation

PM is a form of modulation in which the instantaneous phase of a carrier wave contains the information to be conveyed. Unlike its more popular counterpart, FM, PM is not very widely used for analog communication applications. This is because it tends to require more complex receiving hardware and there can be ambiguity problems with determining whether, for example, the signal has 0° phase or 180° phase.

Suppose that the signal to be sent is as sinusoid with frequency ω_m and phase ϕ_m given by

$$m(t) = M\sin(\omega_\mathrm{m} t + \phi_\mathrm{m}) \tag{2.40}$$

and the carrier onto which the signal is to be modulated is given by

$$c(t) = C\sin(\omega_\mathrm{c} t + \phi_\mathrm{c}) \tag{2.41}$$

The resulting modulated signal is given by

$$y(t) = A_\mathrm{c}\sin(\omega_\mathrm{c} t + m(t) + \phi_\mathrm{c}) \tag{2.42}$$

Thus PM can be viewed as a change to the frequency of the signal and it can be considered a special case of FM where the transmitted signal frequency modulation is the time derivative of the modulating signal.

The precise spectral behavior of PM is difficult to determine, but there are two regions of particular interest:

- Small amplitude signals;
- Single large sinusoidal signal.

For small amplitude signals, PM is similar to AM and exhibits similar doubling of baseband bandwidth and the resulting poor efficiency.

2.3.3.1 PM Modulation Index

As with other modulation indices, in PM the modulation index indicates by how much the modulated variable varies around its unmodulated level. For PM, it relates to the variations in the phase of the carrier signal:

$$h = \Delta\theta \tag{2.43}$$

where $\Delta\theta$ is the peak phase deviation.

2.3.3.2 Carson's Rule

For a single large sinusoidal signal, PM is similar to FM, and its bandwidth is approximately

$$W = 2(h+1)f_m \tag{2.44}$$

where $f_m = \omega_m / 2\pi$ and h is the modulation index given by (2.43). Equation (2.44) is known as the Carson rule for PM.

2.3.4 FM and PM Modulators

Perhaps the most obvious method of FM modulating a carrier is to use a VCO. The amplitude of the modulating signal then determines the frequency of the carrier signal. This technique can be used for both narrowband and wideband FM modulation, although practical limits impose constraints on how narrowband to which the modulation can be limited.

An FM modulator for transmission of narrowband FM (NBFM) consists of an integrator followed by a product modulator, as shown in Figure 2.32.

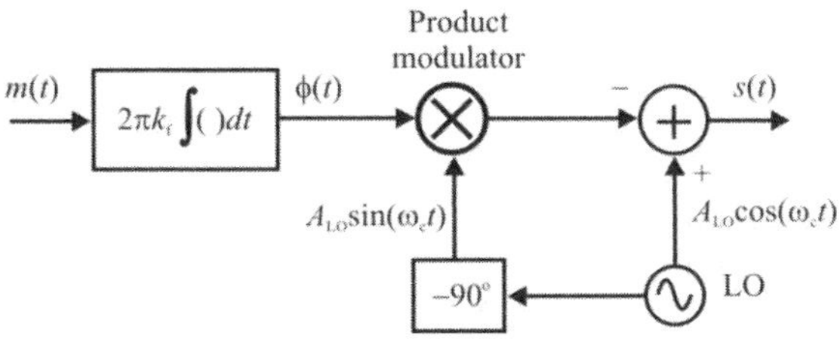

Figure 2.32 NBFM modulator.

$$s_1(t) = A_c \cos[\omega_1 t + 2\pi k_1 \int_0^t m(u)du] \quad (2.45)$$

$$s_2(t) = a_1 s_1(t) + a_2 s_1^2(t) + \cdots + a_n s_1^n(t) \quad (2.46)$$

$$s_2(t) = a_n A_c^n \cos[\omega_1 t + 2\pi k_1 \int_0^t m(u)du] + \text{other terms} \quad (2.47)$$

$$= A\cos[\overset{\omega_c}{n\omega_1}\, t + 2\pi\, \overset{k_f}{nk_1} \int_0^t m(u)du] + \text{other terms} \quad (2.48)$$

2.3.4.1 Voltage-Controlled Oscillators

Any device that can convert between a voltage or current source to a variable frequency can be used as an FM modulator. One popular topology for this purpose is the VCO, one of which is shown in Figure 2.33.

A VCO is a device that varies its oscillating frequency according to a control voltage, hence its name. In this case, a varactor is used in the oscillator's resonant

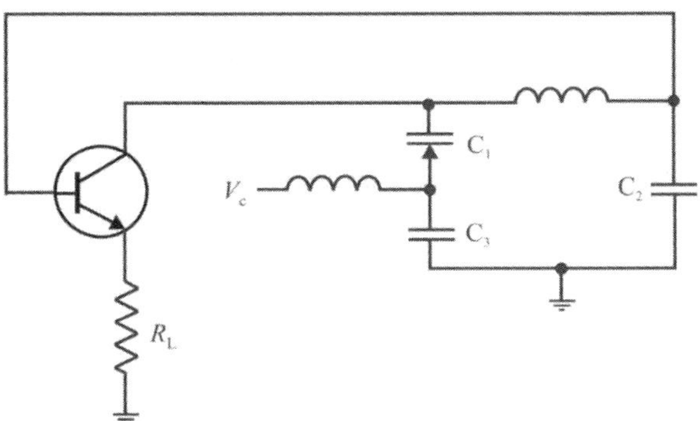

Figure 2.33 A VCO implemented through a π network. The frequency is adjusted by adjusting the reverse bias on the varactor diode, C_1.

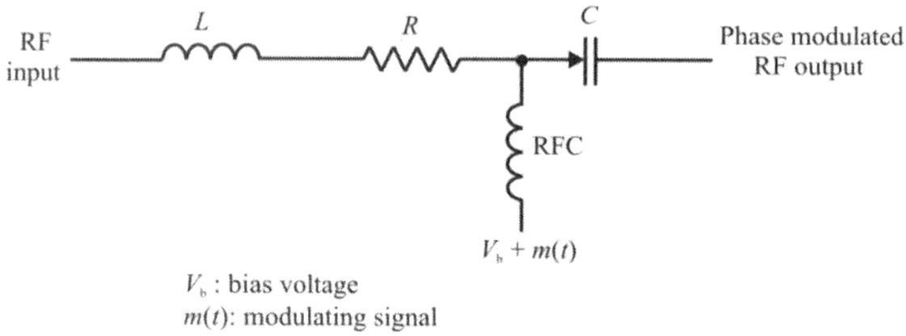

Figure 2.34 LC phase modulator.

circuit (which establishes the operating frequency). VCOs are discussed at length in Chapter 3 where their applications to phase locked loops are presented.

2.3.4.2 Tuned Circuit LC Phase Modulator

The tuned circuit or LC phase modulator is one of several methods used to generate phase modulation for communications signals. Although there is a myriad of designs of LC phase modulators, we describe here a representative circuit and present some of its characteristics [4].

There are various designs of phase modulators that are referred to as reactance modulators, and the LC phase modulator, shown in Figure 2.34, is one form. The LC modulator is a circuit that is resonant at the frequency of the carrier that is to be phase modulated. The circuit can be a parallel or series resonant circuit; however, the series circuit is the most common. Whether the circuit is parallel or series, the operating principle is the same. The value of the inductor and the capacitor are selected to be resonant at the frequency of the RF input. The resistor value sets the Q (and therefore the bandwidth) of the modulator. Since inductors have finite Q, the resistor also includes the resistance of the inductor.

The voltage variable capacitor is a varactor diode described above. The bias voltage (V_b) sets the desired capacitance and inductor to be resonant at the RF input frequency. When the modulation voltage (which is summed with the bias voltage) is applied, the resonant frequency of the circuit varies above and below the center frequency according to the modulation voltage magnitude.

Figure 2.35 shows the amplitude and phase of a single tuned resonant circuit. The plot is normalized so units are independent of Q. The plot shows the phase shift of 45° (0.785 rad) at the –3-dB points on the amplitude curve. The curve for phase goes from +90° to –90° with zero degrees at resonance. The amplitude of the modulating signal applied to the varactor shifts the capacitance of the varactor. This changing capacitance changes the resonant frequency of the resonator (Δf) resulting in a phase shift of the output following the phase response curve, thus producing a phase-modulated carrier at the output.

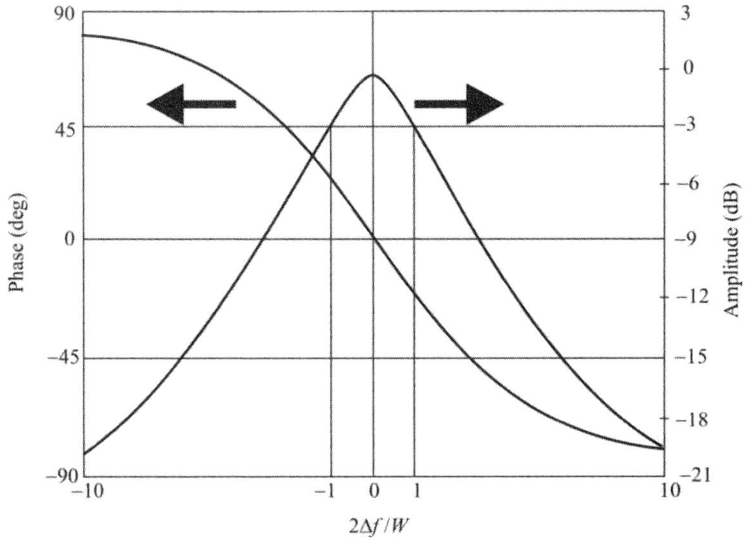

Figure 2.35 Tuned circuit response.

Figure 2.36 shows typical distortion characteristics of the LC phase modulator. The third harmonic is the dominant distortion product from the LC phase modulator. The rising response of the FM discriminator then causes the third harmonic to be 10 dB higher relative to the fundamental than it would be after de-emphasis to restore the modulation of the phase-modulated signal.

2.3.4.3 Armstrong Phase Modulator

FM is normally generated directly, for example, with a VCO. The *Armstrong phase modulator* employs an indirect method of generating FM [5]. In this technique, the

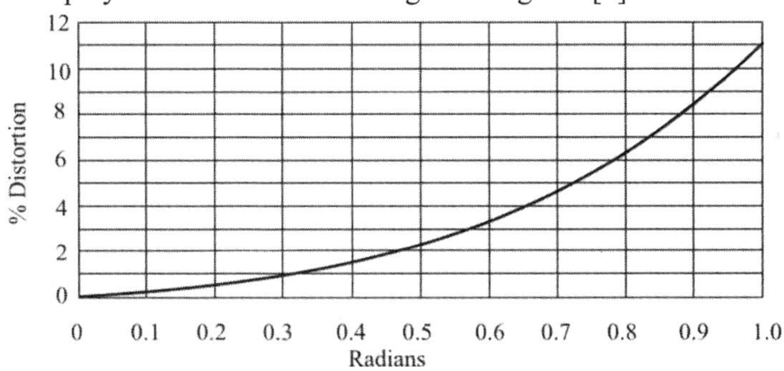

Figure 2.36 LC phase modulator distortion.

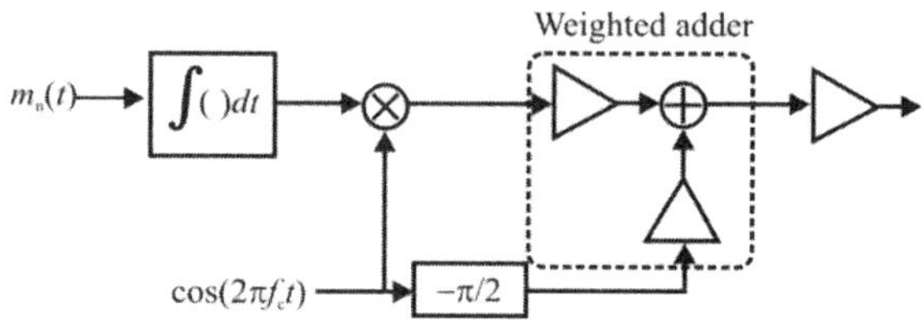

Figure 2.37 Armstrong phase modulator.

message signal is integrated; this produces a new signal whose derivative is the original message signal. Phase modulation of this new signal onto the carrier produces the FM signal

$$y(t) = \cos\left[2\pi f_c t + 2\pi\Delta f \int^{t} m_n(u)du\right] \tag{2.49}$$

where f_c is the nominal carrier frequency, Δf is the frequency deviation, and $m_n(t)$ is the normalized message signal $(|m_n(t)| \le 1)$.. The instantaneous frequency of this carrier is found by taking the derivative of the argument of the cosine and then dividing by 2π. According to the *fundamental theorem of calculus*,

$$\frac{d}{dt}\int^{t} m_n(u)du = m_n(t) \tag{2.50}$$

and we get

$$\frac{1}{2\pi}\frac{d}{dt}\left[2\pi f_c t + 2\pi\Delta f \int^{t} m_n(u)du\right] = f_c + \Delta f m_n(t) \tag{2.51}$$

The instantaneous frequency for the carrier of (2.49) is $f_c + \Delta f m_n(t)$. In other words, the instantaneous frequency is offset from the nominal carrier frequency by an amount that is proportional to the message signal.

The Armstrong phase modulator is only an approximation of a true phase modulator; however, it is a good approximation for small modulation indices. The combination of an integrator and an Armstrong modulator is illustrated in Figure 2.37; this represents an approximation to a frequency modulator. The weighting factors in the weighted adder are shown explicitly as amplifiers.

The output of the integrator is

$$\int^{t} m_{n}(u)du$$

If the message signal is a sinusoid, the integrator output is a sinusoid having the same frequency; but the input and output amplitudes and phases are different.

The Armstrong phase modulator looks similar to a modulator for AM. For both modulators, a DSB term and a residual carrier are added together. Furthermore, the spectrum of the output from the Armstrong modulator looks similar to the spectrum of AM. For a sinusoidal message signal, there are three spectral lines: one from the residual carrier and two from the DSB terms. However, an Armstrong modulator does not produce an AM carrier. In the Armstrong modulator the phase shifter (which delays the residual carrier by $\pi/2$ radians) makes the difference.

In the remainder of this section, the assumption is made that $m_n(t)$ is $\cos(2\pi f_m t)$. Noting that $\cos(2\pi f_m t)$ with a delay of $\pi/2$ radians is $\sin(2\pi f_m t)$, the output of the Armstrong modulator is of the form

$$\begin{aligned} &a\sin(2\pi f_c t) + b\sin(2\pi f_m t)\cos(2\pi f_c t) \\ &\quad = \sqrt{a^2 + b^2\sin^2(2\pi f_m t)}\,\sin\left[2\pi f_c t + \tan^{-1}(b\sin(2\pi f_m t)/a\right] \end{aligned} \tag{2.52}$$

This signal shows both amplitude and angle modulation.

If $|b|$ is small compared with $|a|$ then (2.52) can be approximated as

$$\begin{aligned} &\sqrt{a^2 + b^2\sin^2(2\pi f_m t)}\,\sin\left[2\pi f_c t + \tan^{-1}(b\sin(2\pi f_m t)/a)\right] \\ &\quad \approx |a|\sin\left[2\pi f_c t + \beta\sin(2\pi f_m t)\right] \end{aligned} \tag{2.53}$$

where

$$\beta = \tan^{-1}\left(\frac{b}{a}\right) \tag{2.54}$$

is the modulation index. A comparison of (2.53) and the expression for an FM signal given by

$$y(t) = \cos\left[2\pi f_c t + \beta\sin(2\pi f_m t)\right] \tag{2.55}$$

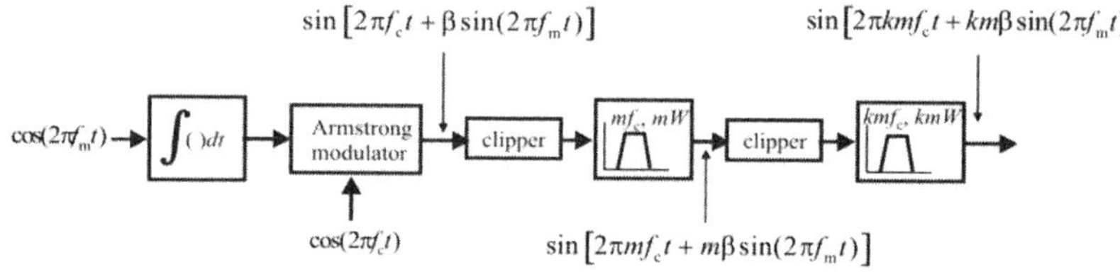

Figure 2.38 Frequency multiplier.

shows that the rhs of (2.53) has the form of FM with a sinusoidal message signal.

The generation of FM by the indirect method, employing an integrator and phase modulator, has an important practical advantage over the direct method. A VCO typically has poor frequency stability, so a carrier generated from a VCO (direct method) exhibits frequency drift. With the indirect method a very stable oscillator (usually a crystal oscillator) can be used as the source of the carrier.

2.3.4.4 Frequency Multiplication

An integrator followed by an Armstrong phase modulator can produce a good approximation of FM when the desired modulation index β is small. This is called NBFM, because the modulation index and therefore the bandwidth is small. However, the well-known performance advantage of FM over AM (for the target signals) occurs only when β is large. However, there is a way to use an Armstrong modulator and yet still achieve a large β. This is illustrated in Figure 2.38.

In other words, the FM is generated with an integrator followed by an Armstrong phase modulator. The carrier frequency f_c and the modulation index β at the output of the Armstrong modulator are both smaller than desired. This NBFM signal is placed at the input of a (memoryless) nonlinear device (such as a clipper), which generates harmonics. Then the mth harmonic is selected by a bandpass filter, having a passband centered at mf_c. This output of this band-pass filter is a carrier frequency of frequency mf_c, having FM with modulation index $m\beta$. The combination of the nonlinear device and the band-pass filter is called a *frequency multiplier*. Actually, *angle multiplier* would be a better appellation, because it is the angle (the argument of the carrier sinusoid) that is multiplied. A second frequency multiplier can be used to further increase both the modulation index and the carrier frequency. If the second multiplier employs a factor k (selects the kth harmonic), then the output of this second frequency multiplier is a carrier of frequency kmf_c, having FM with modulation index $km\beta$. A summary of the effects of frequency multiplication are given by

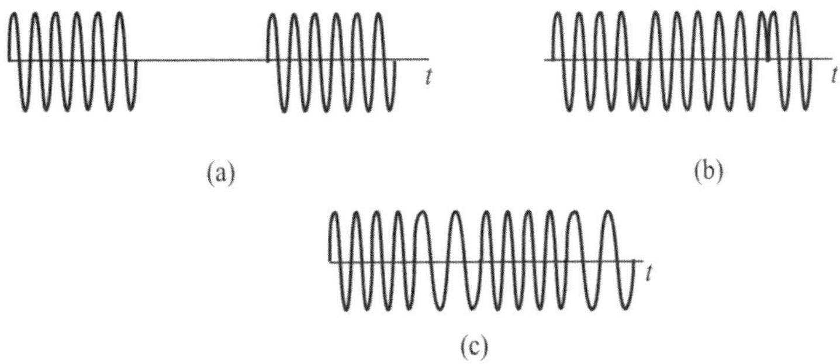

Figure 2.39 Digital modulations: (a) ASK, (b) PSK, and (c) FSK.

$$f_{\text{cfinal}} = kmf_{\text{c}} \tag{2.56}$$

and

$$\beta_{\text{final}} = km\beta \tag{2.57}$$

In this way, quite large modulation indices can be achieved. Because it is the indirect method for generating FM, which permits use of a crystal oscillator for the frequency f_c, the carrier can have excellent frequency stability.

2.4 Digital Modulations

2.4.1 Introduction

In digital RF systems, the carrier is modulated by a digital baseband signal. Digital modulation offers many advantages over analog modulations and is widely used in modern wireless systems.

The digital counterparts of AM, PM, and FM are called *amplitude sift keying* (ASK), *phase shift keying* (PSK), and *frequency shift keying* (FSK), respectively. Figure 2.39 illustrates examples of these waveforms for a binary baseband signal. In RF applications, PSK and FSK find wider usage than does ASK primarily because of their lower sensitivity to amplitude noise.

Analysis of a modem entails quantifying three parameters: signal quality, spectral efficiency, and power efficiency. The quality of digital signaling methods is expressed in terms of the *bit error rate* (BER), defined as the number of erroneous bits observed at the output of the detector divided by the total number of bits received in a unit time. Thus, the goal is to calculate the probability of error in the presence of noise and other interferers.

2.4.2 Amplitude Shift Keying

2.4.2.1 Introduction

The transmission of digital signals is increasing at a rapid rate. Low-frequency analog signals are often converted to digital format pulse amplitude modulation (PAM) before transmission. The source signals are generally referred to as *baseband* signals. Of course, we can send analog and digital signals directly over a medium. From electro-magnetic theory, for efficient radiation of electrical energy from an antenna it must be at least in the order of magnitude of a wavelength in size; $c = \lambda f$, where c is the velocity of light, f is the signal frequency and λ is the wavelength. For a 1-kHz audio signal, the wavelength is 300 km. An antenna of this size is not practical for efficient transmission. The low-frequency signal is often frequency-translated to a higher frequency range for efficient transmission. The process is called modulation. The use of a higher frequency range reduces antenna size. In the modulation process, the baseband signals constitute the *modulating signal* and the high-frequency *carrier signal* is a sinusoidal waveform. There are three basic ways of modulating a sine wave carrier. For binary digital modulation, they are called binary amplitude-shift keying (BASK), binary frequency-shift keying (BFSK) and binary phases shift keying (BPSK). Modulation also leads to the possibility of frequency multiplexing [6].

In a frequency-multiplexed system, individual signals are transmitted over adjacent, nonoverlapping frequency bands. They are therefore transmitted in parallel and simultaneously in time. If we operate at higher carrier frequencies, more bandwidth is available for frequency-multiplexing more signals.

In ASK, the amplitude of the carrier can be one out of M baseband amplitude functions $a_n(t)$, corresponding to one of the M information symbols [7]. The transmitted signal corresponding to the nth symbol is given by

$$s_n(t) = a_n(t)\cos(2\pi f_c t + \phi) \tag{2.58}$$

The carrier frequency $f(t) = f_c$ and the phase shift $\phi(t) = \phi$ are fixed. The amplitude functions $a_n(t)$ are defined over a finite period of time, that is, $t_0 \le t \le t_0 + t$, where T is the *symbol time*. A simple example would be a system using binary symbols where $M = 2$, and square pulses for amplitude functions according to (see Figure 2.40).

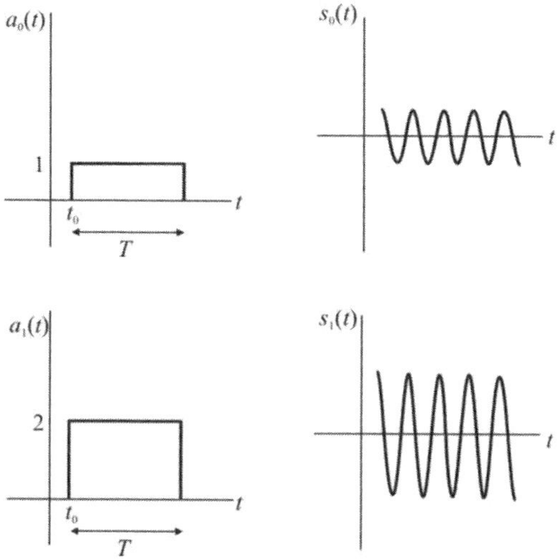

Figure 2.40 Baseband signal and modulated signal for ASK, M = 2.

Symbol "0"

$$a_0(t) = \begin{cases} 1, & t_0 \le t < t_0 + T \\ 0, & \text{otherwise} \end{cases} \Rightarrow s_0(t) = \begin{cases} \cos(2\pi f_c t + \phi), & t_0 \le t < t_0 + T \\ 0, & \text{otherwise} \end{cases} \tag{2.59}$$

Symbol "1"

$$a_1(t) = \begin{cases} 2, & t_0 \le t < t_0 + T \\ 0, & \text{otherwise} \end{cases} \Rightarrow s_1(t) = \begin{cases} 2\cos(2\pi f_c t + \phi), & t_0 \le t < t_0 + T \\ 0, & \text{otherwise} \end{cases} \tag{2.60}$$

2.4.2.2 BASK

We can express the BASK signal as

$$s(t) = Am(t)\cos(2\pi f_c t), \qquad 0 \le t \le T \tag{2.61}$$

where A is a constant, $m(t)$ = 1 or 0, f_c is the carrier frequency, and T is the bit duration. It has a power $P = A^2/2$, so that $A = \sqrt{2P}$. Thus (2.61) can be written as

$$s(t) = \sqrt{2P}\cos 2\pi f_c t, \qquad 0 \le t \le T$$

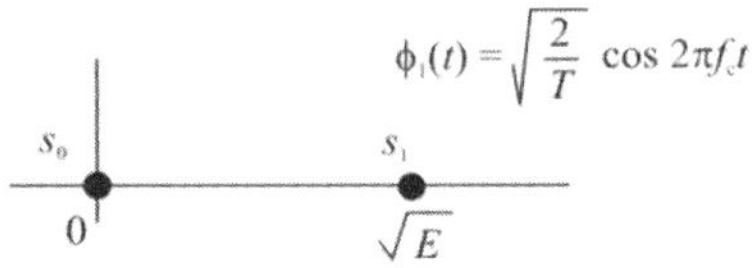

Figure 2.41 BASK signal constellation diagram.

$$= \sqrt{PT}\sqrt{\frac{2}{T}}\cos 2\pi f_c t, \qquad 0 \le t \le T$$
$$= \sqrt{E}\sqrt{\frac{2}{T}}\cos 2\pi f_c t, \qquad 0 \le t \le T \tag{2.62}$$

where $E = P\,T$ is the energy contained in a bit duration. If we take

$$\phi_1(t) = \sqrt{\frac{2}{T}}\cos 2\pi f_c t$$

as the orthonormal basis function, the corresponding constellation diagram of the BASK signals is shown in Figure 2.41.

Figure 2.42 shows the BASK signal sequence generated by the binary sequence 0 1 0 1 0 0 1. The amplitude of a carrier is switched or keyed by the binary

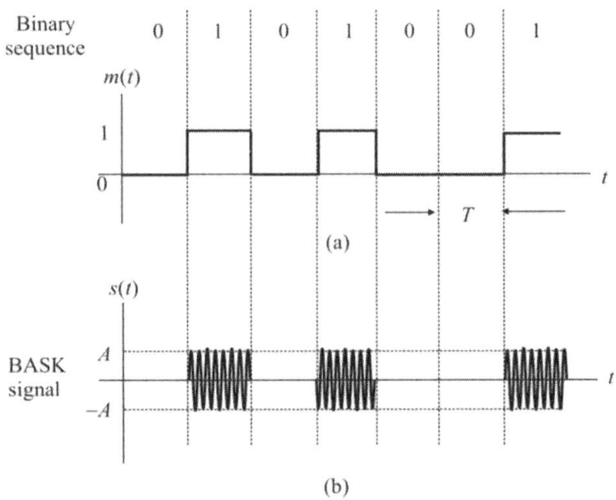

Figure 2.42 (a) Binary modulating signal and (b) BASK signal.

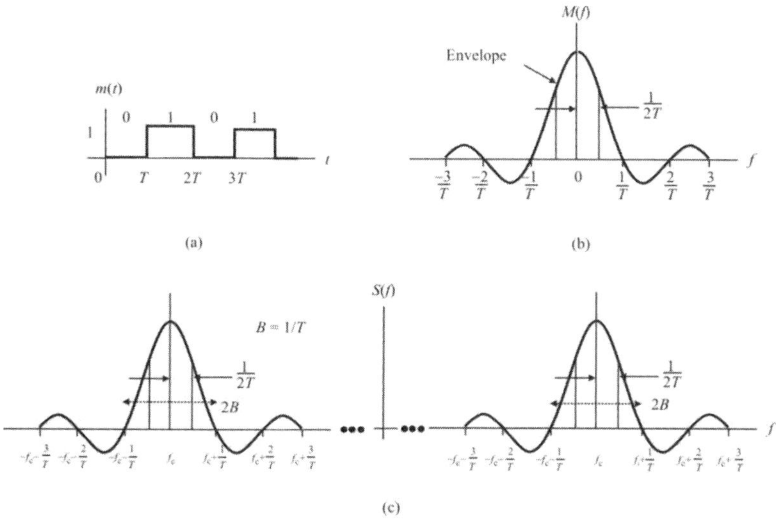

Figure 2.43 (a) Modulating signal, (b) spectrum of (a), and (c) spectrum of BASK signals.

signal *m*(*t*). This is sometimes called *on-off keying* (OOK) since it was the type of modulation originally used for telegraph keying.

The Fourier transform of the BASK signal *s*(*t*) is

$$S(f) = \frac{A}{2}\int_{-\infty}^{\infty}[m(t)e^{j2\pi f_c t}]e^{-j2\pi f t}dt + \frac{A}{2}\int_{-\infty}^{\infty}[m(t)e^{-j2\pi f_c t}]e^{-j2\pi f t}dt$$

$$= \frac{A}{2}M(f - f_c) + \frac{A}{2}M(f + f_c) \tag{2.63}$$

The effect of multiplication by the carrier signal $A\cos 2\pi f_c t$ is simply to shift the spectrum of the modulating signal *m*(*t*) to f_c. Figure 2.43 shows the amplitude spectrum of the BASK signals when *m*(*t*) is a periodic pulse train.

Since we define the bandwidth as the range occupied by the baseband signal *m*(*t*) from 0 Hz to the first zero-crossing point, we have W Hz of bandwidth for the baseband signal and $2W$ Hz for the BASK signal. Figure 2.44 shows the modulator and a possible implementation of the coherent demodulator for BASK signals.

2.4.2.3 M-ary Amplitude-Shift Keying

An *M-ary amplitude-shift keying* (*M*-ASK) signal can be expressed as

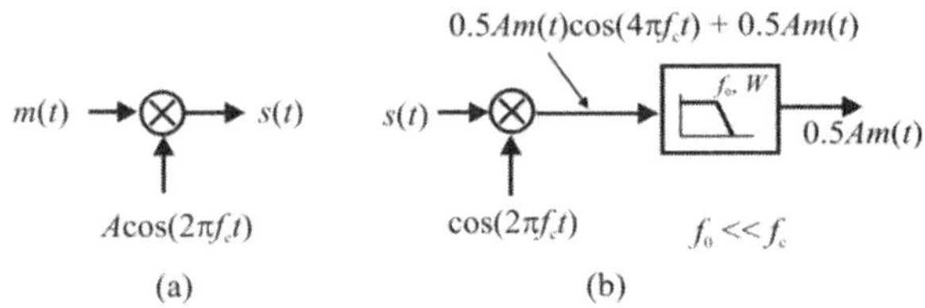

Figure 2.44 (a) BASK modulator and (b) coherent demodulator.

$$s(t) = \begin{cases} A_i \cos 2\pi f_c t, & 0 \le t \le T \\ 0, & \text{otherwise} \end{cases} \tag{2.64}$$

where

$$A_i = A[2i - (M-1)] \tag{2.65}$$

for $i = 0, 1, ..., M-1$ and $M > 2$. Here, A is a constant, f_c is the carrier frequency, and T is the symbol duration. The signal has power $P_i = A_i^2/2$, so that $A_i = \sqrt{2P_i}$. Thus (2.64) can be written as

$$\begin{aligned} s(t) &= \sqrt{2P_i}\sqrt{\frac{2}{T}}\cos 2\pi f_c t, \quad 0 \le t \le T \\ &= \sqrt{P_i T}\sqrt{\frac{2}{T}}\cos 2\pi f_c t, \quad 0 \le t \le T \\ &= \sqrt{E_i}\sqrt{\frac{2}{T}}\cos 2\pi f_c t, \quad 0 \le t \le T \end{aligned} \tag{2.66}$$

where $E_i = P_i T$ is the energy of $s(t)$ contained in a symbol duration for $i = 0, 1, ..., M-1$. Figure 2.45 shows the signal constellation diagrams of M-ASK and 4-ASK signals. Figure 2.46 shows the 4-ASK signal sequence generated by the binary sequence 00 01 10 11. For clarity we assume in Figure 2.46 that the transitions between FSK symbols line up with changes in the data symbols. Figure 2.47 shows

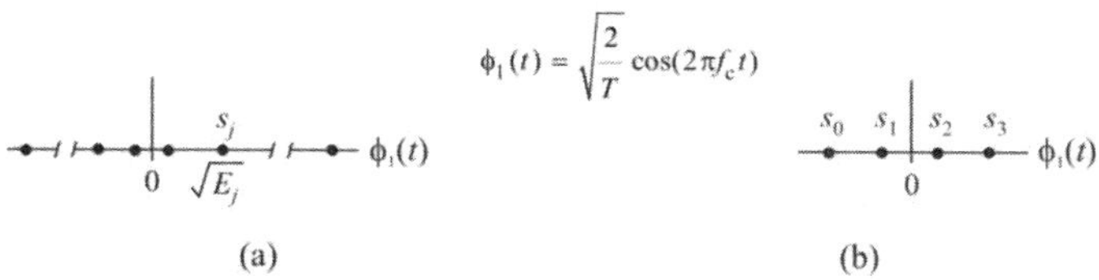

Figure 2.45 Signal constellation diagrams for: (a) *M*-ASK and (b) 4-ASK.

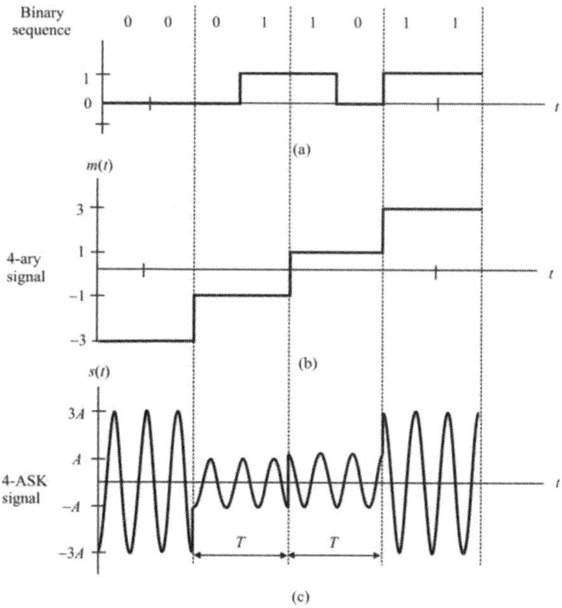

Figure 2.46 4-ASK modulation: (a) binary sequence, (b) 4-ary signal, and (c) 4-ASK signal.

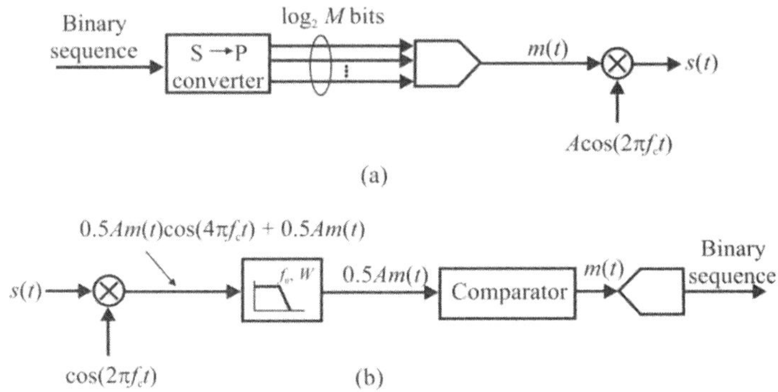

Figure 2.47 *M*-ASK: (a) modulator and (b) coherent demodulator.

the modulator and a possible implementation of the coherent demodulator for *M*-ASK signals.

2.4.3 Pulse Amplitude Modulation: Frequency Shift Keying

M-ASK can be considered to be a kind of PAM system [7, 8]. The data transfer capacity D of such a system is

$$D = \frac{1}{T}\log_2(M) = R\log_2(M) \qquad \text{bits/s} \tag{2.67}$$

where R is the symbol rate. When $M = 2$ for BASK, $\log_2(2) = 1$, and the data transfer capacity becomes $D = 1/T = R$. That is, the symbol time equals the bit time. So, to increase the data transfer capacity we can increase the bit rate R, that is, use a shorter bit (symbol) time T. In so doing, the energy in each symbol is decreased while the background noise remains constant. The result is a lower SNR at the receiver, where estimates of which symbol was transmitted are generated. A lower SNR always corresponds to a higher BER. In addition, higher values to R require more bandwidth to transmit, possibly generating adjacent channel interference when transmit energy from one channel spills over into adjacent channels.

From (2.67) we find that another way of increasing the data transfer capacity is to increase the number of symbols M. In all practical applications, especially tactical EW applications, we only have a limited transmitter power available. Increasing the number of levels M means that the power difference between adjacent levels decreases. Taking the noise into account again, the probability of errors in the receiving process will thereby increase, again increasing the BER. Hence there is a limit to how many symbols M can be used in a transmission system, depending on the level of the noise present.

Another problem with the amplitude functions we have eluded to already—shape of the waveform. A square pulse has a quite wide-frequency spectrum, increasing the risk of causing interference to other users in adjacent frequency bands. For this reason, square pulses are rarely used. Shaping the pulses to generate smoother waveforms is common—a frequently used pulse shape is Gaussian.

2.4.4 FSK

With FSK, the frequency of the signal is determined by the baseband signal, i.e.

$$s_n(t) = A\cos\left[2\pi\int_0^t f_n(u)du + \phi\right] \quad (2.68)$$

The amplitude $a(t) = A$ and the phase shift $\phi(t) = \phi$ are constant. The frequency functions $f_n(t)$ are defined over a finite period of time; that is, $t_0 \le t \le t_0 + t,$ where T is the symbol time, similar to ASK above. A simple example is a system using binary symbols; that is, $M = 2$, and square pulses for frequency functions according to

symbol "0" :

$$f_0(t) = \begin{cases} f_0, & t_0 \le t < t_0 + T \\ 0, & \text{otherwise} \end{cases} \Rightarrow s_0(t) = \begin{cases} a\cos(2\pi f_0 t + \phi), & t_0 \le t < t_0 + T \\ 0, & \text{otherwise} \end{cases} \quad (2.69)$$

symbol "1":

$$f_1(t) = \begin{cases} f_1, & t_0 \le t < t_0 + T \\ 0, & \text{otherwise} \end{cases} \Rightarrow s_1(t) = \begin{cases} a\cos(2\pi f_1 t + \phi), & t_0 \le t < t_0 + T \\ 0, & \text{otherwise} \end{cases} \quad (2.70)$$

(see Figure 2.48). An FSK modulation scheme with $M = 2$ is referred to as a BFSK system. Commonly, the shape of the frequency functions is chosen to be smoother than square pulses, for the spectral occupancy reasons mentioned above.

Figure 2.49 shows a baseband waveform and the corresponding FSK and PSK waveforms associated with it.

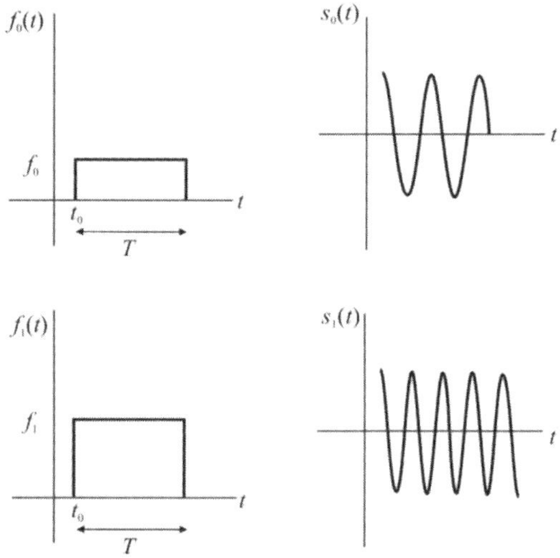

Figure 2.48 Baseband signal and modulated signal for FSK, $M = 2$.

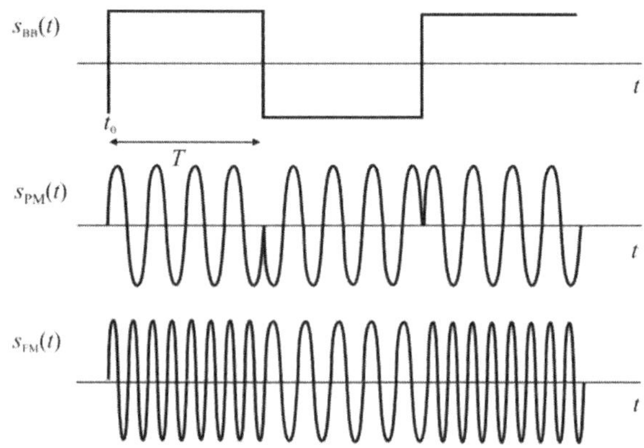

Figure 2.49 PM and FM waveforms (digital).

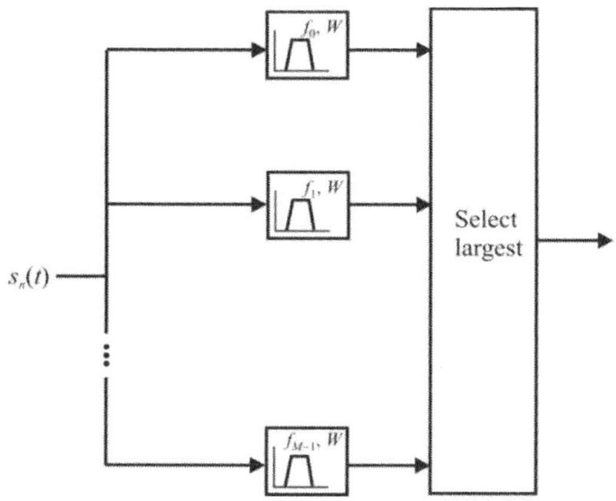

Figure 2.50 Noncoherent FSK detection.

2.4.4.1 FSK Modulators and Demodulators

For the receiver to determine what symbol was transmitted, the frequency of the carrier $s_n(t)$ needs to be determined. This can be accomplished in two ways: noncoherent and coherent detection.

In non-coherent detection, the input signal enters a bank of bandpass filters, where there is a filter for each frequency used; that is, one filter for each symbol used in the system (see Figure 2.50). The outputs of these filters are compared to each other, and the received symbol is assumed to be the one that corresponds to the filter having the strongest output signal magnitude. The phase of the received signal in relation to the phase of the transmitted signal does not matter; the method is noncoherent. This approach is straightforward and quite easy to implement. BFSK systems typically employ noncoherent detection.

In the coherent detection system, the phase of the signal must be known. Hence, phase synchronization between transmitter and receiver is required, which sometimes significantly complicates the system. On the other hand, a coherent system has a greater resistance against noise and interference than a noncoherent system (typically an SNR about 3dB lower is required for the same BER performance). In a coherent system, detection (demodulation) is performed using a cross-correlation technique (see Figure 2.51). The received signal is cross-correlated with locally generated prototype signals $\hat{s}_n(t)$ corresponding to all the M symbols used in the system. Since coherence is needed, the phase of $s'_n(t)$ and the received signal $s_n(t)$ must agree. The prototype signals for our example above would be

symbol "0":

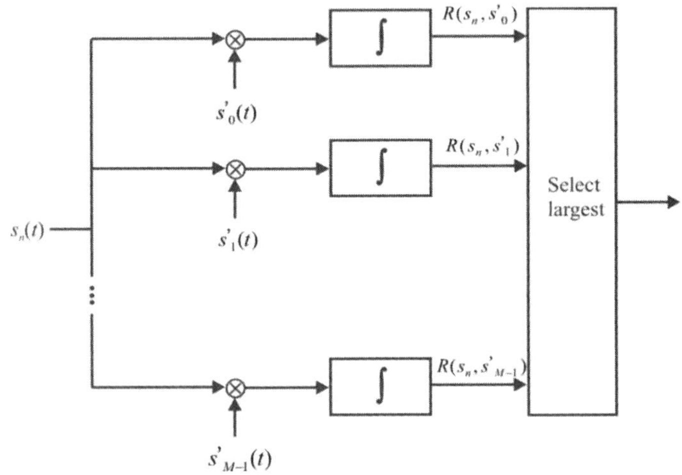

Figure 2.51 Coherent detection of FSK signals using cross-correlation.

$$s'_0(t) = \begin{cases} \cos(2\pi f_0 t + \phi), & t_0 \le t < t_0 + T \\ 0, & \text{otherwise} \end{cases} \tag{2.71}$$

symbol "1":

$$s'_1(t) = \begin{cases} \cos(2\pi f_1 t + \phi), & t_0 \le t < t_0 + T \\ 0, & \text{otherwise} \end{cases} \tag{2.72}$$

The cross-correlation is performed as

$$R_{s_n, s'_n}(\tau) = \int s_n(t) s'_n(t + \tau) dt \tag{2.73}$$

where τ is the delay; that is, phase shift, between $s_n(t)$ and $s'_n(t)$. Since the system is assumed to be coherent, the phase shift between the two signals should ideally be equal to zero; that is, the delay should be an integer number of periods. Therefore, our main interest is the variable $R_{s_n, s'_n}(0)$ where we expect to find maximum correlation. For clarity, the correlation operation in this application will be written as

$$R(s_n, s'_n) = \int_0^T s_n(t) s'_n(t)(t) dt \tag{2.74}$$

The demodulator works as follows. The input signal enters a bank of cross-correlators as in Figure 2.51. The outputs of these correlators are compared to each other, and the received symbol is assumed to be the one that corresponds to the correlator having the strongest output signal.

Coherent detection has a performance advantage over noncoherent detection. This is manifest in the SNR required to achieve a specified BER. Theoretically, and frequently practically, coherent detection can be accomplished with 3-dB lower SNR than noncoherent detection for the same BER.

2.4.4.2 Frequency Shift for Coherent Detection

One obvious question for coherent detection is how large the shift in frequency Δf should be. Obviously, the larger the shift, the more spectrum bandwidth will be occupied, which is typically a detriment, but how small can the shift be for the system to work well? Considering that if the received signal is $s_n(t)$, the correlation to $s'_n(t)$ should be as large as possible, while the (undesired) correlations between $s_n(t)$ and $s_m(t)$ where $m \neq n$ should be a minimum. For $M = 2$ this can be stated as

$$|R(s_0, s'_0)| \to \max, |R(s_1, s'_1)| \to \max, |R(s_0, s'_1)| = 0, |R(s_1, s'_0)| = 0 \tag{2.75}$$

From the last condition in (2.75), we can find the minimum frequency shift required

$$R(s_0, s'_1) = \int_0^T s_0(t) s'_1(t) dt = a \int_0^T \cos(2\pi f_0 t) \cos(2\pi f_1 t) dt = 0 \tag{2.76}$$

where we have assumed $\phi = 0$. Solving (2.76) we find that the frequency shift $|f_0 - f_1| = \Delta f$ needs to be

$$\Delta f = \frac{k}{2T}, \qquad k = 1, 2, 3, \ldots \tag{2.77}$$

If we set $k = 1$ the minimum frequency shift is obtained $\Delta f = 1/2T$. An FSK system using this frequency shift is called a *minimum shift keying* (MSK) system.

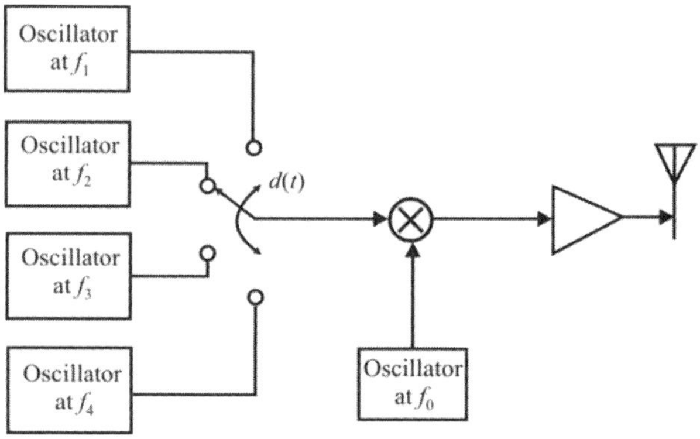

Figure 2.52 4-FSK modulator.

2.4.4.3 MFSK

An MFSK modulator is a switch changing between M carrier oscillators. For example, a 4-FSK modulator is illustrated in Figure 2.52. Although they normally are, the frequencies of these oscillators need not necessarily be adjacent to one another or contiguous. Normal implementations of MFSK incorporate all the data frequencies within the width of one channel, which, in Figure 2.52, is located at f_0. In the military VHF band, for example, the channel width would be 25 kHz, although that is getting narrower as congestion fills up the frequency spectrum.

2.4.5 PSK

In a PSK system, the phase shift of the signal; that is, the phase of a symbol interval compared with the phase of the last symbol interval, is controlled by the baseband signal, [9]:

$$s_n(t) = A\cos[2\pi f_c + \phi_n(t)] \tag{2.78}$$

The carrier frequency $f(t) = f_c$ and the amplitude $a(t) = A$ are constant. The phase functions $\phi_n(t)$ are defined over a finite period of time; that is, $t_0 \le t \le t_0 + t,$ where T is the symbol time as before. A simple example would be a system using binary symbols: $M = 2$ and square pulses for phase functions according to:

symbol “0”:

$$\phi_0(t) = \begin{cases} 0, & t_0 \le t < t_0 + T \\ 0, & \text{otherwise} \end{cases} \Rightarrow s_0(t) = \begin{cases} a\cos(2\pi f_c t), & t_0 \le t < t_0 + T \\ 0, & \text{otherwise} \end{cases} \tag{2.79}$$

symbol "1":

$$\phi_1(t) = \begin{cases} \pi, & t_0 \le t < t_0 + T \\ 0, & \text{otherwise} \end{cases} \Rightarrow s_1(t) = \begin{cases} a\cos(2\pi f_c t + \pi), & t_0 \le t < t_0 + T \\ 0, & \text{otherwise} \end{cases} \tag{2.80}$$

(see Figure 2.53). For spectral occupancy reasons, the shape of the phase functions is normally chosen to be smoother than square pulses. The resulting BPSK signal is as illustrated in Figure 2.54.

Recall that sinusoidal signals can be represented by vectors that have a magnitude and direction. In communication theory these vectors are referred to as *phasors*. If the carrier frequency $f_n = f_c$ is given and constant, the phasor S will represent the signal without any ambiguity, with magnitude

$$a_n = |S_n| \tag{2.81}$$

and phase

$$\phi_n = \angle S_n \tag{2.82}$$

Further, the phasor, being a complex number, can be drawn as a vector starting from the origin and having magnitude a_n and angle ϕ_n in a complex xy-plane.

2.4.5.1 PSK Modulator

A PSK modulator can be implemented as illustrated in Figure 2.34 except in this case $m(t)$ is the digital signal associated with PSK systems. Because it is a phase-based system, however, additional nodes would be required in order to extract the absolute (or relative) phase at the receiver.

Another simple diagram of a BPSK modulator is shown in Figure 2.55 [10]. When there is a positive voltage at the data input port, D_1 and D_3 are ON while D_2 and D_4 are OFF. The carrier signal at the carrier input port will pass through T_1 and couples to the second transformer. The signal then passes through D_1 and D_3 and reaches the primary of T_2. Then the signal will couple to the secondary of T_2. The phase of the waveform at the PSK output terminal will be similar to the phase of the carrier input, as shown in Figure 2.54. On the other hand, when a negative voltage is at the data input port, then D_1 and D_3 are OFF while D_2 and D_4 are ON. Thus, the

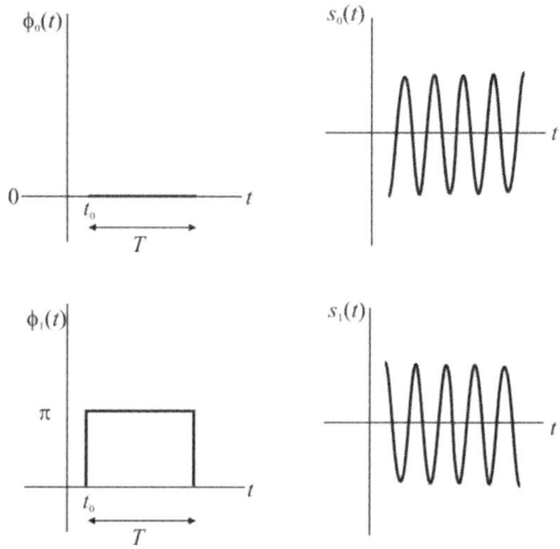

Figure 2.53 Baseband signal and modulated signal for PSK, $M = 2$

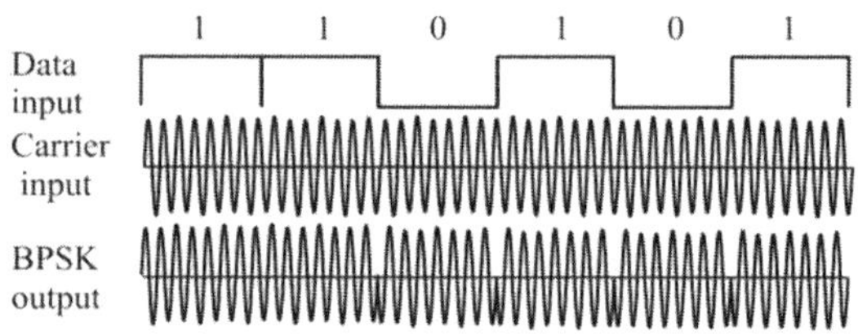

Figure 2.54 BPSK modulation.

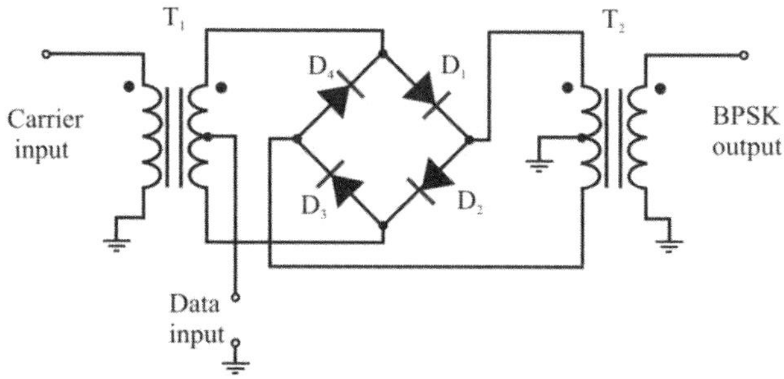

Figure 2.55 A PSK modulator.

phase of the waveform at PSK output terminal is opposite to the phase of the carrier input, thereby implementing a 180° phase shift of the carrier.

2.5 Polar Modulation

Polar modulation is a technique whereby a sinusoidal signal, or carrier, having constant radian frequency ω_c, is time-varied in both magnitude and phase. We can think of polar modulation as transmitting information that both the magnitude (r) and the angle (θ) of a vector simultaneously carry. (We typically reference θ to the positive x-axis). This vector's tail attaches to the origin of the polar-coordinate system, and its head attaches to any other point, P, in the coordinate plane (Figure 2.56).

Any point, P, can be represented in the polar plane by the polar point (r, θ). It should be apparent from Figure 2.56 that we can easily convert a polar coordinate, (r, θ), to a rectangular coordinate, (x, y), by projecting point P onto the x-axis at x_1 and onto the y-axis at y_1. Doing so allows us to convert the polar coordinate, (r, θ), to the rectangular coordinate, (x, y), using

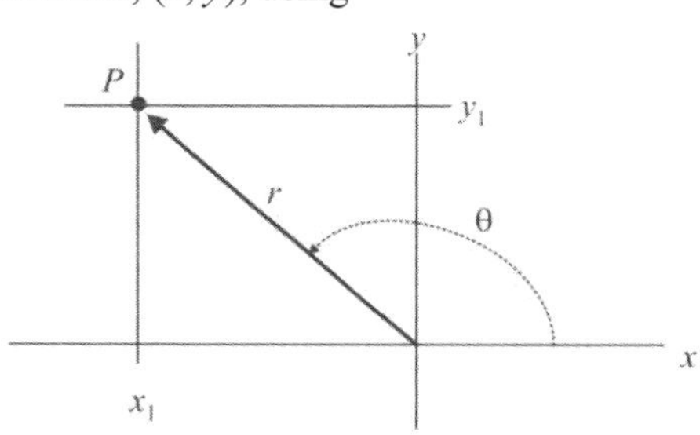

Figure 2.56 We can represent any point, P, in the polar plane by the polar point (r, θ), or convert it to a rectangular coordinate, (x, y), by projecting point P onto the x-axis at x_1 and onto the y-axis at y_1.

$$x = r\cos\theta \tag{2.83}$$

$$y = r\sin\theta \tag{2.84}$$

It is likewise possible to convert a rectangular coordinate (x, y) to a polar coordinate (r, θ) using

$$r = \sqrt{x^2 + y^2} \tag{2.85}$$

$$\theta = \tan^{-1}\frac{y}{x} \tag{2.86}$$

We must take care when performing rectangular-to-polar conversions with (2.86). Most calculators and mathematical-software packages provide a $\tan^{-1}()$ function with only one argument. For example, in $\tan^{-1}(z)$, z is a floating-point number that represents the quotient, y/x, in (2.86), but using the expression "y / x" to provide the z argument for the $\tan^{-1}()$ function presents two problems. First, when the value of x is zero (i.e., on the y-axis) or both x and y are zero (i.e., at the origin), the value of y / x is undefined. Second, a single value of z can represent two angles. For example, $\tan^{-1}(1)$ typically returns a value of $\pi/4$ (0.785 rad). However, $5\pi/4$ (2.927 rad) is an equally valid result.

We can accomplish polar modulation by varying the magnitude of a carrier signal over time, the phase angle of a carrier over time, or both. The varying magnitude and phase carry the information to be transmitted. In polar form, this varying magnitude and phase are time variations in the position of point P in Figure 2.56. That is, the location of P varies over time, which amounts to simultaneously varying r and θ.

For example, suppose the information to be transmitted is a sinusoid of frequency f_0 and peak amplitude A. We can represent this relationship on the polar-coordinate system by making point P follow a circular path. This circle centers on the origin and has a radius of A units. We realize the frequency component by making point P complete one revolution around the circle every $1/f_0$ seconds. In this case, the polar variable, r, has no time variation but is fixed at a constant value, A. The polar variable, θ, on the other hand, varies linearly with time, and its value at any given instant is $\theta = 2\pi f_0 t$. This type of signal, a rotating vector of constant magnitude, is a phasor. The modulation occurs when we impose the information signal—in this case, a phasor—upon the carrier signal.

The most common method of performing polar modulation uses a quadrature modulator, as illustrated in Figure 2.57. $I(t)$ and $Q(t)$ are the input signals to the modulator and carry the information to be transmitted. The quadrature modulator

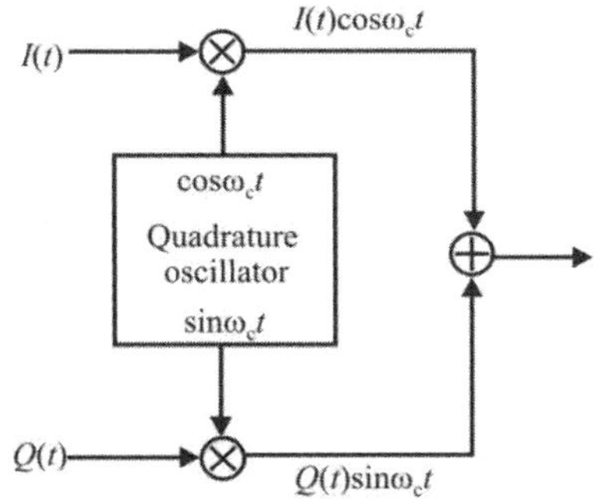

Figure 2.57 The quadrature modulator implements polar modulation by polar-to-rectangular conversion.

implements polar modulation by means of the polar-to-rectangular conversion of (2.83) and (2.84) [11]. That is, we must convert the information signal from polar form (r, θ) to rectangular form (x, y). We can represent the information signal, which is encoded in the time-varying position of P, in polar form as

$$P(t) = [r(t), \theta(t)] \tag{2.87}$$

That is, the position of point P is a function of time—hence, the $P(t)$ designation—and, at any time, t, the location of P is given by the time-dependent values of r and θ [$r(t)$ and $\theta(t)$, respectively]. Equations (2.83) and (2.84) provide the means of converting (2.87) to time-varying x and y signals as

$$x(t) = r(t)\cos\theta(t) \tag{2.88}$$

$$y(t) = r(t)\sin\theta(t) \tag{2.89}$$

In Figure 2.58, input $I(t)$ is identical to $x(t)$ in (2.88), and $Q(t)$ is identical to $y(t)$ in (2.89). Therefore, the information to be transmitted is encoded into the $I(t)$ and $Q(t)$ signals and mixed with the carrier signal. We can think of the process of mixing the $I(t)$ and $Q(t)$ signals with the carrier signal as a multiplication process—hence, the multiplier symbols in Figure 2.57. We add the mixed $I(t)$ and $Q(t)$

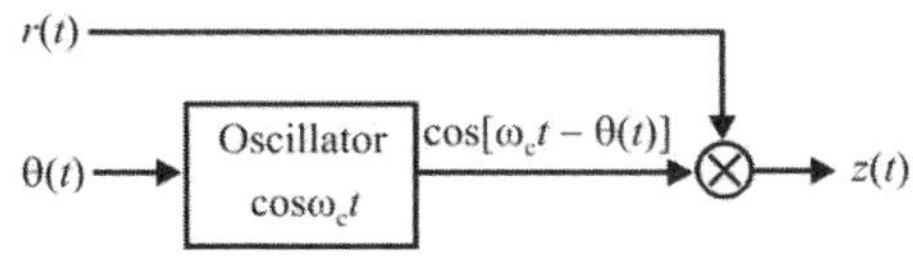

Figure 2.58 Using only a single multiplier, the direct method of polar modulation allows us to apply coding information in polar form.

signals—hence, the summing symbol in Figure 2.57—to produce the final output. It produces the form

$$g(t) = I(t)\cos(\omega_c t) + Q(t)\sin(\omega_c t) \tag{2.90}$$

which is the output signal of the quadrature modulator:

Substitution for $I(t)$ and $Q(t)$ based on (2.88) and (2.89) yields

$$g(t) = \{r(t)\cos[\theta(t)]\}\cos(\omega_c t) + \{r(t)\sin[\theta(t)]\}\sin(\omega_c t)$$

For the moment we drop the function-of-time notation, replacing $g(t)$, $r(t)$, $\theta(t)$, and $\omega_c t$ with variables g, r, θ, and ϕ, respectively. This step allows us to rewrite (2.90) as

$$\begin{aligned} g &= r\cos\theta\cos\phi + r\sin\theta\sin\phi \\ &= r(\cos\theta\cos\phi + \sin\theta\sin\phi) \end{aligned} \tag{2.91}$$

Note that the expression for g contains products of both cosines and sines. We use the following identities for each of these products, as

$$\cos\alpha\cos\beta = \frac{1}{2}\cos(\alpha+\beta) + \frac{1}{2}\cos(\alpha-\beta) \tag{2.92}$$

$$\sin\alpha\sin\beta = \frac{1}{2}\sin(\alpha-\beta) - \frac{1}{2}\sin(\alpha+\beta) \tag{2.93}$$

Substituting (2.92) and (2.93) into (2.91) yields the compact form

$$g = r\cos(\theta - \phi) \tag{2.94}$$

However, since the cosine function is an even function, that is, $\cos(x) = \cos(-x)$, and so (2.94) can be written as

$$g = r\cos(\phi - \theta) \tag{2.95}$$

Restoring the time dependency of the variables in (2.95) yields the output signal of Figure 2.58

$$g(t) = r(t)\cos[\omega_c t - \theta(t)] \tag{2.96}$$

Recall that the input signals in Figure 2.57, I and Q, are the time-varying rectangular coordinates of the information to be transmitted. Equations (2.90)–(2.96) reveal that the output of the quadrature modulator takes the form of a polar equation. That is, the $\cos(\omega_c t)$ portion of (2.96) establishes a phasor (Figure 2.56) that rotates with radian frequency, ω_c (the carrier signal). However, the phasor magnitude varies with time as $r(t)$ prescribes, and its instantaneous angle deviates from $\omega_c t$ as $\theta(t)$ prescribes. Therefore, $r(t)$ and $\theta(t)$ are the polar-form representation of the original rectangular information that $I(t)$ and $Q(t)$ carry.

This result leads to a direct method of polar modulation (Figure 2.58). Instead of coding the information to be transmitted into rectangular form, we can apply it in polar form. One advantage of this technique is that it requires only one multiplier.

By inspection, the output signal, $z(t)$, is

$$z(t) = r(t)\cos[\omega_c t - \theta(t)] \tag{2.97}$$

We see that $z(t)$ is identical to $g(t)$, which is the output signal for the quadrature modulator of Figure 2.57. Hence, we have established the equivalency of the two forms of modulation. Once again, $r(t)$ is the time-varying amplitude, and $\theta(t)$ is the time-varying phase of the information to be transmitted.

Once we establish the two methods for generating a polar-modulated signal (Figure 2.57 and Figure 2.58), we can explore the common QAM (quadrature-amplitude- modulation) case of polar modulation. The name reflects the usual method by which we implement QAM—the quadrature modulator of Figure 2.57. However, we can also accomplish QAM using the direct polar-modulation form of Figure 2.58 by properly controlling the amplitude and the phase of the carrier signal according to (2.97). We use QAM when we want to transmit data that has the form of a serial bit stream but when the spectrum of the output signal must occupy a relatively narrow bandwidth. We accomplish this task by grouping consecutive bits in the original bit stream into M-bit groups, which requires a serial-to-parallel conversion. Each M-bit group, or symbol, can represent 2^M possible binary combinations. For example, if we collect the incoming bits into groups of four with M equal to four, then each symbol has 16 possible binary combinations.

The advantage of grouping bits into symbols is that we can transmit the symbols at a rate that is $1/M^{\text{th}}$ the rate of the original bit stream. The symbol, or baud, rate is always less than the bit rate for $M > 1$, so it proportionally reduces the bandwidth that the transmitted data occupies. Note that we indirectly use the value of M to identify various QAM schemes. For example, the case of $M = 4$ is 16-QAM ($2^M = 16$). Other cases of M—6 and 8, for example—yield 64-QAM and 256-QAM, respectively.

To transmit the QAM data, the value of a symbol takes a unique position in a constellation map. That is, each point in the constellation represents a unique

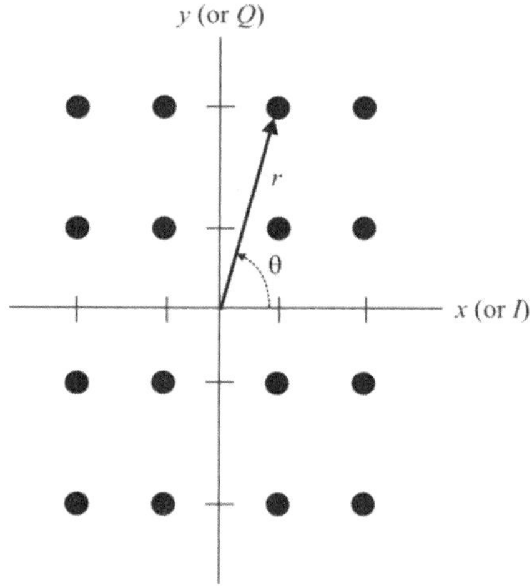

Figure 2.59 Each point in the constellation map for 16-QAM represents a unique combination of bits in the symbol.

combination of bits in the symbol. Figure 2.59 shows the constellation for 16QAM. For example, we could assign the 4-bit combination 0000 to the point at the upper right of Figure 2.59 and assign the 4-bit combination 0001 to the point that the vector indicates. Similarly, each of the 16 possible values of the symbol gets a unique point in the constellation.

If we use QAM, then the (x, y) coordinate of each constellation point embeds the transmitted information. That is, $I(t)$ carries the x information, and $Q(t)$ carries the y information. Alternatively, if we employ direct polar modulation, then the (r, θ) coordinate of each constellation point contains the information. The simplicity of the constellation diagram is one reason that it often appears in the literature on the subject of data transmission. We must remember however, that the vector in Figure 2.59 represents a time-dependent amplitude and phase deviation of the carrier phasor. As such, the vector in Figure 2.59 must move smoothly from point to point and make no instantaneous jump between points. Instantaneous jumps require a broad frequency spectrum, which is contrary to the purpose of using QAM as a data-transmission scheme. This technique is referred to as *pulse shaping* [12, 13].

2.6 Noise Generators

One of the most successful types of modulations for FM EA is FM by modulation with a white Gaussian noise source. It has been shown that with some very reasonable assumptions about the signal to be interfered with (jammed), a Gaussian

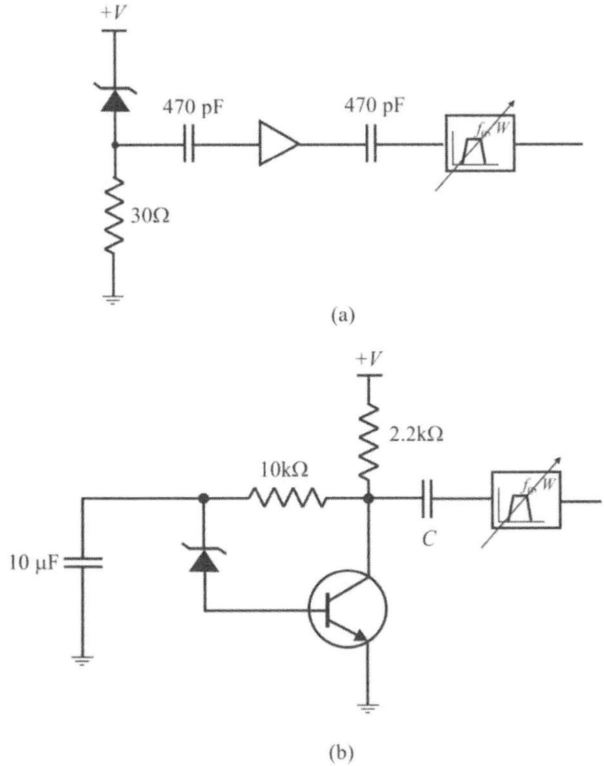

Figure 2.60 These circuits will generate broadband noise from DC up to around 300 MHz with bandwidth dependent on component selection.

noise interfering signal is the optimal waveform from an efficiency point of view [12]. While there are many ways of generating white noise, we will concentrate here on only a few.

Two possible architectures for generating white noise with a reversed-biased Zener diode are shown in Figure 2.60. The variable band-pass filter at the output selects both the bandwidth and center frequency of the noise waveform.

Another analog circuit is shown in Figure 2.61 [14]. With the 50 MHz low-pass filter shown after the npn transistor, the noise spectrum extends from about DC to 50 MHz (see Figure 2.62). The circuit presented here in Figure 2.61, however, is an inexpensive version that produces frequencies up to about 300 MHz with an appropriate low-pass filter. Its operation is based on the noise generated by the Zener breakdown phenomenon in the BJT reverse-polarized base-collector junction. In other words, such shot noise involves the statistical fluctuations of the current flow

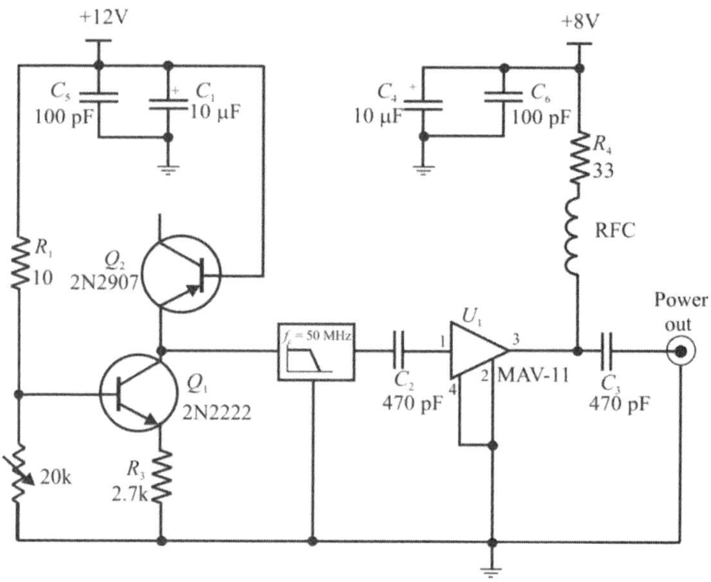

Figure 2.61 White noise generator that is based on the noise generated by the Zener breakdown of Q_2 and is capable of delivering approximately a constant –60dBm in a 0–50 MHz bandwidth.

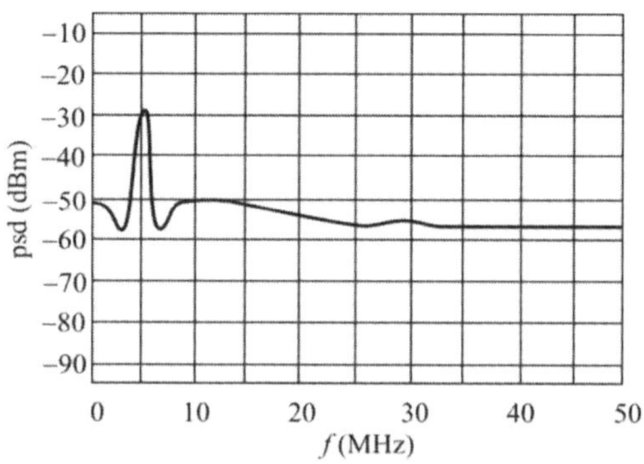

Figure 2.62 White noise spectrum for the circuit shown in Figure 2.64 (measured).

present in the bipolar transistor. The generator shown makes use of a common 2N2907 biased by the constant current source supplied by a 2N2222 (Figure 2.61).

To increase the amount of shot noise attainable, the collector of the 2N2907 is left open and the base-emitter is reverse-biased. That is, the BJT is connected as a Zener diode to exploit the reverse breakdown phenomenon.

With this configuration, the reverse breakdown voltage exhibited by the emitter-base junction can be easily observed using an ordinary spectrum analyzer. The attainable bandwidth is about 300 MHz, and the power output is about –70dBm.

To increase the noise power, one or more amplifiers can be added. The 50 MHz low-pass filter inserted between the generator and U_1 is necessary to maintain the amplifier output power compression at an acceptable value. But, of course, with this configuration, the bandwidth is restricted to the 0–50 MHz range; that is, the power spectrum vanishes outside the cutoff frequency of the filter.

In Figure 2.61, R_4 is needed to limit the current delivered to the amplifier. L_1 provides high impedance to isolate the DC source from the RF signal. C_3 removes any DC content from the output of the generator. The 20 kΩ trimmer connected between the base and ground in Figure 2.61 permits a wide range of the attainable output noise up to –60dBm (Figure 2.62).

Of course, if higher levels of noise are required, broadband amplifiers can be added to any of the outputs to boost the noise to any level.

The waveforms of these analog noise generators are truly random. They are not repeatable as the reversed-biased diodes generate truly random thermal noise.

2.7 Concluding Remarks

The common modulations used for communications were introduced in this chapter along with some methods to apply them to a carrier. Modulating signals such as discussed here are used to modulate a carrier signal that is subsequently sent toward a target receiver node in an attempt to negate successful communications.

In the next chapter we will examine carrier signal generation, which are the RF signals that are modulated in this process of performing EA.

References

[1] Poisel, R. A., *Electronic Warfare Receivers and Receiving Systems*, Norwood, MA: Artech House, 2014, Ch. 2.

[2] Ellis /Ellis%20Mixers%20in%20Communications.html, accessed Jun 15, 2016.

[3] http://whites.sdsmt.edu/classes/ee322/class_notes/322Lecture27.pdf, accessed Aug. 4, 2016.

[4] Leenerts, V., "The Tuned Circuit LC Phase Modulator," 8 Mar. 2005,

http://www.scomcontrollers.com/downloads/lcphasemod.pdf, accessed Sep. 14, 2018.

[5] http://zimmer.csufresno.edu/~pkinman/pdfs/Frequency%20Modulation.pdf, accessed Jan. 5, 2017.

[6] http://www.elec.mq.edu.au/~cl/files_pdf/elec321/lect_mask.pdf, accessed Mar. 13, 2016.

[7] http://www.rfdesignline.com/showArticle.jhtml?articleID=206903486, accessed Nov. 23, 2016.

[8] Stranneby, D., and W. Walker, *Digital Signal Processing and Applications*, 2nd Ed., Amsterdam: Elsevier, Ltd, 2004.

[9] http://www.dspdesignline.com/206904188, accessed Jul. 2014.

[10] http://site.iugaza.edu.ps/mabufoul/files/2010/09/Experiment-9.pdf, accessed Aug 7, 2014.

[11] Gentile, K., "Fundamentals of Digital Quadrature Modulation," *RF Design*, Feb. 2003, p. 40, accessed Feb 15, 2013.

[12] Gentile, Ken, "The Care and Feeding of Digital Pulse Shaping Filters," *RF Design*, April 2002, p. 50, accessed Mar. 23, 2009.

[13] Basar, T., "The Gaussian Test Channel with an Intelligent Jammer," *IEEE Transactions on Information Theory,* Vol. IT-29, No. 1, Jan. 1983, pp. 152–157.

[14] http://electronicdesign.com/test-amp-measurement/wide-band-analog-white-noise-generator, acessed Jan 24, 2015.

Chapter 3

EW Exciters

3.1 Introduction

Exciters in EW systems are the original source of jamming signals. They are comprised of two principal subsystems: (1) an oscillator of some sort, sometimes tunable over a considerable range, and (2) a modulator that imposes the desired type of jamming modulation (typically noise) onto the carrier generated by the oscillator. The oscillator is sometimes referred to as a *synthesizer*, which is the topic of this chapter. Modulators were the subject of Chapter 2.

We provide a general introduction in this chapter. For those desiring to delve more deeply into the subject of synthesizers, [1] is recommended.

The exciter output is typically a fairly low-level signal that must be amplified considerably in order to be useful for EA. It is usually desirable to have the exciter signal clean and as free of harmonics as possible in order to minimize the detrimental effects to friendly communication signals or other signals that are not the target of the EA. We discuss low-level amplifiers for this function extensively in later chapters.

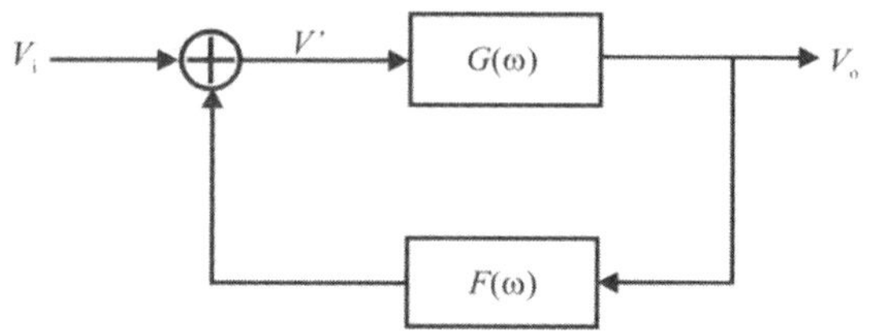

Figure 3.1 Oscillator flow diagram.

This chapter is organized as follows. First, the general considerations for oscillators are presented. Basic notions are covered including the motivation for their use. That is followed by the fundamental principles of operation common to most oscillators. Oscillators that employ LC resonators are covered first. The common types of oscillators (e.g., Clapp) are presented along with discussion of their topology. Oscillators employing crystal resonators are covered next, followed by the introduction of MEMS technology used for oscillators. Phase locked loop synthesizers comes next and that discussion is followed by covering the basics of direct digital synthesis oscillators. The final topic covered in this chapter is oscillator phase noise and how to deal with it.

3.2 Oscillator Basics

Oscillators are amplifiers with a feedback path which contains a resonator in that feedback path; resonator being a term describing a network of inductors, capacitors, and resistors usually tuned to a specific frequency [2]. The resonator can take many forms: crystal, surface acoustic wave (SAW), inductive/capacitive (LC), distributed line, or microelectromechanical (MEM), to mention some. With sufficient gain, noise will be amplified enough to eventually stabilize the amplitude and create an output signal that consists of narrowband noise.

It was long thought that oscillation occurs because the device must be initially powered, and this must create a transient, and this transient jump-starts oscillation; but this is inaccurate. The transient obviously exists, but oscillations are maintained due to initial and continuous noise in the system from passive or active devices; the noise sources being those discussed later in this chapter. An oscillator is a device with very high gain (and Q) that continuously amplifies noise.

The concept of an ideal sinusoidal signal of the form $A\sin(\omega_0 t)$ is convenient for modeling and analysis but does not exist in practice. Oscillators are amplifiers with a resonator somewhere in the positive feedback circuit. Oscillators start oscillating

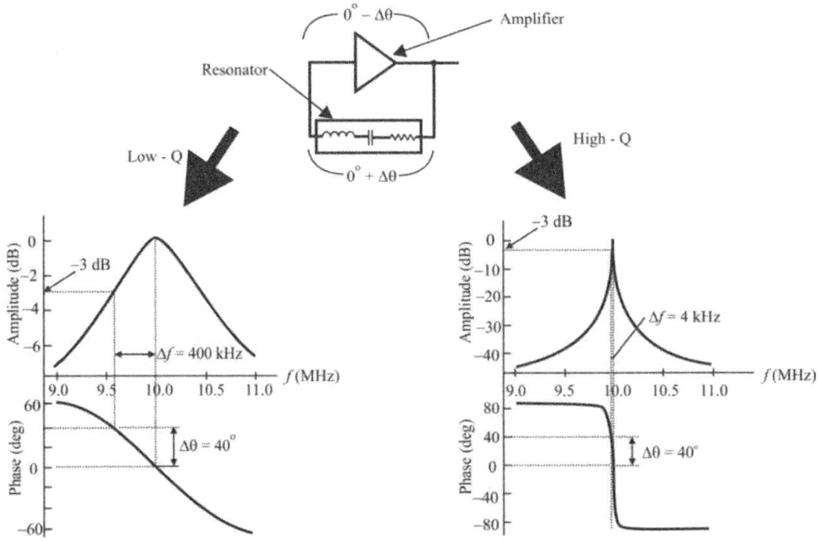

Figure 3.2 Comparing high-Q resonator feedback vs low-Q.

because there is thermal or other noise sources present in the active device. The spectrum is the product of the noise and resonator transfer function.

The resonator in the oscillator with a high Q dictates that the noise spectrum that gets amplified is narrowband. (Q is defined as the ratio of radiated-to-stored energy.) In the limit, narrowband noise with zero bandwidth of necessity is a sinusoid. Therefore, the oscillator generates sinusoidal-like signals. The noise spectrum is centered on an average ω_0 (called the *center frequency*). A general block diagram of an oscillator is shown in Figure 3.1.

The advantage of using a high-Q filter is illustrated in Figure 3.2. Low-Q resonators are possible with LC elements. High-Q resonators are possible by implementation with crystals and even higher with MEMS devices.

Let $G(\omega)$ denote the transfer function of the amplifier. For our purposes it can be assumed that $G(\omega)$ is flat around ω_0 with amplitude A and phase shift $\phi(\omega)$:

$$G(\omega) = Ae^{j\phi(\omega)} \tag{3.1}$$

When oscillating, the signal is narrowband in comparison to amplifier bandwidth and group delay (phase shift rate). The feedback, $F(\omega)$, is a narrowband (high Q) LC resonator.

The transfer function of the system is obtained as follows. From Figure 3.1 we can see that

$$V' = V_i + F(\omega)V_o$$

but

$$V_o = G(\omega)V'$$

so

$$V_o = G(\omega)[V_i + F(\omega)V_o]$$

and

$$V_o[1 - G(\omega)F(\omega)] = G(\omega)V_i$$

so the transfer function of the closed loop (CL) system is

$$CL(\omega) = \frac{V_o}{V_i} = \frac{G(\omega)}{1 - G(\omega)F(\omega)} \tag{3.2}$$

With the aforementioned assumptions, since this is a feedback system and linearity is assumed, the closed loop transfer function is

$$CL(\omega) \approx \frac{Ae^{j\phi(\omega)}}{1 - AF(\omega)e^{j\phi(\omega)}} \tag{3.3}$$

In our case the input voltage, V_i, is noise from one or several internal sources. Oscillations occur when the denominator in (3.3) approaches zero. This happens when $|AF(j\omega_0)| \to 1$ and the phase goes to 0° (or 180°, depending on total phase shift around the circuit, which must be zero).

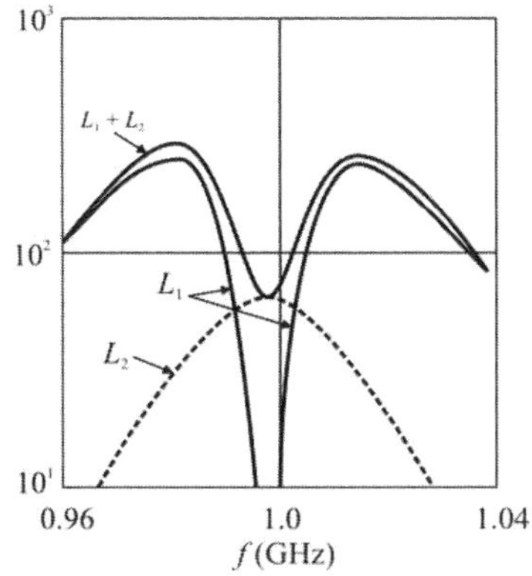

Figure 3.3 Resonator amplitude and phase.

It should be remembered that the output signal from the device is actually amplified noise, not a sinusoid. In addition, the open-loop gain should peak at the same frequency that the group delay peaks (maximum resonator phase rate). Otherwise, additional AM noise will be converted to phase noise as shown in Figure 3.3. The resonator amplitude and phase components contribute to the phase noise. The local minimum in Figure 3.3 is at ω_0, where $dA(\omega)/d\omega = 0$, L_2 is the $d\Phi/d\omega$ term, and L_1 is the $dA/d\omega$ term.

The active, energy restoring device in oscillators act as amplifiers with immense gain resulting in very high Q, amplifying the tiny $k_B T$ signal to a signal in the milliwatt levels with the signal to noise floor dynamic range in the 150- to 160-dB range or more. Active devices also contribute flicker noise, which is a dominant term in practical oscillator design in the region close to the frequency of oscillation.

Consider the circuit shown in Figure 3.4. It is comprised of a tank consisting of R, C, and L, the latter two of which are assumed to be lossless. This tank is augmented with an ideal (noiseless) energy restorer of some kind (a BJT, FET, operational amplifier, TWT).

The circuit (simplified) of one type of oscillator (a Colpitts variety) is shown in Figure 3.5. The energy restorer in this case is an FET. The waveforms of the voltage in this circuit tank and the current through the FET are illustrated by Figure 3.6.

Note that they are not sinusoidal; they are heavily corrupted with noise. The spectrum of the voltage across the tank looks similar to that shown in Figure 3.7.

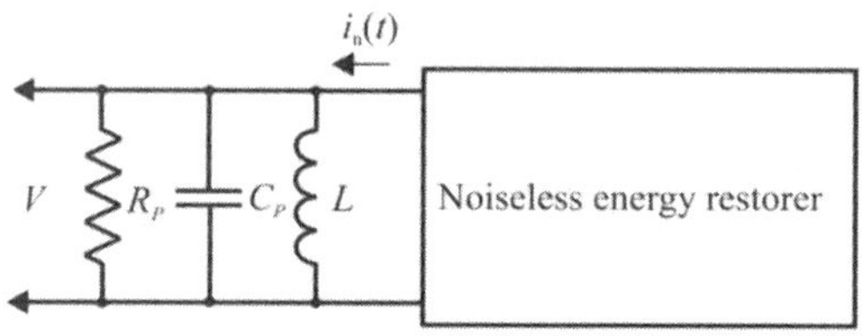

Figure 3.4 Perfectly efficient RLC.

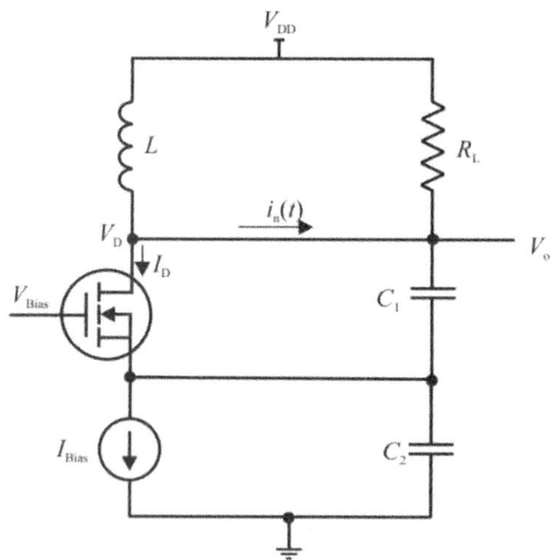

Figure 3.5 Colpitts oscillator.

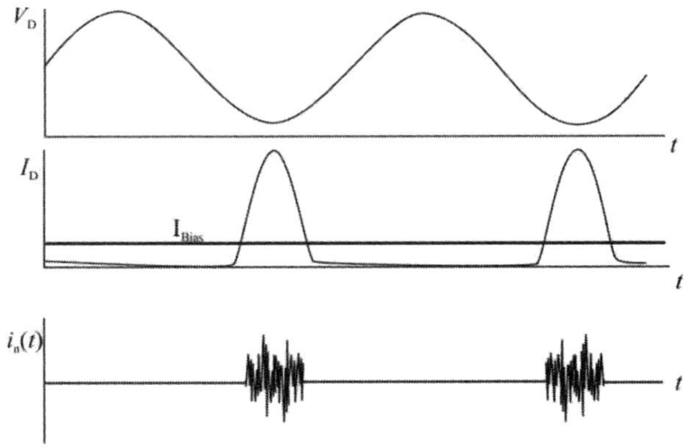

Figure 3.6 Colpitts waveforms. $i_n(t)$ is the current flowing through the tank circuit shown in Figure 3.5.

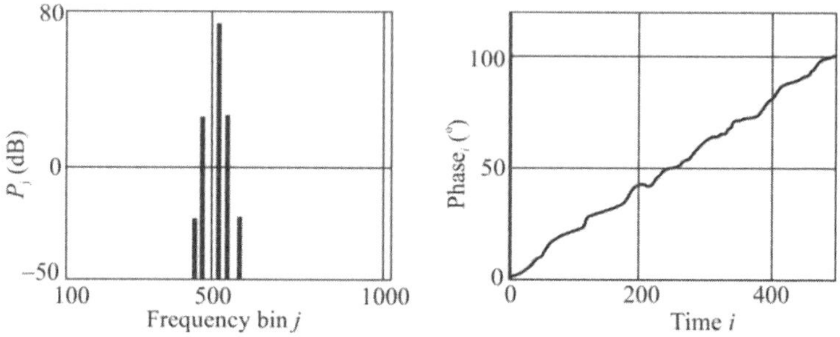

Figure 3.7 Op-amp amplitude and phase.

3.3 Principles of Oscillator Operation

The components added to the simplified oscillator block diagram shown in Figure 3.1 provide a positive feedback path from the output of the active device to the input through a resonator. The resonator must be interfaced to the active device through an appropriate coupling networks (see Figure 3.8).

Figure 3.9 shows a simplified oscillator circuit drawn with only the RF components, no biasing resistors, and no ground connection. The inverting amplifier is implemented with a single transistor and the feedback mechanism depends on which ground reference is chosen. Of the numerous oscillator types, the three most common ones are Pierce, Colpitts, and Clapp configurations, and each consists of common circuitry except that the RF ground points are at different locations.

Operation of an oscillator can be split into two phases: startup and steady state. Oscillation must begin with no external stimulus. When power is applied, signal

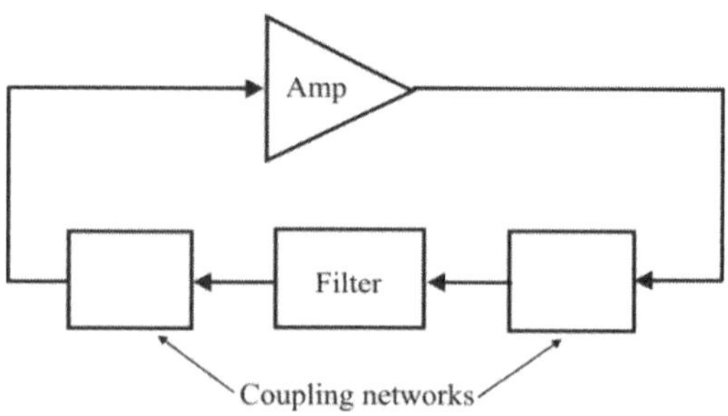

Figure 3.8 A simplified representation of an oscillator.

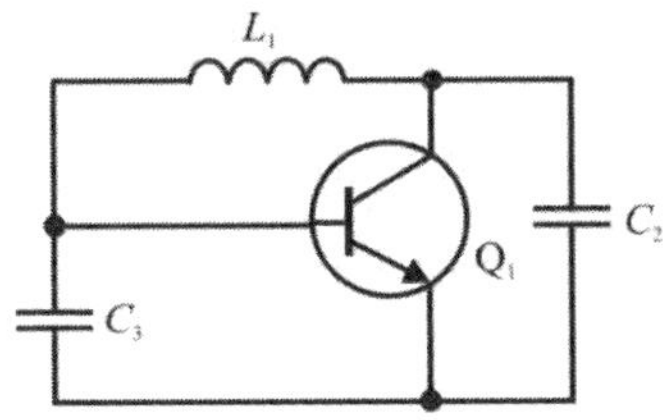

Figure 3.9 This simplified oscillator circuit is designed without an RF ground. Where the ground is attached defines the type of oscillator it is.

buildup begins in the filter network and the signal developed is small. Positive feedback and gain in the amplifier continuously increase the signal until the saturation is reached in the amplifier, limiting the loop gain to unity. At this point, the oscillator enters steady-state operation; the time from power on to steady-state operation is the oscillator startup time.

An oscillator's steady-state operation is governed by the amplifier and the tuned circuit of the filter block—the resonator. Nonlinearity of the amplifier forces unity loop gain at steady state. The Barkhausen phase requirement of 2π rads is met when the tuned circuit reactance adjusts to match. The power output and loading of the tuned circuit are the main concerns at steady state. The amplifier circuit is typically implemented with a BJT, FET, or op-amp. The linear characteristics of the active device determine the starting conditions of the oscillation due to the low signal levels, while the nonlinear characteristics establish an oscillator's operating point.

3.3.1 LC Resonators in Oscillators

The fundamental types of LC resonators are illustrated in Figure 3.10.

The filter block sets the frequency at which the oscillator will operate, which is accomplished by using an inductive-capacitive (LC) tuned circuit or crystal. Figure 3.11 depicts a basic shunt-C coupled LC-series resonator that provides phase shift and a coupling network. Since an inverting amplifier is being used, the filter block must provide a μ-rad (180-deg.) phase shift to satisfy the second Barkhausen criteria.

3.3.1.1 Barkhausen Criteria

The *Barkhausen criteria* establish when the oscillator will oscillate [2]. It states that there are two requirements for oscillation:

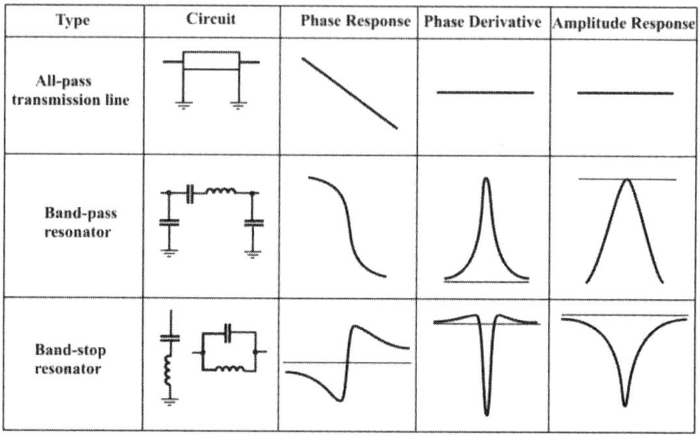

Figure 3.10 Basic types of resonators.

(1) The total magnitude of the gain around the loop is at least unity. That is

$$|AB| \geq 1 \tag{3.4}$$

In practical applications, the total loop gain should be slightly larger than unity to ensure positive feedback is maintained in the presence of device variations. When the gain is larger than unity, the oscillation amplitude will be limited by the onset of device nonlinearity. In the case of the inverter, the nonlinearity acts as a hard limiter. It is clear that the total gain for n inverters is greater than unity.

(2) The total phase shift is zero or an integer multiple of 2π radians at the oscillation frequency. That is

$$\angle AB = 2n\pi, \qquad n = 0, 1, \ldots \tag{3.5}$$

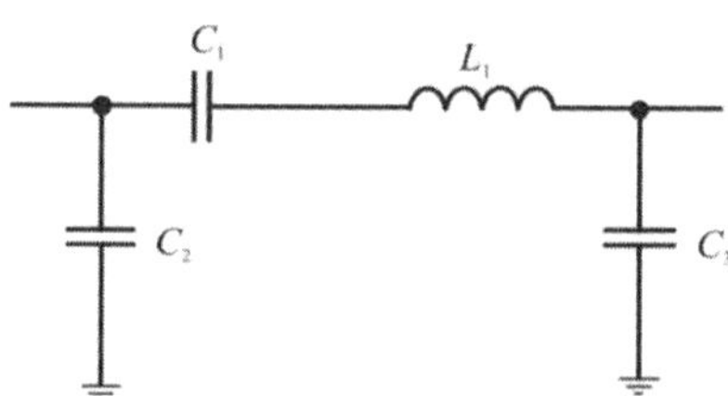

Figure 3.11 This diagram shows a basic LC series resonator.

Each stage provides π/n phase shift and so it requires two passes through the ring to acquire the necessary 2π radian phase shift. Therefore, the frequency of oscillation is [2]

$$f_o = \frac{1}{2nT_d} \tag{3.6}$$

Every oscillator has at least one active device be it a BJT, FET, or vacuum tube (operational amplifiers are made with multiples of these devices); this active device serves as an amplifier. For this first part of the discussion we will confine ourselves to LC oscillators.

At turn on, when power is first applied, random noise is generated within the active device and then amplified [3]. This noise is fed back positively through frequency selective circuits to the input where it is amplified again and so on. Ultimately a state of equilibrium is reached where the losses in the circuit are made compensated by consuming power from the power supply and the frequency of oscillation is determined by the external components, be they inductors and capacitors (LC) or a crystal. The amount of positive feedback to sustain oscillation is also determined by external components.

3.3.1.2 Colpitts Oscillator

An example of an oscillator using a MOSFET as the energy restorer is illustrated in Figure 3.5. In this case the oscillator is one of the most common types called a Colpitts oscillator. Another example of a Colpitts oscillator but with an operational amplifier as the energy restorer is shown in Figure 3.12. Note that neither of these energy restorers is noiseless. When modeling, however, the noise sources in the active devices can be included in the resistances shown. The capacitors and inductors determine the frequency of oscillation, which is given by

$$f_0 = \frac{1}{2\pi\sqrt{\dfrac{LC_1C_2}{C_1 + C_2}}} \tag{3.7}$$

3.3.1.3 Hartley Oscillator

The Hartley oscillator is characterized by two inductors in the feedback path (see Figure 3.13). Its architecture is very similar to the Colpitts oscillator with the inductor and capacitor interchanged. Feedback is applied where the inductors join.

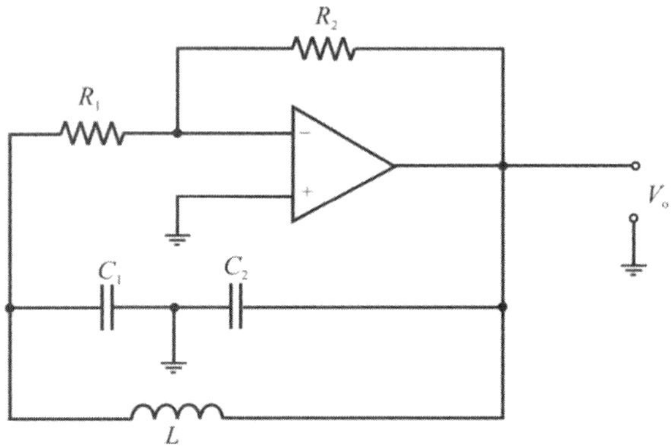

Figure 3.12 Oscillator with op amp.

The two inductors are connected in parallel to a capacitor, forming the frequency-determining resonator.

Since the feedback is provided from the emitter (in this case), the output is not an inverted replica of the transistor input. This configuration does not provide an inversion through the transistor and another 180° phase shift in the resonator, but the total phase shift is 0° (= 360°) nevertheless.

3.3.1.4 Clapp Oscillator

The Clapp oscillator was first documented by James K. Clapp in 1948. Illustrated in Figure 3.14, it is a variation on the Colpitts oscillator and is often preferred over it in *variable frequency oscillators* (VFOs) where the oscillator is designed to be retuned to operate over a range of frequencies. The architecture is comprised of a single inductor and three capacitors, with two capacitors, C_2 and C_3, determining the

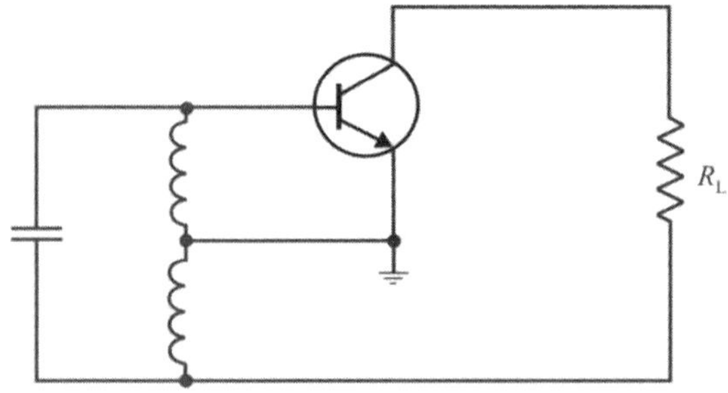

Figure 3.13 Hartley oscillator.

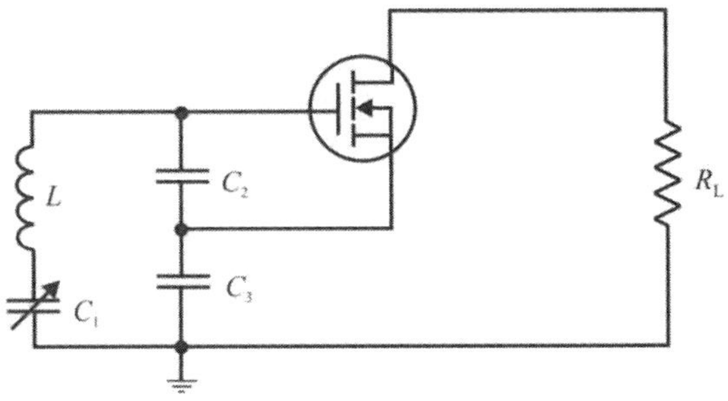

Figure 3.14 Clapp oscillator.

amount of feedback voltage applied to the transistor input. C_1 is often a variable capacitor in a VFO that determines the frequency of oscillation.

The oscillation frequency as determined by the resonator is

$$f_0 = \frac{1}{2\pi}\sqrt{\frac{1}{L}\left(\frac{1}{C_1}+\frac{1}{C_2}+\frac{1}{C_3}\right)} \tag{3.8}$$

To begin oscillation, the circuit must satisfy the Barkhausen criteria mentioned above. We can see that the amplifier provides the gain for the first criteria. The amplifier is inverting, causing a π rad (180°) phase shift to partially meet the requirements of the second criterion. The filter block provides an additional π rad phase shift for a total of 2π rad (360°) around the entire loop. The filter block also sets the frequency of oscillation using a tuned circuit (inductor and capacitor) or crystal [4].

3.3.1.5 Pierce Oscillators

The type of oscillator commonly employed in ICs is the Pierce configuration (Figure 3.15). It has many desirable characteristics, including the capability of operating over a wide range of frequencies with very good short-term stability [3]. Although inductors and capacitors are convenient for use in oscillator-tuned circuits, the primary disadvantage of this type of oscillator is the tendency to drift with changes in temperature, power-supply voltage, or mechanical vibrations. Setting the frequency of an LC oscillator requires precise manual tuning.

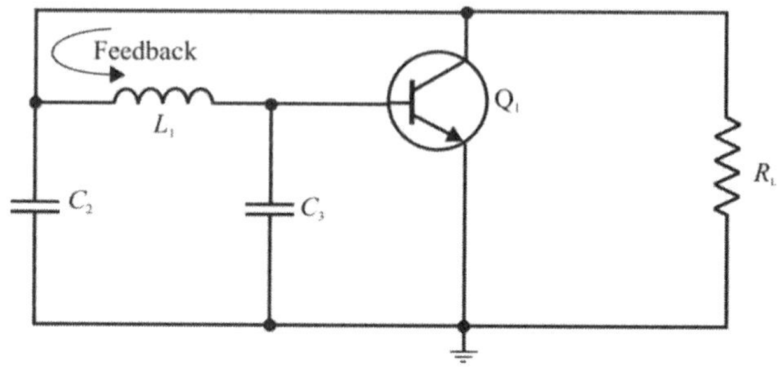

Figure 3.15 Classic Pierce oscillator.

3.3.2 Crystal Resonators

Crystal oscillators use a crystal as the principal resonating element, as opposed to discrete LC elements discussed above. We examine such oscillators in this section.

3.3.2.1 Quartz Crystals

Because their natural oscillation frequencies are very stable, quartz crystals have very desirable characteristics for use in tuned oscillators. The resonance has a very high Q, ranging from 10,000 to several hundred thousand. Ultimately, however, Q values and stability are the principle limitations of crystal-based oscillators.

The practical frequency range for fundamental-mode AT-cut quartz crystals is up to about 30 MHz. Crystals for fundamental frequencies higher than 30 to 40 MHz are very thin and fragile. Higher frequency operation is achieved by utilization of overtones of the fundamental frequency. Ninth-overtone crystals are used up to approximately 200 MHz, the practical upper limit of crystal oscillators. Higher frequencies are possible with phase locked loops, to be discussed later.

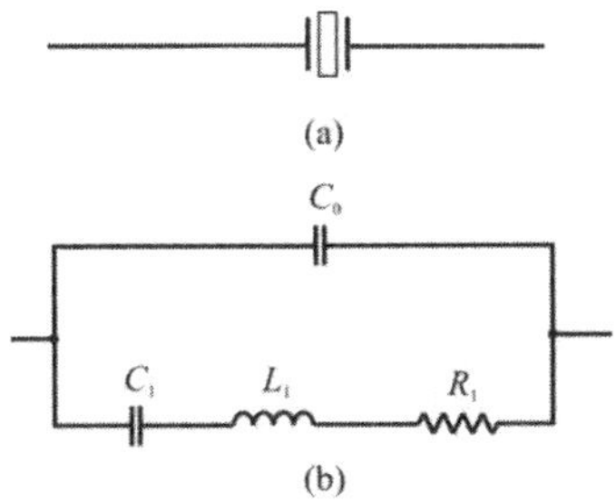

Figure 3.16 Crystal equivalent circuit.

Quartz is a piezoelectric material and when an electric field is placed upon it, a physical displacement occurs. The electronic symbol for a quartz crystal is shown in Figure 3.16(a) and the equivalent circuit for a quartz crystal near fundamental resonance is shown in Figure 3.16(b). The circuit elements C_1, L_1, and R_1 are known as the motional arm and represent the mechanical behavior of the crystal element, while capacitor C_0 represents the electrical behavior of the crystal element and mount.

The equivalent circuit in Figure 3.16(b) represents the fundamental oscillation mode. A more complex model can represent a crystal through as many overtones as desired. For the sake of simplicity, this simple model is usually employed and different values are used to model fundamental or overtone modes. Spurious resonances occur at frequencies near the desired resonance. In high-quality crystals, the motional resistance of spurious modes is at least two or three times the primary resonance resistance and the spurious modes may be ignored. In this model, C_1 represents motional arm capacitance (in farads). It represents the elasticity of the quartz, the area of the electrodes on the face, and the thickness and shape of the quartz wafer. Values of C_1 are in the range of femtofarads (10^{-15} F or 10^{-3} pF).

L_1 represents motional arm inductance (in henrys). It represents the vibrating mechanical mass of the quartz in motion. Low-frequency crystals have thicker and larger quartz wafers and L_1 ranges in a few henrys. High-frequency crystals have thinner and smaller quartz wafers and L_1 ranges in a few millihenrys. R_1 is the resistance (in Ω). It represents the real resistive losses within the crystal. Values range from 10 Ω for 20 MHz crystals to 200 kΩ for 1 kHz crystals. C_0 is the shunt capacitance (in farads). It is the sum of capacitance due to the electrodes on the crystal plate plus stray capacitances due to the crystal holder and enclosure. Values of C_0 range from 3 to 7 pF. Table 3.1 shows some typical values.

Notice that shunt capacitance C_0 is typically listed as a maximum value and not as an absolute value. Notice also that motional parameters C_1, L_1, and R_1 are not

Table 3.1 Example of Crystal Specifications

Parameter	Value
Frequency (f_{XTAL})	8.0 MHz
Load capacitance (C_L)	13 pF
Mode of operation	Fundamental
Shunt capacitance (C_o)	7 pF (maximum)
Equivalent series resistance (ESR)	100 Ω (maximum)

typically provided in a crystal data sheet. They can generally be obtained from a crystal manufacturer, or from measurements. Table 3.2 shows equivalent-circuit values for an example crystal.

A crystal has two resonant frequencies characterized by zero phase shift (see Figure 3.17). The first is the series resonant frequency, f_s, given by

$$f_s = \frac{1}{2\pi\sqrt{L_1 C_1}} \tag{3.9}$$

We recognize this as the basic equation for the resonant frequency of an inductor and capacitor in series. Recall that series resonance is that particular frequency which the inductive and capacitive reactances are equal and cancel: $X_{L1} = X_{C1}$. When the crystal is operating at its series resonant frequency the impedance will be at a minimum and current flow will be at a maximum. The reactance of the shunt capacitance, X_{C0}, is in parallel with the resistance R_1. At resonance, the value of $X_{C0} << R_1$ and, as a result, the crystal appears resistive in the circuit at a value very near R_1. Solving for the example crystal, we find that $f_s = 7{,}997{,}833.8$ Hz.

The second resonant frequency is the anti-resonant frequency, f_a, given by

$$f_a = \frac{1}{2\pi\sqrt{L_1 \dfrac{C_1 C_0}{C_1 + C_0}}} \tag{3.10}$$

Table 3.2 Example of Equivalent Circuit Crystal Values

Equivalent Component	Value
C_o	3.5 pF
C_1	0.018 pF
L_1	22 mH
R_1	30 Ω

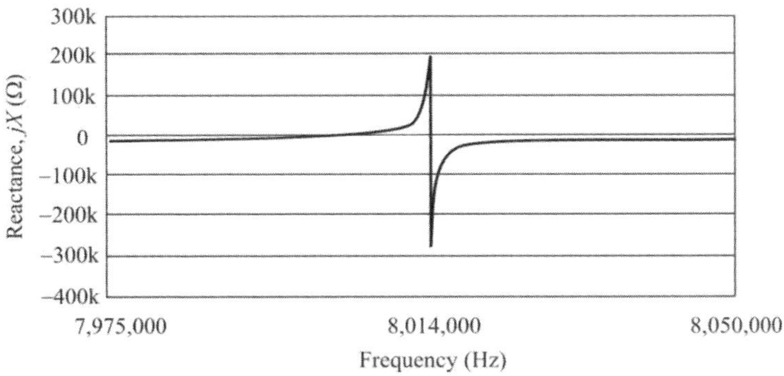

Figure 3.17 Crystal reactance as a function of frequency.

This equation combines the parallel capacitance of C_0 and C_1. When a crystal is operating at its antiresonant frequency, the impedance will be at its maximum and current flow will be at its minimum. Solving for the example crystal, we find that f_a = 8,013,816.5 Hz. Note that f_s is less than f_a and that the specified crystal frequency is between f_s and f_a so that $f_s < f_{XTAL} < f_a$. This area of frequencies between f_s and f_a is known as the *region of parallel resonance*. The crystal has resistance and reactance and therefore impedance. Figure 3.16(b) has been redrawn in Figure 3.18 to show the equivalent circuit. These impedances [2] are defined as:

$$Z_0 = -\frac{j}{2\pi f C_0} \tag{3.11}$$

$$Z_1 = R_1 + j\left(2\pi f L_1 - \frac{1}{2\pi f C_1}\right) \tag{3.12}$$

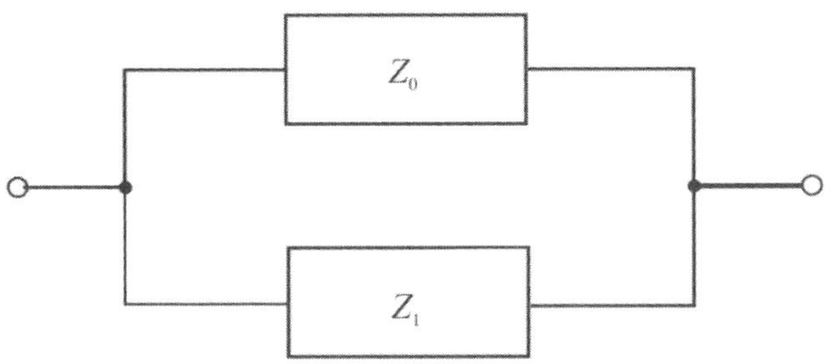

Figure 3.18 Complex impedances.

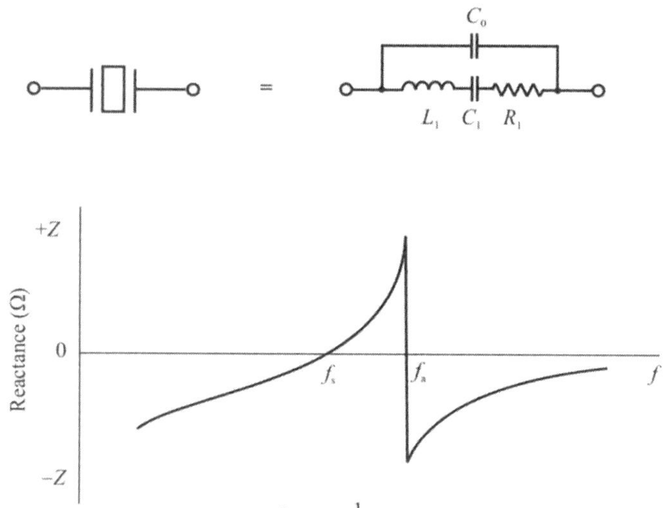

$$f_s = \frac{1}{2\pi\sqrt{L_1C_1}}$$

$$f_a = f_s\sqrt{1+\frac{C_1}{C_0}}$$

$$Q_s = \frac{2\pi f_s L_1}{R_1}$$

Figure 3.19 Crystal response.

Combining Z_1 and Z_2 in parallel yields and using the representative values in Table 3.2, yields of course

$$Z_p = \frac{Z_0Z_1}{Z_0+Z_1} \tag{3.13}$$

and the impedance versus frequency characteristic is illustrated in Figure 3.17. This plot shows where the crystal is inductive or capacitive since positive reactances are inductive and negative reactances are capacitive. The curve in Figure 3.17 is shown in more detail in Figure 3.19. We can see that the reactance is capacitive below f_s, and above f_a while being inductive between f_s and f_a.

There is no difference in the construction of a series resonant crystal and a parallel resonant crystal—they are manufactured exactly alike. However, the desired operating frequency of the parallel resonant crystal is set 100 ppm or so above the series resonant frequency. Parallel resonance means that a small capacitance, known as load capacitance (C_L), of 12 to 32 pF (depending on the

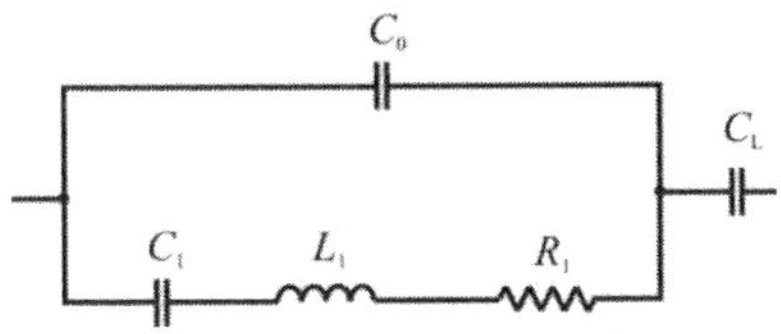

Figure 3.20 This equivalent circuit represents the load capacitance across a crystal.

crystal), should be placed across the crystal terminals to obtain the desired operating frequency [4]. Figure 3.20 depicts load capacitance in series with the crystal equivalent circuit.

A quartz crystal is a tuned circuit with a very-high Q, which, along with many other desirable attributes, makes the crystal an excellent component choice for oscillators. Crystal oscillators are recognizable from their LC oscillator counterparts [6]. For the Pierce oscillator, the crystal replaces the inductor in the corresponding LC-tuned circuit oscillators and not surprisingly, the crystal appears inductive in the circuit.

Upon startup, the amplitude of oscillation builds up to the point where nonlinearities in the amplifier decrease the loop gain to unity. During steady-state operation, the crystal, which has a large reactance-frequency slope (Figure 3.17), is located in the feedback network at a point where it has the maximum influence on the frequency of oscillation. The impedance of the crystal oscillator changes so rapidly with frequency that all other circuit components can be considered to be of constant reactance, this reactance being calculated at the nominal frequency of the crystal. The frequency of oscillation will adjust itself so that the crystal presents a reactance to the circuit, which will satisfy the Barkhausen phase requirement [2].

Figure 3.21 represents a simplified oscillator circuit drawn with only the RF components, with no biasing resistors and no ground connection. The inductor has been replaced by a crystal, which appears as inductive in the circuit. The Pierce crystal oscillator (Figure 3.22), which is designed to look into the lowest possible

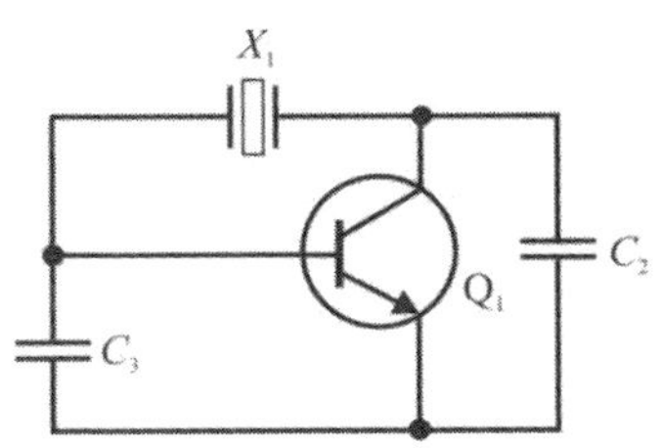

Figure 3.21 This simplified crystal oscillator circuit has no RF ground.

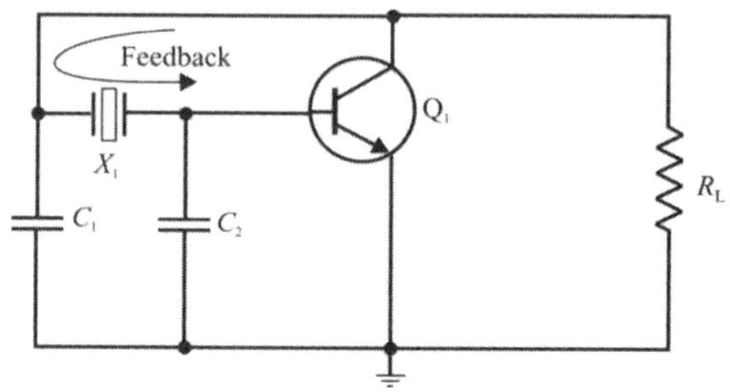

Figure 3.22 Pierce crystal oscillator.

impedance across the crystal terminals [3], oscillates just above the series resonant frequency of the crystal.

In the Pierce oscillator, the ground-point location has a profound effect on the performance. Large phase shifts in the RC networks and the need for large shunt capacitances to ground on each side of the crystal make the oscillation frequency relatively insensitive to small changes in series resistances or shunt capacitances. Additionally, the RC roll-off networks and shunt capacitances to ground provide the circuit with a high immunity to noise, minimizing transient noise spikes [3].

At series resonance, the crystal appears resistive in the circuit (Figure 3.23) and the phase shift around the circuit is 2π rads (360°). If the frequency of the circuit shifts above or below the series resonant frequency of the crystal, it increases or decreases the phase shift so that the total is no longer equal to 360°, thereby maintaining the steady-state operation at the crystal frequency. However, this only happens theoretically. Under actual implementation conditions (Figure 3.24), the phase shift through the transistor is typically more than 180° due to increased delay and the tuned circuit typically falls short of 180°. Therefore, the crystal must appear inductive to provide the phase shift needed in the circuit to sustain oscillation.

Thus, the output frequency of the Pierce crystal oscillator is not at the crystal-series resonant frequency. Typically, a parallel resonant crystal is specified by frequency and load capacitance, C_L. Capacitance C_L is the circuit capacitance required by the crystal for operation at the desired frequency. The circuit-load

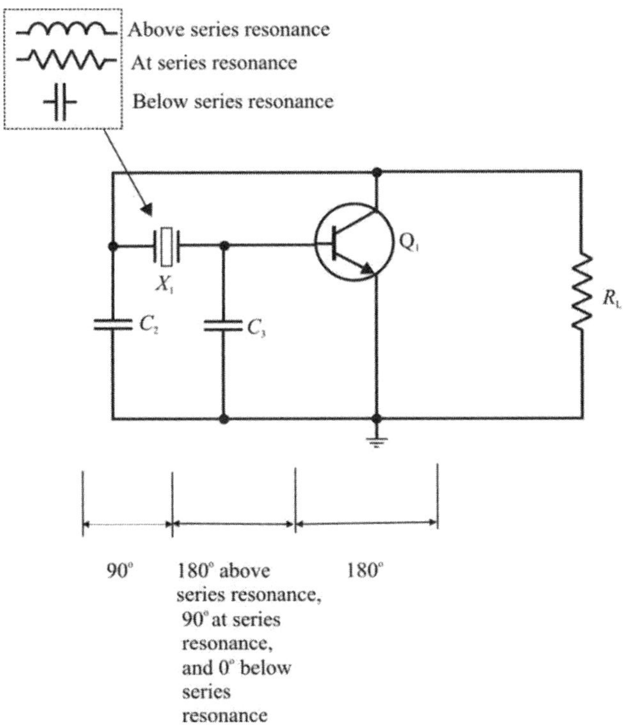

Figure 3.23 Pierce ideal operation.

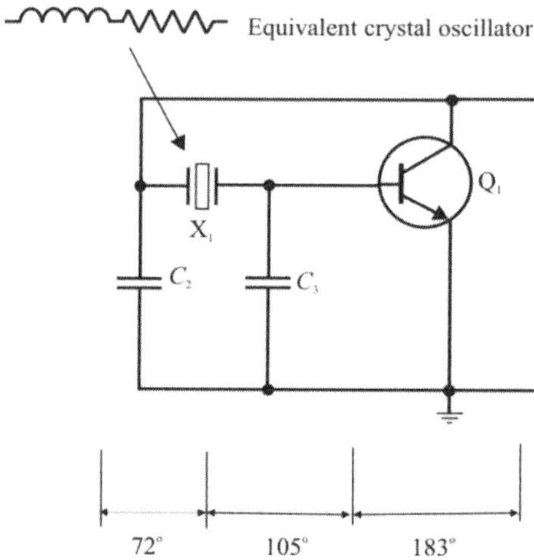

Figure 3.24 Pierce actual operation.

capacitance is determined by external capacitors C_2 and C_3, transistor internal capacitance, and stray capacitance, C_S. The values of capacitors C_2 and C_3 are selected to match the crystal C_L using:

$$C_L = \frac{C_2 C_3}{C_2 + C_3} + C_S$$

The PCB stray capacitance, C_S, can be assumed to be in the range of 2 to 5 pF; it can be minimized by keeping circuit traces on the PCB as short as possible.

The effects of stray reactances and biasing resistors appear across the capacitors C_2 and C_3 in the circuit rather than the crystal, which is a desirable characteristic. If the circuit-load capacitance does not equal the crystal C_L, the operating frequency of the Pierce oscillator will not be at the specified crystal frequency. For example, if the crystal C_L is kept constant and the values of C_2 and C_3 are increased, the operating frequency approaches the crystal-series resonant frequency (i.e., the operating frequency of the oscillator decreases). Note that large values of C_2 and C_3 increase frequency stability but decrease the loop gain and may cause problems during oscillator startup. A trimmer capacitor can be substituted for capacitor C_2 or

C_3 to manually tune the Pierce oscillator to the desired frequency. Capacitors with low temperature coefficients, such as NPO[2], should be selected for this purpose.

3.3.2.2 Frequency Synthesis

In order to make use of the stability of a crystal reference, while generating a carrier frequency of several hundred megahertz it is necessary to place the signal from the crystal oscillator at a different (usually much higher) frequency. Therefore, some form of frequency multiplication or synthesis is required. This section considers a few options for realizing this function.

Harmonic Generation/Frequency Multiplication

Passing a signal through a nonlinear circuit will result in generation of harmonics of the original waveform. Turning a sine wave signal carrier into a square wave will generate a signal rich in odd-order harmonics of the fundamental frequency. Using a tuned circuit to extract one of the harmonics can deliver the output signal at the required frequency. Because the harmonic amplitudes fall with order, it is usual to cascade a number of low-order harmonic generation components, each giving perhaps ×2 or ×3 frequency multiplication.

An alternative to using a nonlinearity to generate harmonics is the use of linear mixing to create signals at multiples of the fundamental frequency. A balanced multiplier will act as a frequency doubler, and when cascaded can generate any integer value of the fundamental source frequency. This approach to frequency synthesis is both relatively expensive and inflexible. Today it has been largely superseded by the fully integrated phase locked loop-based synthesizer.

3.3.3 Microelectromechanical Resonators

3.3.3.1 Introduction

Vibrating mechanical tank components, such as quartz crystals and SAW resonators with *Q*s in the range of 10^3–10^6, are widely used to implement high-*Q* oscillators and band-pass filters in the RF and IF stages of communication devices, including EW receivers and transmitters [7]. Due to orders of magnitude higher *Q*s, filters using such technologies greatly outperform comparable filters implemented using LC technologies in insertion loss, percent bandwidth, achievable rejection, and dynamic range. Oscillators also benefit substantially from high *Q*, as their phase noise at important offsets is often inversely proportional to the square of *Q*.

---

[2] NPO stands for negative-positive 0 ppm/°C, meaning that for any shift in temperature (positive or negative), the capacitance changes 0 parts per million.

Unfortunately, however, the crystal and SAW devices that provide beneficial high *Q*s are off-chip components that must interface with device electronics at the board level, posing a significant bottleneck against the miniaturization of communication/EW devices and trumpeting a need for on-chip replacements.

The rapid growth of IC-compatible micromachining technologies that yield microscale, high-*Q* tank components brings such strategies to reality. Specifically, the high-*Q* RF and IF filters, oscillators, and couplers, currently implemented via off-chip resonators and discrete LC passives now can be realized on the microscale using micromachined equivalents based on a variety of novel devices. These devices include high-*Q* on-chip mechanical resonators. They not only provide an attractive solution, but also enables a paradigm-shift in equipment design where the advantages of high-*Q* (e.g., in filters and oscillators) can be taken advantage of.

In this section we will discuss the fundamental operation of these MEMS resonators. We will also present a few of the popular designs, although many other types are possible.

3.3.3.2. Principles of Operation

The principle of operation of a MEMS resonator is same as that of a guitar string. A guitar string of 25" made from Ni and steel tuned to a note A vibrates at exactly 110 Hz. This guitar string is actually mechanically selecting this frequency, and is doing so with a *Q* on the order of 350, which is ~50 times more frequency selective than a typical on-chip electrical *LC* tank. The RF and IF filters are also doing the same thing but at much higher frequencies. These frequencies can be achieved by scaling (physically downward), similar to transistor technology. IC-compatible materials (such as silicon), supporting it at nodes rather than at its ends (to minimize anchor losses), and exciting it electrostatically rather than plucking it, we can achieve a *free-free beam* (FF-beam) resonator.

Fixed Beam Resonator

Consider the fixed-fixed beam resonator shown in Figure 3.25. Here the moving and fixed part forms two electrodes of a capacitor. A DC actuation voltage (V_p) is applied to the moving part of the beam and an AC signal is applied to the fixed. The DC voltage is needed to create the necessary electrostatic force for the movement of the beam (a bias force). The electrostatic force will bend the beam but the beam will oscillate only if this force oscillates, which is done by applying an AC signal at the fixed electrode. The beam oscillates with maximum amplitude at its resonance frequency

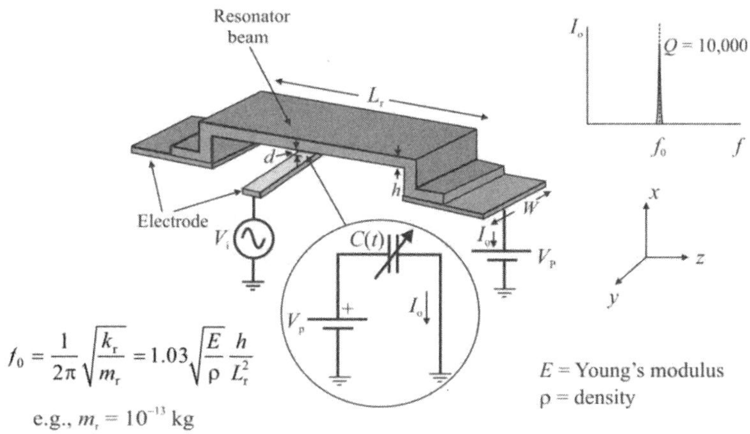

Figure 3.25 MEMS fixed beam resonator.

$$f_0 = \frac{1}{2\pi}\sqrt{\frac{k_r}{m_r}} = 1.03\sqrt{\frac{E}{\rho}\frac{h}{L_r^2}} \tag{3.14}$$

where E is the Young's modulus of the material and ρ is the mass density of the material, h is the height or thickness of the beam and L_r is the beam length as shown in the figure. According to (3.14), the frequency increases as the length decreases. This oscillation of the beam changes the gap width between the electrodes, which changes the capacitance between the two electrodes. The variation in capacitance produces an AC current given by

$$I_o = V_p \frac{dC}{dt} \tag{3.15}$$

The variation in capacitance is maximum at f_0 and hence so is the current.

Disk Resonator

A MEMS disk resonator is shown in Figure 3.26 [8]. Like the fixed beam resonator discussed above, the input voltage is applied to one of the electrodes and a bias voltage is used on the other electrode to set up a static E-field. The disc alternately expands and contracts as shown in Figure 3.26. This mechanical action causes the gap width to vary, which also causes the capacitance to vary in consonance with the

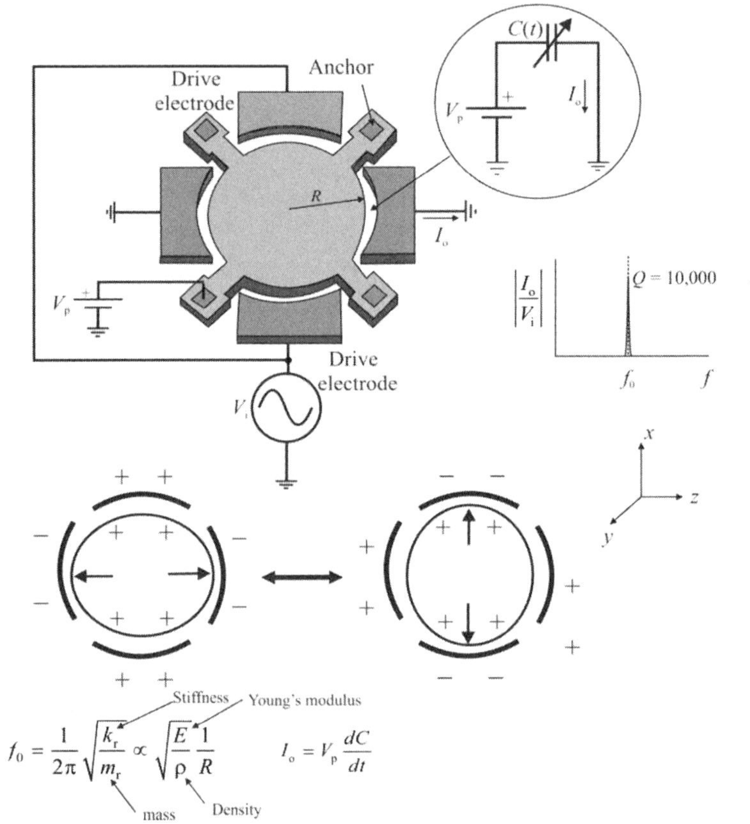

Figure 3.26 MEMS disk resonator.

input voltage. This varying capacitance causes an output current to be created according to

$$I_o = V_p \frac{dC}{dt}$$

which is sensed at the output line and amplified with standard amplifiers. Note that if $V_p = 0$ then the resonator is off.

The resonant frequency of this resonator is given by

$$f_0 = \frac{1}{2\pi}\sqrt{\frac{k_r}{m_r}} \propto \sqrt{\frac{E}{\rho}}\frac{1}{R} \tag{3.16}$$

The transfer function $|I_o / V_i|$ for this resonator is shown in Figure 3.26. The example shown creates a resonator with $Q = 10{,}000$.

We can see from (3.16) that, given the implementation technology so that ρ is fixed, the resonant frequency can be raised by decreasing the radius of the disk, R.

3.3.3.3 Summary

MEMS resonators can be constructed in many forms. We showed two possible configurations, but these were examples. The physical construction is limited only by the designer's imagination. Their biggest single advantage is the extremely high-Q resonators that can be constructed on-chip.

3.4 Phase Locked Loop Synthesisers

3.4.1 Introduction

The *phase-locked loop* (PLL) forms the basis of most modern RF signal generators including the synthesizer in the exciter of an EW system. We can get a high quality (as indicated by good phase noise), high frequency, and flexible (tuneable) oscillator with a PLL. This is obtained by locking the VCO to a stable low-level reference source (often crystal derived) [9 – 12].

The simplest form of single frequency PLL synthesiser contains a VCO operating at the required carrier frequency, a phase comparison circuit, and a loop filter as illustrated in Figure 3.27.

- Phase detector: This is a nonlinear device whose output contains the phase difference between the two oscillating input signals.

- VCO: This is another nonlinear device that produces an oscillation whose frequency is controlled by a lower frequency input voltage (in this application the frequency of this signal is very low—close to DC in a well-designed system).

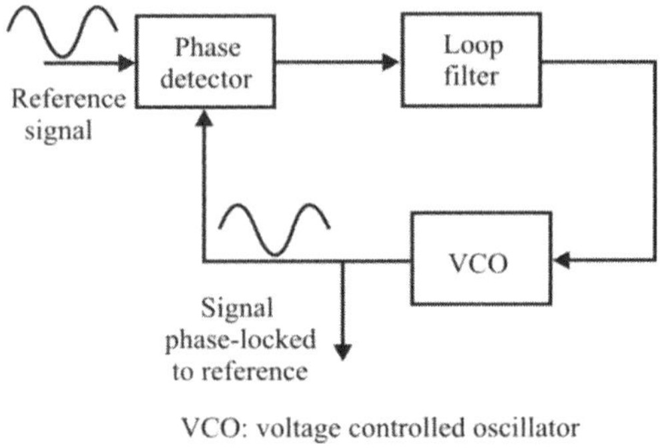

Figure 3.27 A general PLL block diagram.

- A loop filter: While this can conceptually be omitted, resulting in what is known as a first-order PLL, it is always in fact there since PLLs depend on some sort of low-pass filtering in order to function properly.

- A feedback interconnection. The phase detector takes as its input the reference signal and the output of the VCO. The output of the phase detector, the phase error, θ_e, is used as the control voltage for the VCO. The phase error may or may not be filtered.

3.4.1.1 Fractional-*N* Synthesis

In the form shown in Figure 3.27, the PLL is not providing its full functionality, however. If we add a frequency divider (a digital divider circuit) in the feedback loop whose divisor factor N can be adjusted by external control then the PLL acts to lock the reference frequency, f_r, to f_{VCO} / N, resulting in an output frequency of Nf_r. This configuration is shown in Figure 3.28.

By adding an additional frequency divider between the reference frequency source and the PLL reference input as illustrated in Figure 3.29, a signal with a whole range of output frequencies of value $(N / M)f_r$ is created at the output of the VCO. The divider ratio, N/M, provides a great deal of flexibility for an NCO. While we are focused on using PLLs in local synthesizer applications here, it is interesting to note that the PLL has many applications in communication systems and elsewhere (such as control systems) including carrier recovery, clock recovery, tracking filters, frequency and phase demodulation, phase modulation, frequency synthesis, and clock synchronization.

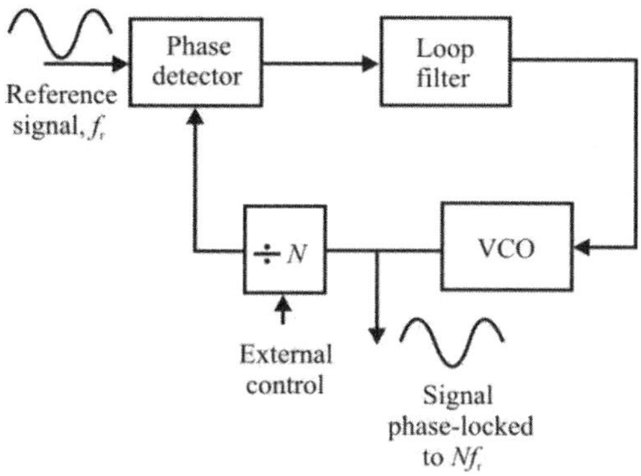

Figure 3.28 Phase-locked loop with fractional N.

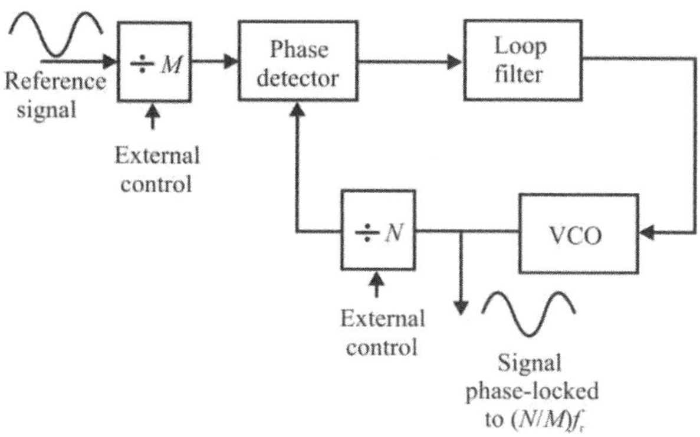

Figure 3.29 Numerically controlled oscillator.

Generation of carrier frequencies that are not constrained to be multiples of the integer divider ratio N/M without compromising on phase noise of the output waveform is accomplished with fractional-N PLLs. These devices allow the generation of pure high-frequency signals with flexible channel spacing from hundreds of hertz to hundreds of kilohertz.

The only current drawback with fractional N synthesis is the somewhat poor level of spurious components that can arise at some output frequencies. Careful design can play a big part in achieving a satisfactory solution.

3.4.2 PLL Basics

The fundamental principal behind a PLL is that if we inject a sinusoidal signal into the reference input, the internal oscillator in the loop will lock to the reference sinusoid in such a way that the frequency and phase differences between the reference sinusoid and the internal sinusoid will be driven to some constant value or 0 (depending on the system type). The internal sinusoid then represents a filtered or smoothed version of the reference sinusoid.

In the realization of the classic mixing phase-locked loop, the loop filter, $F(s)$, is typically low-pass. This filter eliminates the double frequency term arising due to the mixing. It also limits the input noise with minimal impact on the loop dynamics. In addition, typically a band-pass filter would be placed in front of the mixer also to minimize the amount of spurious signals input to the PLL.

A sinusoidal signal at the reference input of a PLL, as shown in Figure 3.29, is given by

$$V_{\mathrm{r}} = A\sin(\omega_{\mathrm{r}}t + \theta_{\mathrm{r}}) \tag{3.17}$$

The output signal from the VCO into the mixer is given by

$$V_{\mathrm{VCO}} = \cos(\omega_{\mathrm{VCO}}t + \theta_{\mathrm{VCO}}) \tag{3.18}$$

The output of the mixer in Figure 3.29 is then given by

$$V_{\mathrm{M}} = AK\sin(\omega_{\mathrm{r}}t + \theta_{\mathrm{r}})\cos(\omega_{\mathrm{VCO}}t + \theta_{\mathrm{VCO}}) \tag{3.19}$$

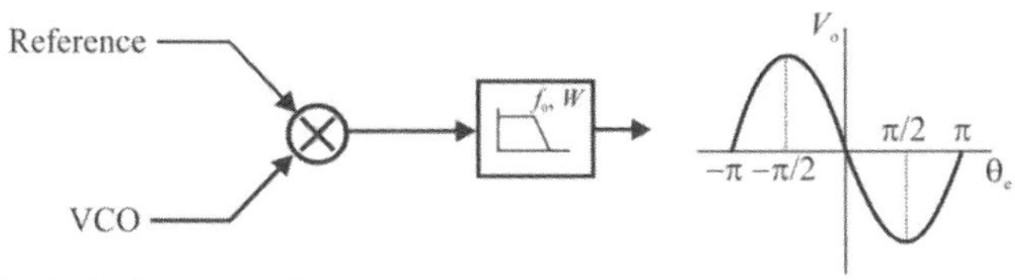

Figure 3.30 Classical mixer phase detector.

where K includes any gain from the mixer. Now we apply the elementary trigonometric identity

$$\begin{aligned}&2\sin(\omega_r t+\theta_r)\cos(\omega_{VCO}t+\theta_{VCO})\\&\quad=\sin[(\omega_r+\omega_{VCO})t+\theta_r+\theta_{VCO}]+\sin[(\omega_r+\omega_{VCO})t+\theta_r+\theta_{VCO}]\end{aligned} \tag{3.20}$$

Let $\theta_d=\theta_r-\theta_{VCO}$. We then assume that:

- The first term in (3.20) is attenuated by the low-pass filter $F(s)$ shown in Figure 3.30 and by the low pass nature of the PLL itself.
- $\omega_r \approx \omega_{VCO}$, so that the difference can be incorporated into θ_d. This means that the VCO can be modeled as an integrator.

Making these assumptions leads to the PLL model shown in Figure 3.31. The problem is that this is still a nonlinear system, and as such is in general difficult to analyze. When the loop is at or near a locked condition, typically a linearized model is assumed so that θ_d is small and slowly varying and $\sin\theta_d \approx \theta_d, \cos\theta_d \approx 1$, and $\theta_d \approx 0$. Note that another model is necessary when the loop is far away from the

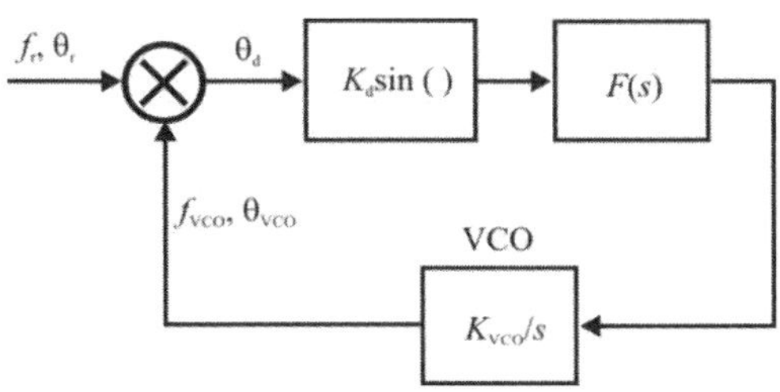

Figure 3.31 Conceptual block diagram of PLL with sine detector. This is a transition stage in the analysis of the classical mixing loop. This model represents the effect of the multiplying detector once the high-frequency component has been attenuated.

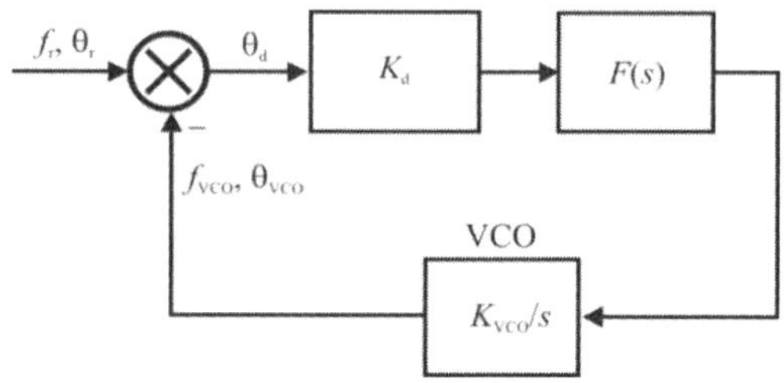

Figure 3.32 Conceptual block diagram of linear PLL. This is derived from the sine detector by assuming that the phase error is small and thus $\sin(\theta_d) \sim \theta_d$.

locked condition that incorporates nonlinear analysis methods. We will not concern ourselves with that condition here.

The linearized model is shown in Figure 3.32. This is derived from the sine detector loop by assuming that the phase error is small and thus $\sin(\theta_d) \approx \theta_d$. This is the model used with most analyses of phase-locked loops.

As seen in (3.19), the amplitude of the phase error is dependent on A, the input signal amplitude. The linearized model has a loop gain that is dependent on the loop components. Thus, the input effects of the amplitude on the loop must be anticipated.

3.4.2.1 PLL Types

Borrowed from control theory, the *type* of a PLL is determined by the number of integrators included. All PLLs are at least type 1, since the VCO acts as an integrator. If the loop contains one additional integrator then it is called a type 2, two additional integrators, it is called a type 3, and so on. So, Figure 3.32 illustrates a type 1 PLL. A type 2 PLL that contains an additional integrator (in the loop filter) is illustrated in Figure 3.33.

3.4.2.2 Linear Analysis of Classical PLLs

The PLL model in Figure 3.32 is a closed-loop feedback system. Simple analysis of Figure 3.32 yields the transfer function from reference phase input to VCO phase output, $T(s)$ as

$$T(s) = \frac{\Theta_{VCO}(s)}{\Theta_r(s)} = \frac{K_d F(s) K_{VCO} / s}{1 + K_d F(s) K_{VCO} / s}$$

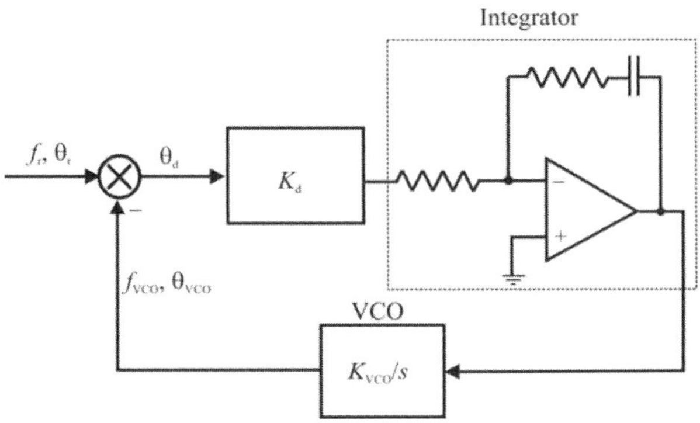

Figure 3.33 Type 2 PLL.

$$= \frac{K_d K_{VCO} F(s)}{s + K_d K_{VCO} F(s)} \tag{3.21}$$

Similarly, the transfer function from the reference phase input to the phase error, $S(s)$, is

$$S(s) = \frac{\Theta_d(s)}{\Theta_r(s)} = \frac{1}{1 + K_d F(s) K_{VCO} / s}$$

$$= \frac{s}{s + K_d K_{VCO} F(s)} \tag{3.22}$$

The loop stability, order, and the type are among the basic properties of interest in this transfer function. The order of the PLL system should be obvious from the denominator of (3.21). A variety of classical methods including root locus, Bode plots, and Nyquist plots can be used to determine stability [13, 14, 18].

Hold Range

The hold range, $\Delta\omega_H$, is that frequency range at which the PLL is able to maintain phase tracking. It is determined by calculating the frequency offset at the reference input that causes the phase error to be beyond the range of linear analysis. For a multiplying phase detector this phase error is $\pi/2$.

For sequential detectors, it will be larger. Loops will be permanently out of lock if the frequency offset at the input is greater than the hold range. Thus, this quantity is more of an academic matter than a practical one, but it can be calculated for a classical PLL (sinusoidal phase detector) as [7, 8, 15]

$$\Delta\omega_H = K_{VCO} K_d F(0) \tag{3.23}$$

Lock Range

The lock range, $\Delta\omega_L$, is defined as that frequency range within which the PLL locks within one single-beat note between the reference frequency and output frequency [7, 8]. A useful approximation if the relative order of numerator and denominator of the PLL are 1, then the loop behaves like a first-order loop at higher frequencies, and the lock range can be estimated as [16]

$$\Delta\omega_L \approx \pm K_{VCO} K_d F(\infty) \tag{3.24}$$

Pull-In and Pull-Out Range

The pull-in range, $\Delta\omega_P$, is defined as the frequency range in which the PLL will always become locked. The pull-out range, $\Delta\omega_{PO}$, is defined as the limit of dynamic stability for the PLL [16, 8]. Unfortunately, there are no simple relationships for these.

Relationships between the Ranges

The hold-in range is the largest of the ranges of a PLL. The smallest range is the lock-in range. In between these two is the pull-in range. These relationships are illustrated in Figure 3.34.

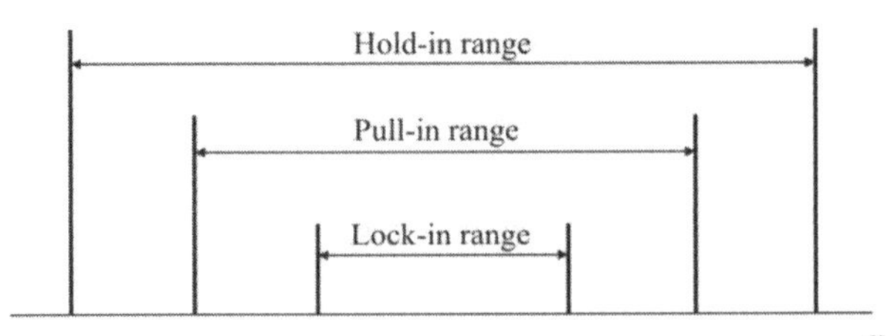

Figure 3.34 PLL ranges.

Steady-State Error

Steady-state errors of PLLs are obtained from the linear analysis via use of the final value theorem from Laplace theory, expressed as [17]

$$\begin{aligned}\lim_{t\to\infty}\theta_{\mathrm{d}}(t) &= \lim_{s\to 0} s\Theta_{\mathrm{d}}(s)\\ &= \lim_{s\to 0} s\Theta_{\mathrm{r}}(s)S(s)\end{aligned} \tag{3.25}$$

As we see from (3.22), the presence of a VCO makes every PLL at least a type 1 system, achieving zero steady-state error to a phase step at θ_{r}. For a phase ramp or (equivalently) a frequency step, there must be another integrator in the forward path, and the natural place for this is the loop filter, $F(s)$.

3.4.2.3 Phase Detectors

The linear analysis in Section 3.4.2.2 allows for determining approximate solutions for PLL performance. This has the property that once an ideal multiplication is done, the analysis of the baseband signal can be more rigorous. It should be kept in mind, however, that these are only linear approximations.

Phase detectors are used to estimate the differences in the phase between two signals of the same frequency. The inputs consist of a reference signal, frequently generated by a crystal oscillator and therefore quite stable, while the other input is the output of the VCO.

Phase detectors can be analog or digital—a popular analog realization is an RF mixer while a popular digital implementation is with an XOR gate.

Mixing

The mixing phase detector shown in Figure 3.30 is a common type of phase detector that is widely used, especially for high frequencies. It has superior noise performance to many of the other detectors [6, 9, 18, 19].

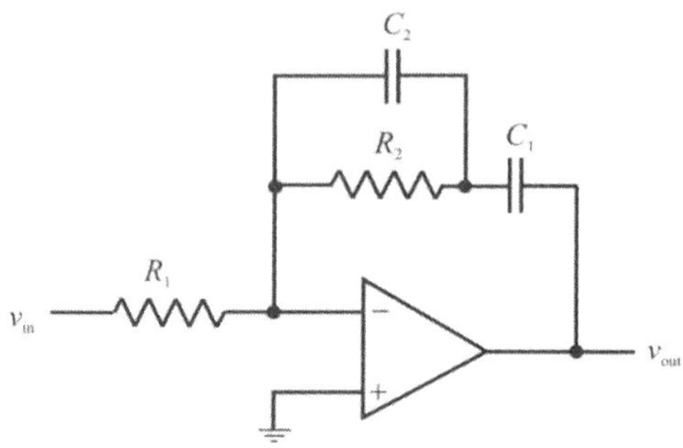

Figure 3.35 An active loop filter.

3.4.2.4 Loop Filters

The loop filter receives the output of the phase detector, frequently after passing through a low-pass filter to remove the double frequency term. The output of this filter is a slowly varying waveform that cause the VCO frequency (phase) to follow that of the reference signal.

The vast majority of PLLs are second order and as the actions of the VCO are modelled as an integrator, loop filters are typically first order. More specifically, since a type 2 system will track a phase ramp, and this corresponds to tracking a step in frequency, the loop filter almost always contains an integrator. For a double integrator system, the loop filter needs a minimum phase zero for stability. This is true whether the filter is implemented as an analog or a digital filter. Higher-order loops (which are rarer) can be obtained by adding extra pole/zero pairs to the filter.

The analog circuit shown in Figure 3.35 shows a form of a loop filter. This particular filter is fairly general with a transfer function of

$$\frac{v_{\text{out}}}{v_{\text{in}}} = -\frac{sR_2C_2 + 1}{sR_1C_1 + 1} \tag{3.26}$$

Two popular and widely used versions of the loop filter are illustrated in Figure 3.36 [15]. In terms of the natural frequency of the loop, ω_n, and the damping ratio, ζ, the transfer function for the passive network in Figure 3.36(a) is given by

$$F_{\text{passive}}(s) = \frac{s\left(2\zeta\omega_n - \dfrac{\omega_n^2}{K_{\text{VCO}}K_d}\right) + \omega_n^2}{s^2 + 2\zeta\omega_n s + \omega_n^2} \tag{3.27}$$

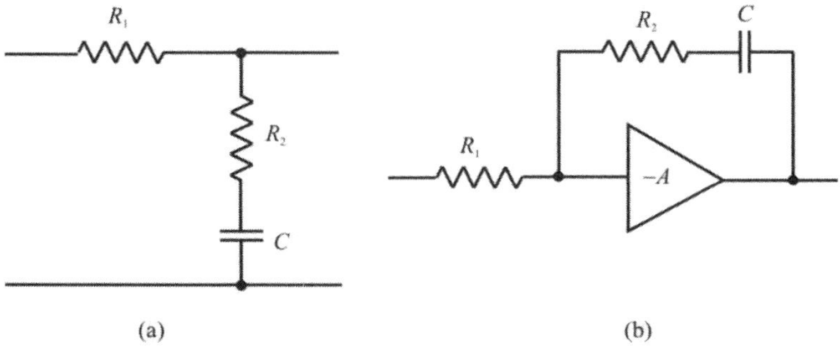

Figure 3.36 Loop filters: (a) passive and (b) active.

where

$$\omega_n = \left(\frac{K_{VCO} K_d}{\tau_1} \right)^{1/2}$$

$$\zeta = \frac{1}{2} \left(\frac{K_{VCO} K_d}{\tau_1} \right)^{1/2} \left(\tau_2 + \frac{1}{K_{VCO} K_d} \right)$$

$$\tau_1 = (R_1 + R_2)C$$

$$\tau_2 = R_2 C$$

The transfer function for the active filter in Figure 3.36(b) is given by

$$F_{active}(s) = \frac{2\zeta\omega_n s + \omega_n^2}{s^2 + 2\zeta\omega_n s + \omega_n^2} \tag{3.28}$$

where

$$\omega_n = \left(\frac{K_{VCO} K_d}{\tau_1} \right)^{1/2}$$

$$\zeta = \frac{\tau_2}{2} \left(\frac{K_{VCO} K_d}{\tau_1} \right)^{1/2} = \frac{\tau_2 \omega_n}{2}$$

$$\tau_1 = R_1 C$$

$$\tau_2 = R_2 C$$

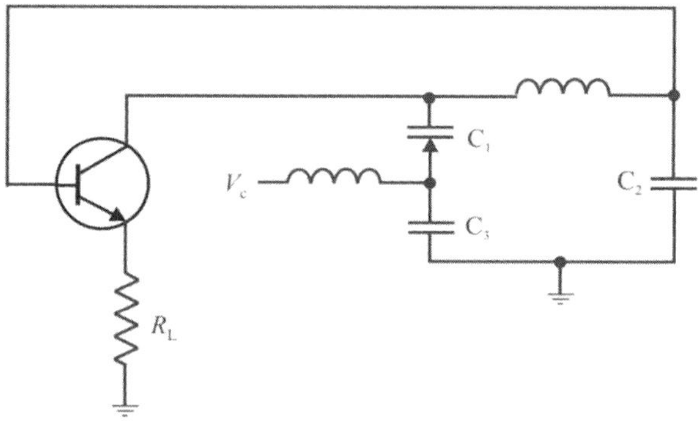

Figure 3.37 A VCO implemented through a π network. The frequency is adjusted by adjusting the reverse bias on the varactor diode, C_1. This image was first presented as Figure 2.33.

3.4.2.5 Voltage-Controlled Oscillators

The VCO was introduced in Chapter 2 as an FM modulator. The example circuit shown there in Figure 2.37 is repeated here as Figure 3.37. A variable capacitor (varactor, C_1) is placed in the tuning loop in the feedback path from the collector to the base and a control voltage is used to tune this varactor, thus varying the frequency of oscillation depending on this voltage.

Another example of a VCO implementation is shown in Figure 3.38 [20]. This version is popular for IC implementation as its properties lend themselves very well for integrated implementation and operation at microwave frequencies using CMOS technology. MOSFETs M_1 and M_2 form a cross-coupled oscillator that implements a negative resistance of $-2/g_m$ across the tank circuits in the drains. This negative resistance cancels the tank resistances R_p at resonance shown in Figure 3.4. At resonance the frequency of the oscillator is given by

$$\omega_0 = \frac{1}{\sqrt{LC_P}} \tag{3.29}$$

M_3 and M_4 in Figure 3.38 are MOSFET configurations of varactors, where the drains and sources are connected together. This configuration is shown in more detail in Figure 3.39. There is an abrupt change in capacitance as an inversion

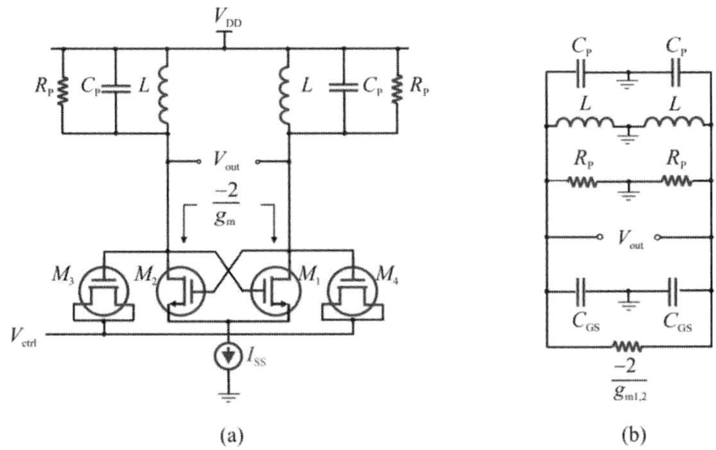

Figure 3.38 Cross coupled VCO: (a) typical LC oscillator with cross coupled pair providing negative resistance and (b) its equivalent model with varactors neglected.

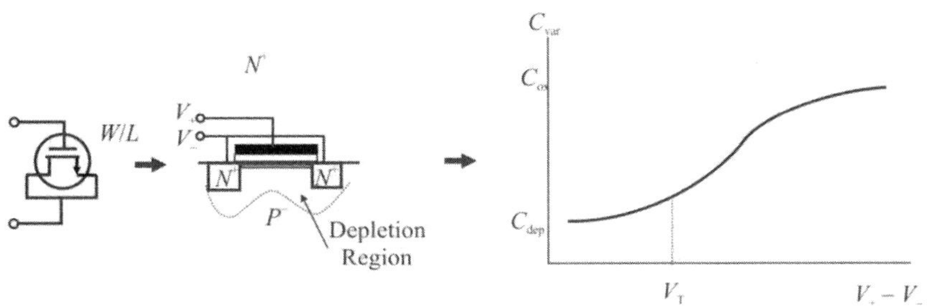

Figure 3.39 MOS varactor. *W* (channel width) and *L* (channel length) are salient parameters in MOS devices.

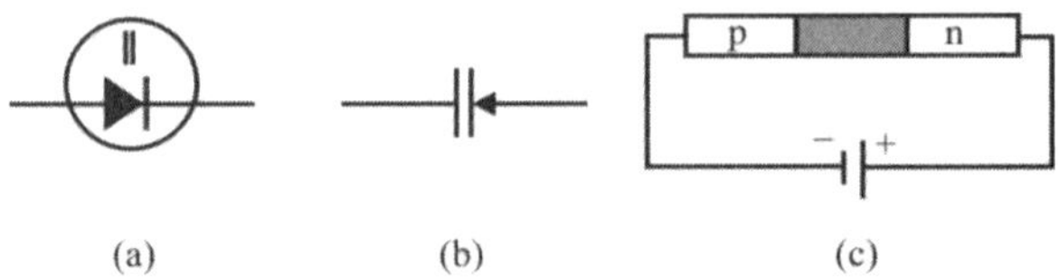

Figure 3.40 Varactor diode. (a) and (b) are symbols and (c) construction with bias.

channel forms. The advantage of this configuration is that it is easily integrated with CMOS. The salient disadvantages are:

- Q is relatively low in the transition region,
- A large signal is applied to the varactor,
- The transition region will be swept across each VCO cycle.

3.4.3 Varactor Diode

One such device is the varactor diode, illustrated in Figure 3.40. Two varactor diode symbols are shown in Figure 3.40. Figure 3.40(b) is the more popular. Also shown is a diagram representation.

When a reverse voltage is applied to a PN junction, the holes in the p-region are attracted to the anode terminal and electrons in the n-region are attracted to the cathode terminal, creating a region where there is little current. This region, the *depletion region*, is essentially devoid of carriers and behaves as the dielectric of a capacitor.

The depletion region increases as reverse voltage across it increases, and since capacitance varies inversely as dielectric thickness, the junction capacitance will decrease as the voltage across the PN junction increases. So, by varying the reverse voltage across a PN junction the junction capacitance can be varied. This is shown in the typical varactor voltage-capacitance curve in Figure 3.41.

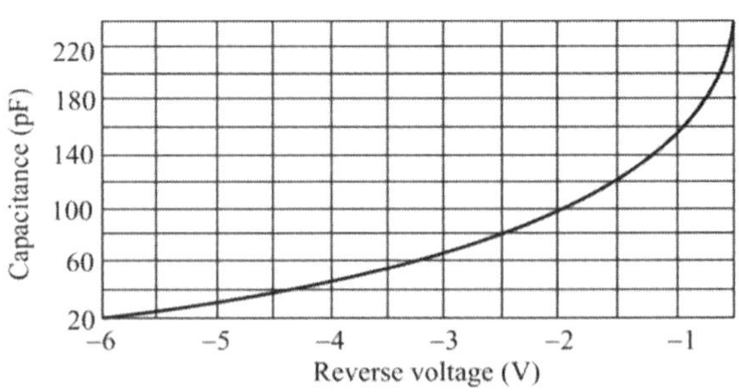

Figure 3.41 Varactor characteristics.

Notice the nonlinear increase in capacitance as the reverse voltage is decreased. This nonlinearity allows the varactor to also be used as a harmonic generator.

Major varactor considerations are:

- Capacitance value
- Voltage
- Variation in capacitance with voltage.
- Maximum working voltage
- Leakage current

3.4.4 MEMS Varactors

3.4.4.1 Introduction

MEMS varactors have considerable advantages compared with other semiconductor devices, including low loss, very high Q at mm-wave frequencies, high power handling capability, low power consumption, and high IIP_2. The RF MEMS varactor can be employed in a phase shifter or true time delay line design to replace the GaAs Schottky varactor diode for low-loss, broadband, and high-frequency applications in modern modulators. It can also be used in low-loss tunable circuits including matching networks, tunable filters, and low noise oscillators [21].

As is well known, the capacitance, C, of any capacitor whose plate area is denoted by A and whose plate separation is denoted by l is given by

$$C = \varepsilon \frac{A}{l} \tag{3.30}$$

where $\varepsilon = \varepsilon_r \varepsilon_0$ is the permittivity of the material between the plates, ε_r is the relative (to a vacuum) permittivity of the material, and ε_0 is the permittivity of a vacuum. Therefore, if we can adjust l by some means, we have a variable capacitor—a MEMS varactor.

3.4.4.2 RF MEMS Extended Tuning Range Varactor

Conventional RF-MEMS varactors usually employ a shunt parallel plate capacitor whose capacitance is determined by the spacing between a fixed-bottom plate and a movable suspended top plate as illustrated in Figure 3.42. Electrostatic actuation occurs when an electrostatic force is created by applying a DC voltage between the capacitor plates, displacing the movable plate toward the fixed plate. However, this

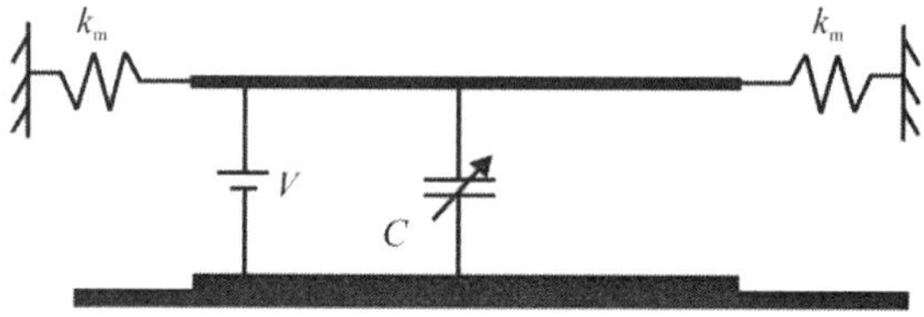

Figure 3.42 MEMS varactor.

shunt capacitance MEMS varactor structure suffers from the so-called *pull-In effect* [22].

It happens when the displacement between the two metal plates exceeds 1/3 of the entire air gap, the electrostatic attraction force loses balance with the mechanical restoring force and that causes the two metal plates to quickly snap into contact. The pull-in effect is the major limitation in MEMS varactor designs. It will cause nonlinearity and mechanical instability of the MEMS varactors. In order to avoid the snap down, the designed capacitance ratio of the conventional MEMS capacitive varactor is usually set to 1.2 to 1.5 [23].

In order to eliminate this pull-in effect, one approach is to employ the so-called MEMS extended tuning range structure [24]. This structure, as shown in Figure 3.43, utilizes a variable-height top metal beam E_1 with separate actuation parts E_2. The airgap between the center part of the top beam E_1 and the bottom plate E_2 has been designed to be less than 1/3 of the airgap between the top beam E_1 and the bottom actuation pads E_2. When DC bias is applied to the actuation parts, the entire top beam E_1 will move down together. Consequently, before the pull-in effect happens at the actuation parts, the center part has already traveled through its entire tuning range, which means that the capacitive ratio of this kind of MEMS varactor can theoretically approach infinity.

The C-V characteristic of a MEMS varactor constructed this way on a GaAs substrate for oscillation at 28 GHz is illustrated in Figure 3.44. In this case, the capacitance is changed by a factor of about 4:1 over the range of control voltage.

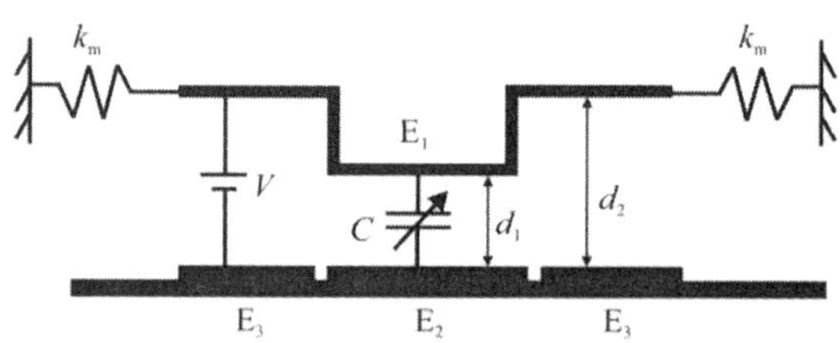

Figure 3.43 MEMS extended tuning range.

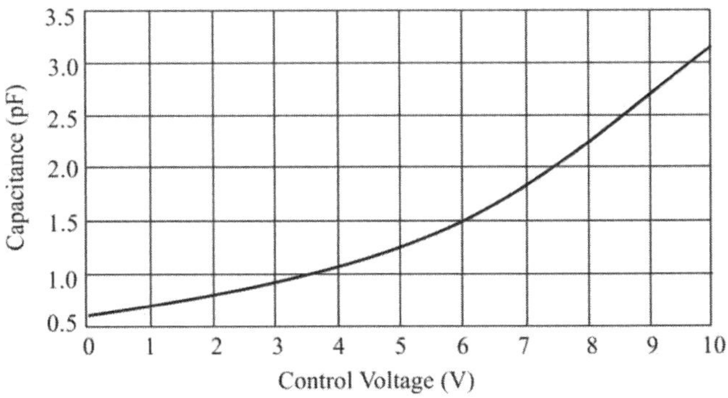

Figure 3.44 Capacitance versus control voltage in a MEMS device.

3.5 Direct Digital Synthesis

3.5.1 Introduction

The long-term frequency stability of an LC oscillator is generally poor. It is difficult to construct a small inductor with stability much better than 0.01%. Its spectral purity can only be increased by increasing the operating *Q* of the tuned circuit, which again is not easy, especially in the case of miniaturized equipment.

Direct digital synthesis (DDS) is a digital method for generating a sinusoidal waveform [25]. The principal advantage of DDS is that we can instantaneously change the frequency of the carrier, an extremely desirable feature in EW systems that have to deal with the newer modulation schemes as well as the older ones.

Instead of having multiple transmitters operating at different carrier frequencies, a single DDS can switch to the desired carriers on demand (Figure 3.45). DDS is a technique for synthesizing signals that is fundamentally different from the analog techniques discussed so far.

With the exception of the output low-pass filter, the whole DDS scheme is entirely digital and is eminently suited for implementation as an SoC.

DDS, as the name suggests, creates the synthesized signal from digital samples of the desired output waveform, which are then passed through *digital-to-analog converters* (DACs). With sufficient processing speed and word-length in the device, it is practical to achieve very fine resolution of the output frequency (millihertz accuracy) with incredibly fast switching times (nanoseconds). Using DDS in conjunction with conventional PLL-based synthesis, it is possible to get the benefits of each method. In the early days, the highest output frequency attainable was no

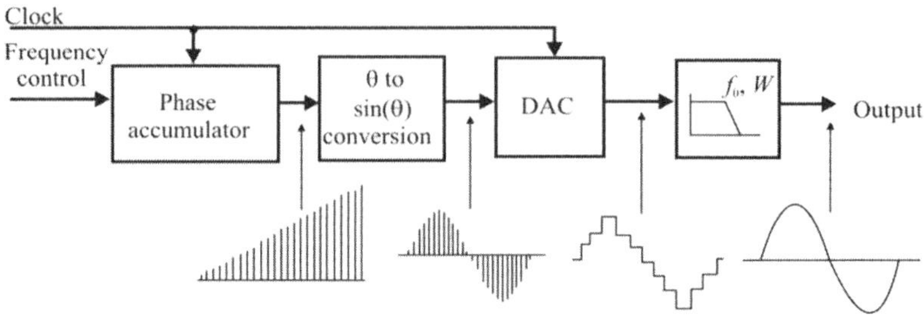

Figure 3.45 A DDS includes an N-bit accumulator, a trigonometric conversion, and an output DAC.

more than a few MHz, limited by the performance of the then available logic ICs and DAC.

The DDS is architecturally simple (see Figure 3.45), yet provides persuasive advantages that are difficult or expensive to achieve with alternative synthesis methods. These advantages include very fast switching, which is important in spread-spectrum or FHSS systems, and no system can change frequency as rapidly as a DDS. Additional advantages include fine steps, excellent phase noise, transient-free (phase continuous) frequency changes, extraordinary flexibility as a modulator, and small size, among others.

With the continued advancement in semiconductor technology, the attainable performance has increased by leaps and bounds. DDS chips realized in GaAs provide signals with frequencies over 400 MHz. Silicon technology delivers output frequencies in excess of 300 MHz. However, when operating at the top of the frequency range, the worst-case levels of spurious outputs have not changed much, still being in the region of 40 to 45dB down from the synthesized frequency.

There are disadvantages as well, of course. Two impose serious restrictions on the synthesizer. The DDS covers an operating range limited by sampling theory [26, 27]. Practically, the output is limited to about 45% of the maximum clock rate at which the logic can be operated. Operating clock rates, however, exceed 1 GHz yielding bandwidths of up to about 450 MHz. The second limitation is spectral purity, which is governed by the density/complexity of the logic circuitry that is attainable at the desired operating speed. Spectral purity and operating bandwidth are inversely correlated, meaning that the lower the bandwidth the better the obtainable spectral purity and vice versa. Despite these limitations, the DDS has evolved into an important tool, and some of its functional capabilities are not attainable with any other signal generation technique.

The conventional DDS architecture shown in Figure 3.45 can be viewed as a simple assembly comprised of only three active components: the phase accumulator, a mapping device, and a DAC.

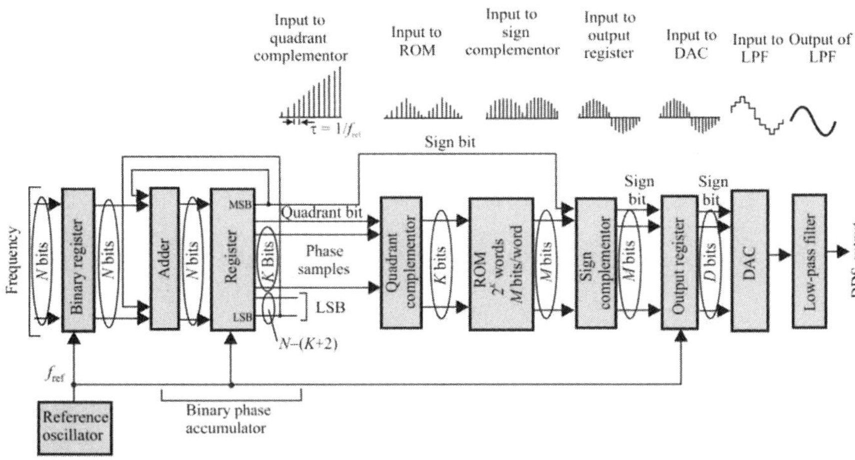

Figure 3.46 Direct digital synthesizer.

3.5.2 DDS Architecture

A more detailed block diagram of the DDS is shown in Figure 3.46. The frequency setting word is given to the adder of a phase accumulator through a binary register. The adder adds the digital signals from the binary register to the current value of the accumulator register and updates the accumulator register with the most recent sum. This register transfers the digital data from the output of the adder to its input at every clock pulse to make the accumulator overflow at regular intervals. The accumulator now addresses the ROM through a quadrant complementor.

Since all the amplitude information for a full 360° sinewave is contained in 90° of information, only 90° of memory mapping is required. The quadrant complementor simply helps the accumulator to clock in both the directions, counting down from 90°, thus giving 180° of the wave. Now the sine complementor changes sign and the process is repeated to get the full sinewave.

A DDS uses an *N*-bit accumulator, an *N*-bit latch and adder, to successively sum an *N*-bit number, θ_0, at the rate of the system clock, operating at frequency F_s. The literature often refers to θ_0 as the frequency-tuning, or frequency-control, word.

3.5.2.1 Phase Accumulator

The *phase accumulator* is an arithmetic device that performs discrete mathematical integration of the form

$$s_n = s_{n-1} + \delta \tag{3.31}$$

where these variables normally take on binary values. In applications there are several variations, including the use of pipelining in high speed circuitry that can degrade the speed or update rate of the DDS. The output of the accumulator is a correlation between the commanded frequency and the clock, in the form of a phase ramp.

The accumulator has an N-bit latch, and the range of binary numbers that its output can represent is $0 \le K \le N-1$, where K is the N-bit binary number at the output of the latch at any time. Therefore, when the value of the accumulator is within the amount θ_0 of its maximum value of $2^N - 1$, the next cycle of the system clock causes the accumulator to overflow. That is, only the amount by which the accumulator exceeds $2^N - 1$ remains in the accumulator. Thus, the value of accumulator output, K, increments by a fixed amount, θ_0, at the rate of F_s. Hence, the average overflow rate of the accumulator (f_c) is

$$f_c = F_s(\theta_0 / 2^N) \tag{3.32}$$

For example, if the fractional value, $\theta_0 / 2^N$ is 0.113, and F_s is 100 Hz, then the accumulator overflows at a rate of $f_c = (100\,\text{Hz})(113/1000) = 11.3\,\text{Hz}$. Note that $\theta_0 / 2^N$ is always a fraction, because θ_0 is by definition less than 2^N. Now, referring to Figure 3.46, we see that the numeric output, K, of the accumulator is linearly advancing in value by θ_0 at a rate of F_s and overflowing at a rate of f_c (3.32).

3.5.2.2 Memory

The second component in a DDS is memory, or a mapping device that performs the nonlinear transformation of $\theta t \to \sin\theta t$. This is usually done with ROM/RAM *lookup tables* (LUTs). However, output signals of high quality require many bits to define both θt and $\sin\theta t$, thus requiring large memories. Furthermore, if the output is to be over a usable bandwidth, the memory must also be fast. Large and fast memories are readily obtainable, however.

Consider the function of the trigonometric conversion block. We interpret the input to this block, K, as an angle. The output is a number that corresponds to the cosine or sine of that angle—that is, the value along the x- or the y-axis that the angle determines. To make this computation, the trigonometric conversion block must map the linear range of the integer, K $(0 \le K \le 2^N)$, that the accumulator generates to the linear range of an angle, $0 \le \phi \le 2\pi$. The relationship $\phi = 2\pi K / 2^N$

easily accomplishes this task. If a trigonometric conversion block implements the cosine function, then its output value is:

$$y = \cos\phi = \cos(2\pi K / 2^N) \tag{3.33}$$

3.5.2.3 DAC

Returning to Figure 3.46, observe that the output of the trigonometric converter drives a DAC. Hence, the value of y in (3.33) must be in the form of an integer that spans a range of 2^D (the binary range of a D-bit DAC). The method of mapping the value of y to the binary range of the DAC is unimportant. The linearly advancing K values at the input of the trigonometric converter represent linearly advancing angle values. The trigonometric converter converts these values, in turn, to amplitude values. Therefore, the output of the trigonometric converter is a numeric sinusoid with a frequency that is the same as the accumulator-overflow rate (3.32).

While it is possible to multiplex both the logic and memory blocks to achieve very fast clock rates, ultimately it is the DAC that limits the performance of the DDC. The traditional DAC consists of a fast switch matrix plus a stack of current sources. Linearity depends on analog (current source) error, as the accuracy of the source corresponding to the MSB must be better than the accuracy of the current flow of the source corresponding to the LSB. In a twelve-bit system, that is a formidable issue.

Further, the settling time of the current sources limits the speed of the system. Overall, the DAC is unquestionably the performance-limiting factor of the DDS. Even when assuming a perfect DAC, with ideal linearity, noise is still introduced by quantization errors. That is, when the ROM calls for an amplitude that falls between two current sources, the output waveform is degraded.

3.5.2.4 Frequency Selection

The frequency of the output of the system is directly proportional to the frequency-setting word. One way of generating the output is to output some points for every cycle of the sinewave. In this case, tuning is affected by changing the rate at which the data is output, as illustrated in Figure 3.47. However, with this method of frequency selection it is difficult to get as fine frequency-resolution as possible. For a second method, the rate at which data is output is kept constant. The basic idea is to store N uniformly spaced samples in the memory. The lowest output frequency contains N distinct samples. Now if we output every other point stored in the ROM at the same rate, we get an output with twice the original frequency. So, if every K^{th} sample is taken, a K times faster waveform will be obtained. The frequency

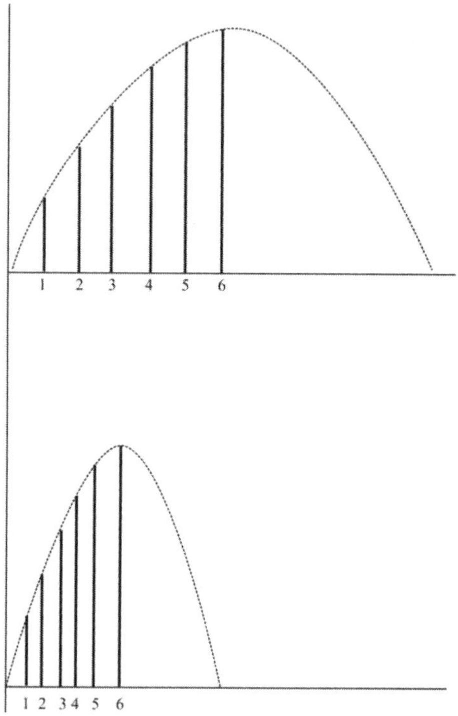

Figure 3.47 Variable sample rate method.

resolution is same as that of the lowest frequency. This method is illustrated in Figure 3.48.

3.5.3 Polar Modulator

Building a polar modulator requires only a few small adjustments to the basic DDS (see Figure 3.49). It should be apparent that we can interpret the output of the accumulator as the instantaneous phase angle of the carrier signal [$\omega_c t$ in Section 2.6]. By inserting a subtraction operation between the accumulator and the trigonometric converter, we can effectively implement the minus $\theta(t)$ part. Furthermore, inserting a multiplier between the trigonometric converter and the DAC scales the output sinusoid by the amount $r(t)$, effectively implementing the $r(t)$ scale factor and constituting the salient features of a DDS-based polar modulator.

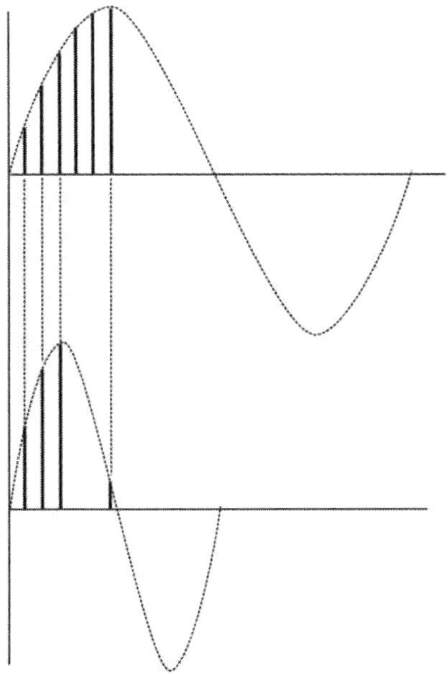

Figure 3.48 Constant sample rate method.

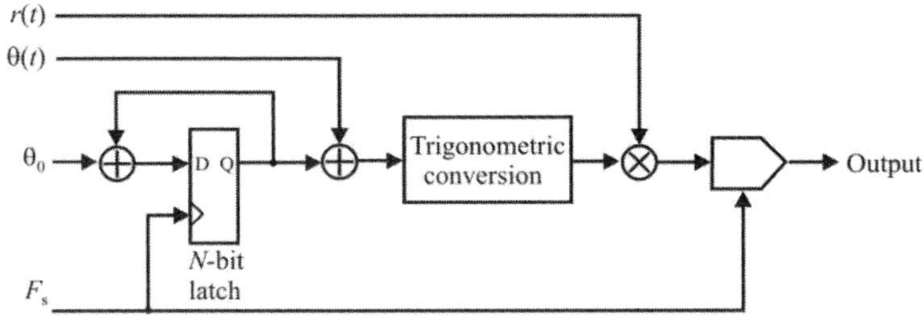

Figure 3.49 For an all-digital modulation approach, DDS blocks replace the oscillator of a conventional polar modulator.

Two caveats are not apparent in the DDS-polar modulator. First, the DDS is a sampled system that operates at the system clock rate, F_S. Therefore, the subtraction element and multiplier in Figure 3.46 must also operate at F_S. However, $\theta(t)$ and $r(t)$ in Figure 3.49 represent the information to be transmitted, which the system samples at a rate that depends on the input data. The problem is that the values of $r(t)$ and $\theta(t)$ occur at the symbol rate, which differs from the system-clock rate, F_S. F_S is usually much higher than the symbol rate, meaning that we must apply a sample-rate conversion to the symbol samples to match the system-clock rate for processing through the multiplier and the subtraction elements of Figure 3.49. If we do not employ a sample-rate conversion, the spectral images of the symbol spectrum, which occur in frequency bands that are multiples of the symbol rate, appear in the final output spectrum—an undesirable consequence of sampling theory. The topic of sample-rate conversion is an important aspect of designing a practical DDS-based polar modulator [28].

The second caveat is that the physical-hardware implementation of the trigonometric-conversion block will likely exhibit some delay. That is, a significant delay may occur between when an input value, K, arrives at the trigonometric-conversion block and when the answer appears at the output. This delay, latency, can span several system-clock cycles. Why is it such a problem? Recall that a polar coordinate, (r, θ), is a coupled pair. The appropriate r must coincide with the appropriate θ to identify the desired point on the polar plane. If the r sample arrives at the multiplier in Figure 3.49 before the converted θ sample, then the coupling of the (r, θ) pair breaks, identifying an incorrect polar coordinate.

We can implement polar modulation in rectangular form using (x, y) coordinates or in direct polar form using (r, θ) coordinates. We can digitally generate the carrier signal to transmit polar encoded information by means of a DDS. More important, the architecture of the DDS provides an elegant means of implementing a direct polar modulator, which can be less hardware-intensive than a traditional quadrature modulator. Furthermore, we can design an entire polar modulator using digital techniques. Thus, we employ all of digital design's positive attributes, including stability, repeatability, and low cost, to their full advantage.

3.5.4 Digital vs Analog Performance

Logic alone (accumulator and ROM) can produce arbitrarily accurate digital signals and the digital portion of the DAC can also be accurate. That is misleading, however, because theoretical digital performance is invariably corrupted by the third active component of the DDS—the DAC.

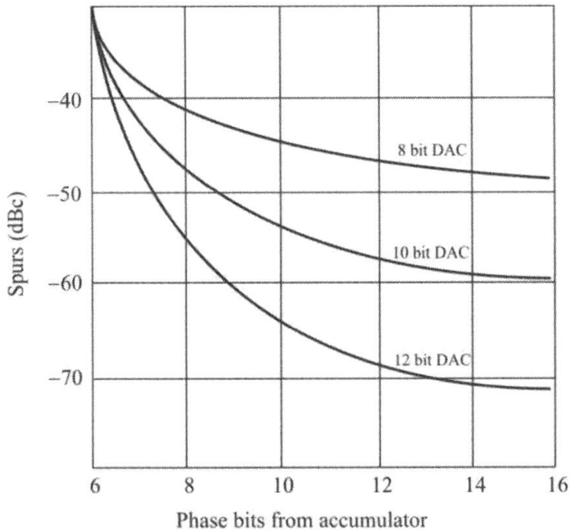

Figure 3.50 Address quantization of sine table.

Most well-designed DDSs will provide excellent phase noise performance (typically that of the clock itself, without degradation). The noise generated by a DDS is therefore almost entirely discrete signals (spurs). Theory permits DDS spurious levels at 6 dB per bit of quantization in the DAC, thus an 8-bit DAC can theoretically attain spurs that are –48dBc in amplitude. If the remaining circuitry has no error whatever, and ignoring RF problems within the DAC (glitch energy, intermodulation, loading, timing, and so forth), there will remain two contributors to spurs.

First is the precision with which the accumulator defines phase, and second is the quantization accuracy of the DAC. Given those two numbers, it is possible to estimate the theoretical limit of spectral purity of any given DDS, as shown in Figure 3.50. As we see, –72-dBc spurious levels can theoretically be achieved by mapping 14 bits of phase information to 12-bits of amplitude. At high clock speeds, that –72-dBc performance will be degraded by glitches during DAC transitions, plus conventional RF problems such as intermodulation distortion, loading, and timing.

Spurs are even more predictable. For instance, assume that the accumulator is controlled by some specific bit pattern of the 32 frequency bits available and outputs 14 phase bits, and the ROM maps the phase data to 12 bits of amplitude. Assume further that the DAC is not a 2's complement device and that 90° of the sinusoidal waveform is mapped (it is flipped and flopped digitally to construct all 360°). Under those circumstances, and for any clock, the dominating spurs can be predicted as to their amplitude and frequency.

In a DDS, spurious signal levels are governed by the accuracy to which the digital circuitry defines the output waveform, plus the effect of glitch energy added by the DAC, plus yet another category of potential problems: typical RF issues such as intermodulation, timing, loading, clock feed-through, etc. Improved digital accuracy adds circuit complexity, hence reducing the maximum clock rate. There is therefore an inverse correlation between the two principal characteristics of the DDS: bandwidth and noise content.

In any signal source, signal degradation follows multiplication at a 6-dB/octave rate (20 log$n$). This means that a DDS that covers 100 MHz with –45-dBc spurs can be made operationally equivalent to one that covers 50 MHz with –51-dBc spurs or 200 MHz with 39-dBc spurs, as shown in Figure 3.50. While it is practical to multiply and divide the output of the DDS to achieve the desired combination of spectral purity and bandwidth, that approach is not suitable for applications where the DDS is used as a multimode modulator.

As an example, consider an accumulator that outputs phase with 12-bits accuracy, memory that maps those data to 10 bits for the DAC, and a 10-bit DAC. According to theory, the best spurious performance that can be achieved is –58 dBc. Similarly, mapping 14 bits of phase data to 12 bits of quantization produces, at best, –72-dBc spurs. The task of achieving this mapping performance is not trivial.

It is relatively simple to design an accumulator and memory combination that produces excellent digital performance. The best designs produce a digital signal with spurious levels of 90 dBc or better. Of course, the spurious output of the DDS must be measured after the DAC, since the DAC by itself degrades the signal more than all the circuitry that precedes it. That –90-dBc digital spurious performance is subject to intermodulation, timing, clock feedthrough, loading, and other common RF problems that are aggravated by higher clock rates. The greatest signal corruption is produced by the fundamental operation of the DAC itself.

In the DAC, quantization error swamps earlier contributors to spurious energy in the DDS output. Assuming a perfect DAC design, quantization error limits spurious suppression to 6 dB per bit of resolution, so an 8-bit DAC is capable of –48-dBc spurious performance, a 12-bit unit can do –72-dBc. Because those limits are so severe in comparison, it is meaningless to consider the digital spurious performance of any logic that precedes the DAC.

Another issue in the DAC is glitch energy, or extraneous signals produced during the transitions of the conversion process (Figure 3.51). Such energy is introduced by the switch matrix as it turns the current sources ON and OFF. As can be seen, when the DAC is operated at high clock rates the glitch takes the same time to settle but the overall step is reduced in the time domain, so there is a corresponding reduction of the percentage of each step that is linear, and a

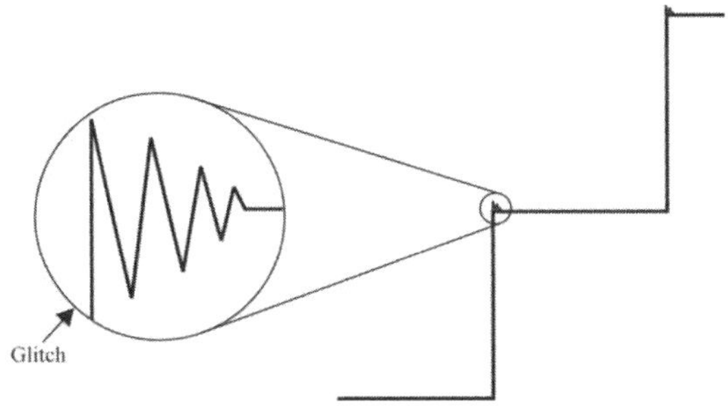

Figure 3.51 After quantification error, the majority of spurious energy from a DDS occurs due to glitches appearing at the transitions in the DAC.

corresponding increase in the percentage that contributes to spurious energy. It is only that linear portion that is useful. This problem can be addressed by a switch that closes to pass the signal only after the DAC has settled to its new level (see Figure 3.52).

3.5.5 DDS Switching Characteristics

The manner in which a DDS digitally constructs its output means that precise and fast manipulation of the digitized data is easy and economical. Switching *mode* and *speed* are parameters of increasing importance: first, consider mode.

A DDS changes frequency in a unique manner when compared to oscillator-based synthesizers. Changes are without significant extraneous transition energy, because a commanded frequency change simply sets a new amplitude of the next bit addressed by the DAC. Frequency changes look like those of a VCO—smooth and without phase discontinuity, as shown in Figure 3.53.

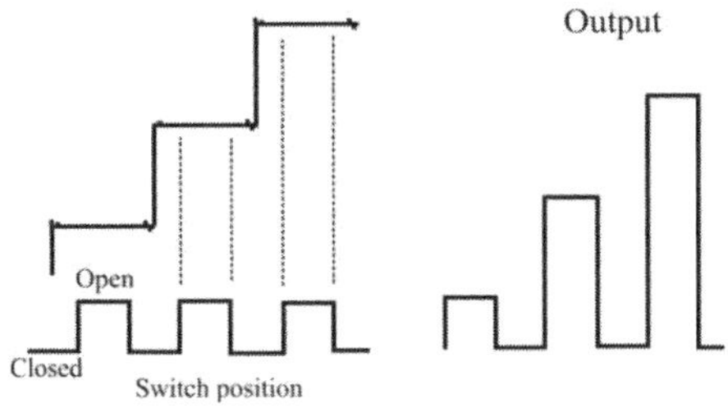

Figure 3.52 Deglitcher.

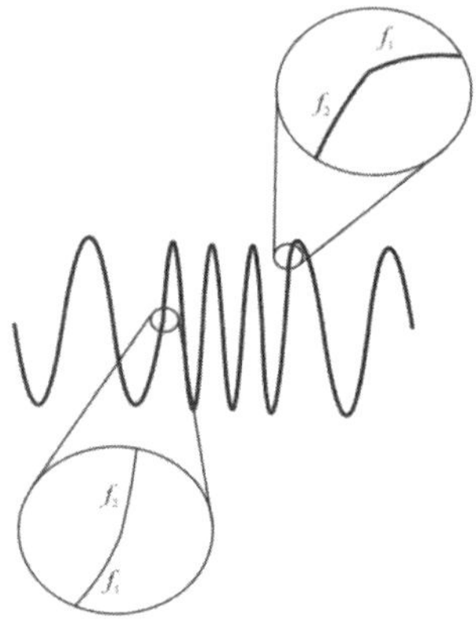

Figure 3.53 Frequency transitions.

As shown in Figure 3.54, DDS frequency changes occur smoothly as the ramp generated by the accumulator changes. Combining that smooth transition with fast switching means that a DDS can operate very much like a synthesized VCO, sweeping across a range with synthesizer accuracy but without the glitches and transients produced by any other synthesizer technique. When the signal is in the digital domain it is economical and easy to manipulate, and the result is a precision not available by any analog means. Of the synthesis techniques, this characteristic is unique to the DDS. It is important in EW applications, where generation of undesired signals during frequency changes demands complex blanking circuitry to avoid several problems. These problems include fratricide of friendly communication networks.

Switching *speed* is an issue in many synthesizer designs. A synthesizer was once considered fast when it could settle at a new commanded frequency within a few milliseconds. Evolution of direct-analog synthesizers allowed the designer to achieve settling within a few microseconds. The DDS is unquestionably the fastest switching technique. The slowest designs include pipelining, which degrades switching performance of some highly complex designs. When switching speed is a critical issue, DDS speed can be as fast as one clock cycle plus the delay added by

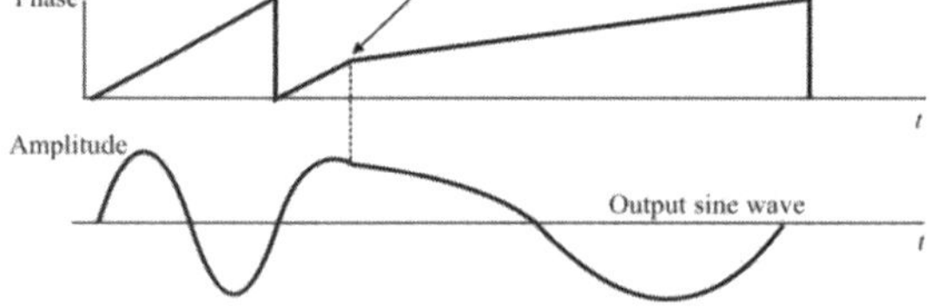

Figure 3.54 Phase transitions.

the output filter. Some DDS designs can switch across any portion of their full bandwidth in one clock cycle.

DSSS and FHSS systems by definition require fast switching. In EW, ultra-fast switching permits a single jammer power generation chain to handle many different targets essentially simultaneously.

3.5.6 Modulation and Complex Waveform Generation

As discussed in Chapter 2, communication systems that transmit, receive, and/or process information require some form of modulation, which is the deliberate modification of a carrier to impress data upon it, in a manner that permits the subsequent recovery of that information. Any signal—including a complex waveform—can be completely defined by only three parameters: $A(t)$ for amplitude, $\omega(t)$ for frequency, and $\phi(t)$ for phase. Modulation requires manipulation of one or more of those parameters, which is ordinarily achieved by adding circuitry to a basic RF source.

The DDS is an optimum modulator because waveform manipulation is simple and inexpensive while the signal is in the digital domain. The addition of a few parts to the DDS will support aggressive combinations of frequency, phase, and amplitude modulation. With DDS technology, minor modifications of the basic architecture will support virtually all modulation schemes because the DDS can impress the required modulation directly on the carrier as an integrated part of the digital construction of the output waveform without the need for a separate modulation subsystem.

We will discuss in this section some of the simple ways that a DDS can be used to modulate the synthesized waveform with analog or digital FM, PM, or AM.

3.5.6.1 FM Modulation

Analog or digital frequency modulation (FM) is achieved by digitizing the information and then adding the digital result to the command word appearing at the control port of the phase accumulator, as shown in Figure 3.55. Changing the phase in this way changes the frequency, thus accomplishing FM.

3.5.6.2 FSK

The combination of fast switching and glitch-free transitions make the DDS an optimum frequency-shift keying (FSK) modulator. FSK can be achieved with a DDS by registering two frequencies in two registers and allowing the modulation data to toggle between them.

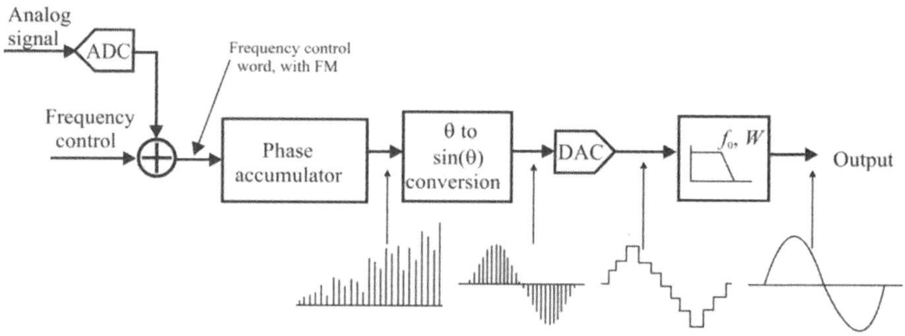

Figure 3.55 The frequency of the signal generated by a DDS can be controlled by adding a digitized version of an analog signal, thus creating FM.

Minimum-shift keying (MSK) is an FSK-derivative used in many classes of systems because of low sidelobes, and a DDS is the only theoretically perfect MSK modulator. Other approaches to MSK produce glitch energy at transition points, which reduces the effectiveness of the technique and complicates system design.

Quadrature (QPSK) or more complex *M*-ary FSK appears in some communication systems, and fast-hopping *M*-ary FSK is used extensively in FHSS anti-jam communications (typically $M = 2$ in this application).

3.5.6.3 Phase Modulation

Phase modulation (PM) is used extensively in communications—in particular satcom and landline telecommunications. BPSK, QPSK, 8-ary and 16-ary phase shifting are frequently encountered. Very high speed BPSK is used in DSSS communications for anti-jam benefits. It is also part of the fourth generation (4G) cellular phone standards. A DDS supports such PM with accuracy/performance not otherwise attainable.

A DDS can shift the phase of the output signal if an adder is placed between the accumulator and the ROM to shift addresses in accordance with the data appearing at that adder, as shown in Figure 3.56. It is possible to achieve nearly any number of phase levels using this technique, and the accuracy will be far better than is possible with analog methods. Excellent analog phase modulators achieve 2° or 3° accuracy; the early phase-control DDS achieved ~1.4°, and modern versions are able to achieve almost any level.

An adder between the accumulator and the ROM will define phase with a precision limited only by the number of bits in the adder. BPSK and QPSK are simple, of course, and an 8-bit adder achieves 256 discrete phase levels (~1.4° steps). Such information density is practical only with DDS.

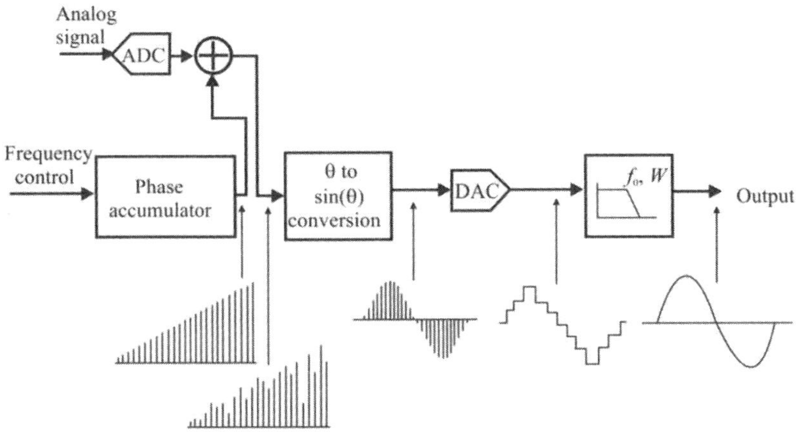

Figure 3.56 PM is achieved by interrupting the data stream between the accumulator and memory, and then inserting the desired digitized information by adding it to the data stream. In a DDS, digital phase control can be extremely precise.

Amplitude modulation (AM) is ordinarily implemented by a multiplier between the memory and the DAC (Figure 3.57), but technology does not yet support multipliers with the speed and circuit density required by high-speed GaAs DDSs today, so this method is limited to the lower frequency ranges.

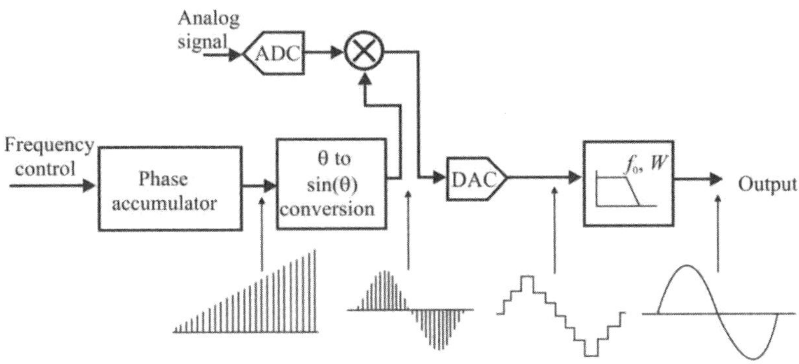

Figure 3.57 Amplitude of a DDS output can be controlled by placing a multiplier between the waveform memory map and the DAC. If the second input to the multiplier is a digitized analog signal, then the result is AM with extreme accuracy.

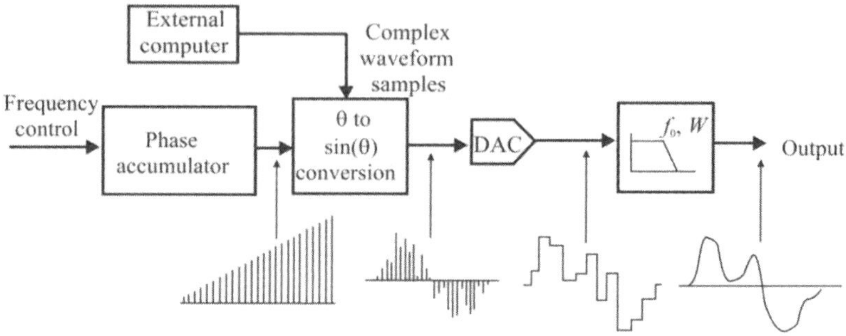

Figure 3.58 Complex waveforms with DDS.

3.5.6.4 Pulse Modulation

Pulse modulation is facilitated by a DDS's ability to execute a high-isolation MUTE function with one digital control bit at the DAC registers. This technique allows preservation of phase coherence from pulse to pulse. Another option is to use a RESET function applied at the accumulator that permits signal resumption at a known point in phase. In both, the ON/OFF ratio will be much better than 100 dB.

Additional logic increases power consumption and complexity, but improvements in device design and basic semiconductor technologies (advanced digital Si and GaAs) support even very aggressive modulation scenarios. Gallium nitride (GaN) is also a new semiconductor technology that shows considerable promise in fast logic applications.

3.5.6.5 Complex Waveform Generation

Complex waveform generation combines frequency synthesis with modulation, usually by the techniques described above. Another technique involves memory. By changing the ROM to RAM in the DDS as shown in Figure 3.58, virtually any desired waveform can be generated. The clock may be fixed or variable; in the latter case, the output waveform can be defined with greater latitude. In effect, the DDS operates like an *arbitrary waveform generator.*

3.6 Oscillator Phase Noise

3.6.1 Introduction

Probably the two most crucial factors affecting an EW signal source (synthesizer) are frequency stability and phase noise [29].

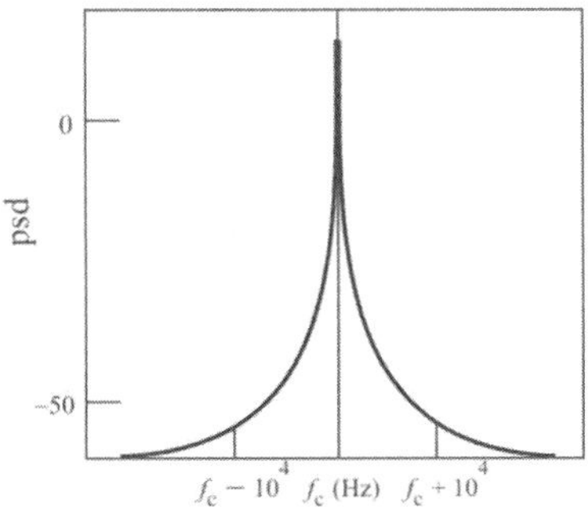

Figure 3.59 Oscillator phase noise.

Frequency stability refers to the drift (or lack of it) of the signal's frequency with temperature, aging, and mechanical factors such as vibration and shock. In EW systems, the frequency stability determines the channel spacing necessary to share the spectrum among several sources. In addition, for a given channel bandwidth (the usual case, as the channel characteristics are normally regulated) the frequency stability determines the maximum data rate that can be used without spilling excessive energy into the adjacent channels.

The oscillator phase noise (also called phase jitter) results in a nonzero bandwidth of the carrier that could also cause energy to spill over into adjacent channels (see Figure 3.59). In addition, if the phase noise is too high the modulation source itself can be corrupted if any of the phase modulations are used. This can limit adjacent channel selectivity due to reciprocal mixing.

In EW systems the targets to be jammed are present in a single channel at a time (by definition, although FHSS signals don't stay there very long). If the exciter drifts around, or if there is excessive phase jitter, jam energy may be broadcast in adjacent channels. The most serious implication of this is electronic fratricide of friendly communications. Another implication is that some of the jammer power is not being used as intended and, in extreme cases, jamming may not occur at all due to insufficient jammer radiated power in the channel where it is intended.

A frequency stable oscillator requires a stable tuned element in the feedback path. A simple LC stabilized oscillator does not generally satisfy this requirement. Using a transmission line in place of the LC circuit usually exhibit less drift with temperature. A ceramic resonator, widely used as a clock source for microprocessor applications, has a frequency tolerance of only approximately 3,000 ppm and hence is not adequate for most radio and EW applications. Coaxial resonators (often ceramic-based) can provide very high-Q elements and hence have excellent phase noise. Their temperature stability is however poor and on their own they are not sufficient to act as the signal source. (Locking a coaxial resonator to a stable (e.g., crystal) source is however a very viable and widely used synthesis process). SAW devices can provide the accuracy needed for wideband operation, giving typically ± 300 ppm stability over –40°C to +60°C for frequencies up to 1 GHz or beyond. They usually have an initial tolerance of approximately ±50 kHz that can be tuned on manufacture. Because they provide a low-cost, on-frequency source, SAW-based oscillators are very popular for wideband synthesizers.

A crystal oscillator is the most widely used stable frequency source, relying on the natural vibration frequency of the quartz crystal. Depending on the quality of the crystal and the type and tolerance of the cut, accurate, temperature-stabilized devices can be fabricated. For a very cheap crystal, a temperature tolerance of ±15 ppm is typical, whereas for a *temperature compensated crystal oscillator* (TCXO), a ±3 ppm tolerance or better is possible. The major drawback with crystal-based oscillators compared with SAW devices is that the oscillation frequency is typically below 30 MHz for a fundamental mode design, and possible up to 100 MHz using a fifth overtone design.

3.6.2 VCO Phase Noise Basics

Consider again the close-loop transfer function (3.2); note that $CL(s)$ at the frequency of oscillation represents a singularity [30]. On paper, the gain is infinite (division by 0). Obviously, this is not possible, so some assumptions will have to be made later in regard to this point. We can, however, analyze what happens close to this singularity as follows.

Expanding the numerator and denominator in Taylor series, close to the oscillation frequency we have

$$CL[j(\omega_0 + d\omega)] \approx \frac{G(j\omega_0) + d\omega \dfrac{d(GF)}{d\omega}}{1 + GF + d\omega \dfrac{d(GF)}{d\omega}}$$

Since $GF = -1$ at resonance we have

$$CL[j(\omega_0 + d\omega)] \approx \frac{G^*}{d\omega} \frac{dF}{d\omega}$$

(G^* is a constant) and

$$|CL[j(\omega_0 + d\omega)]|^2 \approx \frac{G^*}{d\omega^2}\left|\frac{dA}{d\omega}\right|^2$$

because the resonator amplitude response, A, has a close to zero gradient at the frequency of oscillation, but significant phase, ϕ, gradient (see Figure 3.3). This can be further approximated as

$$|CL[j(\omega_0 + d\omega)]|^2 \approx \frac{G^*}{d\omega^2}\left(\frac{d\Phi(\omega_0)}{d\omega}\right)^2$$

Thus, the spectrum around the singularity will be proportional to the resonator group delay. Note that the assumption that $d(FG)/d\omega$ has only a phase component means that at oscillation the resonator amplitude peaks. This is a common simplifying assumption so we can gain some intuitive understanding of the phase noise spectrum. If desired, a more complicated model could be used and then additional components must be added, which results in additional noise.

Note that since $d\Phi / d\omega$ is a measure of bandwidth, we define

$$Q \triangleq \frac{\omega_0}{2}\left(\frac{d\Phi}{d\omega}\right)^{-1}$$

and obtain (ignoring the scaling factor G^*)

$$|CL[j(\omega_0 + d\omega)]|^2 \approx \frac{\left(\dfrac{\omega_0}{d\omega}\right)^2}{4Q^2} \tag{3.34}$$

This is the Leeson phase-noise approximation [31]. Thus, the model shows that flat noise at the input, $k_B TF$ (F is the overall experimental noise factor), will produce a $(2Qd\omega)^{-2}$ spectral behavior at the output terminal of the active element.

Some correction factors have been added to the model, especially to accommodate flicker noise of active devices. Assuming that the noise spectrum of flicker noise is given by $f_c / d\omega$ (f_c is the flicker corner frequency), a general modified Leeson equation [31] at the frequency f_0 is given by

$$\mathcal{L}(f_{\mathrm{m}})=\frac{k_{\mathrm{B}}TF}{P_{\mathrm{S}}}\left(1+\frac{f_{\mathrm{c}}}{|f_{\mathrm{m}}|}\right)\left[1+\left(\frac{f_0}{2Qf_{\mathrm{m}}}\right)^2\right] \tag{3.35}$$

where $\mathcal{L}(f_{\mathrm{m}})$ is the single sideband noise spectrum function, Q is the resonator loaded Q, and F is an experimental correction factor. P_{S} is the power at the input to the active device.

Incorporating the flicker noise multiplier and the experimental noise factor F in this manner makes the Leeson spectral density more of a heuristic than explicitly derived spectrum. F must be determined by measurement as it has no formal definition. The flicker noise factor simply is reasonably incorporated this way.

A general plot of Leeson model prediction, for $f_0 = 2\,\mathrm{GHz}$, $f_{\mathrm{c}} = 10\,\mathrm{kHz}$, and $Q = 20$ is shown in Figure 3.60.

Leeson Equation—Some Comments

The Leeson equation presents certain characteristics that should be addressed for clarity. Some of these areas follow.

- While close to carrier, the phase noise is monotonically declining at –30 dB/decade; it then changes to –20 dB/decade until it hits a noise floor. The equation predicts that when ω_0 doubles, the loss in the phase noise for the same Q is in the 6-dB range. In reality, going from 1 to 2 GHz causes a loss of ~6 dB. Going from 1 to 5 GHz is predicted to cause only 14-dB loss. However, in reality, the number is often significantly higher. In that

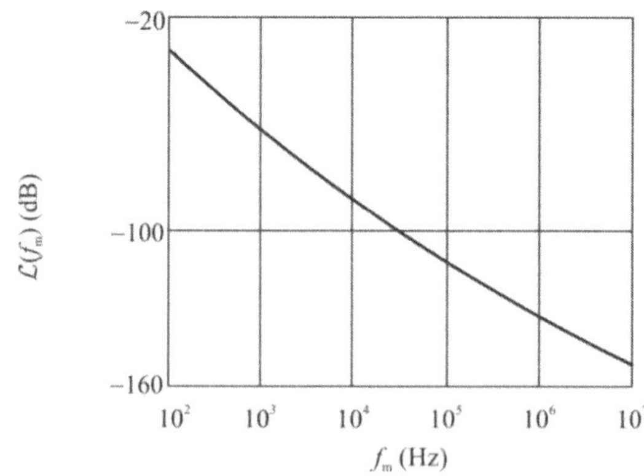

Figure 3.60 Leeson phase noise curve.

regard, the best solution for high-frequency sources is to oscillate at 1 GHz range and use multipliers (multipliers loose only a $20\log_{10}$ multiplication ratio).

- The more power drawn from the device, the better the signal-to-noise ratio, reflected by P_s in the denominator. This, however, reduces the resonator reduces Q.
- No relationship has been established between phase noise performance and VCO constant K_v (MHz / V). In practice, lower K_v allows lower coupling of the resonator, hence higher Q and better phase noise.

Remember that phase noise is always measured by the residual noise as

$$\phi_j^2 = 2\int_{f_1}^{f_2} \mathcal{L}(f_m)df_m \tag{3.36}$$

or

$$SNR = 10\log_{10}(\phi_j^2) \tag{3.37}$$

and f_1, f_2 are defined by the application. Therefore, 2° of RMS phase noise equals $2\pi/180$, which equals 0.035 radians [which is equivalent to $10\log_{10}(1/0.0352)$, or ~ 29-dBc].

3.6.3 Analysis

Assuming such devices exist, a lossless resonator and other noiseless devices would generate a closed-loop transfer function given by

$$CL = \frac{s}{s^2 + \omega_0^2}$$

corresponding to a time domain response given by

$$s(t) = A\sin(\omega_0 t) \tag{3.38}$$

If such a waveform would be generated by the device, it would not be a forced response (by noise, hence a noiseless device) and would generate a perfect, noiseless, undistorted signal. This, however, is not a real-world situation.

Instead, let us specify our model as an amplifier G and the feedback resonator $F(s)$ as a lossy parallel LC circuit with given Q, such that:

$$F(s)=\frac{s\omega_0}{Q\left(\frac{s^2+s\omega_0}{Q+\omega_0^2}\right)^{-1}}$$

The linear closed loop is given by

$$\frac{s\omega_0}{Q}=F_1(s)$$

$$CL(s)=\frac{GF_1(s)}{s^2+F_1(s)(1-G)+\omega_0^2}$$

Because the ideal signal is not possible, $1-G$ must be greater than 0. Hence the poles of a real-world oscillator are not on the imaginary axis of the complex plane; rather they are slightly on the left-hand side.

Oscillator output will actually be forced (by noise); it will always be narrowband noise. The immensity of gain provided by the oscillator is from the $1-G$ term. Hence, oscillators also have large Q and voltage gains of approximately

$$G'=\frac{1}{1-G} \tag{3.39}$$

(on the order of 10^8 to 10^9). This demonstrates, intuitively, how it is possible to build oscillator power from the low value of the noise signal k_BTf.

3.6.4 Equipartition Theorem

The equipartition theorem of statistical mechanics lies at the heart of all calculations of thermal noise [32]. Every state variable (in an electronic system these are comprised of voltages on capacitors and current through inductors) that is not constrained to have a fixed value is free to fluctuate. The thermal fluctuations in the current through an inductor or in the voltage on a capacitor are the ultimate origins of circuit noise. If the energy stored in the system corresponding to state variable x is proportional to x^2, then x is said to be a degree of freedom of the system. Thus, the voltage on a capacitor constitutes a degree of freedom, since the energy stored on it is $CV^2/2$. Statistical mechanics requires that if a system is in thermal equilibrium with a reservoir of temperature T, then each degree of freedom of the system will have a fluctuation energy of $k_B T / 2$. Thus, the mean square fluctuation ΔV^2 in the voltage of a system with a single capacitor must be such that

$$\frac{C\Delta V^2}{2} = \frac{k_B T}{2} \tag{3.40}$$

so that

$$\Delta V^2 = \frac{k_B T}{C} \tag{3.41}$$

This simple and elegant result shows that if all noise is of thermal origin, and the system is in thermal equilibrium, the total noise over the entire bandwidth of the system is determined just by the temperature and capacitance [32]. If we have a large resistance coupling noise to the capacitor, the noise per unit bandwidth is large but the entire bandwidth of the system is small: if we have a small resistance coupling noise to the capacitor, the noise per unit bandwidth is small but the entire bandwidth of the system is large. Thus, the total noise—the product of the noise per unit bandwidth $4k_B TR$ and the bandwidth of the circuit $1/RC$—is constant, independent of R.

We illustrate for the example circuit of Figure 3.61 how the noise from various devices interact to yield a total noise of $k_B T / C$. A circuit with four transistors connected to a common node with capacitance C is shown. By convention the common node is denoted as the source of all transistors, and the forward currents of all transistors are indicated as flowing away from the node while the reverse currents of all transistors are indicated as flowing toward the node. Only the voltage V_s is free to fluctuate; all other voltages are held at fixed values so that the system has only one degree of freedom. The equipartition theorem of statistical mechanics predicts that if we add the noise from all transistors over all frequencies to compute the fluctuation in voltage ΔV_s^2, the total will equal $k_B T / C$ no matter how many transistors are connected to the node or what the other parameters are, so long as all the noise is of thermal origin and the system is in thermal equilibrium.

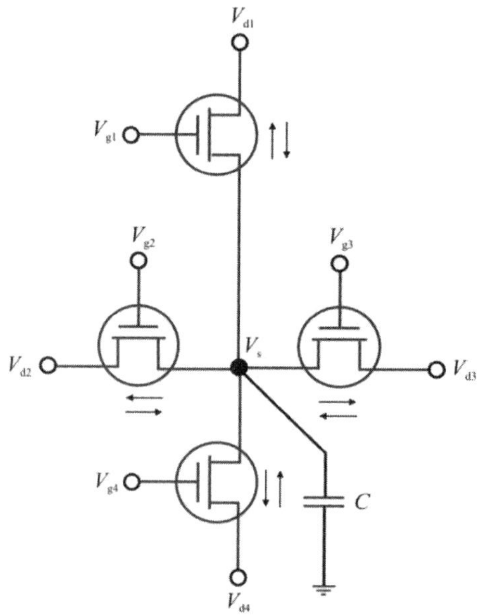

Figure 3.61 Equipartition theorem.

3.7 Concluding Remarks

In this chapter the basic types of oscillators used in the role of exciters in EW systems were discussed. While the LC oscillators, discussed first here, are the oldest types, they have many failings in the functioning as exciters. Chief among them is their relative broad spectral occupancy (low Q). The signals so generated are too wide to be used effectively. Thus, other forms of resonant elements were sought.

Crystal resonators go a long way to addressing this problem. They are fairly stable with relatively (to LC oscillators) high Q.

Phase-locked loops are used almost exclusively in modern oscillator applications due to their flexibility. They are used as reference oscillators in many exciter applications.

Microelectromechanical systems are relatively new but have some very desirable properties: namely small size for applications where SoC are being considered. They also are capable of very high Q.

The last exciter technology discussed was direct digital synthesis where the ultimate in flexibility is possible. Just about any conceivable waveform can be generated with DDS technology.

References

[1] Manassewitsch, V., *Frequency Synthesizers Theory and Design*, 3rd Ed., New York: Wiley, 1987.

[2] Goldberg, B-G, "Oscillator Phase Noise Revisited– a Heuristic Review," www.rfdesign.com, January 2002, pp.62-64, accessed Sep. 14, 2018.

[3] http://my.integritynet.com.au/purdic/oscillators.htm, accessed Aug. 23, 2016.

[4] Matthys, R. J., *Crystal Oscillator Circuits*, Malabar, FL: Krieger Publishing Company, 1992.

[5] Bible, S., "Selecting Crystals for Stable Oscillators," *Microwaves and RF*, Nov. 11, 2002.

[6] Lee, T. H., *The Design of CMOS Radio-Frequency Integrated Circuits,* New York: Cambridge University Press, 1998.

[7] http://www.comsol.com/cd/direct/conf/2012/papers/11386/11509_harshey_paper.pdf, accessed Aug. 15, 2014.

[8] Kshirsagaar, K. S., and P. V. Ingole, "Design Disk Resonator Using Micro-Electromechanical System (MEMS) for GSM and RF Applications*," International Journal of Advanced Innovative Technology in Engineering (IJAITE)*, Vol. 2, No. 5, Sep. 2017, pp. 5–11, http://garph.org/downloads/IJAITE%20Vol.%202,%20Issue%205/2.pdf, accessed Oct. 2018.

[9] Gupta, S. C., "Phase-Locked Loops," *Proceedings of the IEEE*, Vol. 63, pp. 291–306, February 1973.

[10] Lindsey, W. C., and C. M. Chie, "A Survey of Digital Phase-Locked Loops," *Proceedings of the IEEE*, vol. 69, pp. 410–431, April 1981.

[11] Manassewitsch, V., *Frequency Synthesizers Theory and Design*, New York: Wiley, 1987, Chs. 4 and 3.

[12] Abramovitch,D., http://dabramovitch.com/pubs/pll_tutorial.pdf, *Proceedings 2002 Automatic Control Conference*, Agilent Laboratories Communications and Optics Research Lab, Palo Alto, CA, June 24, 2003.

[13] Ogata, K., *Modern Control Engineering*, Prentice-Hall Instrumentation and Controls Series, Englewood Cliffs, NJ: Prentice-Hall, 1970.

[14] Van Valkenburg, M. E., *Network Analysis*, 2nd Ed, Englwood Cliffs, NJ: Prentice Hall, 1964, pp. 284–291.

[15] Hsieh, G-C., and J. C. Hung, "Phase-Locked Loop Techniques — A Survey," *IEEE Transactions on Industrial Electronics*, Vol. 43, pp. 609–615, December 1993.

[16] Best, R. E., *Phase-Locked Loops: Design, Simulation, and Applications* ,3rd Ed., New York: McGraw-Hill, 1997.

[17] Van Valkenburg, M. E., *Network Analysis*, 2nd Ed., Englewood Cliffs, NJ: Prentice Hall, 1964, pp. 207–208.

[18] Wolaver, D. H., *Phase-Locked Loop Circuit Design*. Advanced Reference Series & Biophysics and Bioengineering Series, Englewood Cliffs, NJ: Prentice Hall, 1991.

[19] Crawford, J. A., *Frequency Synthesizer Design Handbook*. Norwood, MA: Artech House, 1993.

[20] Lee, J., "Voltage-Controlled Oscillators and Frequency Dividers," Electrical Engineering Department, National Taiwan University, No. 1, Sec. 4, Taipei, Taiwan, http://cc.ee.ntu.edu.tw/~jrilee/course/COMMIC08/CommIC_Ref1.pdf, accessed Sep. 2018.

[21] Liang, Y., C. W. Domier2, and N. C. Luhmann, Jr., "RF MEMS Extended Tuning Range Varactor and Varactor Based True Time Delay Line Design," *USA PIERS ONLINE*, Vol. 4, No. 4, 2008, pp. 433436.

[22] Senturia, S. D., *Microsystem Design*, Boston: Kluwer Academic Publishers, 2001.

[23] Rebeiz, G. M., *RF MEMS Theory, Design, and Technology*, Hoboken, NJ: John Wiley & Sons, 2002.

[24] Zou, J., C. Liu, et al., “Development of a Wide Tuning Range MEMS Tunable Capacitor for Wireless Communication System," *Technical Digest of International Electron Devices Meeting*, 2000, pp. 403–406.
[25] Manikanandan, K., “Direct Digital Synthesis,” *Electronics for You,* Nov. 2000, pp. 33-38.
[26] Shannon, C. E. "Communication in the Presence of Noise". *Proceedings of the Institute of Radio Engineers*, January 1949, Vol. 37, No. 1, pp. 10–21.
[27] Nyquist, H., "Certain Topics in Telegraph Transmission Theory," *Transactions of the American Institute of Electrical* Engineers, April 1928, Vol. 47, pp. 617–644.
[28] Ifeachor, E. C., B. W. Jervis, *Digital Signal Processing: A Practical Approach*, Reading, MA: Addison-Wesley Publishing Co, 1996, p. 492.
[29] Manassewitsch, V., *Frequency Synthesizers Theory and Design*, 3rd Ed., New York: Wiley, 1987, pp. 104–128.
[30] Niknejad, A. M., “Phase Locked Loops (PLL) and Frequency Synthesis,” http://rfic.eecs.berkeley.edu/ee242/pdf/Module_7_3_PLL.pdf, 2014, accessed Mar. 3, 2015.
[31] Leeson, D. B., “A Simple Model of Feedback Oscillator Noise Spectrum,” *Proceedings IEEE*, Feb. 1966, pp 329-330.
[32] Sarpeshkar, R., T. Delbrock, and C. A. Mead, “White Noise in MOS Transistors and Resistors,” *Circuits and Devices*, Nov. 1993, pp. 23-29.

Chapter 4

Introduction to RF Amplifiers

4.1 Introduction

Amplifiers, which are fundamental elements of circuit design, take a small signal and make it larger. They drive everything from ear buds to antennas. Placed ahead of analog-to-digital converters (ADCs), they reshape signals from sources as diverse as strain gauges to ultrasound probes. By properly selecting feedback passives, they can be configured into high-pass, low-pass, band-pass, and band-elimination filters. Feed them with multiple signals, and they produce harmonics of every component of those inputs—good for some applications, a headache in others.

Extensive coverage of RF amplifiers is provided in many places. For our purposes [1–7] are recommended.

This chapter is structured as follows. Delineation of the amplifier classes is presented first, and how biasing is used to place the operating point of the active device to implement these classes. We then briefly discuss Miller's theorem and its implications in feedback systems. For such a simple result, its impact, especially for one form of amplifier, is tremendous. Next the frequency response of RF amplifiers is discussed and some techniques to quickly estimate the bandwidth are presented. Some wideband amplifier architectures are then presented. The chapter concludes with an introduction to RF amplifiers implemented with digital topologies.

4.2 Amplifier Classes

4.2.1 Amplifier Topology

The general topology for a transistor amplifier is shown in Figure 4.1 (in this case shown with a MOSFET but any properly biased active device that provides gain at the frequency of operation could fill this role). In Figure 4.1, the inductors L_1 and L_2

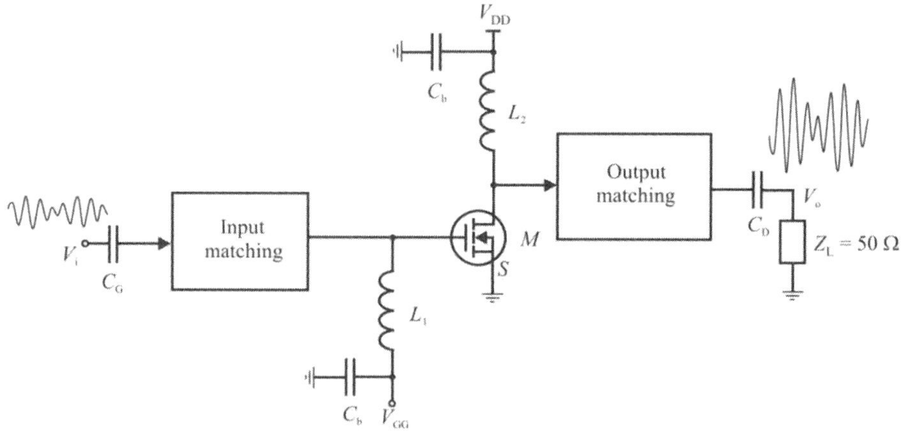

Figure 4.1 Basic topology of an RF amplifier (neglecting most biasing details).

serve as RF chokes which block the RF signals from reaching the power supplies V_{GG} and V_{DD}, respectively, while simultaneously providing DC power to the device. Capacitors C_b, in conjunction with L_1 and L_2, serve as bypass capacitors and provide filtering of the RF signals from the power supplies. The AC coupling capacitors C_G and C_D pass the RF signals while simultaneously isolating the DC bias levels in this stage from the stages prior and subsequent. This is one of the simplest configurations of biasing networks—we discuss more shortly.

The matching networks perform a few different functions depending on how the amplifier is to be used. As a linear amplifier the major concern is interfacing the amplifier properly with stages that precede this stage and the one following. Most of the time these concerns have to do with linearity and minimizing the distortion in the signals being processed, generating as small noise products as possible. When the amplifier is configured as a power amplifier, usually as the output stage of an amplifier chain, it would be feeding a transmission line that leads to an antenna for purposes of radiating the signal toward a target. In that case, the signals are large enough that linearity and noise are less of a concern and stability becomes the major issue. Efficiency is also a major concern in power amplifier stages. These two regimes yield matching networks that can be quite different.

4.2.1.1 Background

Historically, amplifier class designations were related to the biasing of amplifier devices—that is, over how many degrees of each input-signal swing they conducted. It worked for classes A, B, and C. For class AB, it made a certain kind of sense. Today, it's not as clear-cut. In the beginning, classification revealed something about linearity versus efficiency. Class A amplifiers can be made very linear, but with limited efficiency. In theory, a class A amplifier can achieve 50% efficiency with inductive output coupling or 25% with capacitive coupling. Class B amplifiers are subject to crossover distortion, but efficiency runs theoretically as high as 78.5%.

4.2.2 Biasing for RF Devices

Biasing refers to placing the quiescent operating point of the active device so that certain conduction conditions occur [8]. Linear conduction means that changes in the amplified output of the circuit are proportional to changes at the input. All active devices have a limit on the range of input and output voltage and current that are allowed, beyond which the performance becomes nonlinear.

Many RF schematics contain the statement: "bias circuit not shown"; when actually one of the most critical yet often overlooked aspects in any RF circuit design is the bias network. The bias network determines the amplifier performance over temperature as well as RF drive. The DC bias condition of the RF transistors is usually established independently of the RF design.

The principal concerns when considering the biasing configuration are:

- Power efficiency
- Stability
- Noise
- Thermal runway
- Ease of use

Transistor amplifier biasing is required for several reasons:

- Without a DC biasing network, we would require two separate voltage supplies to furnish the desired class of bias for both the emitter-collector and the emitter-base voltages.
- Sometimes, when warranted, this is still done, but biasing was developed so that these separate voltages could be obtained from a single supply.
- Transistor parameters are temperature-sensitive, causing thermal runaway. Thermal runaway will rapidly destroy a BJT, as collector current quickly and uncontrollably increases to damaging levels as the temperature rises, unless the amplifier is temperature stabilized to nullify this effect.

4.2.3 Amplifier Bias Classes of Operation

Specific classes of amplifier bias levels are utilized to achieve different objectives, each with its own distinct advantages and disadvantages. The most prevalent non-

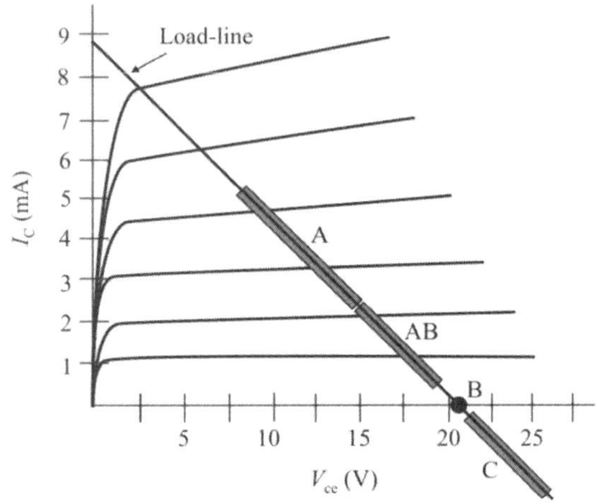

Figure 4.2 Typical BJT load-line characteristics for different bias classes. The resistive load-line is superimposed on the output characteristic of a BJT. I_C is the collector current and V_{CE} refers to the voltage between the collector and emitter. Similar curves apply to FETs and thermionic tubes as well, although the specific shapes of the I_C vs V_{ce} curves can be different.

switching classes of bias operation are Class A, AB, B, and C. All of these classes use circuit components to bias the transistor at a different DC collector operating current, denoted by I_{CQ}, where the Q refers to *quiescent*, and voltage levels. "Q" in this case refers to quiescent point (not to be confused with quality factor used prolifically, nor with heat in Chapter 12). The output characteristics for these classes are summarized in Figure 4.2 in the form of collector current, I_C, versus the collector-emitter current, V_{ce}, as a function of base current.

4.2.4 Amplifier Classes

An analog amplifier class is determined by how it is biased. RF amplifier classes A, B, C, and AB are analog amplifier classifications that are determined by the *number of degrees of current flow during each 360-degree RF cycle* over which plate, collector, or drain current flows. We introduce these classes in this section.

4.2.4.1 Class A

Class A are single-ended amplifiers and are ordinarily used only in small-signal non-power applications. Class A will generally require a constant current bias source to fix the operating point regardless of the RF drive and output. The circuit will have to have some sort of feedback to keep the output current at a fixed level, or a circuit must be created whose current is large compared to the amount of output power required (i.e., it is quasi class A in that the operating point movement is minimal).

The amplifier device is biased in the region of the center of the input signal swing (see Figure 4.2.) In class A, biasing a single active device (a BJT, FET, or thermionic tube) allows it to operate in its linear conduction region during the entire input cycle.

Class A amplifiers operate over a relatively small portion of a tube's plate-current, a transistor's collector-current, or an FET's drain current range and have continuous current flow throughout the entire 360° of each RF cycle. Their efficiency in converting DC-source-power to RF-output-power is poor. DC source power that is not converted to RF output power is dissipated as heat. However, in compensation, Class A amplifiers have higher linearity that is responsible for *lower output-signal distortion* than any other amplifier class. They are most commonly used in small-signal applications where linearity is more important than power efficiency, but also are sometimes used in large-signal applications where the need for high linearity outweighs cost and heat disadvantages associated with poor power efficiency.

4.2.4.2 Class B

A class B active device is biased so that conduction only occurs for ½ of the input signal period. During the other half, the device does not conduct at all. A *push-pull* arrangement ensues when two active devices are included in the amplifier and one active device conducts during half of an input cycle, the other during the other half. The two halves are reassembled at the amplifier's output. These amplifiers are called push-pull because the outputs of the active devices have a 180° phase relationship.

Class B operations require some form of positive biasing—though the operating point will move with RF drive. This will require an open loop circuit with some sort of compensation over ambient conditions.

Class B amplifiers have their tube control grid, transistor base, or FET gate biased near plate, collector, or drain current cutoff, causing plate, or collector, or drain current to flow only during one-half (180°) of each RF cycle. The bias point, as shown in Figure 4.2, is on the V_{ce} axis. That causes the DC source power to RF output-power efficiency to be much higher than with Class A amplifiers, but at the cost of severe output cycle waveform distortion. If the input is a sine wave, for example, only half of that sine wave will appear in the output current waveform.

That waveform distortion is greatly reduced in practical designs by using relatively high-Q resonant output "tank" circuits to reconstruct full RF cycles, but the signal is at higher power levels than the input signal. The quality factor, also denoted by Q, is the ratio of an inductor's reactance at a given frequency to its DC resistance. More generally, though, it reflects how sharply an L-C resonant circuit is tuned (where L implies the presence of one or more inductance and C implies one or more capacitors). Low sinewave distortion results only if the Q of the tank circuit is sufficiently high. Unless the Q is infinite, which, of course, it never can be, the amplitude of one-half cycle will be larger than the other, which is a manifestation of harmonic energy content, *o*f importance to EW amplifier designs. Coupling an antenna system too tightly to the resonant output tank circuit of an amplifier will lower its Q, increasing the percentage of harmonic content in the output.

4.2.4.3 Class AB

For class AB, the active device is biased to conduct over slightly more than half the input signal swing to simplify crossover. When configured in a push-pull arrangement, class AB amplifiers resemble class Bs, except their active devices are biased so both conduct during an overlapping portion of each input cycle. This sacrifices a certain amount of potential gain for better linearity (i.e., there's a smoother transition at the crossover point of the output signal). Class AB sacrifices some of that efficiency for lower distortion.

Simply by decreasing the I_{CQ} of the amplifier by a small amount, class AB operation can be reached. But any class AB single-ended power amplifier will create more output distortion than a class-A type due the output clipping of the signal's waveform.

Like class B, class AB require some form of positive biasing—though the operating point will move with RF drive. This will also require an open-loop circuit with some sort of compensation over ambient conditions. An amplifier will operate in class AB when it is biased in the region shown in Figure 4.2.

As the designation suggests, Class AB amplifiers are compromises between Class A and Class B operation. They are biased so plate, collector, or drain current flows less than 360°, but more than 180°, of each RF cycle. Any bias point between those limits can be used, which provides a continuous selection range reaching from low distortion, low efficiency on one extreme to higher distortion, higher efficiency on the other.

4.2.4.4 Class C Amplifiers

Class C amplifiers are biased well beyond cutoff, so that plate, collector, or drain current flows less than 180° of each RF cycle. That provides even higher power efficiency than Class B operation, but with the penalty of even higher input-to-output nonlinearity, making use of relatively high-Q resonant output tank circuits to restore complete RF sine-wave cycles essential. High amplifying-nonlinearity makes them unsuitable to amplify signals where the amplitude is used to convey the information, such as AM, DSB, or SSB signals.

However, most Class C amplifiers can be amplitude-modulated with acceptably low distortion by varying plate, collector, or drain voltage, because they generally are operated in the region of plate, collector, or drain saturation so that the RF output voltage is dependent upon the instantaneous DC plate, collector, or drain voltage. They also are commonly used in *continuous wave* (CW) and FSK applications and in PM and FM transmitter applications where signal amplitudes remain constant, and the information is conveyed in the instantaneous phase of the carrier signal

The active device in a class C amplifier is biased so that it conducts for less than half the input signal period. As such, they are nonconducting for over half this range. Their response is therefore very nonlinear. See Figure 4.2 for the bias region.

Class C amplifiers offer high efficiency (up to 90%), but the high-Q filters needed for their operation (to return to linear performance) have narrow bands of resonance. Moreover, tank circuits for low (e.g., audio) frequencies are impractical to implement, which ultimately limits them to RF.

For class C amplifiers used in RF applications, the output device drives a resonant tank circuit consisting of an inductor and one or two capacitors. It conducts for only a short portion of each input cycle.

Class C generally requires a negative bias of some kind—or in most cases, the input of the transistor is tied to ground with an inductor or resistor, which is sufficient to keep the conduction angle correct.

Class C amplifiers incorporate a single active device that's biased to conduct during only a small portion of each input-waveform cycle. Energy is driven into a high-Q, L-C tank circuit that continues to ring at its resonant frequency during the times the active device isn't conducting. An analogy would be the continuous tapping of a big bell with a small hammer at a rate equal to the resonant frequency of the bell.

4.2.4.5 Summary

Figure 4.3 summarizes the class of operation that ensues when an amplifier is biased as indicated.

The efficiency of the various amplifier classes is illustrated in Figure 4.4. While there are several different measures of efficiency, in general it refers to how well a particular amplifier topology converts power from the power supply into useful signal power. We will cover several definitions of efficiency in Chapter 9.

4.2.5 Miller's Theorem

Miller's theorem reflects the effects of gain in an amplifier and its interaction with feedback components. We briefly describe it here as we will use it later.

Consider the network in Figure 4.5. An admittance $Y = 1 / Z$ is connected between two nodes. Miller's theorem provides a way for replacing the bridging admittance Y with two admittances Y_1 and Y_2 between node 1 and ground and node 2 and ground.

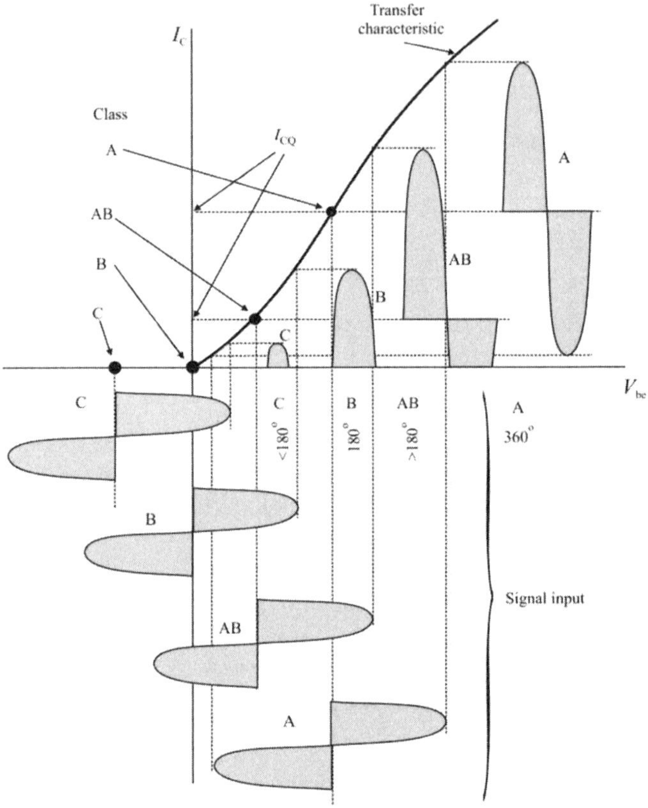

Figure 4.3 Amplifier classes as defined by their bias quiescent point.

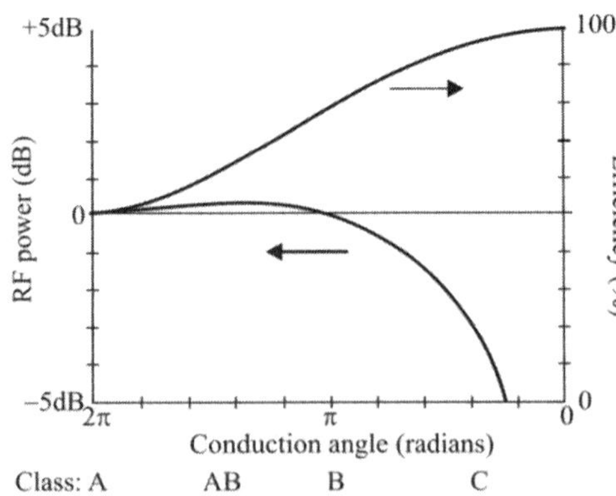

Figure 4.4 Amplifier efficiency.

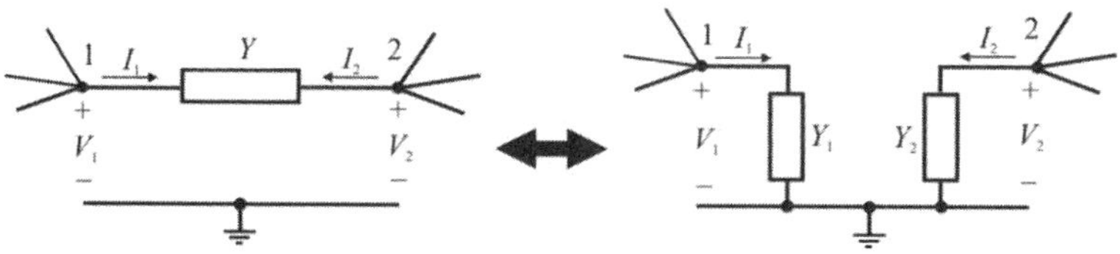

Figure 4.5 Miller's theorem.

The relationship between V_2 and V_1 is given by $G = V_2 / V_1$. To find Y_1 and Y_2

$$\begin{aligned} I_1 &= Y(V_1 - V_2) = YV_1(1 - V_2 / V_1) & I_2 &= Y(V_2 - V_1) = YV_2(1 - V_1 / V_2) \\ I_1 &= YV_1(1 - G) & I_2 &= (YV_2(1 - 1 / K) \\ I_1 &= Y_1 V_1 & I_2 &= Y_2 V_2 \\ Y_1 &= Y(1 - G) & Y_2 &= Y(1 - 1 / G) \end{aligned} \tag{4.1}$$

Equations (4.1) are referred to as Miller's theorem.

We can see from (4.1) that as G gets larger, the admittance represented by Y_1 gets linearly larger (although in the negative direction). Remember that frequently G is negative, making Y_1 grow substantially. The biggest problem arises in active components where Y is actually C_{gd} (MOSFET) or C_{cb} (BJT). The larger these model elements are, the greater their impact, particularly on the frequency response of an amplifier using these components. We can also see that the larger G is, the less of an impact Y has on Y_2, so the effects on the output get smaller.

4.2.6 Frequency Response of Amplifiers

The frequency response of amplifiers consists of three distinct regions (see Figure 4.6).

4.2.6.1 Midband

This is the frequency range normally of most interest for amplifiers. In this region, large capacitors can be treated as short circuits and small capacitors can be treated as open circuits. The gain is constant and can be obtained by linear small-signal analysis.

4.2.6.2 Low-Frequuency Band

At the lower end of the frequency response is the low-frequency band. In this region the gain drops at frequencies lower than f_L. Large capacitors can no longer be treated

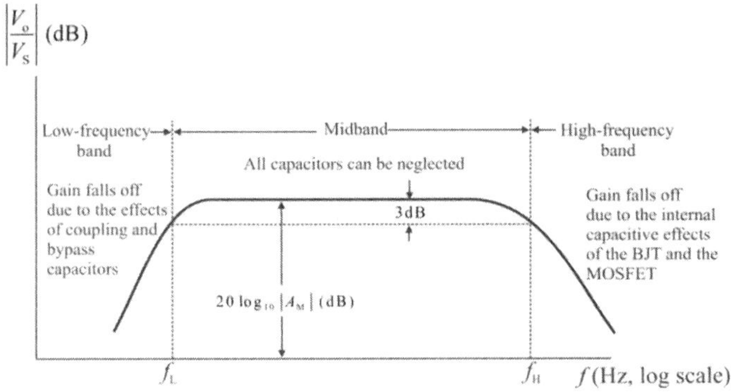

Figure 4.6 Frequency response regions of an amplifier.

as short circuits. However, small capacitors can still be treated as open circuits. The gain roll-off is mainly due to coupling and bypass capacitors.

4.2.6.3 High-Frequency Band

Above the midband lies the high-frequency band. The gain drops at frequencies higher than f_H. Small capacitors can no longer be treated as open circuits while large capacitors can still be treated as short circuits. The gain roll-off is mainly due to parasitic capacitances of the MOSFETs and BJTs.

4.2.7 Some High-Frequency Analysis Techniques

Under certain circumstances, accurate estimates of the frequency performance of amplifiers can be determined essentially by inspection of the circuits [9]. We introduce some of these techniques in this section.

Known as the *short-circuit time constant* (SCTC) and *open-circuit time constant* (OCTC) methods, they are useful for estimating the bandwidth of networks that contain capacitors and resistors. The SCTC method estimates the lower 3-dB cutoff frequency while the OCTC method estimates the high 3-dB cutoff frequency (see Figure 4.7). OCTC is used to find the high-frequency cutoff (DC coupling) and SCTC is used to find the low-frequency cutoff frequency (AC coupling).

4.2.7.1 Estimating the Upper 3-dB Frequency

The general transfer function of an amplifier is given by

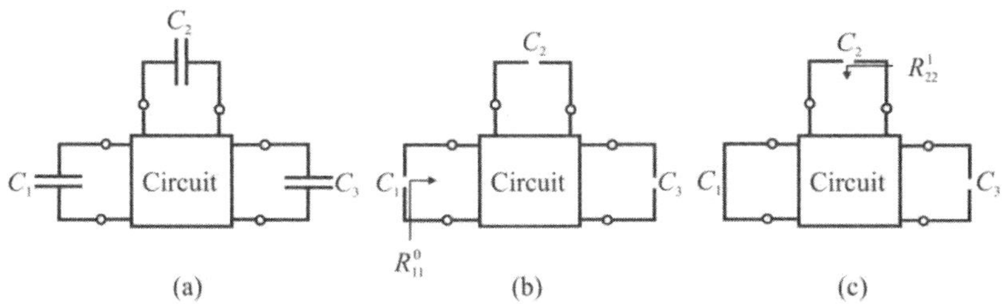

Figure 4.7 Time constant method: (a) the circuit with all capacitors identified, (b) determining R_{ii}^{0}, and (c) determining R_{ii}^{j}.

$$A(s) = A_{\mathrm{M}} F_{\mathrm{H}}(s) = A_{\mathrm{M}} \frac{(1+s/\omega_{z1})(1+s/\omega_{z2})\cdots(1+s/\omega_{zm})}{(1+s/\omega_{p1})(1+s/\omega_{p2})\cdots(1+s/\omega_{pn})} \tag{4.2}$$

$$= A_{\mathrm{M}} \frac{1+a_1 s + a_2 s^2 + \ldots + a_m s^m}{1+b_1 s + b_2 s^2 + \ldots b_n s^n} \tag{4.3}$$

where ω_{zi} represents the transfer function zero frequencies (the frequencies where there is no signal transferred from the input to the output) and ω_{pi} the transfer function pole frequencies (the frequencies at which there is unlimited signal transferred from the input to the output). The upper 3-dB frequency is approximated from (4.2) as follows

$$A(s) = A_{\mathrm{M}} F_{\mathrm{H}}(s) \approx A_{\mathrm{M}} \frac{(1+s/\omega_{z1})(1+s/\omega_{z2})}{(1+s/\omega_{p1})(1+s/\omega_{p2})}$$

$$\left|F_{\mathrm{H}}(j\omega_{\mathrm{H}})\right|^2 = \frac{1}{2} \approx \frac{(1+\omega_{\mathrm{H}}^2/\omega_{z1}^2)(1+\omega_{\mathrm{H}}^2/\omega_{z2}^2)}{(1+\omega_{\mathrm{H}}^2/\omega_{p1}^2)(1+\omega_{\mathrm{H}}^2/\omega_{p2}^2)}$$

$$= \frac{1+\omega_{\mathrm{H}}^2(1/\omega_{z1}^2+1/\omega_{z2}^2)+\omega_{\mathrm{H}}^4\left(\dfrac{1}{\omega_{z1}^2}\dfrac{1}{\omega_{z2}^2}\right)}{1+\omega_{\mathrm{H}}^2(1/\omega_{p1}^2+1/\omega_{p2}^2)+\omega_{\mathrm{H}}^4\left(\dfrac{1}{\omega_{p1}^2}\dfrac{1}{\omega_{p2}^2}\right)}$$

$$\approx \frac{1+\omega_{\mathrm{H}}^2(1/\omega_{z1}^2+1/\omega_{z2}^2)}{1+\omega_{\mathrm{H}}^2(1/\omega_{p1}^2+1/\omega_{p2}^2)} \tag{4.4}$$

$$\omega_{\mathrm{H}} \approx \frac{1}{\sqrt{(1/\omega_{p1}^2+1/\omega_{p2}^2)-2(1/\omega_{z1}^2+1/\omega_{z2}^2)}} \tag{4.5}$$

4.2.7.2 Dominant Pole Response

Usually one of the poles has much lower (high) frequency than any of the other poles and zeros. A dominant pole exists if the lowest (high) frequency pole is at least 4× away from the nearest pole or zero. In that case, from (4.5) we can estimate that frequency with

$$\omega_H \approx \frac{1}{\sqrt{(1/\omega_{p1}^2 + 1/\omega_{p2}^2) - 2(1/\omega_{z1}^2 + 1/\omega_{z2}^2)}} \approx \omega_{p1} \tag{4.6}$$

since squaring the frequency term in the denominator will ensure that the lowest (high) frequency is 16× away.

4.2.7.3 Time Constant Methods

Open Circuit Time Constant Method to Estimate Amplifier Bandwidth

The OCTC is a method to estimate the bandwidth of an amplifier. From (4.2)

$$b_1 = \frac{1}{\omega_{p1}} + \frac{1}{\omega_{p2}} + \ldots + \frac{1}{\omega_{pn}}$$

The exact solution for the open-circuit time constant is given by

$$b_1 = \sum_{i=1}^{n} C_i R_i \tag{4.7}$$

This can be approximated with

$$b_1 = \frac{1}{\omega_{p1}} + \frac{1}{\omega_{p2}} + \ldots + \frac{1}{\omega_{pn}} \approx \frac{1}{\omega_{p1}} \tag{4.8}$$

so that

$$\omega_H \approx \omega_{p1} \approx 1/b_1 = \frac{1}{\sum C_i R_i} \tag{4.9}$$

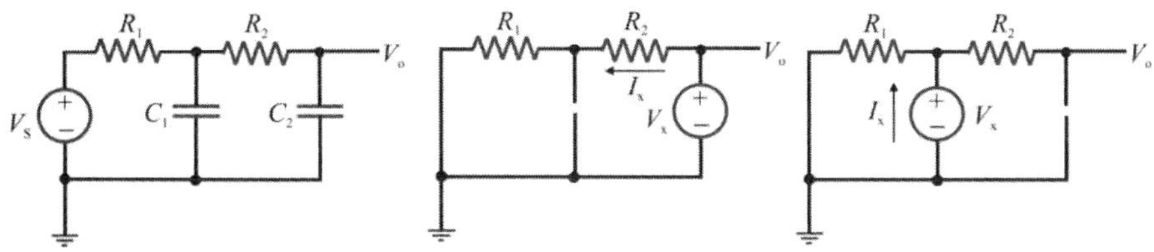

Figure 4.8 Simple RC circuit and auxiliary circuits to find time constants.

If we can find (estimate) the RC time constants in the network we can compute their sum and determine the upper cutoff frequency with (4.9). This is the crux of the OCTCA method. First, we determine b_1 with (4.7) by

$$b_1 = C_1 R_{11}^0 + C_2 R_{22}^0 + C_3 R_{33}^0$$

R_{ii}^0 are the equivalent resistances in parallel with C_i by treating the other capacitors as open circuits.

The method also uses a simplified method for finding the term linear in frequency based on summing the RC products for each capacitor in the circuit, where the resistor R for a selected capacitor is the resistance found by inserting a test source at its site and setting all other capacitors to zero (open circuits) (see Figure 4.8). Hence the name *zero-value time constant technique*. To determine r_E in analysis of a BJT, both a voltage source is used as shown in Figure 4.8, as well as insertion of a current source. The resistance is the ratio of these two results.

Short-Circuit Time Constant Method

The SCTC method recognizes that all the filters in the circuit are of the high-pass type, and tend to yield increasing gain with frequency. It's also worth noting that there are the same number of poles as there are zeros. Under these circumstances, it's expedient to use the *short-circuit time constant approximation* (SCTCA) [10].

With the SCTCA we approximate the combined effect of all the poles and zeros to make a single first order high-pass filter. In other words, we approximate the gain of the overall circuit as

$$\frac{V_o}{V_i} = A_m \frac{s}{s - p_o}$$

where A_m is the midband voltage gain, and p_o is the single pole that approximates the combined effects of the actual poles and zeros of the circuit. To determine p_o,

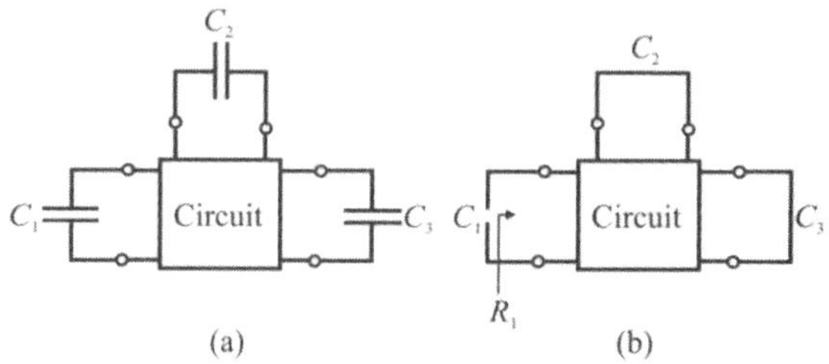

Figure 4.9 Time constant method SCTC: (a) the circuit with all capacitors identified, (b) determining resistances in parallel with all other capacitors short circuited, R_1 for this example associated with C_1. The resulting time constant is given by C_1R_1.

we determine the resistance seen by each capacitor, while all the other capacitors are shorted. We then multiply the resistance seen by the value of the capacitor of interest to obtain the short-circuit time constant for that capacitor (see Figure 4.9). This procedure is then repeated for all capacitors that govern the low-frequency response. Finally, p_o is then given by the sum of the reciprocal of each time constant. Mathematically, this is expressed as:

$$p_o \approx \sum_i -\frac{1}{R_i C_i}$$

Where R_i is the resistance seen by the ith capacitor with all other capacitors shorted.

4.3 Amplifier Parameters

The most prominent parameters associated with amplifier analysis and design are given in Table 4.1.

4.3.1 Determining the Lower 3-dB Frequency

The coupling and bypass capacitors produce a low-frequency high-pass frequency response with three poles (and three zeros where $s = j\omega = 0$, that is, at DC). If the poles are sufficiently separated then a Bode plot can be used to estimate the response (see Figure 4.10). The lower 3-dB frequency is the highest frequency pole. ω_{p2} is typically the highest frequency pole due to the small resistance represented by $1/g_m$. However, if the poles are located closely together, the lower 3-dB frequency has to be evaluated by the transfer function, which is more complicated.

Table 4.1 Definitions of Amplifier Parameters

Parameter	Definition
Current Gain, A_I	$A_I = I_o / I_i$
Voltage Gain, A_V	$A_V = V_o / V_i$
Power Gain, G_P	$G_P = P_o / P_i$
Input Impedance, Z_i	$Z_i = V_i / I_i$
Output Impedance, Z_o	$Z_o = V_o / I_o$

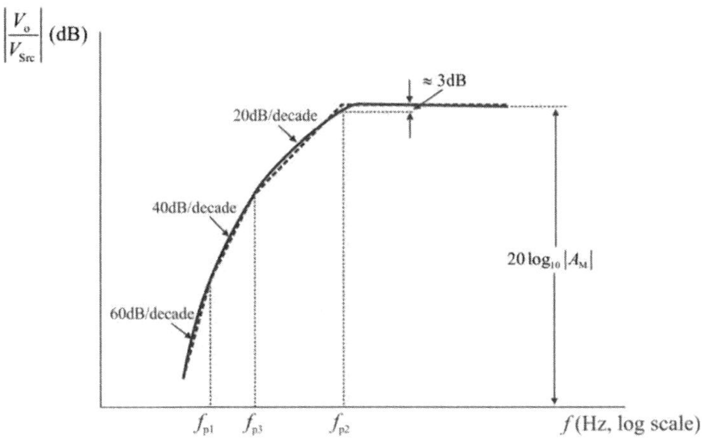

Figure 4.10 Bode plot.

4.3.2 Selecting Values for the Coupling and Bypass Capacitors

These capacitors are typically required for discrete amplifier designs. C_S is determined to satisfy the needed f_L. C_{C1} and C_{C2} are chosen such that the poles are 5 to 10 times lower than f_L.

4.4 Amplifier Topologies

Analog amplifier classes are categorized by which terminal is common to both the input and output circuits. Thus, we have for the BJT: common emitter, common collector, and common base, and for the FET: common source, common drain, and common gate. Each of these configurations has its own attributes, advantages, and disadvantages.

4.4.1 Bipolar Transistor Configurations

As the BJT is a three-terminal device, there are basically three possible ways to connect them within an electronic circuit with one terminal being common to both the input and output. Each method of connection responding differently to its input signal within a circuit as the static characteristics of the devices vary with each circuit arrangement:

- Common base configuration – voltage gain but no current gain.
- Common emitter configuration – both current and voltage gain.
- Common collector configuration – current gain but no voltage gain.

4.4.2 MOSFET Configurations

Likewise, the MOSFET is a three terminal device that can be connected in three ways:

- Common gate configuration – both voltage and current gain.
- Common source configuration – voltage gain; current gain is meaningless for the MOSFET since the input current is zero.
- Common drain configuration – current gain but no voltage gain.

4.4.3 Summary

The topologies mentioned above will be discussed extensively in the material contained in the chapters to follow. They are mentioned here simply to list the configurations to be explored.

4.5 Active Loads

An active load replaces the collector or drain resistor with a current source consisting of a BJT or MOSFET network. Resistors (especially big ones) and capacitors take up excessive space on IC real estate. Active loads take up less.

Consider the complementary MOSFET amplifier shown in Figure 4.11(a). M_2 and M_3 form a PMOS current mirror. In addition to help form a current mirror, M_2 also functions as a current source; that is, an active load, for M_1. For M_2 to be a current source, it must operate in the saturation region.

We show an output characteristic curve for M_2 in Figure 4.11(b) (that is, V_{GS2} is a constant value). Consider when $I = I_{ref}$, then $V_{GD2} = 0$ due to symmetry with M_1. This implies that $V = V_{SG}$, which is the Q-point shown in Figure 4.11(b). This characteristic curve is the *load curve* for M_1 formed by the active load M_2. (Recall that a resistive, passive load curve is a straight line—the concept here is the same except that the very small slope on the saturation region of the characteristic curve implies a very large value for r_{o2}, a desirable characteristic for a current source.)

We can graphically construct the transfer function V_o/V_i for this amplifier. The drain currents of M_1 and M_2 are the same since they are in series. The operating point of the amplifier is found from the intersection of the M_1 characteristic curve with the load curve of M_2 for a particular V_{GS1} (see Figure 4.12).

Collecting these intersections from Figure 4.12 as V_{GS1} (= V_i) changes, we can construct point-by-point the transfer characteristic curve for this amplifier. The results are shown in Figure 4.13. From this chart we can see that Region 3 exhibits a linear relationship between V_o and V_i. This is the region where the circuit of Figure 4.13 can be used as a linear amplifier.

The configuration shown in Figure 4.11(a), which uses both PMOS and NMOS devices, is used extensively in *complementary MOS* (CMOS) integrated circuits. As with other IC technologies, the advantage of active loads in CMOS IC designs is that an active load is comprised of just another transistor, which consumes considerably less real estate than large resistors. The same topology, of course, can be implemented with BJTs. They must be biased in their active regions and the transistors must be matched as well as possible. The current mirror on the left side of Figure 4.14 is equivalent to the current source shown on the right side. Ideally

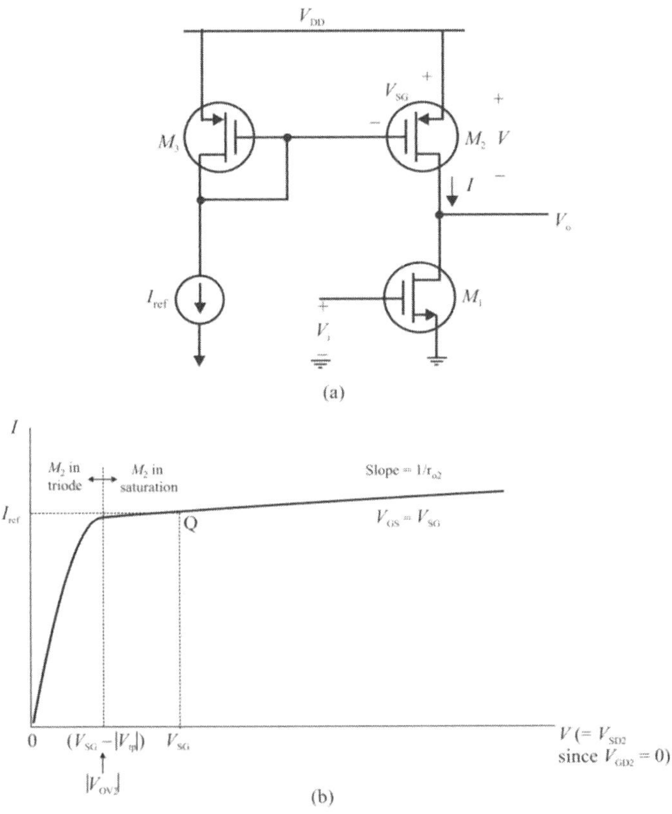

Figure 4.11 CS amp active load.

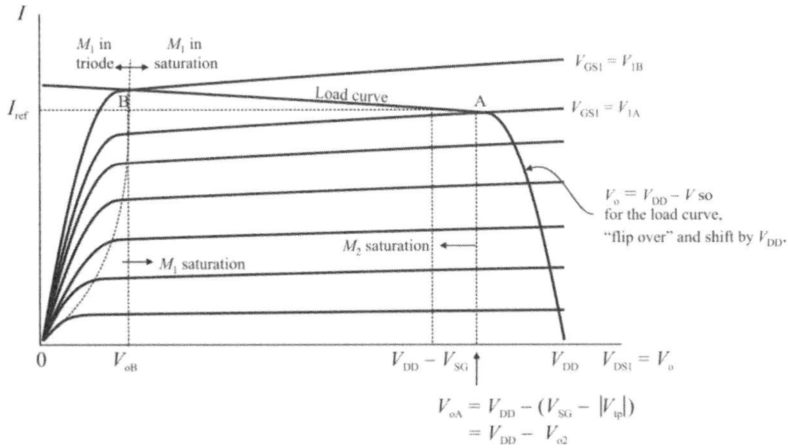

Figure 4.12 M_2 load curve with M_1 characteristic.

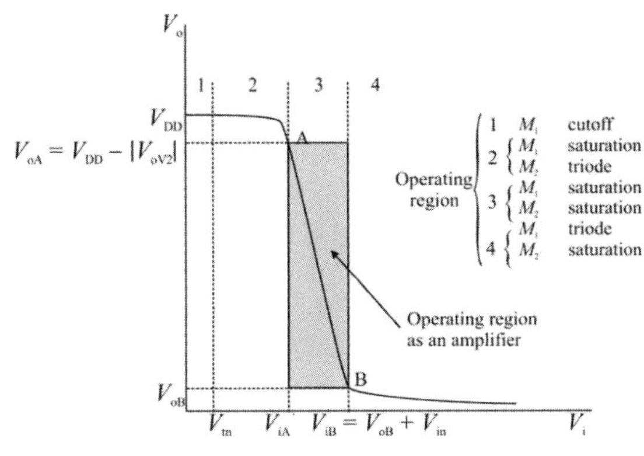

Figure 4.13 Active load transfer characteristic.

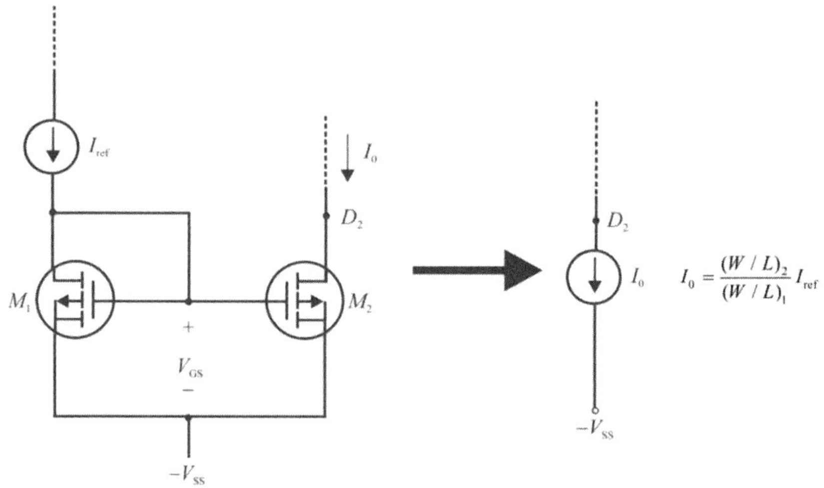

Figure 4.14 Current mirror.

the current source resistance is infinite. The current from the source is governed by

$$I_0 = \frac{(W/L)_2}{(W/L)_1} I_{\text{ref}} \tag{4.10}$$

where, in this case, W_i is the width of the channel in MOSFET i and L_i is the length of the channel in MOSFET i, $i = 1, 2$. I_{ref} is typically established with a (smaller) resistor.

The current mirror as an active load is the most prolific way to bias the active components in integrated circuits.

4.6 Gain

The maximum available power gain for unconditionally stable two-ports is given by

$$G_{\max} = \left|\frac{S_{21}}{S_{12}}\right|(K_{\text{R}} - \sqrt{K_{\text{R}}^2 - 1}) \tag{4.11}$$

where K_{R} is the Rollett stability parameter to be discussed in Chapter 10.

The *bilateral transmission power gain* of a two-port can be split into three parts

$$G = G_S G_0 G_L \tag{4.12}$$

where

$$G_S = \frac{(1-|\Gamma_S|^2)(1-|\Gamma_i|)^2}{|1-\Gamma_S\Gamma_i|^2} \tag{4.13}$$

$$G_0 = |s_{21}|^2 \tag{4.14}$$

and

$$G_L = \frac{1-|\Gamma_L|^2}{|1-\Gamma_L s_{22}|^2 (1-|\Gamma_i|^2)} \tag{4.15}$$

where Γ_i is the reflection coefficient of the two-port input.

The curves of constant gain are circles in the reflection coefficient plane. The circle for the load-mismatched two-port with gain G_L is centered at

$$\Gamma_{center} = \frac{(s_{22}^* - s_{11}\Delta_S^*)G_L}{G_L(|s_{22}|^2 - |\Delta_S|^2)+1} \tag{4.16}$$

with radius

$$R = \frac{\sqrt{1-G_L(1-|s_{11}|^2-|s_{22}|^2+|\Delta_S|^2)+G_L^2|s_{12}s_{21}|^2}}{G_L(|s_{22}|^2-|\Delta_S|^2)+1} \tag{4.17}$$

The circle for the source-mismatched two-port with gain G_S is centered at

$$\Gamma_{center} = \frac{G_S\Gamma_1^*}{1-|\Gamma_1|^2(1-G_S)} \tag{4.18}$$

with radius

$$R = \frac{\sqrt{1-G_S}\,(1-|\Gamma_1|^2)}{1-|\Gamma_1|^2\,(1-G_S)} \tag{4.19}$$

with

$$\Gamma_1 = s_{11} + \frac{s_{12}s_{21}\Gamma_L}{1-\Gamma_L s_{22}} \tag{4.20}$$

The available power gain, G_A, of a two-port network is reached when the load is conjugate matched to the output port:

$$G_A = \frac{|s_{21}|^2\,(1-|\Gamma_S|^2)}{|1-s_{11}\Gamma_S|^2 - |s_{22}-\Delta_S\Gamma_S|^2} \tag{4.21}$$

The curves with constant gain G_A are circles in the source reflection coefficient (Γ_S) plane with the center at

$$\Gamma_{center} = \frac{g_A C_1^*}{1+g_A\,(|s_{11}|^2 - |\Delta_S|^2)} \tag{4.22}$$

and radius

$$R = \frac{\sqrt{1-2K_R g_A\,|s_{12}s_{21}| + g_A^2\,|s_{12}s_{21}|^2}}{\left|1+g_A\,(|s_{11}|^2 - |\Delta_S|^2)\right|} \tag{4.23}$$

where $C_1 = s_{11} - s_{22}^*\Delta_S$ and $g_A = G_A / |s_{21}|^2$.

The operating power gain, G_P, of a two-port is given by the ratio of the power delivered to the load to that of the input of the amplifier. In terms of S-parameters it is given by

$$G_P = \frac{|s_{21}|^2\,(1-|\Gamma_L|^2)}{|1-s_{22}\Gamma_L|^2 - |s_{11}-\Delta_S\Gamma_L|^2} \tag{4.24}$$

The curves of constant gain G_P are circles in the load reflection coefficient (Γ_L) plane. The center of the circles is at

$$\Gamma_{\text{center}} = \frac{g_P C_2^*}{1+g_P(|s_{22}|^2 - |\Delta_S|^2)} \tag{4.25}$$

with radius

$$R = \frac{\sqrt{1-2K_R g_P |s_{12}s_{21}| + g_P^2 |s_{12}s_{21}|^2}}{\left|1+g_P(|s_{22}|^2 - |\Delta_S|^2)\right|} \tag{4.26}$$

where

$$C_2 = s_{22} - s_{11}^* \Delta_S$$

and

$$g_P = G_P / |s_{21}|^2$$

Example 4.1

We will examine a single-stage CE BJT amplifier that is to operate at 1 GHz [11]. The particular BJT we use is the BFG193 from Infineon whose S-parameters at 1 GHz are shown in Table 4.2. The bias parameters are V_{CE} = 10V and I_C = 10mA.

From Table 4.2 we calculate that

$$|\Delta| = |s_{11}s_{22} - s_{12}s_{21}| = |0.1346 - j0.2539| = 0.2873$$

$$K_R = \frac{1-|s_{11}|^2 - |s_{22}|^2 + |\Delta|^2}{2|s_{12}s_{21}|} = 1.0144$$

$K_R > 1$ (but not by much) and $|\Delta| < 1$ so the circuit is stable at 1 GHz. We assume that the input and output are conjugate matched so that

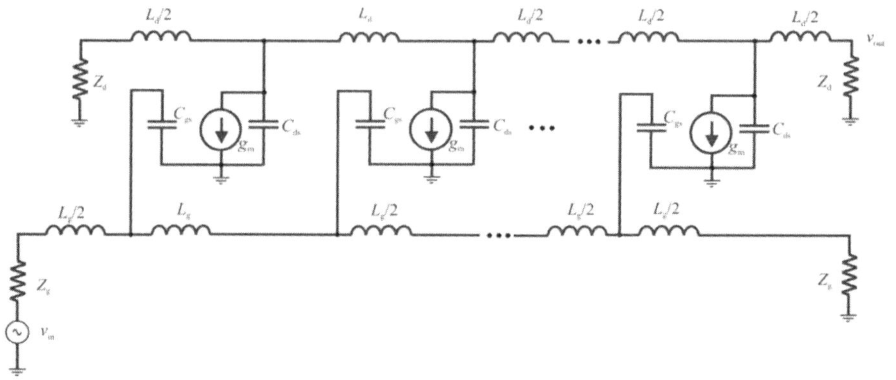

Figure 4.15 A distributed MESFET amplifier using a very simplified unilateral FET model.

$$\Gamma_S = s_{11}^*$$
$$\Gamma_L = s_{22}^*$$

4.7 Wideband Amplifiers

One of the common requirements for amplifiers in EW applications is broad bandwidth since the target set is rarely completely known in advanced. The amplifiers involved with generating the EW signal must be able to cover the entire frequency range of concern—preferably instantaneously so that retuning is not necessary.

There are some methods that can be used to widen the operating bandwidth of both small signal amplifiers as well as power amplifiers. We will discuss a few of these methods in this section.

4.7.1 Distributed Amplifiers

A technique that achieves extremely broadband operation is the *distributed amplifier* (DA), shown in Figure 4.15. In this configuration, instead of trying to tune out the transistor capacitances, they are used as part of a lumped-element approximation to a transmission line on both sides of the active device. On the input, inductors L_g are placed between the gate-to-source capacitances C_{gs} of the adjacent transistors, and the familiar lumped-element artificial transmission line with a characteristic impedance of $Z_g = \sqrt{L_g / C_{gs}}$ is formed. Z_g is nearly frequency independent. The phase velocity of a wave traveling along this line is $v_{\mathrm{g}} = 1/\sqrt{L_{\mathrm{g}}C_{\mathrm{gs}}}$. This transmission line can be resistively terminated at the end with little loss of input signal. On the output, inductors L_{d} are placed between drain-to-source capacitances of the adjacent devices, and a transmission line with a characteristic impedance $Z_{\mathrm{d}} = \sqrt{L_{\mathrm{d}}/C_{\mathrm{ds}}}$ and phase velocity $v_{\mathrm{d}} = 1/\sqrt{L_{\mathrm{d}}C_{\mathrm{ds}}}$ is formed. This is an active transmission line and the signal builds up along it with each successive amplification stage. The phases of the outputs of the individual stages will only be appropriate for left-to-right propagation, so little power will be lost at the resistive termination at the left end of the line. In effect, the two transmission lines are coupled lines with a coupling coefficient greater than unity. The inductors L_{g} and L_{d} can be chosen to equalize the phase velocities of the two coupled lines. Terminating resistors Z_{g} and Z_{d} are placed to minimize destructive reflections. Note that this discussion is valid only for a unilateral transistor approximation.

The transconductive *gain* of each device is g_{m} and the output impedance seen by each transistor is half the characteristic impedance of the transmission line. The resulting overall voltage gain of the DA is:

$$A_{\mathrm{V}} = \frac{1}{2} N g_{\mathrm{m}} Z_0 \tag{4.27}$$

where N is the number of stages.

From the above discussion, we might surmise that an arbitrarily large gain can be achieved by employing a large number of sections. In the equivalent circuit for the transistor, however, there are some resistors as well (not shown in Figure 4.15), and they make the transmission lines lossy. As a result, a limited number of stages can be included before the losses overtake the gain. Gain as a function of the number of sections is shown in Figure 4.16 for one example. It shows that after 5 sections, there is no appreciable increase in gain, whereas the flatness of the gain is reduced. Distributed amplifiers have been reported with flat gain from 1 to 40 GHz, and into the 100 GHz range.

Distributed amplifiers are normally monolithically integrated so that the devices are very small compared to the guided wavelength. In practice, it is difficult to make good inductors in monolithic circuits at high microwave frequencies. Therefore, short sections of transmission lines are used instead between the stages of a distributed amplifier.

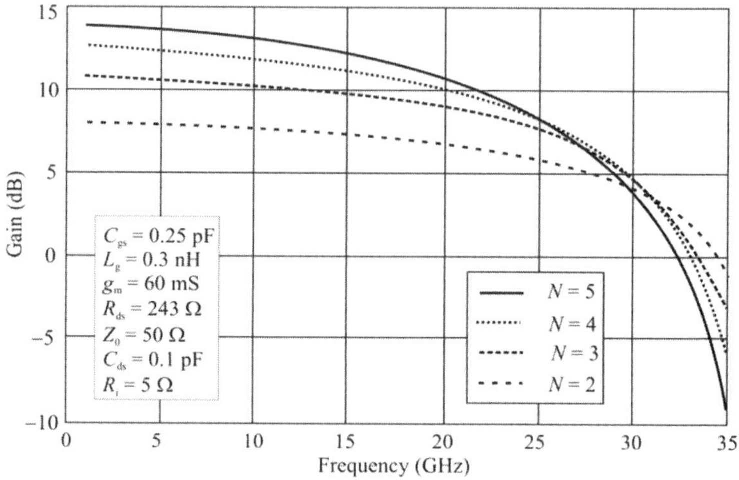

Figure 4.16 Dependence of the gain versus frequency as a function of the number of sections of a MESFET distributed amplifier.

4.7.2 Balanced Amplifiers

A common approach to the problem of broadband amplifier design is a balanced amplifier topology as shown in Figure 4.17. It consists of a pair of 3-dB couplers, acting as splitters (input) and combiners (output) of two matched amplifiers. For example, if the couplers are ideal 90° hybrids (such as a branch line coupler), the scattering matrix for the balanced amplifier is:

$$\mathbf{S} = \begin{bmatrix} (s_{11,1} - s_{11,2})/2 & j(s_{12,1} + s_{12,2})/2 \\ j(s_{21,1} + s_{21,2})/2 & (s_{22,1} - s_{22,2})/2 \end{bmatrix} \tag{4.28}$$

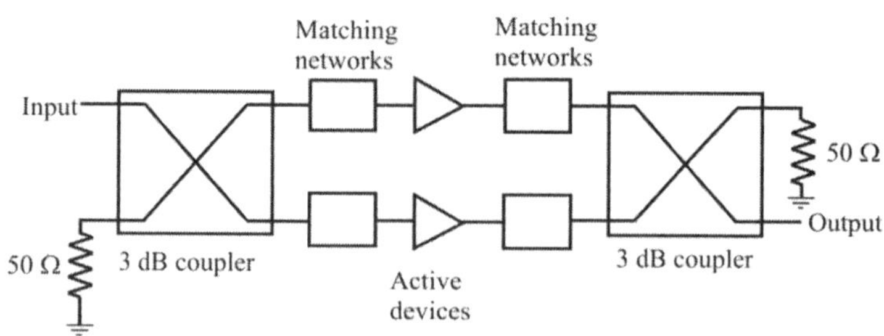

Figure 4.17 Balanced amplifier configuration using directional couplers.

and if the two amplifiers are identical, the scattering matrix becomes:

$$\mathbf{S} = \begin{bmatrix} 0 & js_{12} \\ js_{21} & 0 \end{bmatrix} \tag{4.29}$$

This means that, as long as the amplifiers are identical, they can be whatever we wish, and the amplifier still has input and output matching. The two amplifiers can individually be tuned for gain, noise, or flatness of frequency response.

Another popular balanced amplifier configuration uses Wilkinson combiner/dividers instead of the branch-line couplers shown in Figure 4.17. In this case, quarter-wave sections in the two lines provide a 180° phase between the two waves reflected from the inputs of the amplifiers, and the reflections are canceled. The bandwidth of the amplifier is obviously limited by the bandwidth of the directional coupler or Wilkinson splitter, both of which rely on quarter-wave sections for proper operation. Therefore, a hybrid is not the best choice (it has about 15% bandwidth). Instead, a Lange coupler based on coupled line sections is most commonly used, which can have a bandwidth of 2 octaves. It is also possible to design broadband multisection branch line and Wilkinson combiners with over a decade bandwidth. See Appendix 4A to examine how a Wilkinson power divider works.

Balanced amplifiers ideally have the same gain and twice the output power as compared to a single amplifier. When the signal becomes large, each of the transistors receives only half the power, so balanced amplifiers can handle more power with less signal distortion. However, twice the input signal is required, as is two times more DC power. An additional disadvantage is the size of the circuit and the fact that a large part of the real estate is taken by passive elements. This is very costly in MMIC implementations.

An important factor in balanced amplifier design is the amplitude and phase mismatch between the coupler output ports as a function of frequency, as well as the sensitivity of this mismatch to load impedance variations. This is normally verified with simulation prior to fabrication.

4.7.3 Resistive Feedback Amplifiers

Resistive feedback can also be used for implementing a broadband amplifier. The effect of a feedback resistor between the gate and drain of a MESFET is to lower the input and output impedance and to broaden the gain curve. The biggest drawback is resistive coupling between the bias circuits, as well as overall lower gain than for reactively matched amplifiers.

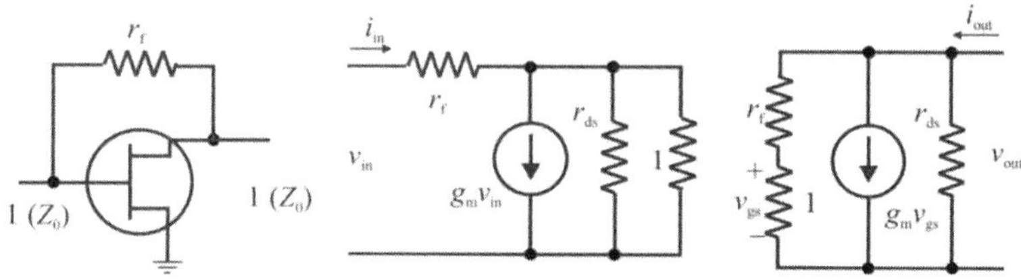

Figure 4.18 Simplified equivalent circuits for input and output MESFET circuits in a series feedback amplifier.

By observing the approximate circuit model for the MESFET with normalized resistance values (to 50 Ω), shown in Figure 4.18, the equations for the current and voltage in the input circuit are found to be:

$$i_{in} = g_m v_{in} + i \tag{4.30}$$

$$v_{in} = r_f i_{in} + \frac{r_{ds}}{r_{ds}+1} i \tag{4.31}$$

Solving for the input impedance we obtain:

$$Z_{in} = \frac{v_{in}}{i_{in}} = \frac{r_f + \dfrac{r_{ds}}{r_{ds}+1}}{1 + g_m \dfrac{r_{ds}}{r_{ds}+1}} \tag{4.32}$$

For the output circuit, the current and voltage can be expressed as:

$$i_{out} = \frac{v_{out}}{1+r_f} + g_m v_{gs} + \frac{v_{out}}{r_{ds}} \tag{4.33}$$

$$v_{out} = (1+r_f) v_{GS} \tag{4.34}$$

and the expression for the output impedance is:

$$Z_{out} = \frac{v_{out}}{i_{out}} = \frac{r_{ds} \dfrac{1+r_f}{1+g_m}}{r_{ds} + \dfrac{1+r_f}{1+g_m}} \tag{4.35}$$

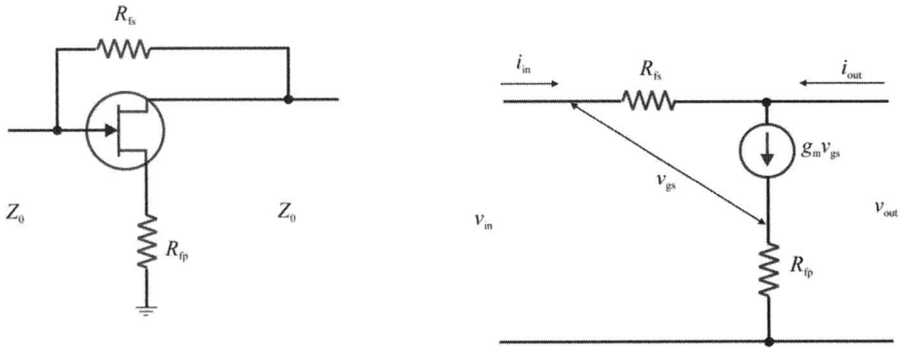

Figure 4.19 Simplified circuit for series-shunt resistive feedback amplifier.

In both cases, the impedance is reduced and can be controlled by the amount of feedback resistance.

The above simplified analysis is an example of the more general case of series-shunt resistive feedback shown in Figure 4.19, where in addition to the series feedback between gate and drain, there is a parallel feedback resistor placed in the source. The admittance matrix for this network (for a very simplified FET model) can be written as

$$\begin{bmatrix} i_{in} \\ i_{out} \end{bmatrix} = \begin{bmatrix} \dfrac{1}{R_{fs}} & -\dfrac{1}{R_{fs}} \\ \dfrac{g_m}{1+g_m R_{fp}} - \dfrac{1}{R_{fs}} & \dfrac{1}{R_{fs}} \end{bmatrix} \begin{bmatrix} v_{in} \\ v_{out} \end{bmatrix} \tag{4.36}$$

The admittance matrix can be converted to the S-matrix using standard conversion formulas and we get

$$\mathbf{S} = \begin{bmatrix} \dfrac{1}{\Delta}\left(1 - \dfrac{g_m Z_0^2}{R_{fs}(1+g_m R_{fp})}\right) & \dfrac{2Z_0}{\Delta R_{fs}} \\ \dfrac{1}{\Delta}\left(\dfrac{-2g_m Z_0}{1+g_m R_{fp}} + \dfrac{2Z_0}{R_{fs}}\right) & \dfrac{1}{\Delta}\left(1 - \dfrac{g_m Z_0^2}{R_{fs}(1+g_m R_{fp})}\right) \end{bmatrix} \tag{4.37}$$

where

$$\Delta = 1 + \frac{2Z_0}{R_{fs}} + \frac{g_m Z_0^2}{R_{fs}(1 + g_m R_{fp})}. \tag{4.38}$$

When we match the input and output (i.e. $s_{11} = s_{22} = 0$), then the resistor values are related to the transconductance by

$$1 + g_m R_{fp} = \frac{g_m Z_0^2}{R_{fs}} \tag{4.39}$$

or

$$R_{fp} = \frac{Z_0^2}{R_{fs}} - \frac{1}{g_m} \tag{4.40}$$

From the above equations, s_{21} and s_{12} can be found to be

$$s_{21} = \frac{Z_0 - R_{fs}}{Z_0} \tag{4.41}$$

and

$$s_{12} = \frac{Z_0}{R_{fs} + Z_0} \tag{4.42}$$

We can see that the gain of the amplifier depends only on the characteristic impedance and the value of the series feedback resistor, not on the device parameters, a very desirable characteristic since the device parameters can vary greatly. This means that we can obtain flat gain over a frequency range with such feedback.

4.8 Switched-Mode RF Amplifiers

4.8.1 Introduction

Switching-mode power amplifiers are widely used in different frequency ranges and output power levels beginning from several kilowatts at low frequencies and up to

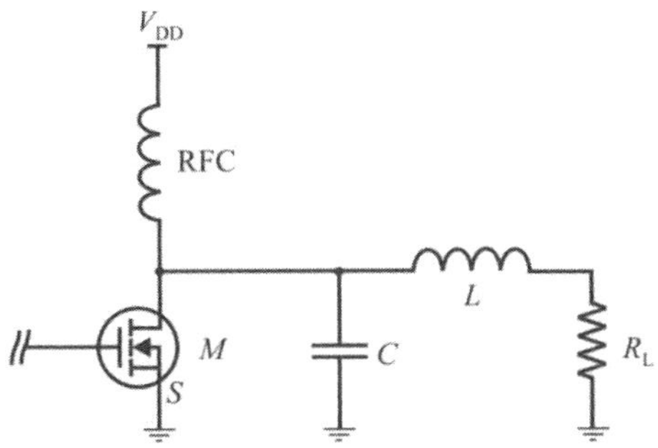

Figure 4.20 Switched-mode PA shunt.

one to several watts at microwaves. In these power amplifiers, the transistors operate as ON-OFF switches and the waveforms of the output current and the output voltage (ideally) do not overlap at any given time. Therefore, the power dissipation is minimized and the efficiency is maximized. Such an operation mode can be realized by an appropriate choice of the values for the reactive elements in the output load network.

The first possibility of increasing the efficiency of the single-ended PA by modifying the output matching circuit was described by Lohrmann in 1966 [12]. Three years later Artym [13] and Gruzdev [14] provided the theoretical analysis of the single-ended switching-mode power amplifiers with the calculation of their circuit parameters. Analysis was provided for switching-mode power amplifiers not only with the shunt capacitance but also with the resonant circuit tuned to the fundamental to provide the sinusoidal signal flowing into the load. Later, the generalized analysis of the electrical performance and circuit parameters of the single-ended switching-mode power amplifiers with shunt capacitance C and series inductance L, as well as with a parallel LC circuit was presented by Popov [15] and Kozyrev [16]. These topologies are shown in Figure 4.20 and Figure 4.21, respectively. These two general design models are still in use today.

The salient benefit of a switching-mode PA is the high efficiency possible (up to 100%). But the transistors are used as switches and the largest disadvantage of a switching-mode PA is the resulting nonlinearity. Therefore, the signal spectrum can be degraded and severe adjacent channel interference can occur.

The typical switching-mode power amplifiers are class D and class E power amplifiers. The class F power amplifiers are also popular and are often used as the driving stages of the class E power amplifiers.

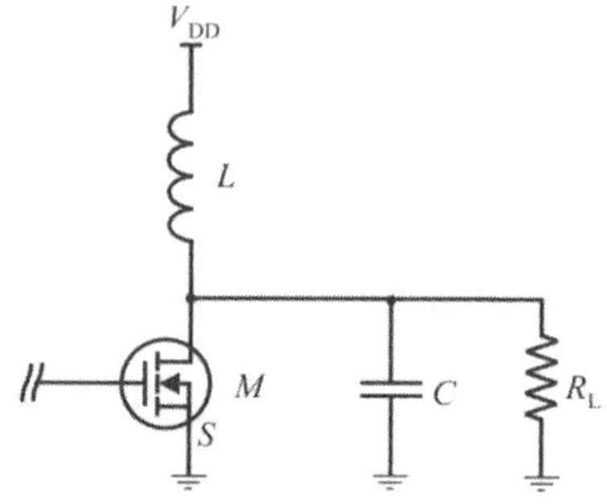

Figure 4.21 Switched-mode PA shunt output.

4.8.2 Class D PA

The original class D PA is the voltage mode class D power amplifier. Using a MOSFET device, the simplified circuit of this PA and its waveforms of the drain voltage and the drain current are shown in Figure 4.22. Two transistors are connected in parallel and a series filter comprising the inductor *L* and the capacitor

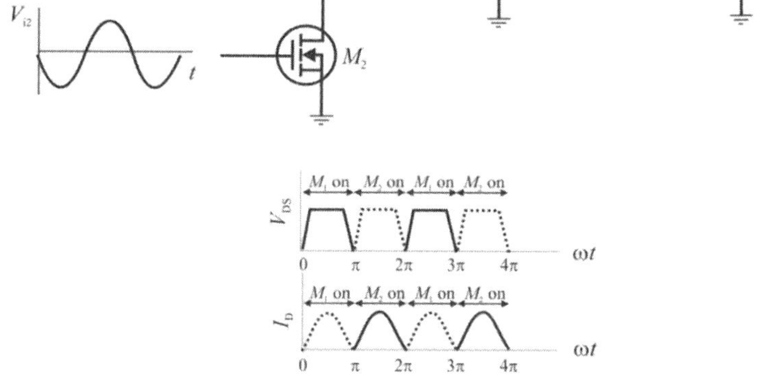

Figure 4.22 The voltage mode class D power amplifier and its waveforms.

C is employed. This filter is tuned to the fundamental frequency. The two transistors M_1 and M_2 are driven 180° out-of-phase. That means the input connection guarantees that only one transistor is driven ON at a given time, with one transistor handling the positive half-cycles and the other one the negative half-cycles. It works just like a push-pull class B PA (covered later in Chapter 9). The difference here is that the transistors are driven hard enough to act like switches. The voltage across the transistors is a square wave and the transistor current becomes a half-wave rectified sine wave. *Zero-current-switching* (ZCS) is implemented, which means that the drain current is zero, just when the transistor is turned ON. Ideally, there is no overlap of the drain voltage and the drain current. Therefore, the DC power consumption is theoretically zero.

Unfortunately, a real transistor always has parasitic capacitors (e.g., the drain-source capacitor in an FET). The transistor must be charged or discharged to the supply voltage V_{DD} or ground through the transistor. That means the voltage waveform cannot have a perfect square wave shape and some transient current spikes occur when the transistor turns ON—the overlap of voltage and current cannot be totally avoided. The energy loss per cycle, E_c, is given by [17]

$$E_c = \frac{1}{2} C_{DS} V_{DS\text{-}ON}^2 \tag{4.43}$$

where C_{DS} is the drain-source capacitance and $V_{DS\text{-}ON}$ is the drain-source voltage when the transistor is turned ON. Interestingly, this energy loss on the output capacitance is independent of the channel resistance of the transistor and becomes the dominant loss mechanism at high frequencies.

To overcome the problem of energy loss the current-mode class D PA was developed, which is shown in Figure 4.23. Two transistors are again connected in parallel, but instead of a voltage source a current source is used. Instead of a series filter a shunt filter is employed, which is also tuned to the fundamental frequency. The resulting voltage across the transistors is a half-wave rectified sine wave and the transistor current becomes a square wave. Due to the filter resonance, there is no voltage across the transistors when a transistor is turned ON; therefore, a so-called *zero voltage-switching* (ZVS) is realized where the transistor has no voltage across it (theoretically) when it is turned ON. The output capacitance of the transistors can be included as a part of the parallel filter, so that an overlap of the drain voltage and the drain current can be avoided, at least theoretically.

4.8.3 Class E PA

Another type of switching-mode PA is the class E PA, which was introduced by Socal in 1975 [18]. The topology of a class E PA is shown in Figure 4.24. A shunt

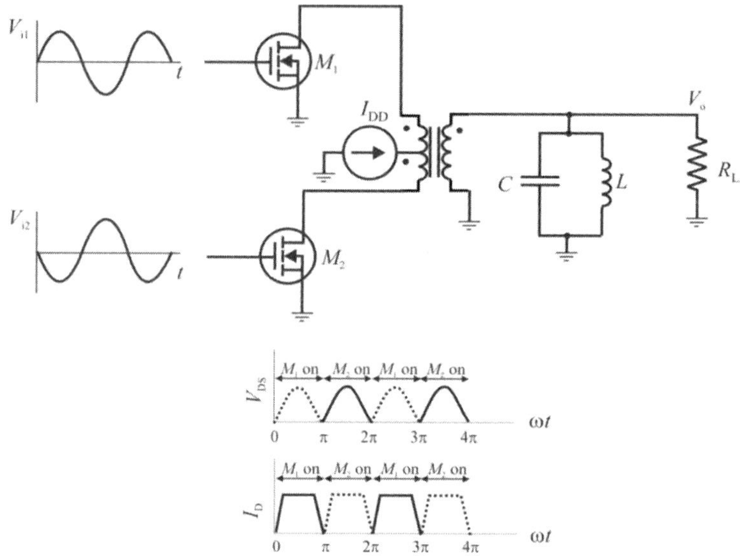

Figure 4.23 The current mode class D power amplifier and its waveforms.

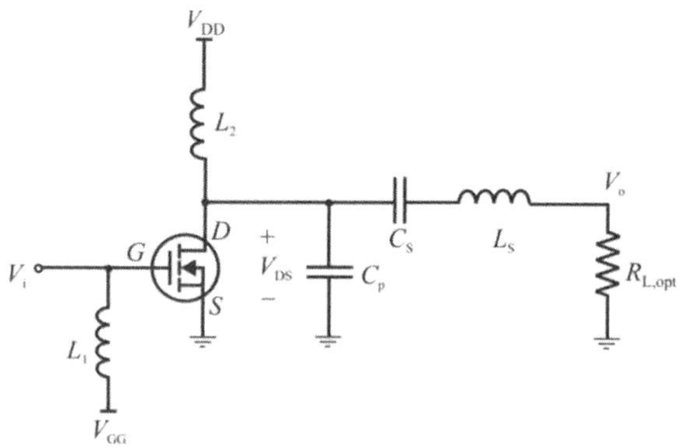

Figure 4.24 The schematic of a class E power amplifier.

capacitor C_p from the drain to ground is used to provide the charge-discharge function when switching. It ensures that the voltage across the switch remains relatively low, when the switch is turned OFF, until the drain current reaches zero.

A series filter consisting of a series inductor L_s and a series capacitor C_s is employed between the drain and the load. This filter is tuned at the fundamental frequency to ensure a sine waveform on the load. $R_{L,opt}$ is the optimum load of the class E power amplifier, to which the 50 Ω reference load should be transformed.

The waveforms of the voltage across the transistor and the current flowing through it are shown in Figure 4.25.

The switch alternately opens and closes at the operating frequency, and the common choice of the duty cycle is 50%. During the ON state, the transistor is overdriven and should provide a resistance as low as possible while sustaining the current running through it. The voltage across the switch is zero and the current flows totally through the switch. Conversely, during the OFF state the transistor is in the cutoff region and should provide very high impedance. It should also be able to sustain the voltage rise across its terminals. The entire current is flowing through the shunt capacitor C_p, charging and discharging it, and the switch voltage has a characteristic asymmetric waveform. The voltage and the current waveforms never simultaneously have nonzero values, so that no power is dissipated in the switch and the efficiency of this operation can theoretically reach 100%.

To realize the so-called *soft switching*, which means that the shunt capacitor will not be discharged through the switch resulting in power loss, two conditions must be fulfilled [19, 20]. The first is that the voltage returns to zero at the switch turn-on. The second one is zero voltage slope at the switch turn-on. These two conditions are described as

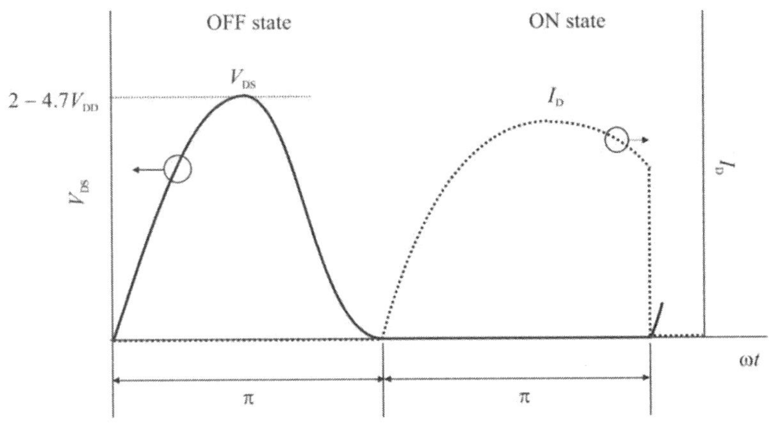

Figure 4.25 Voltage and current waveforms for an optimum class E PA.

$$V_{\mathrm{DS}}(t=\tau/2)=0 \tag{4.44}$$

$$\left.\frac{dV_{\mathrm{DS}}}{dt}\right|_{t=\tau/2}=0 \tag{4.45}$$

where V_{DS} is the drain-source voltage of the transistor as shown in Figure 4.24; the time constant is

$$\tau=\frac{2\pi}{\omega} \tag{4.46}$$

These two conditions are met by an appropriate choice of the reactive elements in the output network. The value of these components can be determined with [21]

$$R_{\mathrm{L,opt}}=0.577\frac{V_{\mathrm{DD}}^{2}}{P_{\mathrm{out}}} \tag{4.47}$$

$$C_{\mathrm{p}}=\frac{0.1837}{\omega\times R_{\mathrm{L,opt}}} \tag{4.48}$$

$$L_{\mathrm{s}}=\frac{Q\times R_{\mathrm{L,opt}}}{\omega} \tag{4.49}$$

and

$$C_{\mathrm{s}}=\frac{1}{\omega^{2}\times L_{\mathrm{s}}} \tag{4.50}$$

Therefore, the reference impedance, normally 50 Ω, should first be transformed to $R_{\mathrm{L,opt}}$ given in (4.47), considering the required output power. The other components can then be calculated in turn. The values of L_{s} and C_{s} depend on the chosen quality factor Q. The shunt capacitance C_{p} in (4.48) includes the output capacitance of the active device.

Unfortunately, the transistors used as switches always have a finite ON-resistance, and the transition times from the OFF-state to the ON-state and vice versa are not negligible. In order to avoid power loss from the finite ON-resistance of the switch, the FET is designed with a large gate width. On the other hand, doing so increases the transition time of the input signal due to the larger parasitic capacitance. To obtain the shortest transition time from one state to the other, it is necessary to use a driving stage in the power amplifier, which feeds a square-wave-like signal to the input of the class E PA. Class F power amplifiers are good candidates to generate such square-wave-like periodic signals, which we discuss next.

4.8.3 Class F PA

The class F PA was introduced by Tyler in 1958 [22]. In this topology, appropriate terminations at fundamental and harmonic frequencies are placed in the drain network to improve the efficiency of the active device output. Open- and short-circuit terminations at odd and even harmonics are presented at the output of the transistors to shape the drain waveforms. This can be done with an output network that includes the load as well as series and shunt harmonic resonators. Ideally an infinite number of resonators should be used, which is obviously not feasible in practice.

Figure 4.26 shows the schematic of a class F PA using an FET as the active device. This circuit has a band rejection filter tuned to the second harmonic in the output matching network, which consists of the inductors L_1, L_2, L_3, and the capacitors C_1, C_2, and C_3. Such a configuration provides a close approximation to a square wave for the output voltage [23], and hence can be used as the driving stage for a class E PA. L_4 and L_5 serve as RF chokes while C_3 at the drain is the DC blocking capacitor.

The waveforms of the drain voltage and the drain current are illustrated in Figure 4.27 for the ideal case [24] (i.e., the output matching network includes an infinite number of resonators). The drain-source voltage has a square waveform, while the drain current presents a half-sine waveform. The quiescent current, I_Q, is defined using

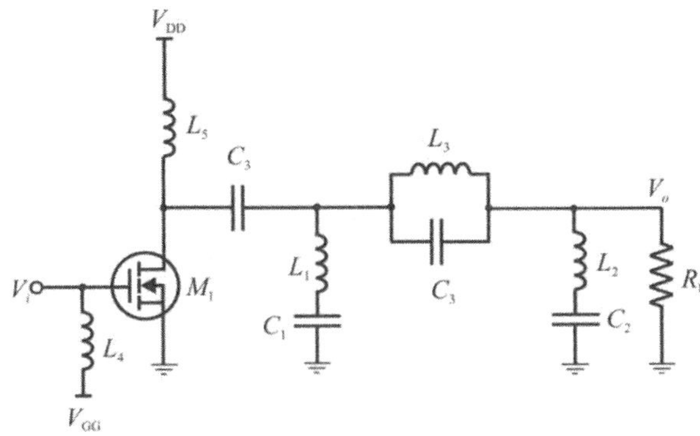

Figure 4.26 The topology of a class F power amplifier.

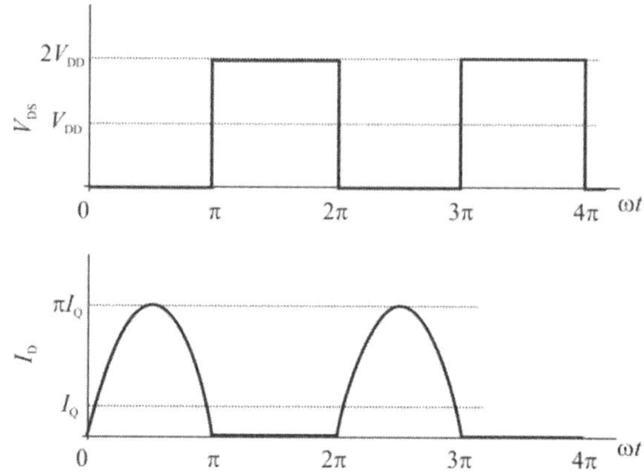

Figure 4.27 Ideal waveforms of a class F power amplifier.

$$A_S = A_Q \tag{4.51}$$

where A_S is the area covered by the half-sine waveform and A_Q is the area covered by the quadrate shown in Figure 4.28. The area A_S can be calculated as

$$\begin{aligned} A_S &= \int_0^{\pi} I_m \sin x dx \\ &= 2I_m \end{aligned} \tag{4.52}$$

On the other side, the quadrate area A_Q can be calculated as

$$A_Q = I_Q \times 2\pi \tag{4.53}$$

Using (4.52) and (4.53) in (4.51), we obtain

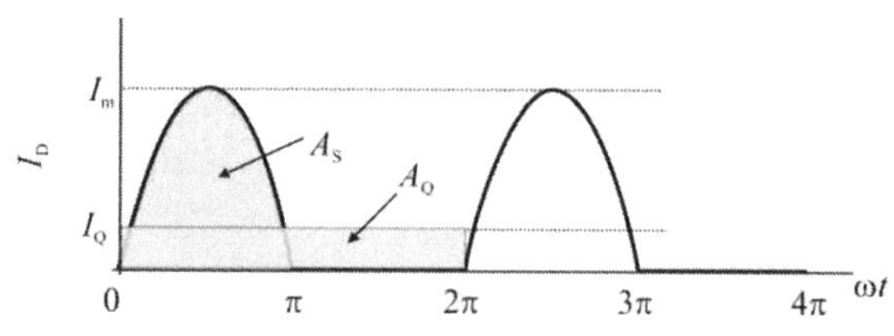

Figure 4.28 Calculation of the quiescent current.

$$I_{\mathrm{m}} = \pi \times I_{\mathrm{Q}} \tag{4.54}$$

for the maximum value of drain current.

In practice a class F PA has only a limited number of resonators; hence the drain-source voltage waveform includes one or more odd harmonics and only approximates a square-wave-like waveform. The current waveform includes even harmonics and approximates a half-wave rectified sine wave. Theoretically, the efficiency of a class F PA increases as more harmonics are included. It approaches the ideal value of 100% when all the harmonics are rejected.

4.9 Concluding Remarks

We provided an introduction to RF amplifiers in this chapter. This information will be expanded upon in the chapters to follow.

There are two fundamental types of RF amplifiers that a generally used: analog and digital. The former is considerably older than the latter, and were discovered in the early twentieth century largely due to the requirements of war efforts. The latter were discovered later in the century largely due to the advancement in semiconductor technology, and small, efficient amplifiers became available.

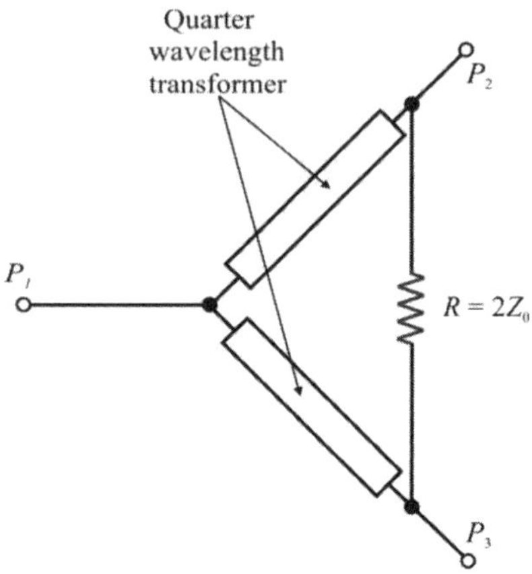

Figure 4.29 Wilkinson power divider, splitter / combiner circuit.

Appendix 4A Wilkinson Power Divider/Combiner

The Wilkinson power divider / Wilkinson combiner uses quarter-wave transformers to split the input signal to provide two output signals that are in phase with each other (see Figure 4.29).

The resistor between the two output ports enables the two outputs to be matched while also providing isolation. The resistor does not dissipate any power, and as a result the Wilkinson power divider can theoretically be lossless. In practice there are some losses, but these are generally low.

The values within the two-way Wilkinson divider / combiner can be calculated:

$$R = 2Z_0 \tag{4.55}$$

$$Z_{\text{match}} = \sqrt{2}Z_0 \tag{4.56}$$

where:

R = the value of the terminating resistor connected between the two ports
Z_0 = the characteristic impedance of the overall system
Z_{match} = the impedance of the quarter-wave transformers in the legs of the power divider/combiner.

To see how the Wilkinson divider works, consider a signal entering the left-hand port, P_1, in Figure 4.29. The signal reaches the physical split and passes to both outputs, ports P_2 and P_3 of the Wilkinson divider. As the two legs of splitter / divider are identical, the signals appearing at the outputs will have the same phase. This means that ports P_2 and P_3 will be at the same potential and no current will flow in the resistor.

As the power is being split, it is necessary to ensure that the impedances within the Wilkinson divider are maintained. To achieve this, the two output ports must each appear as an impedance of $2Z_0$—the two output ports of $2Z_0$ in parallel will present an overall impedance of Z_0. The impedance transformation is achieved by placing a quarter-wave transmission line between the star point and the output— the transmission line has an impedance of $\sqrt{2}Z_0$. In this way, the impedance within the system is maintained.

The Wilkinson power divider operates in both directions and can also be used as a combiner. In this mode signals entering ports P_2 and P_3 will emerge at port P_1, and none at port P_3 (i.e., it is isolated).

Consider power entering port P_2. It will split equally between the resistor and port P_1. Thus, half the power passes to port P_1 and half is dissipated in the resistor—a fact that is important for power applications.

As the signal enters port P_2, half passes through the resistor and the other half passes through the first quarter-wave transformer. It then appears at the star point. Any power passing through the other quarter-wave transformer to port P_3 will be out of phase with that appearing via the resistor as it will have passed through two quarter-wave lines. As a result, there is isolation between the two ports, half the power is dissipated within the resistor and half appears at port P_1.

The Wilkinson power divider or combiner is used for many RF applications. Some of its key advantages: its low level of loss and its high level of isolation between the output ports means that the Wilkinson power divider is a particularly attractive proposition. A further advantage at microwave frequencies is that it can often be made very cheaply because the transmission line elements can be printed on the PCB. This means that the only component required for the Wilkinson divider is a resistor, although it may mean that additional board space is required.

The Wilkinson power divider/splitter/combiner is widely used in many areas and is able to provide excellent performance.

References

[1] Millman, J. M., and C. C. Halkias, *Electronic Devices and Circuits*, New York: McGraw-Hill, 1964.

[2] Razavi, B., *RF Microelectronics*, Upper Saddle River, NJ: Prentice Hall, 1998.

[3] Krauss, H. L., C. W. Bostain, and F. H. Raab, *Solid State Radio Engineering*, New York: Wiley, 1980.

[4] Gonzalez, G., *Microwave Transistor Amplifiers, Analysis and Design*, 2nd Ed., Upper Saddle River, NJ: Prentice Hall, 1994.

[5] Rizzi, P. A., *Microwave Engineering Passive Circuits*, Upper Saddle River, NJ: Prentice Hall, 1988.

[6] White, J. F., *High Frequency Techniques*, New York: Wiley, 2004.

[7] Pozar D. M., *Microwave Engineering*, 3rd Ed., New York: Wiley, 2004.

[8] Rosu, J., YO3DAC / VA3IUL, http://www.qsl.net/va3iul/, accessed Sep. 2008.

[9] Lu, L. H., "Frequency Response of Amplifiers," NTUEE Electronics, http://libvolume3.xyz/electrical/btech/semester3/analogelectroniccircuits/transistorfrequencyresponse/transistorfrequencyresponsenotes1.pdf, accessed Nov. 2014.

[10] Goldsman, N., "Short Circuit Time Constant Approximation," http://www.ece.umd.edu/~neil/man/node70.html, accessed Sep. 14, 2018.

[11] http://www.emagtech.com/wiki/index.php?title=RF Tutorial Lesson 9: Designing a Distributed RF BJT Amplifier, accessed June 2010.

[12] Lohrmann, D. R., "Amplifier Has 85% Efficiency While Providing Up to 10 Watts Power Over a Wide Frequency Band," *Electronic Design*, Vol. 14, March 1966, pp. 38–43.

[13] Artym, A. D., "Switching Mode of High Frequency Power Amplifiers," *Radiotekhnika*, Vol. 24, pp. 58–64, 1969.

[14] Gruzdev, V. V., "Calculation of Circuit Parameters of Single-Ended Switching-Mode Tuned Power Amplifiers," *Trudy MEI*, Vol. 2, 1969 pp. 124–128.

[15] Popov, I. A., "Switching Mode of Single-Ended Transistor Power Amplifier," *Poluprovodnikovye pribory v tekhnike svyazi*, Vol. 5, 1970, pp. 15–35.

[16] Kozyrev, V. B., "Single-Ended Switching-Mode Tuned Power Amplifier with Filtering Circuit," *Poluprovodnikovye pribory v tekhnike svyazi*, Vol. 6, 1971, pp. 152–166.

[17] Kobayashi, H., J. M. Hinrichs, and P. M. Asbeck, "Current-Mode Class D Power Amplifiers for High-Efficiency RF Applications," *IEEE Transactions on Microwave Theory and Techniques*, Vol. 49, No. 12, Dec. 2001, pp. 2480–2485.

[18] Socal, N. O., and A. D. Socal, "Class E – A New Class of High-Efficiency Tuned Single Ended Switching Power Amplifiers," *IEEE J. Solid-State Circuits*, vol. SC-10, June 1975, pp. 168–176.

[19] Grebennikow, A. V., and H. Jaeger, "Class E with Parallel Circuit – A New Challenge for High-Efficiency RF and Microwave Power Amplifiers," *IEEE MTT-Digest*, 2002, pp. 1627–1630.

[20] Grebennikov, A. V., "Switched-Mode Tuned High-Efficiency Power Amplifiers: Historical Aspect and Future Prospect," *IEEE Radio Frequency Integrated Circuits Symposium*, June, 2002, pp. 49–52.

[21] Milosevic, D., J. Tang, and A. Roermund, "Explicit Design Equations for Class-E Power Amplifiers with Small DC-Feed Inductance," *European Conference on Circuit Theory and Design*, Vol. 3, August, 2005, pp. 111-113.

[22] Tyler,V. J., "A New High-Efficiency High Power Amplifier," *Marconi Review*, Vol. 21, 1958, pp. 96–109.

[23] Tu, S., and C. Toumazou, "Highly Efficient CMOS Class E Power Amplifier for Wireless Communications," *Proceedings of the 1998 IEEE International Symposium on Circuits and Systems*, Vol. 3, June 1998, pp. 530–533.

[24] Fortes, F., and M. Rosario, "A Second Harmonic Class F Power Amplifier in Standard CMOS Technology," *IEEE Transactions on Microwave Theory and Techniques*, Vol. 49, No. 6, June. 2001, pp. 1216–1220.

Chapter 5

Semiconductor Technologies

5.1 Introduction

While many active devices can be used to generate and amplify RF signals, we concentrate here on the use of *bipolar junction transistors* (BJTs) and *metal oxide semiconductor field effect transistors* (MOSFETs), and their close derivatives.

This chapter is structured as follows. After a brief introduction to microwave semiconductors in general, we discuss as amplifying devices. We cover their structure, operation, and high-level characteristics. That is followed by a discussion of the same properties for MOSFETs. Power amplifier (PA) device technology is covered next, followed by a discussion of JFETs as amplifiers. GaAs semiconductor technology is next compared with GaN, as they are both used in the role of amplifiers. GaN on SiC is then discussed by comparing that approach to GaAs and Si MOSFETs. The characteristics are compared. After that, extensive discussions of the characteristics of GaN are presented. Lastly, we briefly mention MMIC approaches to high-frequency amplifier designs.

5.2 Semiconductor Devices

Many types of devices can be used for amplification of RF signals, including vacuum tubes, diodes of many varieties, and transistors.

Here we will focus on the BJT and MOSFET and their closely related families of devices. We exclude vacuum tubes because their use is minimal in the primary frequency ranges in which we are interested (less than 10 GHz).

RF/microwave transistors offer many different performance capabilities by type of device, with some better for lower noise and some better for higher power.

Transistors for RF/microwave applications are based on several different elements and compounds, with each substrate material bringing its own strengths and weaknesses. Not long ago, the choice for a microwave transistor was essentially between silicon and *gallium arsenide* (GaAs). But the last few decades have seen the emergence of additional high-frequency semiconductor substrates, including indium phosphide (InP); silicon carbide (SiC); silicon germanium (SiGe); gallium nitride (GaN); and even combinations of the materials, such as GaN on SiC. Transistors fabricated on these semiconductor materials offer a wide range of performance capabilities, from low noise figures to high output powers, from the high-frequency (HF) range through millimeter-wave frequencies.

The comparison of silicon BJTs to GaAs transistors, such as *metal-epitaxial-semiconductor field-effect transistors* (MESFETs), has long been simply the differentiation of the two technologies' operating frequency ranges. In addition to using different substrate materials, silicon BJTs and GaAs MESFETs differ in structure, although both are three-terminal semiconductor devices.

Semiconductor devices have been in use for several decades for RF (HF, VHF, UHF, and microwave) applications. However, their extension to higher and higher frequencies has been rather slow. In the lower frequency regions, vacuum tubes are being gradually replaced by a variety of solid-state devices. However, until recently, the frequency range above 1 GHz remained the preserve of vacuum tubes. Microwave semiconductor devices and their associative circuits can perform a variety of functions, such as, generation, amplification, detection, switching modulation and limitation. Harmonics of the microwave signals can be generated; mixing of two microwave signals can be accomplished, and so on. In each function, the design of device and circuit is different.

The first solid-state amplifiers for microwave applications were negative-resistance diodes. For example, a tunnel diode. This was followed by the development of parametric amplifiers that used as variable capacitance diode (varactor) and an oscillator to vary the junction capacitance. The main feature of parametric amplifiers was the low noise that could be achieved by cooling the diode to liquid nitrogen temperatures. Power outputs greater than 10 mW for tunnel diodes were difficult to obtain. On the order hand, the varactor multiplier source was more successful, since it uses a high-power high frequency generating power at a few hundred megahertz. Here, the transistor drives a nonlinear reactance, harmonics are generated, and the power transfer from the fundamental to the harmonics can be quite efficient.

Parametric amplifiers became the prominent and most widely used solid-state amplifiers from 1958 to about 1970. By 1970, improvements in materials preparation and processing technology had resulted in development of NPN silicon BJTs with a maximum frequency of oscillation greater than 10 GHz. During the next two decades further progress in the design and manufacture of high-frequency microwave BJTs and FETs was dramatic. The key to successful microwave

Table 5.1 Reactance of a Few Components at Microwave Frequencies

	Frequency (GHz)		
	1	**10**	**100**
L = 0.1 nH	0.628 Ω	5.28 Ω	62.8 Ω
L = 1 nH	5.28 Ω	62.8 Ω	628 Ω
C = 0.1 pF	1592 Ω	159 Ω	15.9 Ω
C = 1 pF	159 Ω	15.9 Ω	1.6 Ω

transistors design is miniaturization, which is a necessity in order to reduce device and package parasitic capacitances, lead inductances, as also to overcome the lead inductances and the finite transit time of the charge carriers. An appreciation for the need to reduce parasitic capacitances and inductance can be obtained by referring to Table 5.1. For example, as seen from Table 5.1, an inductance of 0.1 nH at 10 GHz represents a reactance of 5.28 Ω, which is not negligible in a 50 Ω system. Similarly, a capacitance of 0.1 pF at 10 GHz has a reactance of 159 Ω and would be a significant shunt reactance across a 50 Ω transmission line.

Transit times are dependent on the election mobility and saturation velocity in the semiconductor material. In this regard, GaAs is better than silicon for high frequency devices. By 1980, the design and fabrication of MESFETs were well established in the frequency range above 5 GHz and MESFET devices are widely used.

In order to achieve high-frequency performance in transistors, it was necessary to develop the technology that would enable key device dimensions to be less than 1 μm. For example, gate lengths with submicron dimensions.

5.3 Microwave Semiconductor Devices

There are mainly two types of devices used for the design of RF amplifiers for the EW applications we are concerned with:

- Bipolar junction transistors;
- Field effect transistors.

5.3.1 Bipolar Transistors

Microwave bipolar transistors are similar to low-frequency transistors but are fabricated so as to have improved performance characteristics, such as smaller transit time. To make the transit time a small fraction of a cycle, the base layer must be thin. At the same time, the base resistance must be small so that the inner collector capacitance can have a charging time, which is almost the same as the period of the

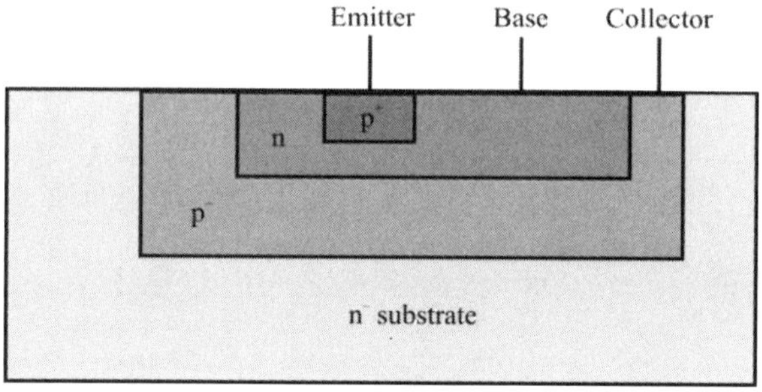

Figure 5.1 PNP BJT cross section.

operating frequency. The base doping cannot be made too high because then the efficiency with which the emitter injects minority carriers will be lowered. Hence, the charging time can be kept low only by making the diameter narrow. Therefore, microwave transistors are now made in the form of many base and emitter strips that are very narrow and close to each other.

A cross section of a PNP BJT is sketched in Figure 5.1. The collector base depletion layer is made thin so that the transit time from the base across the depletion layer to the collector is small. The emitter charging time is also made very short. In the frequency range below 5 GHz silicon bipolar transistors are preferred except for very low noise amplifies design.

5.3.2 Field-Effect Transistors

Though FETs, including *metal insulator semiconductor* (MIS) devices, have a large number of uses, their speed at microwave frequency is not very high. Hence, they are inferior to BJTs. There are two main characteristics of FETs that make them superior to BJTs in RF amplifiers: the lower noise characteristics and the higher frequency of operation (due to the higher electron mobility in GaAs). The higher electron mobility and the absence of shot noise that result in low noise.

5.3.3 Molecular Beam Epitaxy Technique

By means of *molecular beam epitaxy* (MBE), it has been possible to grow high-quality epitaxial layers and controlled doping profiles in highly localized regions. MBE techniques also led to the development of hetero structures, which in turn, led to the development of the *high electron mobility transistor* (HEMT), which can operate at frequencies as high as 100 GHz.

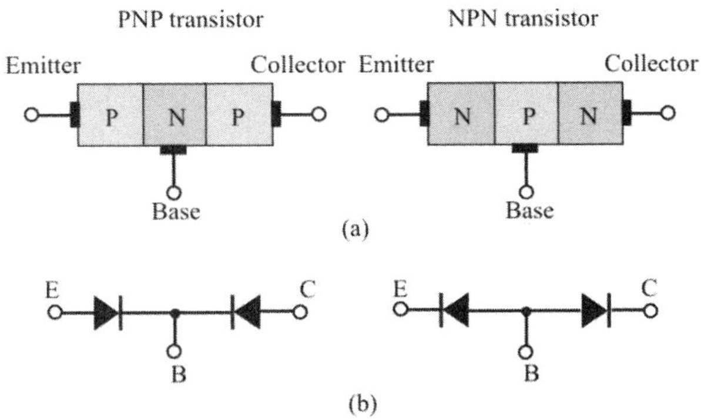

Figure 5.2 BJT construction: (a) physical construction and (b) two-diode models.

5.3.4 Bipolar Technology vs MOSFET Technology

The BJT has the advantage over the MOSFET of a much higher transconductance (g_m) at the same value of DC bias current. Thus, much higher values of voltage gain can be realized in a BJT amplifier stage than in the corresponding MOSFET circuit. Also, BJT amplifiers have superior high-frequency performance than their MOS counterparts.

On the other hand, practically infinite input resistance (R_{gs}, where $Z_i = R_{gs} + X_{gs}$) at the gate of a MOSFET makes it possible to design MOS amplifiers with extremely high input resistance and an almost zero input bias current. Also, the MOSFET provides an excellent implementation of a switch: while a saturated BJT exhibits an offset voltage of few tenths of a volt, the I-V characteristics of the MOSFET pass through the origin, resulting in zero offset. The availability of good switches in CMOS technology makes possible a host of analog circuit techniques that are employed in the design of, among other things, data converters and filters.

It can thus be seen that each of the two circuit technologies, bipolar and MOS, has its distinct and unique advantages and disadvantages.

5.4 Bipolar Junction Transistors

The terminals in a silicon BJT are the base, collector, and emitter; a small current at the base terminal can switch or control a much larger current between the collector and emitter terminals.

When two diodes are formed on the same substrate back-to-back a three-layer, two-junction, three-terminal device is formed that is the basis of a BJT. Depending on the particular type of semiconductor, N-type, doped predominantly with extra electrons, or P-type, doped predominantly with excess holes (lack of electrons), two configurations emerge as possible (see Figure 5.2).

The different semiconductor materials can act as either an insulator or a conductor by the application of a small signal voltage. The transistor's ability to change between these two states enables it to have two basic functions: switching (digital electronics) or amplification (analog electronics).

Bipolar transistors operate within three different regions:

- Active region – the transistor operates as an amplifier and $I_C = \beta I_B$;
- Saturation – the transistor is operating as a switch, is fully-ON, and $I_C = I_{saturation}$;
- Cut off – the transistor is operating as a switch, is fully-OFF, and $I_C = 0$.

We will see in Chapter 9 that there are some configurations of switching amplifiers that have very desirable properties, partially blurring these distinctions.

The word *transistor* is a combination of the words *transfer varistor* used to describe their mode of operation. There are two basic types of bipolar transistor construction, PNP and NPN, which describes the physical arrangement of the P-type and N-type semiconductor materials from which they are made. Bipolar transistors are current regulating devices where the amount of current flowing between the emitter and collector is in proportion to the amount of biasing current applied to the base terminal acting like a current-controlled switch. The principle of operation of the two transistor types PNP and NPN is exactly the same, the only difference being in their biasing and the polarity of the power supply for each type. There are differences in performance between the two types, however, which is due primarily to the difference in the speed at which holes travel versus the speed at which electrons travel through the devices. The latter travels at a speed that is about twice as fast as the former (on average).

5.4.1 Bipolar Transistor Construction

The circuit symbols for both the PNP and NPN BJTs are shown in Figure 5.3 with the arrow in the circuit symbol always showing the direction of "conventional current flow" between the base terminal and its emitter terminal. The direction of the arrow always points from the positive P-type region to the negative N-type region for both transistor types, exactly the same as for the standard diode symbol.

Figure 5.3 BJT circuit symbols: (a) NPN and (b) PNP.

The silicon (Si) BJT is the original solid-state RF power device. Since the BJT is a vertical device, as seen in the cross-section diagram in Figure 5.4, the power density can be very high. Si BJTs typically operate from 28V supplies and remain usable at frequencies up to 5 GHz and sometimes higher, especially in high-power (1 kW) pulsed applications. While Si RF power devices have higher gain at high frequencies, their fundamental properties are basically those of low power RF BJTs.

The positive temperature coefficient of BJTs tends to allow thermal runaway. The high power dissipated at the collector junction causes heating at that junction, which causes further temperature rise. If this process continues unabated, thermal runaway occurs. To avoid thermal runaway the bias circuits have to be carefully designed.

Power BJTs are typically composed of multiple, smaller BJTs essentially connected in parallel. Adding emitter resistance is generally employed to force even division of the current among these BJTs.

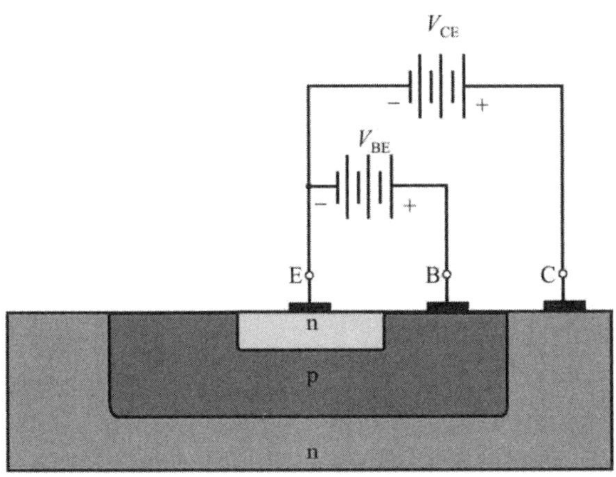

Figure 5.4 NPN BJT cross section shown with biasing power supplies.

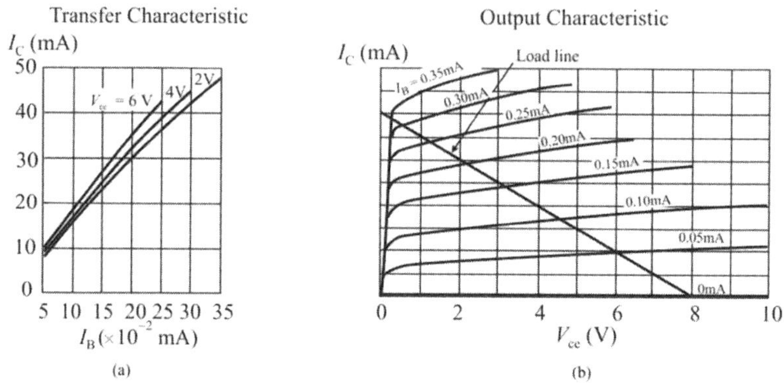

Figure 5.5 Typical operating characteristics of a BJT: (a) base current to collector current (transfer) and (b) collector current versus collector to emitter voltage (output).

5.4.2 Operating Characteristics of BJTs

Typical operating characteristics of a BJT are shown in Figure 5.5. The transfer characteristics in the form of I_C as a function of the base current I_B are shown in Figure 5.5(a). In the active region, where $I_C \approx \beta I_B$, the collector current is approximately a linear function of the base current as we see. The output characteristics are shown in Figure 5.5(b) along with a typical load line. The slope of the load line is given by the inverse of the output resistance as seen by the transistor. The region in the output characteristic that is essentially vertical, close to the ordinate axis represents a nonlinear region of the transistor. Where I_C is maximum, the BJT is saturated (fully ON, approximately 40 mA in this example). Significantly to the right of this region, approximately where the load line cuts across the operating characteristic curves of the transistor, we can see that as the base current is varied the collector current is varied, as is V_{ce}. Thus, we achieve current gain with the BJT. Where I_C goes to zero, about 8V in this example, the transistor is OFF.

5.5 Field Effect Transistors

Like the BJT, an FET is also a three-terminal device (actually they are available with more than three but for our purposes assuming only three suffices). An FET's terminals are the gate (G), source (S), and drain (D)—the voltage at the gate controls the current between source and drain. We can therefore consider the FET as a

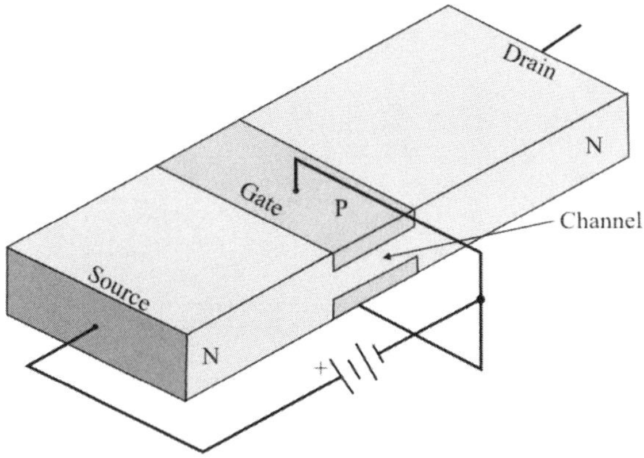

Figure 5.6 Field effect transistor.

transadmittance amplifying device ($y = I_D / V_{gs}$). A diagram of an FET is shown in Figure 5.6.

The output characteristics of MOSFETs, as indicated by charts of I_D versus V_{ds} with V_{gs} as a parameter, are the usual way of considering MOSFETs. One such characteristic is illustrated in Figure 5.7. In the linear region the device acts like a (fairly linear) resistor with I_D increasing as V_{ds} increases—that is, with a positive linear slope. Past this region, as V_{ds} is increased further (with a constant value of V_{gs}) is the saturation region. As V_{ds} increases, the output current remains relatively constant, behaving much like a current source.

An FET is a unipolar transistor because it uses only one form of conductor, electrons, or holes. The current that flows between the drain and the source is

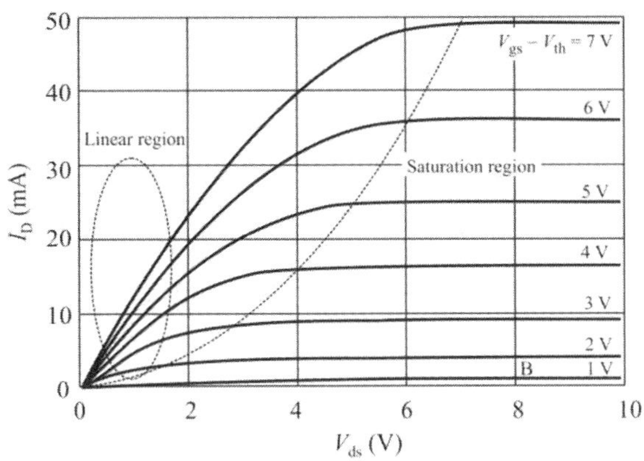

Figure 5.7 NMOSFET output characteristic.

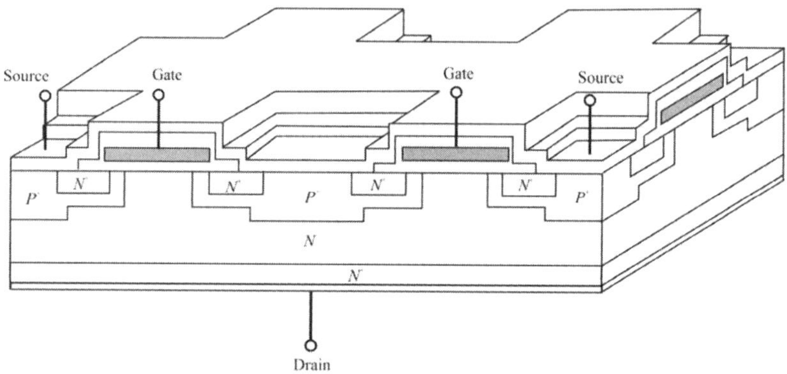

Figure 5.8 Cross section of a power MOSFET, with square cells. A typical power transistor consists of several thousand cells like these.

controlled by the voltage between the gate and the source. FETs, in both silicon and GaAs forms, can be used as switches (as they often function in digital circuits or power supplies) or as amplifiers (providing gain). As mentioned above, switching amplifier architectures are available using FETs that are extremely efficient (compared with other topologies). More on this in Chapter 9.

Silicon MOSFETs are capable of high power-handling capabilities at lower (such as audio) frequencies and in switching power supplies. GaAs FETs—although not capable of the high-power levels of silicon MOSFETs—can operate through microwave and millimeter-wave frequencies, typically in solid-state low-noise or power amplifiers. FETs can provide current gain, voltage gain, or both.

To construct MOSFET topologies that can handle high power levels, many (thousands) of basic cells illustrated in Figure 5.8 are placed in parallel, with each cell handling only a small portion of the total power.

5.5.1 CMOS

Si served for many years as the favored substrate for many high-power silicon bipolar transistors at submicrowave frequencies. It has also been the principal basis for many high-frequency *complementary MOS* (CMOS) integrated circuits (ICs). When fabricated with sufficiently small dimensions, such as with 90-nm processes, silicon CMOS transistors are capable of low-power, high-gain operation at frequencies to 60 GHz and beyond. Digital and computer applications have been the primary use for CMOS devices; however, their low cost and capabilities for operation in the millimeter-wave region make them attractive candidates for use in higher-frequency applications. It is also possible to combine digital and analog functions on a single chip with this technology.

Silicon also works with other elements as part of compound materials substrates, including SiGe and SiC, providing higher frequency and higher power capabilities. SiGe substrates provide for higher-frequency applications while SiC materials offer higher power levels, whether as the main semiconductor substrate or as a base for other types of semiconductors.

High-frequency SiGe BiCMOS devices provide high-frequency analog and mixed-signal ICs. Some of this SiGe technology is based on 130-nm silicon CMOS foundations, and is suitable for applications at 60 GHz and higher frequencies.

Heterojunction bipolar transistors (HBTs) for low-noise applications combines a blend of SiGe and carbon. Developed principally for 5-GHz wireless local area-network (WLAN) applications, this transistor technology features a noise figure of only 0.75 dB with about 14-dB gain in the 5.8-GHz band. This performance represents a significant improvement in noise-figure performance and reasonable gain. The addition of carbon has made it possible to optimize the device's collector-base profile, increasing the range of supply voltages that can be used.

SiC discrete devices and modules are capable of high-voltage, high-power operation through lower microwave frequencies. SiC MOSFET die and packaged devices are available at voltages to 1200V. Targeted for motor drives and switched-mode power supplies applications, they are available that can handle 24A at +1200 VDC. We discuss power supplies for RF amplifier applications in Chapter 13.

Combining the electrical capabilities of GaN and the thermal capabilities of SiC yields a powerful combination, fabricating GaN-on-SiC transistors with excellent high-power, high-frequency capabilities [1]. One transistor is available as a discrete GaN-on-SiC HEMT that is usable from DC to 5.5 GHz. It provides 15-dB linear gain at 5. 5 GHz with 55W output power at that frequency when operating from a +28 VDC supply. It is built with 0.25μm production GaN processes and is suitable for commercial and military applications in communications, radar, and avionics systems. There is also a discrete GaN-on-SiC HEMT, with 100W output power from DC to 18 GHz. It offers maximum power-added efficiency of 52%, with 15.5-dB power gain at 3 GHz. These are, of course, only current-day examples of what is possible, as research reveals new capabilities continually.

The thermal conductivity of GaN is much higher than that of GaAs, about 170 W/m-K for GaN versus about 50 W/m-K for GaAs, which allows for much higher power levels for GaN devices compared to GaAs transistors. GaN also offers more than twice the bandgap energy of InP and GaAs. Quite simply, GaAs and InP transistors are limited in output power at higher frequencies, especially compared to GaN, making GaN an attractive semiconductor material into the terahertz region.

5.6 PA Device Technologies and Architectures

There are a variety of solid-state devices that are used for RF amplifiers, each with its own characteristic advantages and limitations. These devices offer amplification to 100 GHz or more and are, in many cases, grown to order in MMIC form. A summary of the most common types of power transistors is presented in this section [2 – 6].

These active devices are currently available for use in RF power amplifiers, and RF power transistors are available in packaged, die, and grown-to-order forms. Packaged devices are used at frequencies up to X band, and are dominant for high power at VHF and lower frequencies. A given package can contain one or more module connected in parallel and can also include internal matching for a specific frequency of operation. In MMICs, the RF-power device is grown to order, allowing its size and other characteristics to be optimized for a particular application. This form of construction is essential for upper-microwave and millimeter-wave frequencies to minimize the effects of strays and interconnects. Virtually all RF power transistors are npn or n-channel types because the greater mobility of electrons (vs. holes) results in higher frequency capabilities.

GaAs MESFETs have dominated higher frequency applications since their discovery in the 1970s. GaAs MMICs brought considerable integrated functionality to high frequency applications. CMOS manufacturing capability reached the 3-μm level in the 1980s, which brought CMOS into the mainstream, competing head-to-head with bipolar technology. CMOS integration levels have reached the sub 1-μm level with considerable functionality possible on a single chip. CMOS is the technology of choice for prolific applications because of its inexpensive manufacturing costs and widespread installed base. Silicon *laterally diffused MOS* (LDMOS) devices began to be used in PAs for cellular phone systems in the 1990s.

5.6.1 Metal-Oxide-Silicon Field-Effect Transistor

The topology of a MOSFET includes an insulated gate. Structures with both vertical and lateral current flow are used in RF applications. Because the insulated gate conducts essentially no DC current, MOSFETs are very easily biased.

The drain current decreases with temperature in a MOSFET because the temperature coefficient is negative. This prevents thermal runaway and allows multiple MOSFETs to be connected in parallel. The absence of base-charge storage time allows fast switching and also eliminates a mechanism for subharmonic oscillation. A saturated MOSFET can conduct drain current in either direction, which is very useful in switching mode operation with reactive loads and also allows them to be used as voltage-controlled resistors.

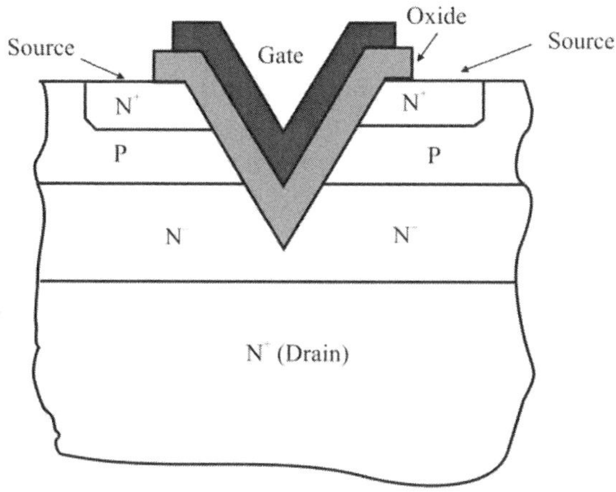

Figure 5.9 VMOS cross-section structure.

Vertical RF power MOSFETs are useable through VHF and UHF. These devices can deliver up to 1 kW at HF and 100s of watts at VHF. VMOS devices typically operate from 12, 28, or 50V supplies, although some devices are capable of operation from 100V or more.

5.6.2 Vertically Diffused MOSFET

The cross section of the *vertically diffused MOSFET* (VDMOSFET) is shown in Figure 5.9 [7]. A unique feature of the VDMOSFET is the shaped groove. We can see that the source is at the top of the device while the drain is at the bottom. Thus current flows vertically in the device instead of horizontally as in standard FETs.

The V-shaped gate increases the cross-sectional area of the source-drain path. This reduces the ON resistance of the device allowing it to handle much higher powers. The gate consists of a metalized area over the V-groove and this controls the current flow in the p region.

The main drawback to the VMOS FET is that the structure is more complicated than the traditional FET and this makes it slightly more expensive.

5.6.2.1 Summary

VDMOSFETs differ from lateral MOSFETs because of the vertical structure of the VMOS. These are used in a variety of applications that require high switching speeds and high frequency operation, along with a variety of voltage levels. The doping and channel lengths contribute to the characteristics of these MOSFETs.

5.6.3 Laterally Diffused MOSFET Transistors

Direct grounding of the source that eliminates bond-wire inductance that produces negative feedback and reduces gain at high frequencies make LDMOS transistors particularly useful at UHF and lower microwave frequencies. Such grounding also eliminates the need for a beryllium oxide (BeO) insulating layer commonly used in other RF-power MOSFETs. LDMOS devices typically operate from 28V supplies and are currently available with power outputs of 120W at 2 GHz. They are relatively low in cost compared to other devices for this frequency range.

Up until recently, radio frequency power amplifiers (RFPA) in the VHF band could only be fabricated using bipolar transistors, because the devices designed in vertical MOS technology (VMOS), which were the only MOS power devices available at that time, were not able to amplify signals of more than a few hundred megahertz. There were two types of parasitic parameters caused by the vertical transistor layout: the source inductor and the output to input feedback capacitor (gate-drain capacitor). Both of these effects limited the frequency range to less than approximately 300 MHz [8].

In RF power amplifiers, MOS transistors must be operated in the common-source configuration since the package flange to which the source is electrically connected must be attached to the cooling heat sink (that is the ground) to dissipate heat. In the same way, BJT transistors are connected in the common-emitter configuration since for them the emitter is attached to the flange.

For MOS transistors, the common-source configuration (which is analogous to the common emitter of the bipolar transistors) magnifies the capacitance existing between the gate and the drain, C_{GD}, because of the Miller effect. In BJT transistors, the capacitance is the one between the base and the collector, C_{BC}. Even a small value of these capacitances produces a strong feedback between the output and input, which decreases the gain. We examined the Miller effect in Chapter 4 and will encounter it many more times in the subsequent chapters.

The main difference observed between bipolar and MOS transistors used in RF power applications is the gain: for the same output power level, a BJT has a typical gain value of approximately 8 to 9 dB, while an MOS transistor is able to provide gains up to 14 dB. Thus if bipolar devices are used as the power amplifiers driving the antenna, eight or nine amplifier stages may be necessary. If it is possible to use MOS power transistors, two to three intermediate stages could be eliminated, thanks to their higher gain. This obviously has many advantages.

5.6.4 Technological Development

To avoid the frequency limitations described previously it is possible to combine lateral diffusion (LD) with the VMOS transistor (vertical) process. The reductions of the feedback capacitor and the source inductor have allowed the devices to

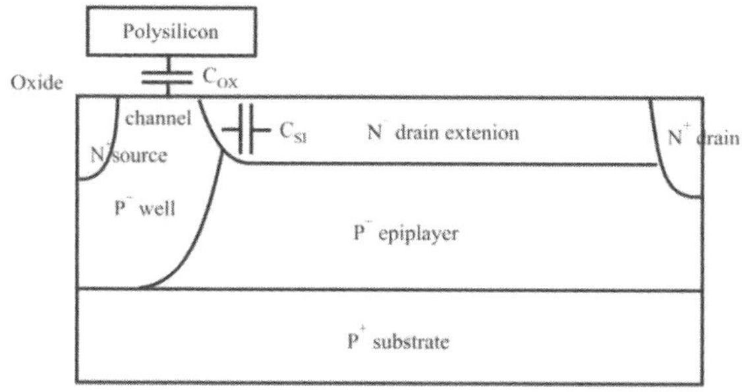

Figure 5.10 LDMOS substrate capacitors.

support high power applications in the frequency range from approximately 300 MHz to 3 GHz.

There are two developments in the layout of the transistors that have allowed improving the frequency performance of the devices. The first reduces the feedback capacitor C_{gd}, comprised of the two series capacitors C_{OX} and C_{SI}, where C_{OX} is the overlap capacitor between the gate and the diffused N^- drain extension region (see Figure 5.10). This capacitance is proportional to the area of such overlap and depends on the oxide thickness. C_{SI} represents the capacitor associated with the junction P^- well and N^- drain extension, the value of which depends on the depth of the junction and the doping of the n^- drain extension region (see Figure 5.10). The C_{SI} capacitor is reduced by the addition of the diffusion region n^- drain extension. This is the larger of the two capacitors in the series, allowing the reduction of the total capacitor C_{GD}.

It is possible to cancel the source inductor by means of the diffusion of one p+ via that forms the contact between the source of the device and its substrate, which has direct contact with the package flange (see Figure 5.11). In such a way, the flange of the device represents the source contact that, in common-source configuration (which is the only one possible in RF and microwave power applications), must be connected directly to the ground, attaching the same flange of the transistor to the heat sink without an electrical insulator, which worsens the heat transfer. Avoiding the use of a bond wire between source and the ground cancels the associated inductor responsible for the gain decrease that is proportional to frequency.

A cross section of a LDMOS transistor is shown in Figure 5.11. With this layout it is possible to realize with two devices (in push-pull) utilizing interdigitated structures in order to optimize the matching with the resulting power of up to 250W. It is also possible to implement the impedance matching networks required for external matching to the standard 50 Ω.

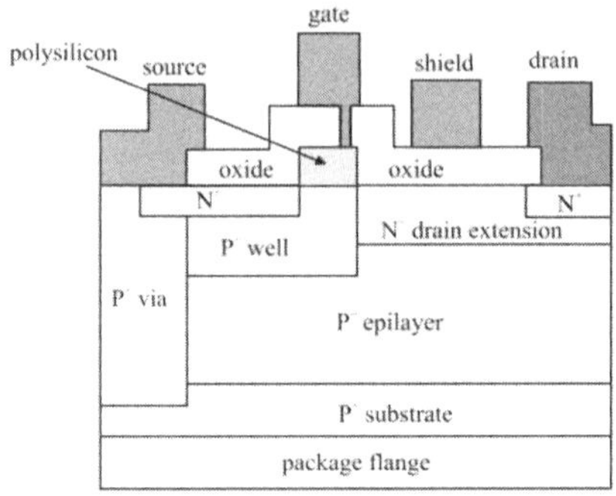

Figure 5.11 LDMOS cross section.

The LDMOS transistor is simply a special version of the standard enhancement-mode MOSFET. Designed for high power dissipation, it can achieve hundreds of watts at frequencies to about 6 GHz. This format is used primarily with silicon devices.

5.6.5 Advantages Compared to Bipolar Technology

The smaller intermodulation distortion and the better linearity in class A and class AB operations are characteristics that are intrinsic to MOS transistors compared to BJTs. LDMOS transistors have higher efficiency and superior thermal stability, particularly if the bias current for class A is next to the crossing point characteristics. The superior quality of such devices is their gain: their average value is approximately 4 to 5 dB higher than for bipolar transistors for equal output power levels. LDMOS transistors have a typical gain of 13 dB in wideband UHF applications; therefore, they could show gains of over 20 dB in narrowband conditions.

It is possible to obtain RF power amplifiers capable of providing from 5 to 250 W or more output power with lateral diffusion technology. In EW applications good linearity is required to minimize adjacent channel interference. These amplifiers require very good linearity that is necessary to minimize the interference. The main consideration to achieve that linearity is the DC biasing for optimal drain current for a given power output.

5.6.6 Junction FET

In many applications, the BJT has the disadvantage of low input impedance because the base of the transistor is the signal input and the base-emitter diode is forward-biased. A semiconductor device that operates with the input diode junction reversed-biased, called a *junction field effect transistor* (JFET) can serve as an amplifier as

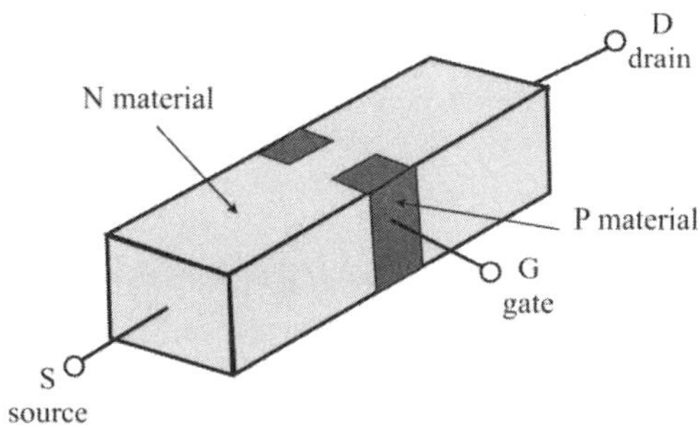

Figure 5.12 N-channel JFET.

well. With the reverse-biased input junction there is little input current so it has very high input impedance. Having high input impedance minimizes the loading effect on the signal source. Impressive power and efficiency have been obtained from RF JFETs based on Si, SiGe, and SiC at frequencies through UHF [9].

For an N-channel FET, JFET, or MOSFET, the device is constructed from a bar of N-type material, with the shaded areas composed of a P-type material as a gate as illustrated in Figure 5.12. Between the source and the drain, the n-type material acts as a resistor. The current flow consists of the majority carriers (electrons for n-type material). The difference between the MOSFET and JFET is that the current flow is controlled by the E-field established at the gate in the former while there is a junction in the latter. The amount of voltage on the junction controls the amount of current flowing from the source to the drain.

The transfer characteristic is given as the drain current (which is also the source current since there is no gate current to speak of) at a constant value of V_{DS}, and an example is shown in Figure 5.13. The transfer function is typically nonlinear, but there is a small region around $V_{gs} \sim$ 0V that can be considered linear. The control functioning of the JFET comes from depletion of charge carriers from the n-channel that reduces the drain current through the channel. When the gate is made more negative, it depletes the majority carriers from a larger depletion zone around the gate. This reduces the current flow for a given value of V_{ds}. Modulating V_{gs} modulates the current flow through the device. Characteristic curves for the JFET are shown in Figures 5.13 and 5.14.

The output characteristics for the JFET are illustrated in Figure 5.14, which are graphs of I_D versus V_{ds} at given values of V_{gs}. We can see that for a given value of V_{gs}, I_D is very nearly constant over a wide range of V_{ds}. The amount of output current supplied is controlled by the gate voltage. This characteristic makes the JFET, like MOSFETs, excellent current sources since the output voltage varies little with the current being drawn. Note that the inverse of an I_D vs V_{ds} curve is the output impedance of the JFET and is quite large because these curves are very nearly flat.

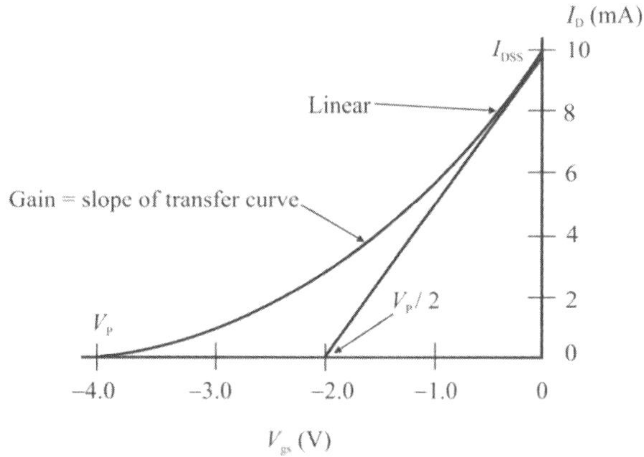

Figure 5.13 JFET transfer characteristic at a constant V_{ds}

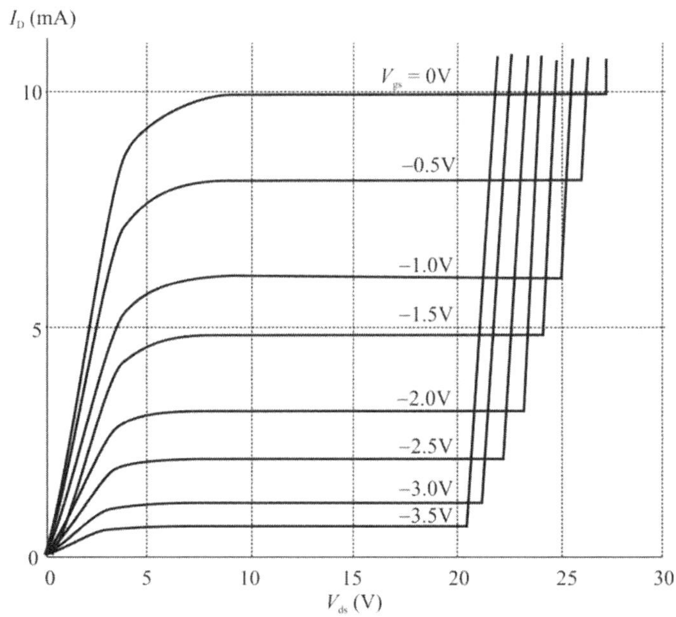

Figure 5.14 JFET output characteristic.

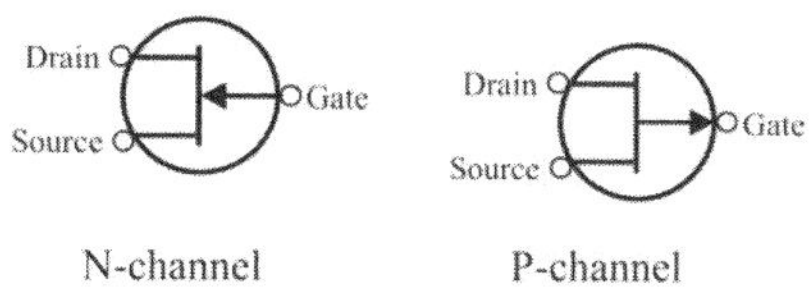

Figure 5.15 JFET symbols.

The transfer characteristic for the JFET is useful for visualizing the gain from the device and identifying the region of linearity. The gain is proportional to the slope of the transfer curve. The current value I_{DSS} represents the amount of drain current when the gate is shorted to ground, the maximum current for the device. The gate voltage at which the current reaches zero is called the *pinch voltage*, V_P (–4V in Figure 5.13). Note that the line representing the gain in the linear region of operation strikes the zero current line at about half the pinch voltage; this is a common characteristic of a JFET.

The circuit symbols for JFET devices are shown in Figure 15.15.

5.6.7 High Electron Mobility Transistors

The HEMT is a heterostructure FET. Its operating principle is based on a heterojunction that consists of two different semiconducting materials brought into contact. Because of the different band gaps and their relative alignment to each other, band discontinuities occur at the interface between the two semiconducting materials.

These discontinuities are referred to as the *conduction and valence band offsets* ΔE_c and ΔE_v. If the proper materials in the right combinations are selected, the conduction band offset can form a potential well confining electrons in the horizontal direction only. Within the well the electrons can only move in a two-dimensional plane parallel to the interface and the region is referred to as a *two-dimensional electron gas* (2DEG). To determine the exact shape of the conduction and valence bands, the Schrödinger and Poisson equations must be solved self-consistently.

5.6.7.1 Indium Phosphide (InP) HEMT

Taking advantage of the fact that the 2DEG offers exceptional high carrier mobilities compared to bulk material, a typical InP HEMT having the layer structure

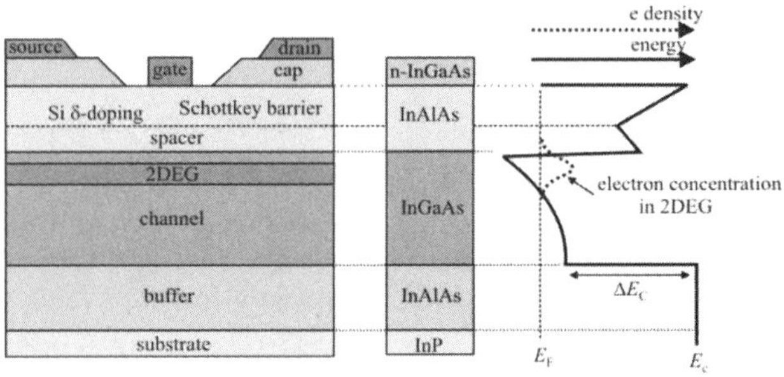

Figure 15.16 InP HEMT structure.

is shown in Figure 15.16. It consists of:

- Silicon δ-doping layer. Highly doped layer with only few atomic layers thickness. Located between the Schottky-Barrier and spacer layer, it acts as a donor of charge carriers—it provides electrons to the channel. Since electrons tend to occupy the lowest allowed energy state, they drain into the potential well and form the confined 2DEG in the channel. A high δ-doping level provides high electron densities in the channel and therefore results in high transconductances, current densities, and cutoff frequencies.
- The spacer layer assures the separation between the electrons and their positively charged Si-donors, reducing impurity scattering and hence enhancing electron mobility.
- A highly N-doped cap layer helps minimize the contact resistance of the source and drain contacts. The cap also provides protection from oxidation for the sensitive InAlAs layer beneath.
- The Schottky-barrier layer, in contrast to the ohmic source and drain contacts, provides a so-called Schottky contact between gate-metal and semiconductor material with a rectifying characteristic. It prevents large currents from flowing through the gate and limits tunneling to the channel.
- Channel properties have a major impact on the device performance. This is why InGaAs, with its excellent electron mobility properties at room and cryogenic temperatures, is the material of choice.
- The gate has a special T-shape that helps minimize the gate resistance by enlarging the cross section while maintaining a small footprint and thus a small gate length.

InP-HEMTs show excellent noise and gain performances at microwave frequencies. At cryogenic temperatures, these properties improve further. Thus InP-HEMTs are a prime choice for receiver systems, which have the most stringent requirements for low noise and high sensitivity due to the expense of cooling the device to such low temperature.

5.6.7.2 Gallium Nitride HEMT

Another HEMT is based on GaN with GaN/AlGaN heterojunctions [10]. Instead of using InP substrates as above, the substrates are based on sapphire (Al_2O_3) or silicon carbide (SiC). These semiconductors are both wide bandgap materials (5.4 eV and 5.3 eV compared to 1.3 eV for InP) and therefore have high electric breakdown fields, which enables applications at high supply voltages. Furthermore, this allows the material to withstand high operating temperatures and provides improved radiation hardness.

To achieve high currents and high frequency operation, high carrier mobilities and high saturation velocities are desirable. Typically, wide bandgap semiconductors attain only relatively low mobilities but high saturation velocity values. Compared to the InP-HEMT structure the main differences are:

- No doping in the AlGaN barrier layer is required. Built-in polarization fields, due to spontaneous polarization and piezopolarization help induce the 2DEG.
- Higher 2DEG concentrations are achievable (above $10^{13}/cm^2$) due to the very large conduction band discontinuity.

The direct bandgap of GaN and its alloys enables the material to be used for both optical and electronic applications. At 300 K the bandgap of GaN is 5.44 eV, which corresponds to a wavelength in the near ultraviolet region of the optical spectrum. This enables the fabrication of high-power optical devices as LEDs.

In EW applications, GaN is an excellent option for high-power/high-temperature microwave and lower applications because of its high electric breakdown field and high electron saturation velocity ($\sim 1.5 \times 10^7$ cm/s). The former is a result of the wide bandgap (5.44 eV at room temperature) and enables the application of high supply voltages, which is one of the two requirements for high-power device performance.

5.7 GaAs and GaN RF Power Amplifiers

5.7.1 III-V Semiconductors

GaAs and GaN are known as III-V compound semiconductors and they fall into categories of elements with either three or five valence electrons. Boron, aluminum, gallium, indium, and thallium are Category III elements. Nitrogen, phosphorus, arsenic, antimony, and bismuth are Category V elements [11].

Combining a Category III element with a Category V element produces a covalent bond with eight electrons, yielding unique semiconductors. Such semiconductors have higher electron mobility than silicon, so their high frequency performance is better. GaAs and GaN are particularly useful for microwave power amplifiers.

5.7.2 Other Transistor Types

GaAs and GaN can be used to make any type of transistor including BJTs and MOSFETs. But other types are also possible, such as HBTs, MESFETs, HEMTs, and LDMOS transistors. These transistors take advantage of the materials to produce the best amplifying and power handling capability.

HBTs are similar to the standard BJT configuration but use different materials for the base and emitter. One popular combination, a GaAs emitter and an aluminum-gallium-arsenide (AlGaAs) base, yields very high gain at microwave frequencies up to 150 GHz.

A MESFET is essentially a JFET with a metal gate that's used to form a Schottky junction with the main conducting channel. The operation is depletion mode, where the device is normally on and is turned off by a applying a negative gate voltage. MESFETs are usually made with GaAs and can have significant gain at microwave frequencies.

The HEMT is a variation of the MESFET. Also called a hetrostructure FET (HFET) or modulation-doped FET (MODFET), it is usually made with GaAs or GaN with extra layers and a Schottky junction. It provides exceedingly high levels of performance at microwave frequencies.

The HEMT offers a combination of low noise figure combined with the ability to operate at the very high microwave frequencies. Accordingly the device is used in areas of RF design where high performance is required at very high RF frequencies. It is a field-effect transistor incorporating a junction between two materials with different bandgaps (i.e., a heterojunction) as the channel instead of a doped region (as is generally the case for a MOSFET).

The depletion mode is the most common configuration. The pseudomorphic or pHEMT version improves performance by using extra layers of indium to further speed electron movement. These devices work at frequencies up to 20 GHz or so.

5.7.3 GaAs RF Power Amplifiers

Most GaAs products are ICs, specifically monolithic microwave ICs (MMICs). These MMICs are usually low-signal-level gain blocks. Such amplifier blocks use HBTs or MESFETs, but some use pHEMTs. Frequency ranges up to 30 GHz are achievable. The actual high-frequency cutoff (f_T) or the unity gain-bandwidth for GaAs devices is in the range of 150 GHz. Such MMICs are widely used in most microwave equipment designs including EW systems.

GaAs power amplifiers are available with power levels up to about 5W. That's roughly the upper power limit for GaAs devices, since they cannot withstand the high voltages, currents, and heat levels of silicon or GaN. But by using multiple devices in push-pull or parallel or combining amplifier outputs in transformers or networks, power levels up to about 20 to 40W are possible. They are commonly configured as class AB linear amplifiers with maximum PAEs in the 20% to 50% range.

5.7.4 GaN RF Power Amplifiers

GaN has emerged as a technology of choice for millimeter-wave electronics including electronic warfare systems [12]. GaN's high power density, or its ability to dissipate heat from a small package, makes it attractive for such applications. While GaAs has a basic power density of about 1.5 W/mm, GaN has a power density ranging from 5 to 12 W/mm. It also has high electron mobility, meaning it can amplify signals well into the upper-gigahertz ranges—typical transistor f_{Ts} are up to 200 GHz. Furthermore, GaN can do all this at relatively high breakdown voltages levels, up to 80V or so.

Although not normally a first-order concern for EW system applications, the downside to GaN is its high cost. The materials are expensive, and the processes to make devices are costly. As volume increases, production costs will come down but will still remain way above the process costs of bulk CMOS or even higher-cost GaAs production. We must remember, however, while it is on the list, cost is not normally a first-order concern for military equipment. Performance is almost always more important.

GaN technology mainly focuses on microwave and millimeter-wave power amplifiers. Individual amplifiers can achieve power levels of tens of watts. In other parallel/push-pull/Doherty configurations, power levels of hundreds to thousands of watts are possible.

Also, GaN amplifiers replace traveling wave tube (TWT) amplifiers in many applications. TWTs are capable of hundreds and even thousands of watts.

GaN has greatly increased power density compared to other technologies. Increasing power density results in smaller transistor die with higher output power. Compared to Si LDMOS FETs and GaAs MESFETs of similar output power, GaN HEMTs have lower parasitic capacitances. The result is transistors that have higher gain with larger input and load impedances. The matching networks have fewer sections and exhibit broader bandwidth than those required for other technologies.

5.7.5 Advantages and Disadvantages

Gallium nitride is a wide bandgap material, which enables operation at higher drain voltages. This further increases the transistor's impedances relative to LDMOS and GaAs. Another benefit is tolerance of open- and short-circuited outputs, as well as operation into high VSWR loads. This is advantageous for applications where use of a circulator is not feasible such as when a multioctave amplifier is used to directly drive a broadband antenna, typical of EW applications.

Currently, the principal disadvantage of gallium nitride technology is cost, which again is normally not a first order concern for EW applications. The majority of GaN HEMTs are produced on silicon carbide substrates, which are both costly and limited in size. In spite of the smaller die, they cost five to ten times more than LDMOS transistors of comparable output power. GaN HEMTs grown on silicon substrates potentially offer substantial cost savings over ones grown on silicon carbide. The drawback is reduced thermal conductivity of silicon substrates, which limits the power density of the transistor.

The increased transistor cost in a GaN HEMT amplifier is somewhat offset by several factors. In applications where multiple lower power LDMOS FETs or MESFETs would need to be used in parallel to achieve the required power and bandwidth, a single pair of higher power GaN HEMTs could be used. For example, an 80W amplifier covering 800 MHz to 2500 MHz using GaAs MESFET technology would require eight 10W transistors configured in parallel. Three stages of power combining would also have to be implemented. The GaN topology would require only two 45W transistors in parallel. There are considerable savings in size, complexity, and tuning time, which lower the cost of the GaN amplifier, particularly as the power level and bandwidth are increased.

GaN HEMTs exhibit soft power compression characteristic. LDMOS FETs and GaAs MESFETs typically exhibit gains identical to their small signal gains over most of their drive range. When producing the rated output power, the gain typically drops by 1 dB. This is called the P_{1dB} output power. Output power will saturate at a power approximately 1 dB greater than P_{1dB}, while the gain will have typically decreased by 3 dB compared to the small signal gain. By comparison, GaN HEMTs driven even 10 dB below their rated output power will exhibit a decrease in gain compared to the small signal gain. This results in a lower value for P_{1dB} than the

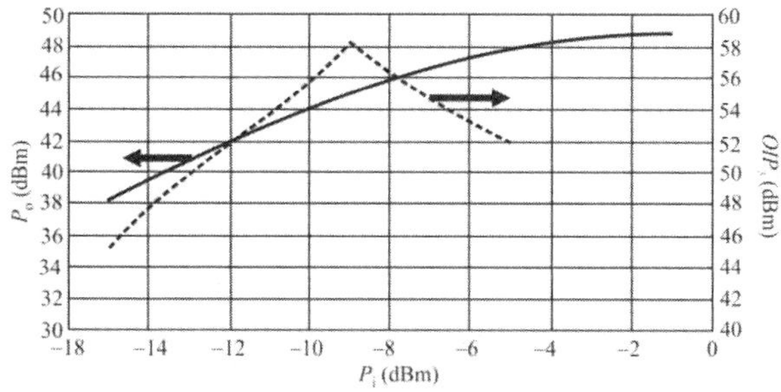

Figure 5.17 GaN L-band amplifier output power and OIP3 versus input power.

rated power might indicate. GaN HEMT output powers are therefore usually specified where the gain drops by 3 dB or P_{3dB}. When GaN HEMTs are used in multiple stages of a power amplifier, greater levels of gain compression can occur before the maximum output power is reached.

The soft compression characteristic also impacts the linearity of GaN HEMT amplifiers. LDMOS and MESFET amplifiers usually exhibit third-order output intercepts (OIP_3) 10 dB greater than their P_{1dB} when measured at output powers 10-dB or greater backed off from P_{1dB}. At output powers closer to P_{1dB}, OIP_3 generally degrades due to the compression of the transistors. There will be further discussion of this compression in Chapter 9.

Figure 5.17 illustrates typical performance of a L-band GaN HEMT showing P_o and OIP_3 versus input power. P_o peaks at about 49 dBm at maximum input power of –1 dBm (gain = 50-dB). On the other hand, the OIP_3 peaks at midband at about 58 dBm with –9-dBm input power.

Figure 5.18 illustrate the performance of a typical amplifier implemented with GaN technology and compares that performance to one implemented in LDMOS

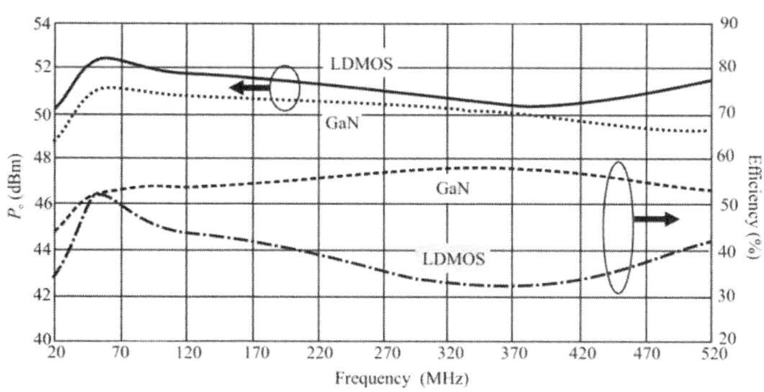

Figure 5.18 LDMOS vs GaN power output and efficiency.

technology. We can see that the LDMOS amplifier exhibits somewhat better output power performance over the frequency range considered while the GaN amplifier shows a considerable improvement in efficiency, approaching a factor of two in mid-range. This efficiency performance brings several advantages including increased reliability, the possibilities of using a smaller power supply for the same output power, and a reduction in the amount of heat that needs to be dissipated.

5.7.6 Summary

The increased power density of GaN HEMTs enables the creation of power amplifiers with higher output power, smaller size, and greater efficiency than other technologies. A high efficiency class F power amplifier at L-band with over 60% efficiency was discussed. A broadband VHF/UHF power amplifier with over 50% efficiency has also been achieved over a decade bandwidth.

GaAs continues to dominate the microwave space with small-signal MMICs and low-noise amplifiers (LNAs) as well as low-level power amplifiers. That will continue.

5.8 Comparing GaN-on-SiC Power Transistor Technology with GaAs and Si

Much has been written about the potential of GaN power devices to improve gain, bandwidth, linearity, and efficiency at VHF through microwave frequencies. In this section we compare the performance of GaN technology with the most common alternative technologies based on GaAs and Si [13].

Among these technologies, Si LDMOS power devices have established a strong presence in lower-frequency, high-power systems, as well as in wireless amplifiers operating in the 2-GHz range, while GaAs FET devices are commonly used at higher frequencies for space and terrestrial microwave applications.

5.8.1 Silicon LDMOS

Silicon LDMOS is the lateral type of power MOSFET. As we see in Figure 5.19, the lateral structure creates an inversion channel under the gate, over the laterally diffused P-well. Vertical power FETs are excellent at handling high power and high currents. However, they have very high input and output capacitances that severely limit high-frequency performance. LDMOS was developed to solve this problem, and the technology has proven to be effective with devices in the 100W range, available for operation up to 3 GHz and much higher power devices for lower frequency applications.

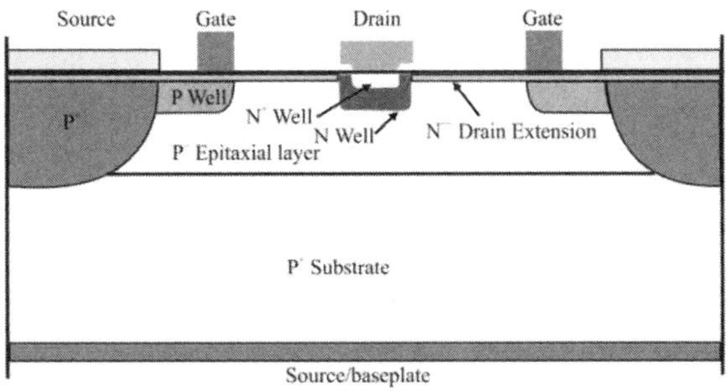

Figure 5.19 Simplified die structure of a Si LDMOS power transistor. LDMOS devices have multiple gate channels; two are shown here.

Si LDMOS features include high linearity, especially when compared to other silicon technologies, low input and output impedances, requiring internal matching for the desired frequency band, lower drain-gate capacitance than GaAs, higher breakdown voltage than GaAs, and low-cost, established silicon fabrication processes.

5.8.2 GaAs FET Power Devices

Gallium arsenide can be used up to at least 250 GHz. These devices include a semi-insulating substrate with a dielectric constant of approximately 12.5 and high-bulk resistivity, which makes it feasible to fabricate microstrip or coplanar waveguide (CPW) structures on-chip. GaAs is radiation resistant for space applications and provides low-noise, high linearity, and very good broadband performance. Limitations include relatively poor thermal performance and modest breakdown voltage (12 to 20V). In addition, GaAs MESFETs and PHEMTs require a negative supply.

GaAs is a mature technology, and several transistor structures have been developed including MESFETs and MMICs, PHEMT (Figure 5.20) and HBTs that offered a different set of advantages for low-noise, high-frequency operation, as well as reliable fabrication._Higher-voltage GaAs devices with V_{ds} (V_{ce} in HBTs) breakdown voltages of 28 to 50V and are available. These devices are primarily intended for satellite communication and point-to-point microwave equipment, provided in prematched configuration for the various microwave bands. They are useful for EW applications in this portion of the spectrum as well. Salient characteristics are indicated in Table 5.3.

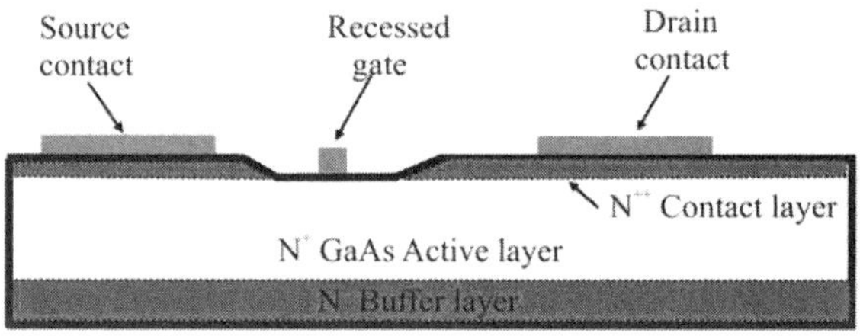

Figure 5.20 The basic GaAs HEMT structure.

Table 5.3 Salient Characteristics of Selected Transistor Types

MESFET	Excellent microwave substrate Photonic properties (i.e., light-sensitive gate) 16 to 20V breakdown Dual positive and negative supplies Relatively high channel temperature
PHEMT	Typical operation to 50 GHz High power and efficiency compared to other GaAs types 12V breakdown, low operating voltage Dual positive and negative supplies E-beam fabrication increases cost
HBT	Single supply All-optical fabrication process Limited heat dissipation Requires emitter resistors for stability (Chapter 10)
HV GaAs	Reduces current requirement for GaAs amplifiers Typically, available as prematched devices for specific bands Breakdown voltages of 28, 40, 50 are available Available using MESFET and HBT technologies

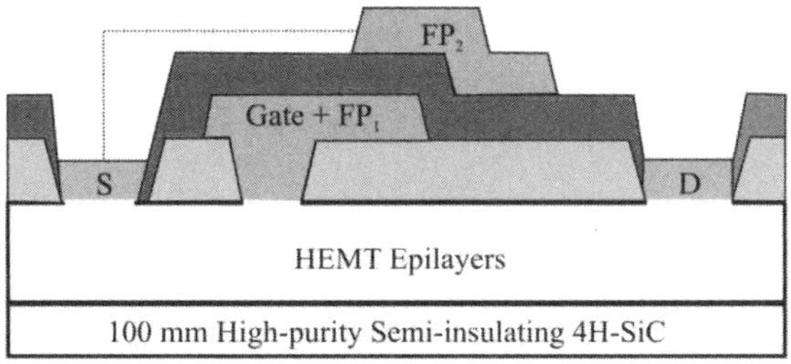

Figure 5.21 Schematic cross section of the RF GaN HEMT structure.

5.8.3 GaN-on-SiC Technology

Figure 5.21 is a simplified sketch of the GaN-on-SiC die structure, depicting GaN devices fabricated using a *metal-organic chemical vapor deposition* (MOCVD) in a high-volume reactor on 100-mm semi-insulating 4H SiC substrates. As shown in the figure, ohmic contacts are formed directly on the top AlGaN layer, with the gate connection extended toward the drain side to create a field shaping plate (FP).

Among power transistor technologies, GaN has a smoother transition into saturation, which means that amplifiers can operate further into the saturated region where efficiency is higher. Although GaAs and Si LDMOS also have excellent linearity, power amplifiers using GaN devices can provide the same performance with high *power-added efficiency* (PAE). This behavior is achieved by several different characteristics. Compared to LDMOS or GaAs, GaN-on-SiC technology offers high electron mobility, higher breakdown voltage, higher power density, compact gate area with shorter interconnects, lower input and output capacitances, higher input and output impedances, and higher PAE.

Each of these characteristics leads to improved linearity, both within the device itself and by enabling simpler matching with wider bandwidth. With greater inherent linearity, the remaining distortion can be more easily reduced by feedback, fixed predistortion, or adaptive predistortion (more in Chapter 11). We examine each of the items in this list.

5.8.3.1 High Electron Mobility

GaN has an electron mobility approaching 2000 $cm^2/V{\cdot}s$, which is higher than LDMOS but lower than GaAs. GaAs' higher electron mobility enhances operation at higher frequencies; however, GaN structures and processing are still improving in the area of f_{max}.

5.8.3.2 Higher Breakdown Voltage

GaN devices can operate with V_{ds} of 28 and 48V, compared to 12V for GaAs devices. LDMOS technology is available with the same operating voltages as GaN. For a given power level, higher operating voltages result in lower drain current and lower resistive losses. Along with GaN's internal structure, high voltage operation contributes to higher impedance and lower capacitance, thus higher operating frequency possibilities.

5.8.3.3 Higher Power Density

The GaN structure enables effective thermal conduction pathways, but the SiC substrate is also a primary contributor to high thermal performance. SiC has far better thermal performance than sapphire or silicon substrates that are also used in GaN devices. High power density allows operation at higher heat sink temperatures, simplifying cooling requirements. An obvious benefit is the safety factor for systems that are able to maintain typical operating temperatures. We discuss this in more depth in Chapter 12.

5.8.3.4 Compact Die Size

The structure of a GaN HEMT (see Figure 5.21) allows straightforward implementation of passive components: thin film resistors, metal-insulator-metal (MIM) capacitors, and slot vias.

5.8.3.5 Lower Input and Output Capacitances

The small size also contributes to lower input and output capacitances as do the gate structure, high breakdown voltage, and straightforward interconnections. AM-PM conversion, a significant distortion factor in wireless PAs, results primarily from variation in C_{gs} with voltage. The smaller gate periphery of GaN results in lower C_{gs}, less change versus voltage, and reduced AM-PM conversion than other devices.

5.8.3.6 Higher Input and Output Impedances

Low capacitance and the higher resistive impedance resulting primarily from higher operating voltage result in higher impedances. This greatly simplifies matching, allowing simpler, lower loss circuits and enables wide bandwidth matching networks. In many cases, internal matching is not required, which is important for broadband EW applications.

5.8.3.7 Higher PAE

GaN's slower rise in distortion as the device approaches saturation allows amplifiers to operate closer to saturation; that is, with less backoff from P_{sat}. This property helps effective implementation of linearization techniques like predistortion (discussed in Chapter 11).

5.8.3.8 Summary

GaAs, Si, and GaN power devices all have high linearity, along with a spectrum of characteristics that affect maximum power output, frequency range, bandwidth, thermal performance, and ease of circuit implementation. When all factors are taken into consideration, GaN has the overall highest performance for many applications, especially when priorities include high PAE, excellent thermal performance, and easy implementation of broad bandwidth matching, a primary requirement for many EW applications.

5.8.4 GaN Technology Overview

GaN has at least ten times the power density per power cross-section millimeter of transistor gate periphery, higher operating voltages (reducing impedance transformation issues), higher efficiency, and the ability to combine high RF power output over broad bandwidths at high frequencies.

We now review the different configurations of GaN: on silicon, silicon carbide (SiC), or diamond substrates. The basic characteristics of these topologies will be presented and compared.

5.8.4.1 GaN on Silicon

This approach yields lower performance than either of the alternatives but has the potential advantage of using the world's low-cost, large-wafer silicon CMOS and power silicon foundries. As a result, it is price-competitive with existing silicon and GaAs technologies.

5.8.4.2 GaN on SiC

The high-end version of GaN for RF, GaN on SiC will deliver GaN's highest power levels and other performance metrics whose fit will be those applications where the highest performance is required irrespective of cost. While not independent of cost parameters, military EW applications usually do not put cost very high on the list of trade-offs so this technique is particularly applicable for applications of interest to us.

5.8.4.3 GaN on Diamond

While the benefits are significant, it is not easy to combine these two materials. Industrial diamond has the highest thermal conductivity of any material known (more on this in Chapter 12). It therefore can remove heat better than any other material. Replacing SiC, silicon, or any other substrate materials with diamond allows the thermal conductivity benefits to be realized very close to the active area of the device.

HEMT GaN transistors first started appearing in about 2004 with *depletion-mode* (d-mode) RF transistors. The architecture employed used GaN on SiC substrates [14]. The HEMT structure was based on the phenomenon first described in 1975 by T. Mimura et al. [15] and again in 1994 by M. A. Khan et al. [16] A 2DEG near the interface between an AlGaN and GaN heterostructure produces unusually high electron mobility. Adapting this phenomenon to GaN grown on silicon carbide, benchmark power gain in the multigigahertz frequency range was produced. The first d-mode RF HEMT transistor made with GaN grown on silicon wafers was introduced in 2005.

GaN RF transistors have continued to make inroads in RF applications. Outside this market, however, penetration has been limited by device cost as well as the inconvenience of depletion-mode operation. In June 2009 the first enhancement-mode (e-mode) GaN on silicon FET designed specifically as power MOSFET replacements was introduced. In addition to enabling enhancement mode devices in lieu to the more difficult to work with depletion mode, one of the goals was to produce them in high volume at low cost using standard silicon manufacturing technology and facilities.

5.8.4.4 Gallium Nitride Characteristics

Silicon has been the dominant material for power management since the late 1950s. All of the advantages of Si stemmed from the basic physical properties of silicon combined with a huge investment in manufacturing infrastructure and engineering. Table 5.4 shows four key electrical properties of the three principal semiconductor materials contending for the power management market [17, 18]. Silicon has been the dominant material primarily because of its availability.

One way of translating these basic crystal parameters into a comparison of device performance in a power transistor is to calculate the best theoretical performance that could be achieved in each of the three candidates. For power devices there are many characteristics that matter in the variety of power conversion

Table 5.4 Material Properties of GaN, SiC, and Silicon at 300 Kelvin

Property	GaN	Si	SiC
E_G (eV)	3.4	1.12	3.2
E_{BR} (MV/cm)	3.3	0.3	3.5
V_s ($\times 10^7$ cm/s)	2.5	1.0	2.0
μ (cm^2/V_S)	990-2000	1500	650

From: [17. 18].
E_G is the bandgap energy
E_{BR} is the critical electric field for breakdown in the crystal
V_s is the saturated drift velocity
μ is the mobility of electrons
GaN gains an additional boost to the mobility (μ) by virtue of the ability to form a 2DEG at the GaN/AlGaN interface. This 2DEG more than doubles the mobility of electrons (from 990 cm^2/V_s to about 2000 cm^2/V_s [17]).

systems available. Five of the most important characteristics are conduction efficiency, breakdown voltage, switching efficiency, size, and cost.

Using the data from Table 5.4, we can calculate the theoretical minimum device on-resistance (the inverse of conductivity) as a function of breakdown voltage and as a function of material [18]. We can see in Figure 5.22 that SiC and GaN both have a superior relationship between on-resistance and breakdown voltage due to their higher critical electric field strength compared to Si [19]. Devices can thus be smaller and the electrical terminals closer together for a given breakdown voltage requirement. GaN has an extra advantage compared with SiC as a result of the enhanced mobility of electrons in the 2DEG. GaN devices with a smaller size for a given on-resistance and breakdown voltage can therefore be constructed.

e-GaN FETs result from a relatively new technology and, as such, remain somewhat more expensive to produce than their silicon counterparts. This, however, is a temporary situation. There are no known insurmountable barriers to achieving an even lower cost for an equivalent performance e-GaN FET compared with a power MOSFET or IGBT.

5.8.4.5 Basic GaN FET Structure

The basic GaN d-mode transistor structure is shown in Figure 5.23. As with any power FET, there are gate, source, and a drain electrodes. The source and drain electrodes penetrate through the top AlGaN layer to form an ohmic contact with the underlying 2DEG. A short circuit between the source and the drain is thus created until the 2DEG pool of electrons is depleted and the semi-insulating GaN crystal can block the flow of current. In order to deplete the 2DEG, a gate electrode is placed on top of the AlGaN layer (see Figure 5.23). In many of the early GaN transistors, this gate electrode was formed as a Schottky contact to the top surface (as opposed to utilizing the field effect as in other MOSFETs). By applying negative

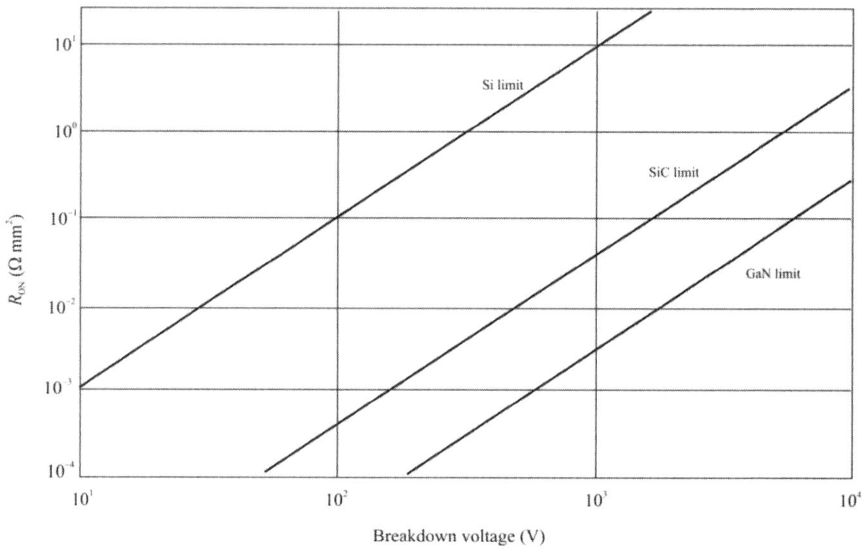

Figure 5.22 Theoretical on-resistance vs blocking voltage capability for silicon, silicon-carbide, and gallium nitride [19].

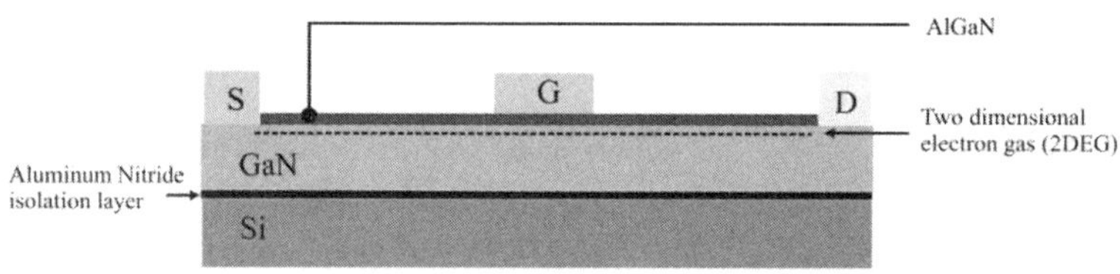

Figure 5.23 Typical AlGaN/GaN HFET structure with three metal-semiconductor contacts for the source, gate and drain.

voltage to this contact, the Schottky barrier becomes reverse-biased and the electrons underneath are depleted (rejected). Therefore, to turn this device OFF, a negative voltage relative to both drain and source electrodes is needed.

In power conversion applications, d-mode devices are inconvenient because at the startup of a power converter, a negative bias must first be applied to the gate of power devices or a short circuit will result. An *enhancement mode* (e-mode) device, on the other hand, does not suffer this limitation. With zero bias on the gate, an e-mode device is OFF and does not conduct current.

Characteristics between a d-mode and e-mode GaN FET are shown in Figure 5.24.

5.8.4.6 e-GaN FET Structure: Enhancement Mode

The enhancement mode manufacturing process begins with silicon wafers. A thin layer of aluminum nitride (AlN) is grown on the silicon to provide a seed layer for the subsequent growth of a GaN heterostructure. A heterostructure of aluminum gallium nitride (AlGaN) and then GaN is grown on the AlN. This layer provides a foundation on which to build the e-GaN FET (see Figure 5.25).

A thin AlGaN layer is then grown on top of the highly resistive GaN. It is this thin layer that creates a strained interface between the GaN and AlGaN crystal layers. This interface, combined with the intrinsic piezoelectric nature of GaN, creates the 2DEG, which is filled with highly mobile and abundant electrons [20]. Further processing of a gate electrode forms a depletion region under the gate.

To cause conduction in the FET, a positive voltage is applied to the gate in the same manner as turning on an n-channel, enhancement mode power MOSFET. A cross section of this structure is depicted in Figure 5.25. Additional layers of metal are added to route the electrons to gate, drain, and source terminals (see the cross section in Figure 5.25). This structure is repeated many times to form a power device. Conveniently, e-GaN FETs behave similarly to silicon MOSFETs with some exceptions that will be explained in the following sections.

On-Resistance

On-resistance ($R_{DS(ON)}$) versus gate-source voltage (V_{gs}) curves are similar to MOSFETs. e-GaN transistors are designed to operate with 4-5 V gate drive. The temperature coefficient of $R_{DS(ON)}$ of the e-GaN FET is also similar to the silicon MOSFET as it is positive, but the magnitude is significantly less. At 125°C the $R_{DS(ON)}$ is 1.45 times the $R_{DS(ON)}$ at 25°C compared to 1.7 for silicon. This advantage increases with increasing device voltage rating.

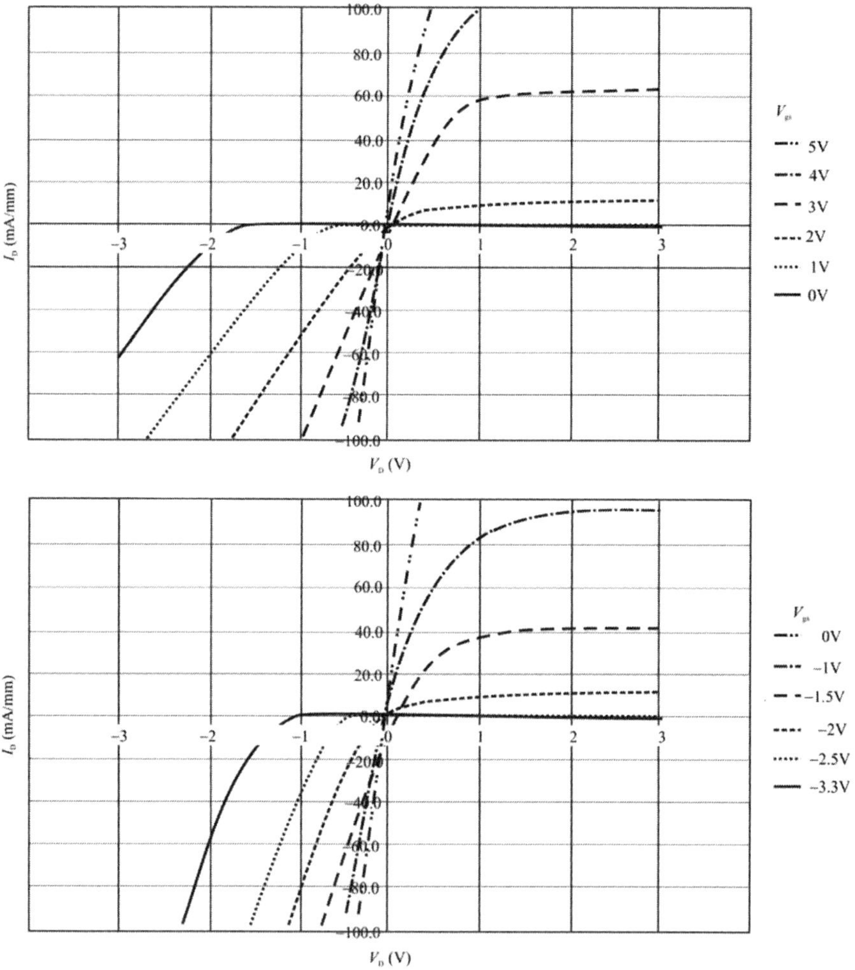

Figure 5.24 Comparison of output characteristics between an enhancement mode GaN FET (top) and a depletion mode GaN FET (bottom).

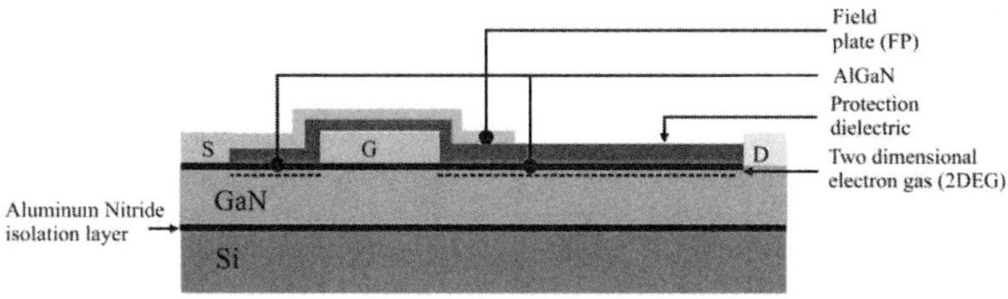

Figure 5.25 e-GaN FET structure.

Gate Threshold Voltage

The threshold of e-GaN FETs is lower than that of silicon MOSFETs. This is made possible by the almost flat relationship between threshold and temperature along with the very low gate-to-drain capacitance (C_{gd}). Since the device starts to conduct significant current at 1.6V, care must be taken to ensure a low impedance path from gate to source when the device needs to be held off during high speed switching.

Capacitance

In addition to the low $R_{DS(ON)}$, the lateral structure of the e-GaN FET makes it a very low capacitance device as well. It has the capability of switching hundreds of volts in nanoseconds, giving it high power and multiple megahertz capability. Most important in switching is C_{gd}. With the lateral structure, C_{gd} comes only from a small corner of the gate and is much lower than the same capacitance in a vertical MOSFET.

Gate-to-source capacitance (C_{gs}) consists of the junction from the gate to the channel, and the capacitance of the dielectric between the gate and the field plate. C_{gs} is large when compared with C_{gd}, giving e-GaN FETs good *dv*/*dt* immunity, but still small when compared with silicon MOSFETs. The drain-to-source capacitance C_{ds} is also small, being limited to the capacitance across the dielectric from the field plate to the drain. Capacitance versus voltage curves for e-GaN FETs are similar to those for silicon except that, for a similar resistance, its capacitance is significantly lower.

Body Diode

The parasitic bipolar junction common to silicon-based MOSFETs is known as the body diode. As seen in Figure 5.25, the GaN transistor structure is a purely lateral

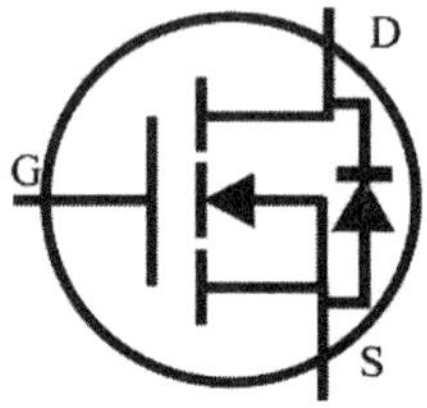

Figure 5.26 Standard symbol for an enhancement mode MOSFET or e-GaN FET. (As a memory trigger the broken line for the channel indicates that the gaps need to closed for current to flow, thus, enhancement.)

device, absent the body diode of MOSFETs. As such, reverse bias or diode operation has a different mechanism but similar function.

With zero bias gate to source there is a minimum of electrons under the gate region. As the drain voltage is decreased, a positive bias on the gate is created relative to the drift region, injecting electrons under the gate. Once the gate threshold is reached, there will be sufficient electrons under the gate to form a conductive channel.

The benefit to this mechanism is that there are no minority carriers (holes) involved in conduction, and therefore no reverse recovery losses. While Q_{RR} is zero, output capacitance (C_{OSS}) has to be charged and discharged with every switching cycle. For devices of similar $R_{DS(ON)}$, e-GaN FETs have significantly lower C_{OSS} than silicon MOSFETs. As it takes a bias on the gate greater than the threshold voltage to turn on the e-GaN FET in the reverse direction, the forward voltage of the diode is higher than silicon transistors.

Since the fundamental operation of e-GaN FETs is similar to that of silicon MOSFETs, they can be represented schematically the same way as shown in Figure 5.26.

Summary

In this section we introduced the technology for making power transistors using gallium nitride grown on top of silicon. Enhancement mode transistors have characteristics very similar to the power MOSFET, but with improved high speed switching, lower on-resistance, and a smaller size than their silicon predecessors. These capabilities, combined with high-density packaging, enable reduced power losses, reduced system size, improved efficiency, and ultimately, reduced system costs.

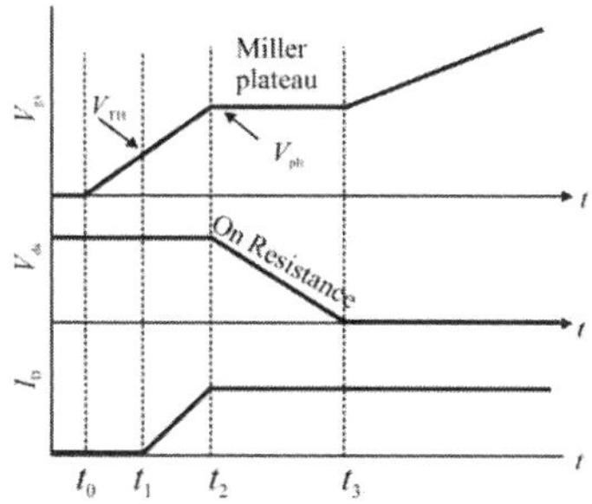

Figure 5.27 Turn-on characteristic for MOSFET devices.

5.9 Miller Plateau Region

MOSFETs (as well as IGBJTs) are controlled by the various charges distributed throughout the device. The capacitors internal to a MOSFET device are responsible for holding these charges.

The turn-on characteristic for a MOSFET is illustrated in Figure 5.27. The gate driver is responsible for providing sufficient current to charge the capacitor between the gate and the source, C_{GS}. A pulse provided to the gate at t_0 causes the gate voltage, V_{GS}, to increase as the charge provided by the current increases. At t_1 the charge provided by the gate driver is sufficient to cause V_{GS} to attain a V_{GS} threshold and the MOSFET begins to turn on and I_D begins to rise. At t_2 another V_{GS} threshold, V_{plt}, is reached but the charge provided on the gate by the driver is inadequate to allow the voltage on the gate to rise. The region from t_1 to t_2 is known as the *Miller plateau region.* During this time V_{GS} is essentially flat. At t_3 the charge provided to CG_S by the gate driver, Q_{GD}, is sufficient to cause V_{GS} to again increase causing V_{DS} to fall to zero and the MOSFET is fully on.

There is a similar characteristic plateau when the MOSFET turns off. This is illustrated in Figure 5.28.

The Miller plateau region is a concern for when the MOSFET is used in high-power switching applications such as power supplies. It is also a concern when using power MOSFETs in switching amplifiers, such as D, E, F, F^{-1}, G, and S architectures.

5.10 MICs and MMICs

Microwave amplifiers usually constructed either as *hybrid microwave integrated circuits* (MICs) or as *monolithic microwave integrated circuits* (MMICs). In hybrid

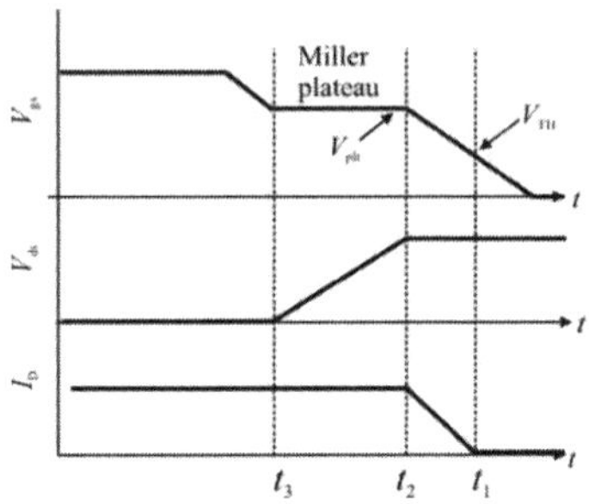

Figure 5.28 Turn-off characteristic for MOSFET devices.

construction, the transmission lines and matching networks are usually realized as microstrip circuit elements on a suitable substrate material and then the discrete components, such as, chip capacitor, resistors, and transistors are connected in place by soldering or using wire-bonding techniques. Discrete device are available with beam leads for easy insertion into the hybrid circuit.

A MMIC is a circuit where all active devices, for example transistors and passive circuit elements, such as transmission lines, capacitors, resistors, and inductors are fabricated on a single semiconductor crystal. The substrate material used has typically been GaAs, because of its high resistivity in the undoped state.

In the frequency range below 10 GHz, where distributed circuit elements are relatively large, the hybrid form of construction is often less costly than monolithic construction. However, in the frequency range of 1.0 to 10 GHz, the ability to produce miniature inductors and capacitors has led to the development and production of many MMIC systems using lumped circuit elements instead of distributed circuit elements. The millimeter wavelength band monolithic microwave integrated circuit construction promises to be more cost effective and to yield circuits with greater reliability and uniformity.

5.10.1 Monolithic Microwave Integrated Circuits

MMICs integrate RF power devices and matching/decoupling elements such as on-chip inductors, capacitors, resistors, and transmission lines. The proximity of these elements to the RF-power devices is essential for input, output, and interstage matching at microwave and millimeter-wave frequencies.

5.11 Concluding Remarks

We introduced the predominate semiconductor technologies used for low power as well as power amplifiers in EW systems in this chapter. These topics will be further expanded in the chapters to follow, as this is the principal topic of this material.

We discuss the general characteristics of BJT amplifiers in the next chapter and MOSFET amplifiers in the chapter following that. Power amplifiers are then examined as that is where the heavy lifting in EW systems occurs.

References

[1] Manz, B., "How Far Can We Take GaN Technology?" *The Journal of Electronic Defense*, Vol. 38, No. 3, Mar. 2015, pp. 41–45.

[2] Raab, F. H., P. Asbeck, S. Cripps, P. B. Kenington, Z. B. Popovic, N. Pothecary, J. F. Sevic, and N. O. Sokal, "RF and Microwave Power Amplifier and Transmitter Technologies—Part 1," *High Frequency Electronics*, Vol. 2., No. 3, May 2003, pp. 22–35.

[3] Raab, F. H., P. Asbeck, S. Cripps, P. B. Kenington, Z. B. Popovic, N. Pothecary, J. F. Sevic, and N. O. Sokal, "RF and Microwave Power Amplifier and Transmitter Technologies—Part 2," *High Frequency Electronics*, Vol 2., No. 4, July 2003, pp. 22–35.

[4] Raab, F. H., P. Asbeck, S. Cripps, P. B. Kenington, Z. B. Popovic, N. Pothecary, J. F. Sevic, and N. O. Sokal, "RF and Microwave Power Amplifier and Transmitter Technologies—Part 3," *High Frequency Electronics*, Vol. 2, No. 5, Sep. 2003, pp. 34–48.

[5] Raab, F. H., P. Asbeck, S. Cripps, P. B. Kenington, Z. B. Popovic, N. Pothecary, J. F. Sevic, and N. O. Sokal, "RF and Microwave Power Amplifier and Transmitter Technologies—Part 4," *High Frequency Electronics*, Vol. 2, No. 7, Nov. 2003, pp. 38–49.

[6] Raab, F. H., P. Asbeck, S. Cripps, P. B. Kenington, Z. B. Popovic, N. Pothecary, J. F. Sevic, and N. O. Sokal, "RF and Microwave Power Amplifier and Transmitter Technologies—Part 5," *High Frequency Electronics*, Vol. 3, No. 1, January 2004, pp. 46–55.

[7] www.d.umn.edu/~sburns/EE4611Spring2013/, accessed May, 2015.

[8] Giuseppe Vacca Bari Polytechnic Bari, Italy June 1, 2006.

[9] http://hyperphysics.phy-astr.gsu.edu/hbase/electronic/fet.html, accessed June 2010.

[10] Mishra, U. K., L. Shen, T. E. Kazior, and Y. F. Wu, "GaN-Based RF Power Devices and Amplifiers," *Proceedings of the IEEE*, Vol. 96, No. 2, Feb. 2008, pp. 287–305.

[11] Frenzel, L., *Electronic Design,* Oct 18, 2012.

[12] Baker, R., " Comparing GaN-on-SiC Power Transistor Technology with GaAs and Si," https://www.ecnmag.com/article/2017/01/comparing-gan-sic-power-transistor-technology-GaAs-and-Si, accessed Aug. 2018,

[13] Ambacher, O. J. Smart, J.R. Shealy, N. G. Weimann, K. Chu, M. Murphy, W. J. Schaff, L. F. Eastman, R. Dimitrov, L. Wittmer, M. Stutzmann, W. Rieger, and J. Hilsenbeck, "Two-Dimensional Electron Gases Induced by Spontaneous and Piezoelectric Polarization Charges in N- and Ga-Face AlGaN/GaN Heterostructures," *Journal of Applied Physics*, Vol. 85, No.6, 1999, pp. 322–335.

[14] Mitani, E., H. Haematsu, S. Yokogawa, J. Nikaido, Y. Tateno, "Mass Production of High Voltage GaAs and GaN Devices," CS Mantech Conference, Apr. 24-27, 2006, Vancouver B.C., Canada.

[15] Mimura, T., N. Tokoyama, H. Kusakawa, K. Suyama, M. Fukuta, "GaAs MOSFET for Low-Power High-Speed Logic Applications," the 37th Device Research Conference, 25-27 June 1979, University of Colorado, Boulder, CO.

[16] Khan, M., J. N. Kuznia, and D. T. Olson, *Applied Physics Letters*, Vol. 65, No. 9, pp. 1121-1123, 29 Aug. 1994, http://www.businesswire.com/portal/site/home/permalink/? ndmViewId=news_view&newsId= 20051005005600&newsLang=en, accessed Sep. 2010.
[17] Gaska, R., J. W. Yang, Q. Chen, A. Khan, A. O. Orlov, G. L. Snider, and M. S. Shur, "Electrontransport in AlGaN-GaN Heterostructures Grown on 6H-SiC Substrates," *Applied Physics Letters*, Vol. 72, No.6, pp.707–709.
[18] Kaminski, N., "State of the Art and Future of Wide Band-Gap Devices," *Proceedings EPE* 2009, Barcelona, 2009.
[19] Baliga, B. J., "Semiconductors for High-Voltage, Vertical Channel Field-Effect Transistors," *Journal of the Applied Physics Laboratory* (1982), Vol. 53, pp. 1759–1764.
[20] Lidow, A., J. Strydom, A. Ferencz, and R. V. White, "Driving e-GaN FETs in High Performance Power Conversion Systems," PCIM Asia, June 2011, http://epc-co.com/epc/documents/presentations/PCIM_Asia_2011.pdf, accessed Apr. 2013.

Chapter 6

BJT Amplifiers

6.1 Introduction

In this chapter we discuss the basics of amplifiers constructed with BJT devices.

Most analysis of amplifier circuits, whether constructed with BJT or other active devises, other than with computer simulation, use models of the active device. In the case of the BJT there are several models that can be used; however, we will use what is called the hybrid-π model shown in Figure 6.1.

This chapter is structured as follows [1 – 5]. First we examine the principal small signal models for BJTs that are used to perform linear analysis. Then we have a section were the CE topology is discussed, followed by a section on the CC configuration, and lastly the CB structure. Next the BJT cascode amplifier is discussed and last the Darlington pair topology is examined.

6.2 BJT Models

Models of BJTs are frequently employed for circuit analysis purposes. We introduce two such models in this section. Which of these models to use is circuit-dependent. However, they are equivalent and both lead to the same results.

6.2.1 BJT Hybrid-π High-frequency Model

The BJT high-frequency hybrid-π model is shown in Figure 6.1. Resistor $r_{\mathrm{Bb'}}$ (sometimes referred to as r_{x}, and much smaller than r_π), and C_π are added to the input of the low-frequency hybrid BJT model along with two feedback components

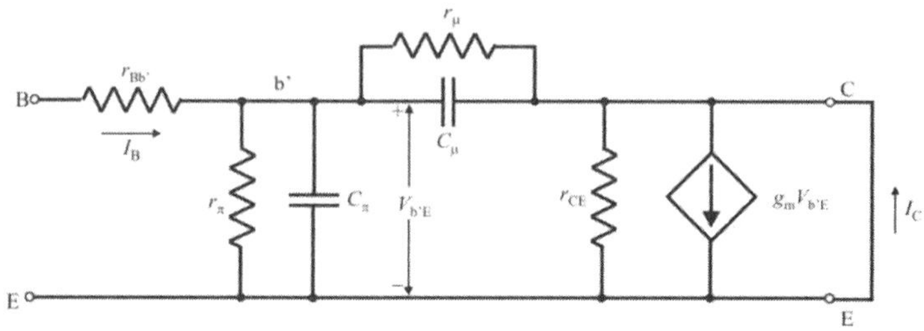

Figure 6.1 CE BJT hybrid-π high-frequency model.

r_μ and C_μ. This model is generally used when the CE and CC amplifier topologies are being considered.

The unity-gain, or cutoff frequency, denoted by f_t, is where the current gain of the device becomes 0 dB. It is given by

$$f_t = \frac{1}{2\pi}\frac{g_m}{C_\pi + C_\mu}$$

The cutoff frequency is typically used as an indicator to evaluate the high-frequency capability of the device.

6.2.2 BJT T-Model

To model the CB configuration of a BJT amplifier, the modified T-model shown in Figure 6.2 is normally used. α is the short-circuit current gain of the device.

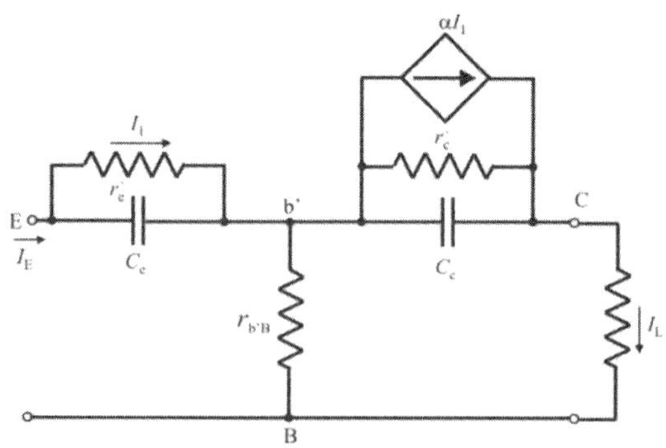

Figure 6.2 CB BJT high-frequency T-model.

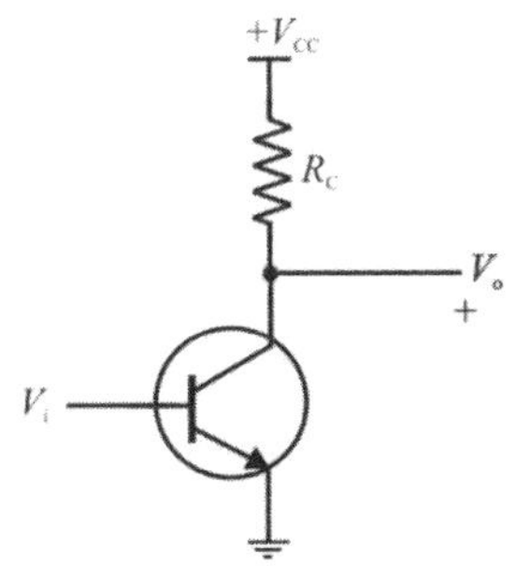

Figure 6.3 Basic NPN common-emitter circuit (neglecting most biasing details).

6.3 Common Emitter

In the *common emitter* (CE) or grounded emitter configuration, the input signal is applied between the base and emitter, while the output is taken from between the collector and the emitter as shown in Figure 6.3. This type of configuration is the most commonly used circuit for transistor-based amplifiers and represents the normal method of bipolar transistor connection.

The common emitter amplifier configuration produces the highest current and power gain of all the three bipolar transistor configurations. This is mainly because the input impedance is LOW as it is connected to a forward-biased PN-junction, while the output impedance is HIGH as it is taken from a reverse-biased junction.

In a CE amplifier, the NPN configuration of which is shown in Figure 6.3, the emitter is typically at RF ground. This is accomplished by bypassing the emitter biasing resistor with an appropriately sized capacitor. The capacitor maintains the correct DC biasing level. The CE amplifier is typically used as a voltage amplifier. The input signal is applied to the base and the (inverted) output signal is extracted from the collector.

6.3.1 Common Emitter Amplifier

A common emitter amplifier configuration is shown in Figure 6.4 where one form of biasing is included (we cover other biasing techniques later). We will first determine its low-frequency characteristics by considering the effects of each capacitor separately.

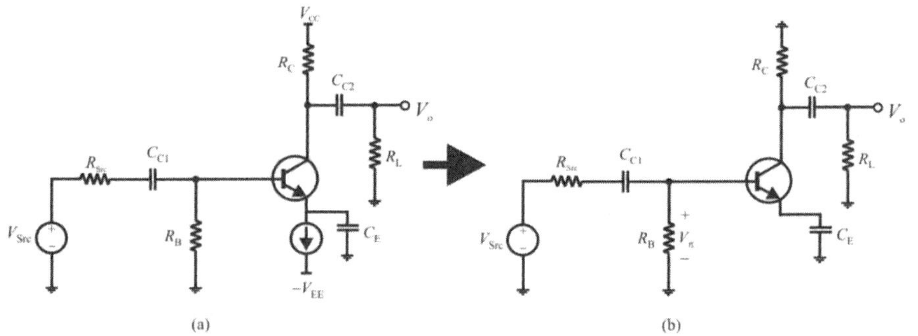

Figure 6.4 Common-emitter: (a) circuit and (b) equivalent high-frequency circuit.

6.3.1.1 C_{C1} Only

When we consider only C_{C1} we get the equivalent circuit shown in Figure 6.5. Now[3]

$$V_\pi = V_{Src} \frac{R_B \| r_\pi}{R_B \| r_\pi + R_{Src} + \dfrac{1}{sC_{C1}}}$$

$$V_o = -g_m V_\pi (R_C \| R_L)$$

$$\frac{V_o}{V_S} = \underbrace{-\frac{g_m (R_B \| r_\pi)(R_C \| R_L)}{R_B \| r_\pi + R_{Src}}}_{A_M} \frac{s}{s + \underbrace{\dfrac{1}{C_{C1}(R_B \| r_\pi + R_{Src})}}_{\omega_{p1}}}$$

$$\omega_{p1} = \frac{1}{C_{C1}(R_B \| r_\pi + R_{Src})}$$

$$A_M = -\frac{g_m (R_B \| r_\pi)(R_C \| R_L)}{R_B \| r_\pi + R_{Src}}$$

6.3.1.2 C_E Only

When we consider only C_E with the other capacitors shorted we get the equivalent circuit shown in Figure 6.6. Small signal analysis yields

[3] Note: Parallel lines (||) indicate components in parallel.

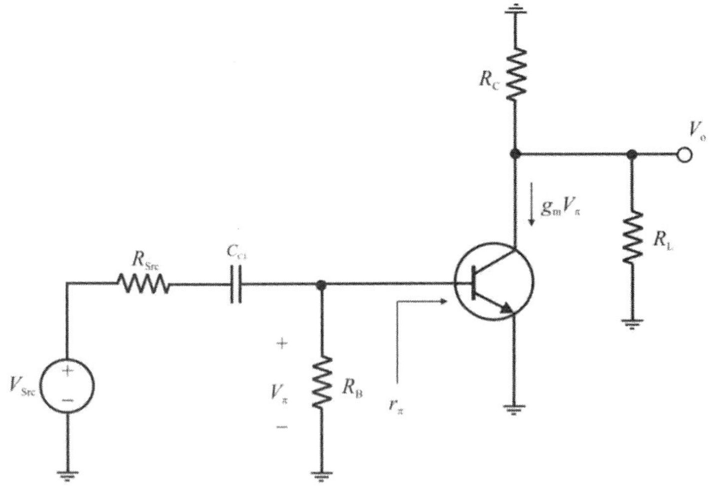

Figure 6.5 C_{C1} only.

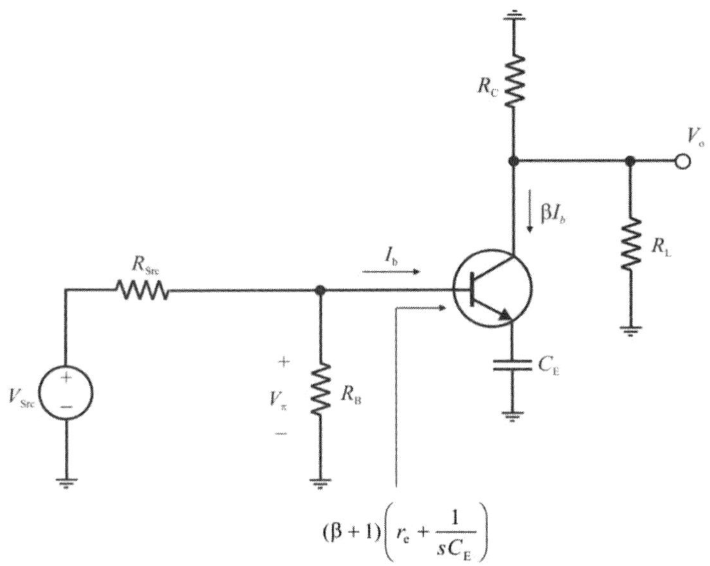

Figure 6.6 C_E only.

$$I_b = V_{Src}\frac{R_B}{R_B + R_{Src}}\frac{1}{R_B \| R_{Src} + (\beta+1)\left(r_e + \frac{1}{sC_E}\right)}$$

$$V_o = -\beta I_b (R_C \| R_L)$$

$$\frac{V_o}{V_S} = -\frac{R_B}{R_B + R_{Src}}\frac{\beta(R_C \| R_L)}{R_B \| R_{Src} + (\beta+1)r_e}\frac{s}{C_E\left(\frac{1}{r_e + \frac{R_B \| R_{Src}}{\beta+1}}\right)}$$

thus

$$\omega_{pE} = \frac{1}{C_E\left(r_e + \frac{R_B \| R_{Src}}{\beta+1}\right)}$$

6.1.3.3 C_{C2} Only

Last, we consider C_{C2} only with the other capacitors shorted. The equivalent circuit is shown in Figure 6.7. Small signal analysis yields

$$V_\pi = V_{Src}\frac{R_B \| r_\pi}{R_B \| r_\pi + R_{Src}}$$

$$V_o = -g_m V_\pi \frac{R_C}{R_C + \frac{1}{sC_{C2}} + R_L} R_L$$

$$\frac{V_o}{V_{Src}} = -\frac{g_m (R_B \| r_\pi)(R_C \| R_L)}{R_B \| r_\pi + R_{Src}}\frac{s}{s + \frac{1}{C_{C2}(R_C + R_L)}}$$

yielding

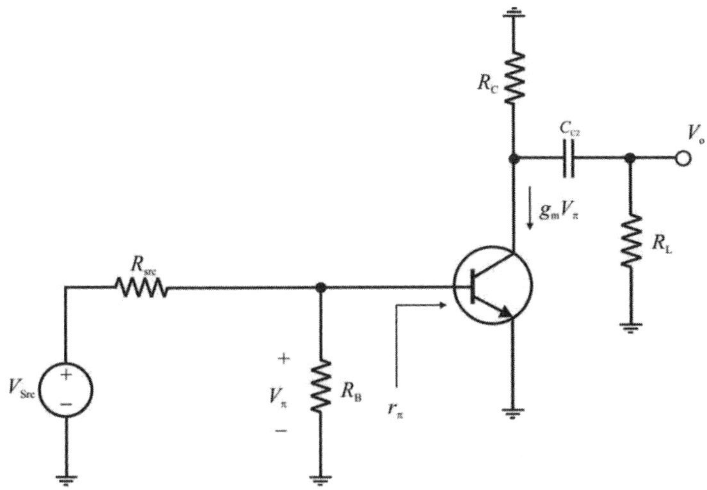

Figure 6.7 C_{C2} only.

$$\omega_{p2} = \frac{1}{C_{C2}(R_C + R_L)}$$

6.3.1.4 The Lower 3-dB Frequency

The coupling and bypass capacitors result in a high-pass frequency response with three poles:

$$\frac{V_o}{V_{Src}} = -A_M\left(\frac{s}{s+\omega_{p1}}\right)\left(\frac{s}{s+\omega_{p2}}\right)\left(\frac{s}{s+\omega_{p3}}\right)$$

The lower 3-dB frequency is the highest frequency if the poles are sufficiently separated. The highest frequency pole is typically ω_{p2} due to the small resistance of R_E. An approximation of the lower 3-dB frequency is given by

$$f_L \approx \frac{1}{2\pi}\left(\frac{1}{C_{C1}R_{C1}} + \frac{1}{C_E R_E} + \frac{1}{C_{C2}R_{C2}}\right)$$

The coupling and bypass capacitors are typically required for discrete amplifier designs. C_E is determined first to satisfy the needed f_L. C_{C1} and C_{C2} are chosen such that the poles are 5 to 10 times lower than f_L.

6.3.1.5 Characteristics of the CE BJT Amplifier

The salient characteristics of the CE BJT amplifier shown in Figure 6.3 are discussed in this section using the hybrid-π model shown in Figure 6.1. These characteristics are the short-circuit current gain, the open-circuit voltage gain, the output impedance, and the input impedance.

Short-Circuit Current Gain

The short circuit current gain is given by

$$A_I = \left.\frac{I_o}{I_i}\right|_{R_L=0} = \beta \tag{6.1}$$

Typically, β has a value between 20 and 200 for most general-purpose transistors.

Open-Circuit Voltage Gain

The open-circuit voltage gain is given by

$$\begin{aligned} A_V &= \left.\frac{V_o}{V_i}\right|_{R_L=\infty} = -\frac{\beta R_C}{r_\pi + (\beta+1)R_E} \\ &\approx -\frac{R_C}{R_E} \qquad \beta >> 1 \end{aligned} \tag{6.2}$$

Input Resistance

The input resistance is given by

$$R_i = r_\pi + (\beta+1)R_E \tag{6.3}$$

$$\approx \beta R_E \qquad \beta >> 1 \tag{6.4}$$

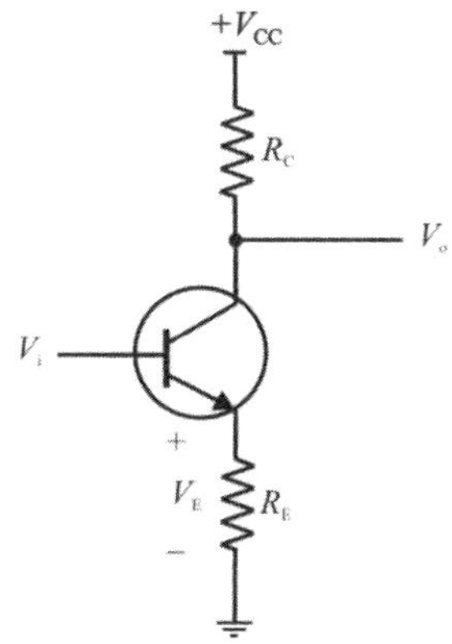

Figure 6.8 Common emitter degeneration. Adding an emitter resistor, R_E, decreases gain, but increases linearity and stability (discussed in Chapters 10 and 11).

Output Resistance

The output resistance is given by

$$R_o = R_C \tag{6.5}$$

6.3.2 Emitter Degeneration

The gain in CE amplifiers depends strongly on both transistor temperature and bias current, making the resulting gain somewhat unpredictable. Stability can also be an issue with such high gain circuits. Instability can arise because of unintentional positive feedback that may arise in implementation. One way of addressing these issues is to use *emitter degeneration*, which is a form of negative feedback. An unbypassed resistance is added between the emitter and ground as shown in Figure 6.8. The result of adding this resistance is to reduce the transconductance $G_m = g_m$ of the amplifier by a factor of $G_m R_E + 1$; that is

$$A_V = \frac{-g_m R_C}{1 + g_m R_E} \tag{6.6}$$

$$\approx -\frac{R_C}{R_E} \qquad (g_m R_E >> 1) \tag{6.7}$$

which, of course, reduces the voltage gain to $-g_m R_C$ if $R_E = 0$. Adding R_E effectively makes the voltage gain more dependent on the not-so-sensitive bias

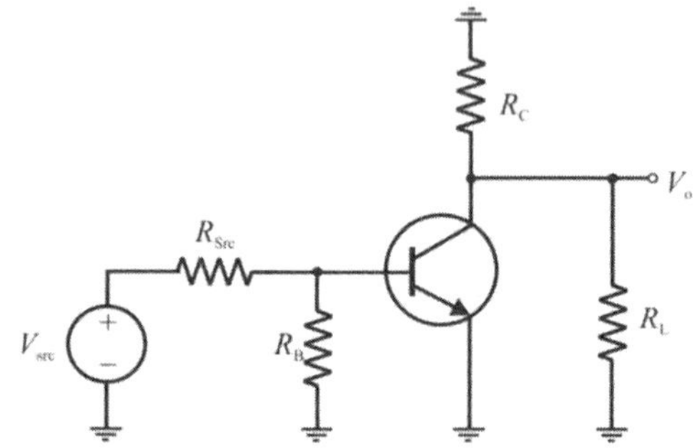

Figure 6.9 CE midband equivalent where the capacitors have no effect on performance.

resistances than on the BJT's characteristics. What this benefit costs is a reduction in gain by the factor of $g_m R_E + 1$.

6.3.3 Bandwidth

The bandwidth of the common emitter amplifier tends to be low due to high input capacitance resulting from the Miller effect. The base-collector capacitance shown in Figure 6.1 added to the basic model to account for operation at RF, is effectively multiplied by the factor

$$1 + |A_V| \tag{6.8}$$

thus substantially increasing the total input capacitance and which results in reduction of the overall bandwidth. This bandwidth reduction problem is less severe with emitter degeneration than without it partially due to the reduction in gain, which reduces the multiplying factor given by (6.8).

6.3.4 Common-Emitter Amplifier High-Frequency Response

The AC model of a simple BJT CE amplifier is shown in Figure 6.9. The equivalent circuit showing the discrete components is shown in Figure 6.10 while the simplified equivalent circuit incorporating the Miller effect capacitors is shown in Figure 6.11. The high-frequency voltage gain of the CE BJT amplifier based on Figure 6.11 is given by

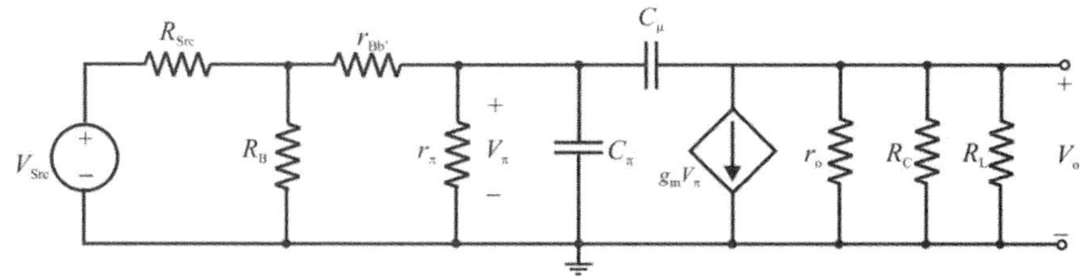

Figure 6.10 CE equivalent circuits using the hybrid-π model of the BJT.

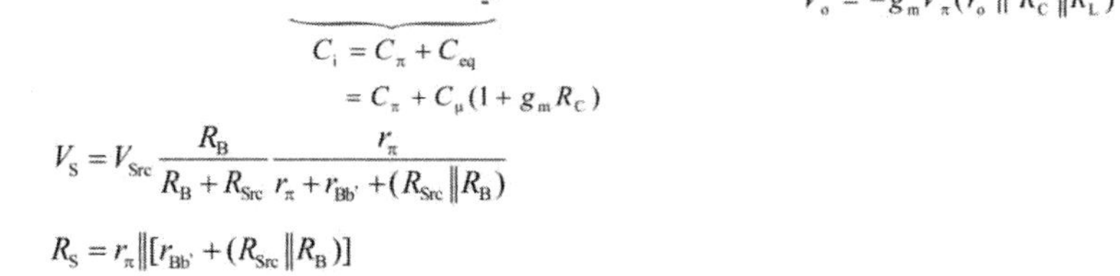

Figure 6.11 CE equivalent circuits htat shows the Miller effect capacitance. C_o is the Miller effect output capacitance which is minimally affected by the feedback path.

$$A_V = \left[\frac{-g_m R_B r_\pi (r_o \parallel R_C \parallel R_L)}{R_{Src}(R_B + r_{Bb'} + r_\pi) + R_B(r_{Bb'} + r_\pi)} \right] \times \left[\frac{1}{1 + j\omega C_o (r_o \parallel R_C \parallel R_L)} \right] \left\{ \frac{1}{1 + j\omega C_i (r_\pi \parallel R_S + r_{Bb'})} \right\} \tag{6.9}$$

where

$$R_S = R_{Src} \parallel R_B$$

and the Miller effect capacitances are

$$C_{eq} = C_\mu [1 + g_m (r_o \parallel R_C \parallel R_L)]$$

$$C_o = C_\mu \frac{1 + g_m (r_o \parallel R_C \parallel R_L)}{g_m (r_o \parallel R_C \parallel R_L)}$$

$$C_i = C_\pi + C_{eq} = C_\pi + C_\mu [1 + g_m (r_o \parallel R_C \parallel R_L)]$$

So, the high-frequency poles for the CE BJT amplifier are given by

$$\omega_{p1} = \frac{1}{1 + j\omega C_o (r_o \parallel R_C \parallel R_L)} \tag{6.10}$$

and

$$\omega_{p2} = \frac{1}{1 + j\omega C_i \left[r_\pi \parallel (R_S + r_{Bb'}) \right]} \tag{6.11}$$

We can see from (6.9) that the high-frequency voltage gain is just the product of the midband gain and the two pole terms.

6.3.5 Summary

The CE is probably the most prolific of the BJT amplifier forms. It provides high-voltage gain and moderate impedance levels. It is also a very flexible configuration, and can be used in many applications. Stability and temperature performance must be taken into consideration in most cases. We consider these in Chapters 10 and 12.

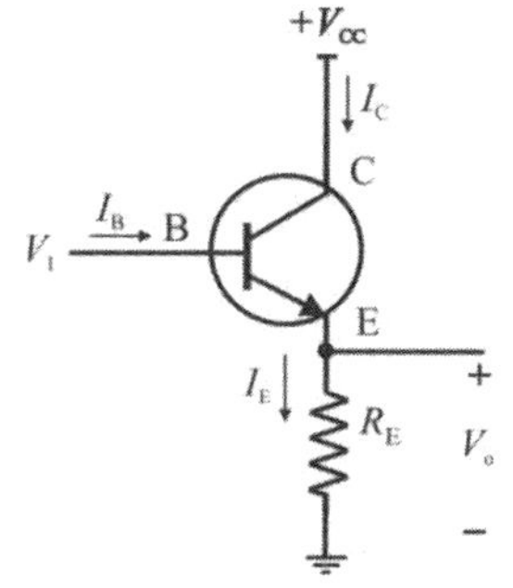

Figure 6.12 Basic NPN common collector circuit (neglecting most biasing details).

6.4 Common Collector

In a common collector BJT amplifier configuration, the collector is held at RF ground, and the output is taken from the emitter, which is connected to ground through the emitter resistor, R_E. The input is applied to the base of the BJT. Such a configuration is also referred to as an *emitter follower* and *voltage follower*. The most common use for this amplifier is as a voltage buffer.

A simplified configuration for the emitter follower is shown in Figure 6.12 while a topology for a common collector amplification stage is shown in Figure 6.13 with biasing details included with specific component values. The power supply has also been shorted to reflect AC operation.

The major advantage of a voltage follower is for transformation of impedance levels. The input impedance is relatively high, making coupling to the previous stage easier by reducing the loading effects. This is particularly useful in the load

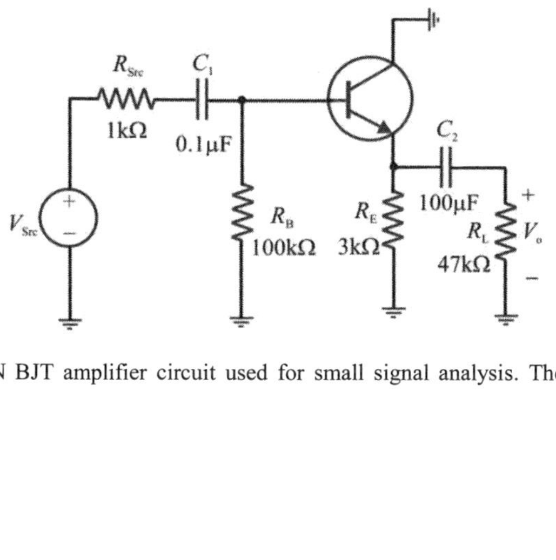

Figure 6.13 CC NPN BJT amplifier circuit used for small signal analysis. The voltage sources are removed (shorted).

impedance (without the voltage follower) is low. Likewise, the output impedance is low, making the voltage follower a good voltage source for the following stage.

6.4.1 Common Collector Characteristics

The salient characteristics of the common collector (*emitter follower*) topology are given in this section. At low frequencies we use a simplified hybrid-π model.

6.4.1.1 Current Gain

Referring to Figure 6.12

$$
\begin{aligned}
I_E &= I_C + I_B \\
A_I &= \frac{I_E}{I_B} = \frac{I_C + I_B}{I_B} \\
&= \frac{I_C}{I_B} + 1 \\
&= \beta + 1 \qquad (6.12) \\
&\approx \beta \qquad \beta >> 1 \qquad (6.13)
\end{aligned}
$$

6.4.1.2 Voltage Gain

The voltage gain of the CC amplifier is given by

$$
\begin{aligned}
A_V &= \frac{V_o}{V_i} = \frac{g_m R_E}{g_m R_E + 1} \\
&\approx 1 \qquad g_m R_E >> 1 \qquad (6.14)
\end{aligned}
$$

This type of bipolar transistor configuration is noninverting in that the signal voltages of V_i and V_o are in-phase. It has a voltage gain that is always less than 1. The load resistance of the common collector transistor receives both the base and collector currents giving a large current gain (as with the common emitter configuration), therefore good current amplification with no voltage gain are provided with this topology.

6.4.1.3 Input Resistance

The input resistance to the emitter follower is given by

$$R_i = \frac{V_i}{I_i} = r_\pi + (\beta+1)R_E$$
$$\approx \beta R_E \qquad\qquad g_m R_E >> 1 \text{ and } \beta >> 1 \qquad (6.15)$$

so it minimizes the loading effects on the previous stage by increasing the (otherwise) low impedance as given by R_E, by a factor β.

6.4.1.4 Output Resistance

The output resistance of the emitter follower is given by

$$R_o = R_E \left\| \left(\frac{r_\pi + R_{Src}}{\beta+1} \right) \right.$$
$$\approx \frac{1}{g_m} + \frac{R_{Src}}{\beta} \approx \frac{R_{Src}}{\beta} \qquad\qquad \beta >> 1 \text{ and } R_i >> R_{Src} \qquad (6.16)$$

The second approximation is because, typically, $R_{Src}/\beta >> 1/g_m$.

Using (6.15) and (6.16) we see that placing such an amplification stage between a source with large output impedance and a load with small impedance allows for voltage buffering.

The emitter follower is commonly used in the output stages of class-B and class-AB amplifiers. The base is biased to the correct point to cause the transistor to operate in class-B or -AB mode.

The common collector, or emitter follower, configuration is very useful for impedance matching applications because of the very high input impedance, in the region of hundreds of thousands of ohms, while having relatively low output impedance.

The AC model of a BJT common-collector amplifier is shown in Figure 6.14. The midband equivalent circuit is shown in 6.14(a) while the high-frequency equivalent is shown in Figure 6.14(b). For simplicity we will assume that r_o is very large compared to R_E (and since r_o appears in parallel to R_E at the output, it can be neglected).

The Miller effect capacitances are

$$C_i = C_\pi + C_\mu \left(1 + g_m R_E\right)$$

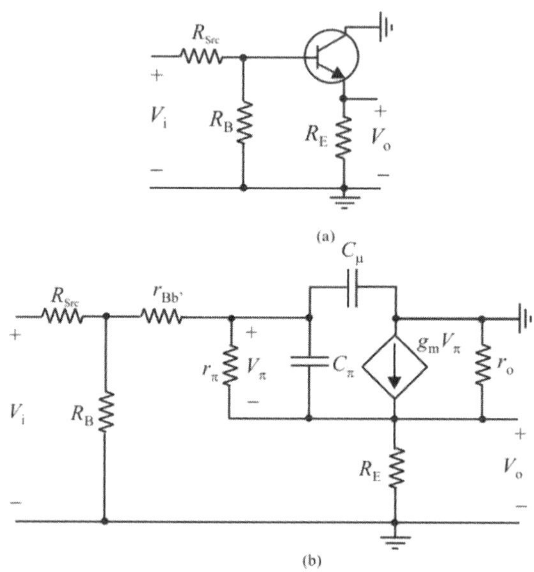

Figure 6.14 CC high-frequency equivalent circuits: (a) midband AC equivalent (all capacitors are neglected) and (b) high-frequency equivalent (neglecting r_μ).

$$C_o = C_\mu \left(\frac{1 + g_m R_E}{g_m R_E} \right) \approx C_\mu$$

The pole introduced by C_i is of primary interest since C_o is significantly less than the input capacitance. Therefore, assume that the pole introduced by C_o is sufficiently large so that C_o can be replaced by an open circuit. The resulting high-frequency voltage gain of the common-collector amplifier is

$$A_V \approx \left[\frac{g_m R_B r_\pi}{R_{Src} \parallel R_B + r_{Bb'} + (1 + g_m R_E) r_\pi + j\omega C_i r_\pi (R_{Src} + r_{Bb'})} \right] \frac{R_{Src} \parallel R_B}{R_{Src}} \qquad (6.17)$$

The dominant high-frequency pole is

$$\omega_{p1} \approx \frac{1}{C_i \left(r_\pi \parallel \dfrac{R_{Src} + r_{Bb'}}{1 + g_m R_E} \right)} \qquad (6.18)$$

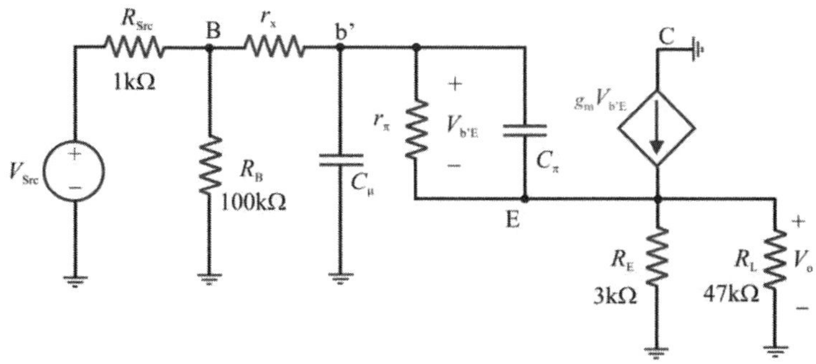

Figure 6.15 The CC amplifier in Figure 6.13 redrawn with the transistor replaced with its high-frequency small signal model.

6.4.2 Common-Collector Amplifier Approximate High Cutoff Frequency ω_H

In this section we will estimate the high cutoff frequency of the CC amplifier. We will derive an expression for the high cutoff frequency using the OCTC method.

Example 6.1: In this example we will determine the bandwidth of the amplifier shown in Figure 6.14 assuming that $V_{CC} = V_{EE} = 5\,\text{V}$, and assuming the transistor parameters are $f_T = 500\,\text{MHz}, \beta_0 = 100, C_\mu = 0.5\,\text{pF}$, and $r_x = 250\,\Omega$.

The CC amplifier is redrawn in Figure 6.15 by replacing the transistor in Figure 6.13 by its high-frequency small signal model. All the coupling and bypass capacitors are considered short circuits at high-frequency.

Figure 6.15 can be simplified as shown in Figure 6.16 where

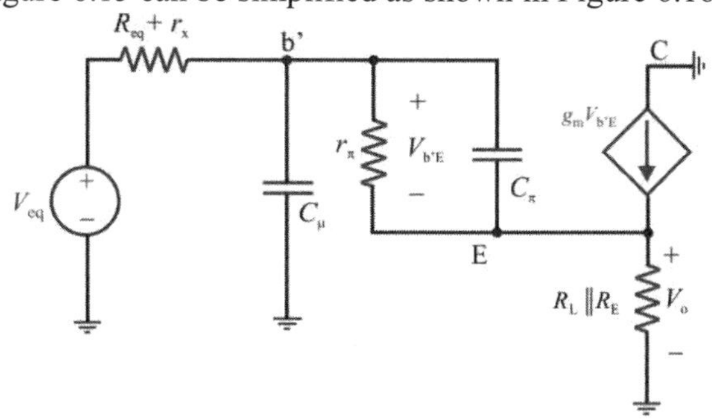

Figure 6.16 Simplified circuit to calculate the high-frequency response of the CC amplifier.

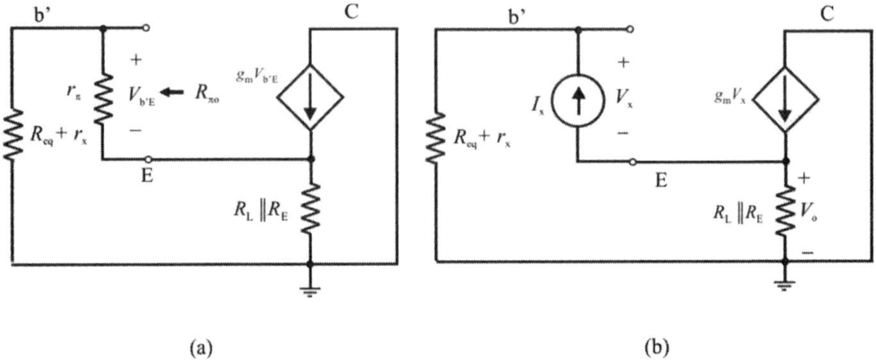

Figure 6.17 Circuit used to calculate the open-circuit time constant associated with C_π.

$$V_{eq} = V_{Src} \frac{R_B}{R_{Src} + R_B}$$

and

$$R_{eq} = \frac{R_{Src} R_B}{R_{Src} + R_B}$$

For $C\pi$, the resistance seen across the capacitor terminals is determined using Figure 6.17(a). The calculation may be simplified by considering the equivalent circuit in Figure 6.17(b)

$$R_{\pi o} = r_\pi \| (V_x / I_x)$$

$$V_x = V_{b'} - V_e = I_x (R_{eq} + r_x) - (g_m v_x - I_x)(R_L \| R_E)$$

$$V_x [1 + g_m (R_L \| R_E)] = I_x [R_{eq} + r_x + (R_L \| R_E)]$$

$$\frac{V_x}{I_x} = \frac{R_{eq} + r_x + (R_L \| R_E)}{1 + g_m (R_L \| R_E)}$$

$$R_{\pi o} = r_x \left\| \left(\frac{R_{eq} + r_x + (R_L \| R_E)}{1 + g_m (R_L \| R_E)} \right) \right.$$

$$R_{\pi o} \approx \frac{R_{eq} + r_x + (R_L \| R_E)}{1 + g_m (R_L \| R_E)}$$

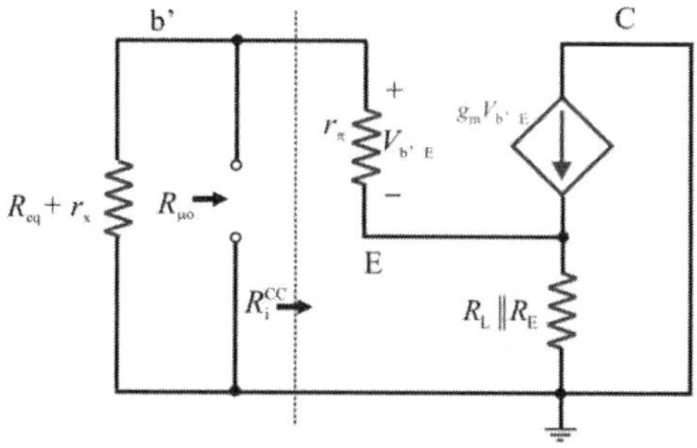

Figure 6.18 Circuit used to calculate the open-circuit time constant associated with C_μ.

and the time constant associated with C_π is $C_\pi R_{\pi o}$.

For C_μ the resistance across the capacitor terminals may be calculated from Figure 6.18

$$R_{\mu o} = (R_{eq} + r_x) \| R_{in}^{CC} = (R_{eq} + r_x) \| [r_\pi + (\beta_0 + 1)(R_L \| R_E)]$$
$$\approx R_{eq} + r_x$$

so the time constant associated with C_μ is $C_\mu R_{\mu 0}$.

The approximate high cutoff frequency is:

$$\omega_H \approx \frac{1}{\sum_{i=1}^{2} R_{io} C_i} = \frac{1}{R_{\pi o} C_\pi + R_{\mu o} C_\mu}$$

$$= \frac{1}{[R_{eq} + r_x + (R_L \| R_E)] \left(\frac{C_\pi}{1 + g_m (R_L \| R_E)} \right) + (R_{eq} + r_x) C_\mu}$$

If we assume that $R_L \| R_E << (R_{eq} + r_x)$ in the expression of $R_{\pi 0}$ we get the estimate:

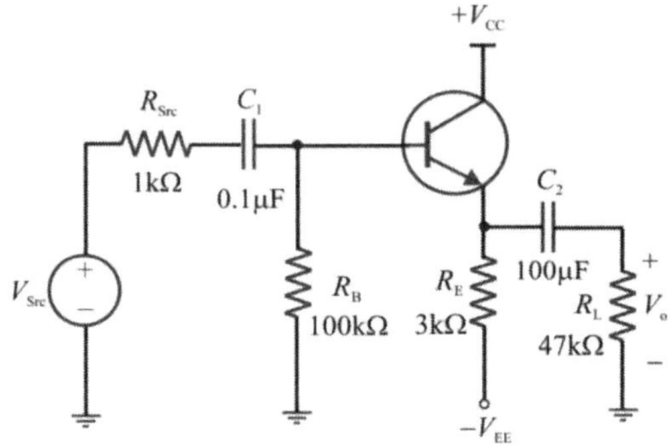

Figure 6.19 Common collector NPN BJT amplifier circuit.

$$\omega_H \approx \frac{1}{(R_{eq} + r_x)\left(\dfrac{C_\pi}{1 + g_m (R_L \| R_E)} + C_\mu\right)}$$

The DC circuit of the amplifier may be obtained from the CC amplifier circuit in Figure 6.19 by considering all the capacitors as open circuits and the AC sources equal to zero yielding the DC circuit shown in Figure 6.20.

Considering the B-E loop and assuming that the transistor is in the active mode, we have

$$I_B R_B + V_{BE} + I_E R_E - V_{EE} = 0$$

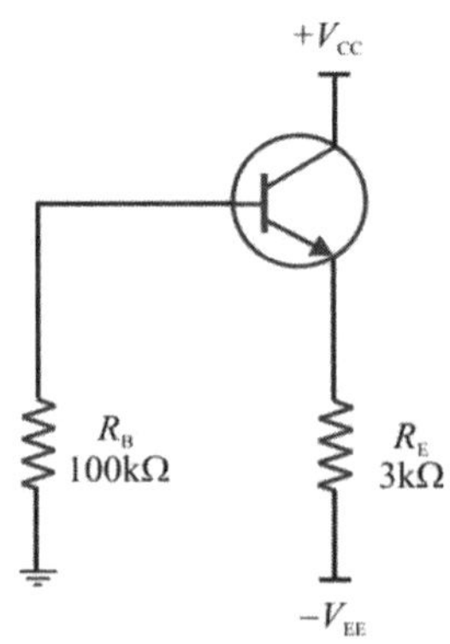

Figure 6.20 Common collector NPN BJT DC equivalent.

$$I_E\left(\frac{R_B}{1+\beta_0}+R_E\right)+V_{BE}-V_{EE}=0$$

$$I_E=\frac{V_{EE}-V_{BE}}{\frac{R_B}{1+\beta_0}+R_E}=\frac{5-0.7}{\frac{100}{101}+3}=1.08\text{mA}$$

$$I_C=\alpha_0 I_E\approx 1.08\text{mA}$$

In the C-E loop

$$V_{CC}=V_{CE}+I_E R_E+V_{EE}$$

so

$$V_{CE}=5-1.08\times 3+5=6.76\text{V}$$

Since $V_{CE}>V_{CESat}$, then the transistor is in the active region as assumed, with Q-point = (1.08 mA, 6.76V).

So

$$g_m=\frac{I_C}{V_T}=40I_C=43.2\,\text{mS}$$

$$r_\pi=\frac{\beta V_T}{I_C}=\frac{100\times 25\,\text{mV}}{1.08\,\text{mA}}=2315\,\Omega \tag{6.19}$$

$$R_{eq}=R_{Src}\|R_B=1\|100=\frac{100}{101}=0.99\,\text{k}\Omega \tag{6.20}$$

$$C_\pi=\frac{g_m}{2\pi f_T}-C_\mu=13.25\,\text{pF}$$

$$R_L\|R_C=47\|3=2.82\,\text{k}\Omega$$

$$\omega_H=\frac{1}{[R_{eq}+r_x+(R_L\|R_E)]\frac{C_\pi}{1+g_m(R_L\|R_E)}+(R_{eq}+r_x)C_\mu}$$

$$=\frac{1}{(438+620)\times 10^{-12}}=945\,\text{Mrad/sec}$$

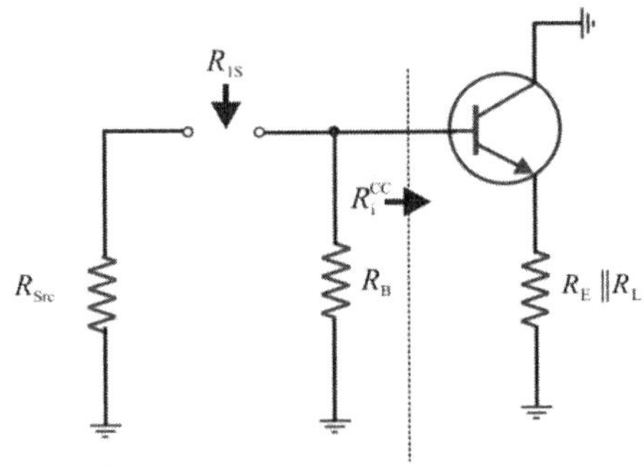

Figure 6.21 Circuit used to calculate the SCTC associated with C_1.

$$W = \frac{\omega_H - \omega_L}{2\pi} \approx \frac{\omega_H}{2\pi} = 150\,\text{MHz}$$

By comparing the bandwidth of the CC amplifier to that of the CE amplifier we find the bandwidth of the CC is much larger. This is because of the absence of the Miller's capacitance associated with the common emitter configuration.

6.4.3 Common-Collector Amplifier Approximate Low Cutoff Frequency ω_L

In this section we will determine low cutoff frequency of the CC amplifier shown in Figure 6.19. We employ the SCTC technique discussed above.

Example 6.2: We again assume that $\beta_0 = 100$ and $V_{CC} = V_{EE} = 5\,\text{V}$. The small signal circuit may be obtained with a short circuit on the DC supplies as shown in Figure 6.13. For C_1 the resistance seen across the capacitor terminals may be calculated from Figure 6.21:

$$R_{1S} = R_{Src} + (R_B \| R_i^{CC}) = R_{Src} + \{R_B \| [r_\pi + (\beta_0 + 1)(R_E \| R_L)]\}$$

and the time constant associated with C_1 is $C_1 R_{1S}$.

For C_2 the resistance across the capacitor terminals may be calculated with Figure 6.22

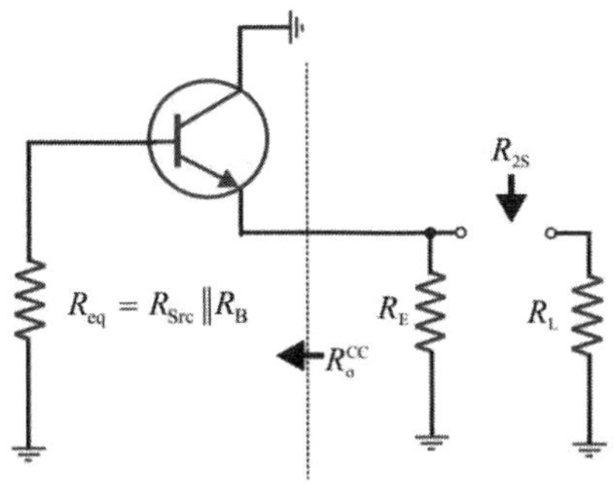

Figure 6.22 Circuit used to calculate the SCTC associated with C_2.

$$R_{2S} = R_L + (R_E \| R_o^{CC}) = R_L + \left(R_E \| \frac{r_\pi + R_{eq}}{\beta_0 + 1} \right)$$

$$R_{eq} = R_{Src} \| R_B$$

So the time constant associated with C_2 is $C_2 R_{2S}$.

The low cutoff frequency is calculated as:

$$\omega_L \approx \sum_{i=1}^{2} \frac{1}{R_{iS} C_i}$$
$$= \frac{1}{C_1 \left\{ R_{Src} + \left[R_B \| [r_\pi + (\beta_0 + 1)(R_E \| R_L)] \right] \right\}} + \frac{1}{C_2 \left[R_L + \left(R_E \| \frac{r_\pi + R_{eq}}{\beta_0 + 1} \right) \right]} \tag{6.21}$$

To calculate ω_L we need the value of r_π that was found at (6.19) at the DC operating point of the amplifier, as well as R_{eq} that was found at (6.20). Substitution of (6.19) and (6.20) in (6.21) yields

$$\omega_L = 133 + 0.21 = 133.2 \text{ rad/sec}$$

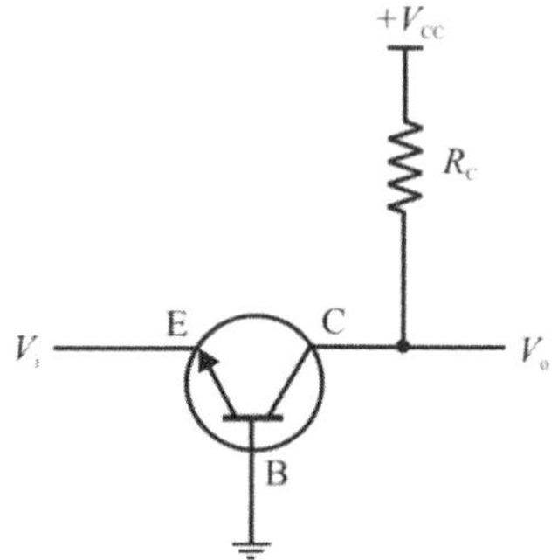

Figure 6.23 Basic NPN common base circuit (neglecting most biasing details).

6.5 Common Base

As its name suggests, in the common base or grounded base configuration, the base connection is common to both the input signal and the output signal with the input signal being applied between the base and the emitter terminals. The corresponding output signal is taken from between the base and the collector terminals with the base terminal grounded or connected to a fixed reference voltage point (which is an equivalent AC ground).

The common base topology is characterized by a large voltage gain, small current gain (less than one, so actually a gain loss), low input resistance, and high output resistance.

The common-base amplifier is typically used as a current buffer or voltage amplifier. In this circuit the emitter terminal of the transistor serves as the input, the collector the output and the base is common to both (see Figure 6.23).

This arrangement is popular in high-frequency amplifier applications (VHF and UHF) because its input capacitance does not suffer from the Miller effect, which degrades the bandwidth of the common emitter configuration, and because of the relatively high isolation between the input and output. This high isolation means that there is less feedback from the output back to the input, yielding better stability.

This configuration is useful as a current buffer since it has a current gain of approximately unity. A cascode topology ensues when a CE stage precedes a CB stage. The cascode has several useful properties and is discussed later.

6.5.1 Low-Frequency Characteristics

The circuit in Figure 6.24 can be represented by that in Figure 6.25 at low frequencies and under small-signal conditions; the hybrid π model for the BJT has

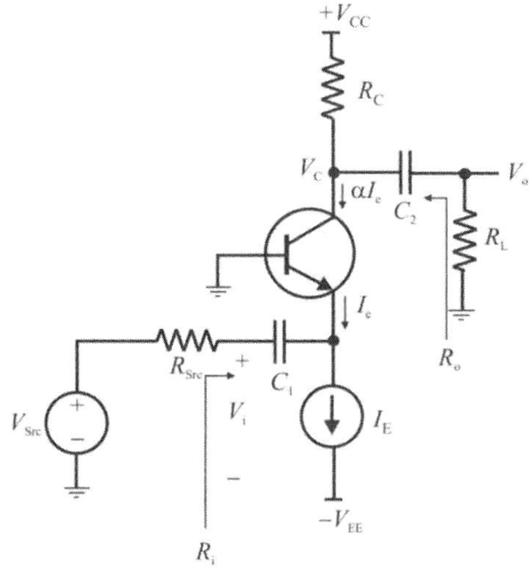

Figure 6.24 Common-base amplifier.

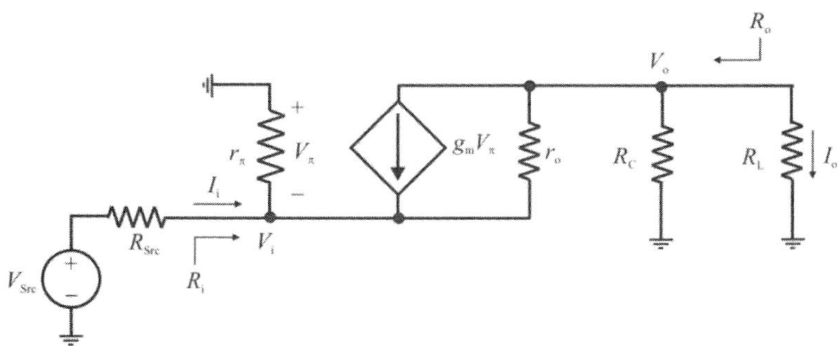

Figure 6.25 Low-frequency small-signal common-base amplifier.

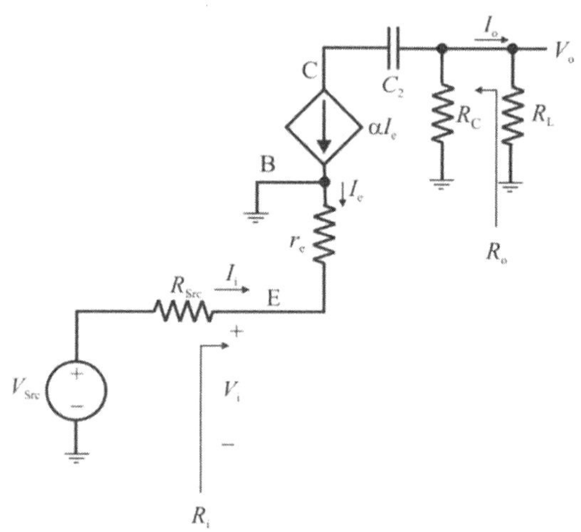

Figure 6.26 Common-base amplifier midfrequency AC equivalent based on the modified T-model of the BJT shown in Figure 6.2.

been employed. The input signal is represented by a source voltage source, V_{Src}, with a series resistance R_{Src} and the load is a resistor R_L. We use this circuit to derive the characteristics of the common-base amplifier.

A common-base amplifier circuit is shown in Figure 6.24. The small signal equivalent circuit for this amplifier is shown in Figure 6.26 ignoring r_o by assuming that $r_o >> R_C$ and $r_o >> R_C$.

6.5.1.1 Small-Signal Current Gain

The short-circuit current gain; that is, with no load at the output, is given by

$$A_I = \left.\frac{I_o}{I_i}\right|_{R_L=0} = \frac{r_\pi + \beta r_o}{r_\pi + (\beta+1)r_o} \approx 1 \qquad \beta >> 1 \tag{6.22}$$

6.5.1.2 Small-Signal Voltage Gain

The open-circuit voltage gain; that is, with an open circuit for the load, is given by

$$A_V = \left.\frac{V_o}{V_i}\right|_{R_L\to\infty} = \frac{(g_m r_o + 1)R_C}{r_o + R_C}$$

$$\approx g_m R_C \qquad\qquad r_o >> R_C \qquad (6.23)$$

6.5.1.3 Input Resistance

The input resistance is given by

$$R_i = \frac{V_i}{I_i} = \frac{(r_o + R_C \| R_L) r_e}{r_o + r_e + \dfrac{R_C \| R_L}{\beta + 1}}$$

$$\approx r_e \quad \left(\approx \frac{1}{g_m}\right) \qquad r_o >> R_C \| R_L \qquad (\beta >> 1) \qquad (6.24)$$

Since $r_e = V_T / I_E$, where V_T is the thermal voltage and I_E is the DC emitter current, the input resistance is extremely low. For example, for $V_T = 26$ mV and $I_E = 10$ mA, typical values, then $R_i = 2.6\Omega$. If I_E is reduced to increase R_i, there are other consequences like lower transconductance (g_m), higher output resistance, and lower β that also must be considered. A practical solution to this low input impedance problem is to place a common emitter stage at the input to form a cascode amplifier, which we discuss below.

Typically, small R_i is an undesirable trait, although in EW/RF systems where impedance values of 50 Ω are the norm, low impedance is of significant value due to the high powers involved, especially in PA discussions. Low impedances lead to lower losses.

6.5.1.4 Output Resistance

The output resistance is given by

$$R_o = \left.\frac{V_o}{-I_o}\right|_{V_S=0} = R_C \| \left\{ \left[1 + g_m (r_\pi \| R_{Src})\right] r_o + (r_\pi \| R_{Src}) \right\}$$

$$\approx \begin{cases} R_C \| r_o, & R_{Src} << r_e \\ R_C \| \left[(r_\pi \| R_{Src})(1 + g_m r_o)\right], & R_{Src} >> r_e \end{cases} \qquad (6.25)$$

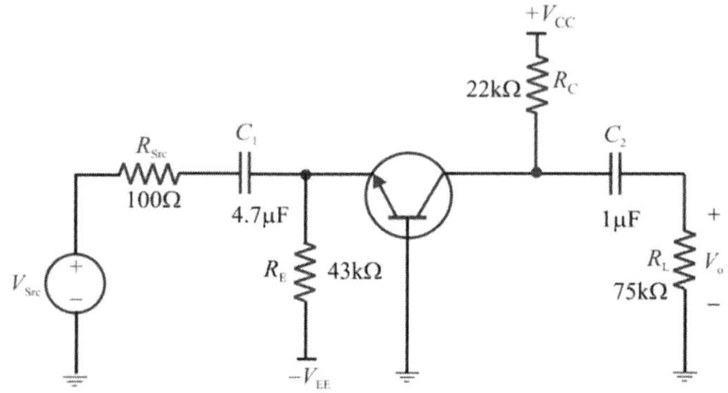

Figure 6.27 Common base amplifier circuit.

6.5.1.5 Summary

In summary, the seminal characteristics of the CB amplifier are:

- Very low input resistance
- A_V can be large, although R_{Src} is critical
- $A_{is} = \alpha$
- Output resistance can be large (dependent on R_C)

Depending on the source and load shown in Figure 6.25 and with a voltage source as signal, generally the overall voltage/current gain may be substantially less than the open/short-circuit gains shown above.

6.5.2 Common-Base Amplifier Low Cutoff Frequency, ω_L, Estimation

In this section the low cutoff frequency of the CB amplifier will be determined.

Example 6.3 The CB amplifier circuit in Figure 6.27 is the topology under consideration. We will assume that the numerical values are $\beta_0, V_{CC} = V_{EE} = 5\text{V}$.

The small-signal circuit may be obtained by shorting the DC supply as shown in Figure 6.28. For C_1 the resistance seen across the capacitor terminals may be calculated from Figure 6.29 as

$$R_{1S} = R_{Src} + (R_E \| R_i^{CB}) = R_{Src} + (R_E \| r_e)$$

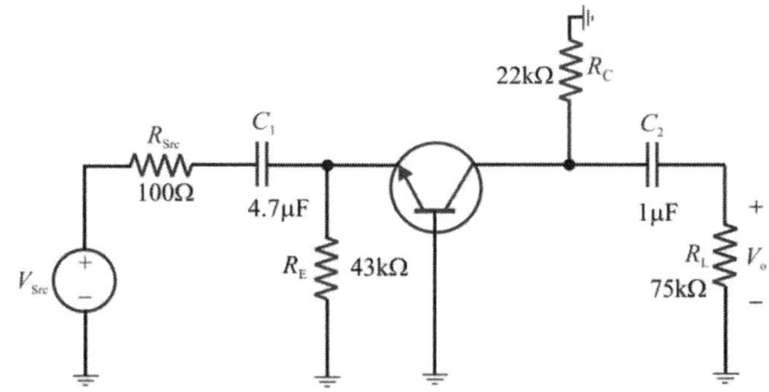

Figure 6.28 CB amplifier circuit used for small-signal analysis.

Figure 6.29 Circuit used to calculate the short-circuit time constant associated with C_1.

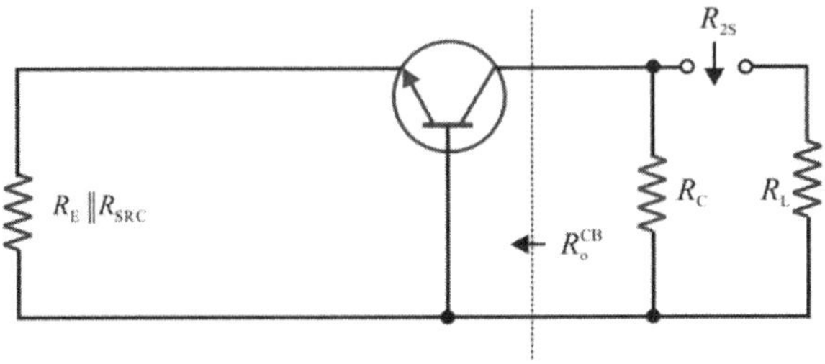

Figure 6.30 Circuit used to calculate the short-circuit time constant associated with C_2.

so the time constant associated with C_1 is $C_1(R_{Src} + R_E \| r_e)$.

For C_2 the resistance seen across the capacitor terminals may be calculated from Figure 6.30.

$$\begin{aligned} R_{2S} &= R_L + (R_C \| R_o^{CE}) \approx R_L + (R_C \| r_o) \\ &\approx R_L + R_C \end{aligned}$$

using Miller's theorem and assuming high gain. So the time constant associated with C_2 is $C_2(R_L + R_C)$.

The low cutoff frequency is estimated as:

$$\omega_L \approx \sum_{i=1}^{2} \frac{1}{R_{iS}C_i} = \frac{1}{C_1[R_{Src} + (R_E \| r_e)]} + \frac{1}{C_2(R_L + R_C)} \tag{6.26}$$

We can see that to calculate ω_L we need to find r_e; this value may be found from the DC operating point of the amplifier. The DC circuit of the amplifier may be obtained from the CE amplifier circuit in Figure 6.27 by considering all the capacitors are open circuits and the AC sources are zero. The DC circuit is shown in Figure 6.31.

From the B-E loop and assuming that the transistor is in the active mode we get

$$V_{BE} + I_E R_E - V_{EE} = 0$$

$$I_E = \frac{V_{EE} - V_{BE}}{R_E} = \frac{5 - 0.7}{43\text{k}} = 0.1\,\text{mA}$$

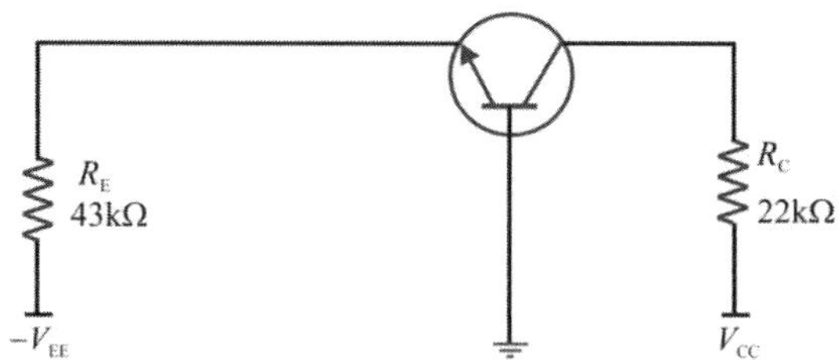

Figure 6.31 CB DC circuit.

$I_C = \alpha_0 I_E \approx 0.1\text{mA}$

For the C-E loop

$$V_{CC} = I_C R_C + I_E R_E + V_{EE}$$
$$V_{ce} = 5 - 0.1 \times (22 + 43) + 5 = 3.5\text{V}$$

The transistor is in the active region at Q-point = (0.1 mA, 3.5V) as we assumed since $V_{ce} > V_{CESat}$. Also

$$r_e = \frac{V_T}{I_E} = \frac{25\,\text{mV}}{0.1\,\text{mA}} = 250\,\Omega \tag{6.27}$$

Substituting (6.27) in (6.26) yields

$$\omega_L = 610 + 10 = 620\,\text{rad/sec}$$

6.5.3 Common-Base Amplifier High Cutoff Frequency, ω_H, Estimation

In this section the high cutoff frequency of the CB amplifier will be determined.

Example 6.4 We will derive an expression for the high cutoff frequency for the circuit in Example 6.3 shown in Figure 6.27. We will also calculate the amplifier bandwidth assuming that $V_{CC} = V_{EE} = 5\text{V}$ given the following transistor parameters: $f_T = 500$ MHz, $\beta_0 = 100$, $C_\mu = 0.5$ pF, and $r_x = 250\,\Omega$.

The CB amplifier is redrawn for small-signal analysis as shown in Figure 6.32 by replacing the transistor in Figure 6.28 by its high-frequency small-signal model. All the coupling and

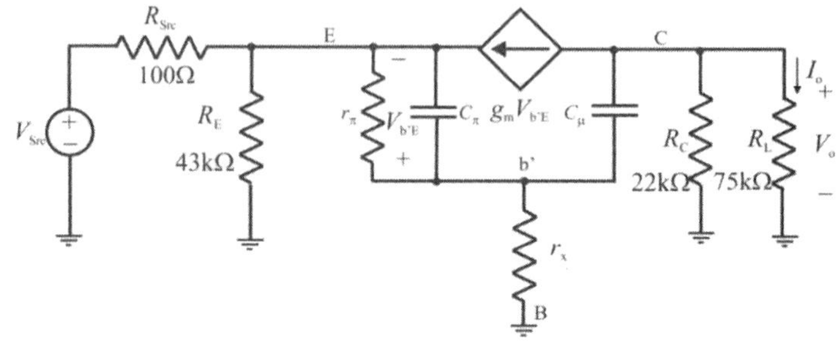

Figure 6.32 The CB amplifier in Figure 6.27 redrawn with the transistor replaced by its high-frequency small-signal model.

bypass capacitors are considered short circuit at high-frequency as well.

The small-signal model can be simplified by using Thevenin's theorem. The simplified circuit is shown in Figure 6.33.

$$V_{eq} = V_{Src}\frac{R_E}{R_{Src} + R_E}$$

$$R_{eq} = \frac{R_{Src}R_E}{R_{Src} + R_E}$$

For C_π the resistance seen across the capacitor terminals may be calculated from Figure 6.34(a). The calculation may be simplified by considering the circuit in Figure 6.34(b).

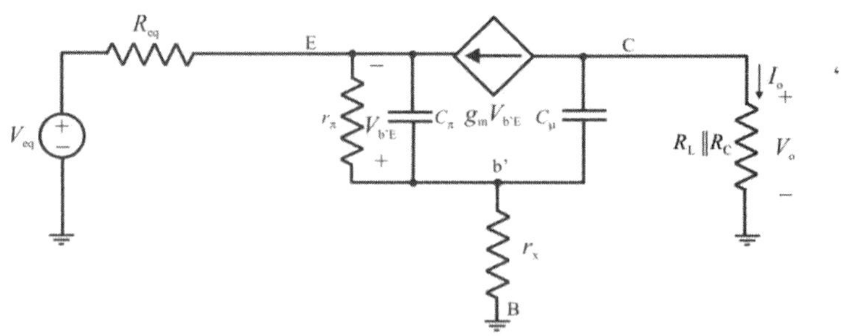

Figure 6.33 Simplified circuit to calculate the high-frequency response of the CB amplifier.

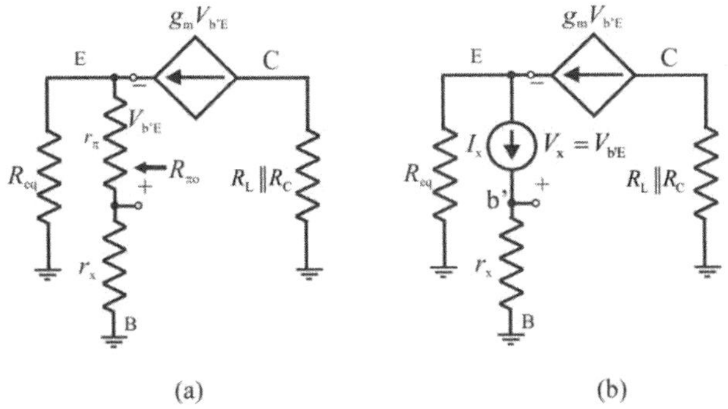

Figure 6.34 Circuit used to calculate the open-circuit time constant associated with C_π.

$$R_{\pi o} = r_\pi \| \frac{V_x}{I_x}$$

$$V_x = V_{b'} - V_E = I_x r_x - (g_m V_x - I_x) R_{eq}$$

$$V_x (1 + g_m R_{eq}) = I_x (R_{eq} + r_x)$$

$$\frac{V_x}{I_x} = \frac{R_{eq} + r_x}{1 + g_m R_{eq}}$$

$$R_{\pi o} = r_\pi \| \frac{R_{eq} + r_x}{1 + g_m R_{eq}}$$

Assuming that $\beta_0 >> 1$ and $r_x << r_\pi$

$$R_{\pi o} = \frac{R_{eq} + r_x}{1 + g_m R_{eq}}$$

so the time constant associated with C_π is $C_\pi R_{\pi 0}$.

For C_μ the resistance seen across the capacitor terminals may be calculated from Figure 6.35. Note that split-source transformation is used to ease the calculation. Assuming that $\beta_0 >> 1$ and $r_x << r_\pi$. These assumptions imply that $I_B = 0$ or $I_E = I_C = g_m V_{b'E}$ as well as all the current sourced from I_x that shunt r_x will pass in r_x only.

For the loop containing r_x, r_π, and R_{eq}

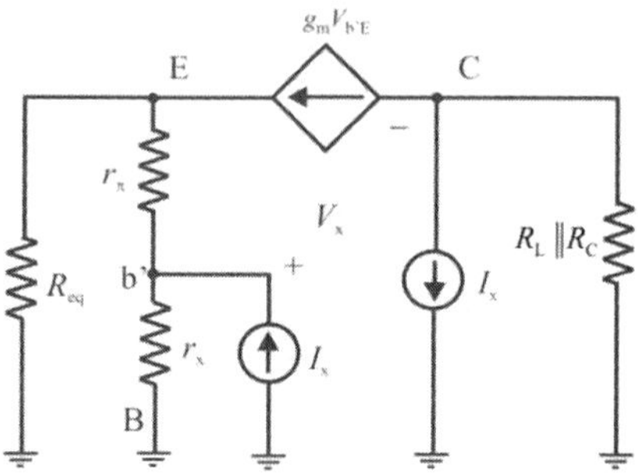

Figure 6.35 Circuit used to calculate the open-circuit time constant associated with C_μ.

$$V_{b'} = I_x r_x \approx V_{b'E} + g_m V_{b'E} R_{eq} = V_{b'E}(1 + g_m R_{eq})$$

$$V_{Eb'} = \frac{I_x r_x}{1 + g_m R_{eq}}$$

Summing the current at the collector node

$$V_C = -(g_m V_{Eb'} + I_x)(R_L \| R_C) = -I_x \left(1 + \frac{g_m r_x}{1 + g_m R_{eq}}\right)(R_L \| R_C)$$

$$R_{\mu o} = \frac{V_x}{I_x} = \frac{V_{b'} - V_C}{I_x} = r_x \left(1 + \frac{g_m (R_L \| R_C)}{1 + g_m R_{eq}}\right) + R_L \| R_C$$

so the time constant associated with C_μ is $C_\mu R_{\mu 0}$.

The high cutoff frequency is

$$\omega_H = \frac{1}{\sum_{i=1}^{2} R_{io} C_i} = \frac{1}{R_{\pi o} C_\pi + R_{\mu o} C_\mu}$$

$$= \frac{1}{r_x \left[\frac{C_\pi}{1 + g_m R_{eq}} \left(1 + \frac{R_{eq}}{r_x}\right) + C_\mu \left(1 + \frac{g_m R_L \| R_C}{1 + g_m R_{eq}}\right) \right] + C_\mu R_L \| R_C}$$

Neglecting the first term which is of order $1/\omega_T$ and hence the last term is dominant.

$$\omega_H \approx \frac{1}{C_\mu (R_L \| R_C)}$$

$$R_L \| R_C = 75 \| 22 = 17\text{k}\Omega$$

$$\omega_H = 117.6\text{Mrad/sec}$$

$$W = \frac{\omega_H - \omega_L}{2\pi} \approx \frac{\omega_H}{2\pi} = 18.73\text{MHz}$$

By comparing the bandwidth of the CB amplifier to that of the CE amplifier we see that the bandwidth of the CB is much higher. This is due to the absence of the Miller's capacitance associated with the common emitter configuration.

6.5.4 Active Loads for the CB Amplifier

For voltage amplification, the range of allowed output voltage swing in this amplifier is tied to its voltage gain when a resistor load R_C is employed, as in Figure 6.24. That is, as indicated by (6.23), large voltage gain requires large R_C, and that in turn implies a large DC voltage drop across R_C. For a given supply voltage, the larger this drop, the less output swing is allowed before saturation of the transistor occurs, with resultant distortion of the output signal. To avoid this situation, an active load can be used; for example, a current mirror. If this choice is made, the value of R_C in (6.23) is replaced by the small-signal output resistance of the active load, which is generally at least as large as the r_o of the active transistor in Figure 6.24. On the other hand, the DC voltage drop across the active load is a fixed low value (the so-called *compliance voltage* of the active load), much less than the DC voltage drop incurred for comparable gain using a resistor R_C.

However, active load or not, large AC gain still is coupled to large AC output resistance, which leads to poor voltage division at the output except for large loads $R_L >> R_o$.

For use as a current buffer, as we can see from (6.22) gain is not affected by R_C, but output resistance is [see (6.25)]. Because of the current division at the output, it is desirable to have an output resistance for the buffer much larger than the load R_L being driven so large signal currents can be delivered to a load.

6.5.5 Common-Base Amplifiers

Because the input impedance is so low, most signal sources have larger source impedance than the input impedance of the common-base amplifier, R_i. The consequence is that the source essentially delivers a *current* to the input rather than a voltage, even if it is a voltage source, given by $I_i = V_{Src} / R_{Src}$. If the output signal is also a current, the amplifier is a current buffer and delivers the same current as its input. If the output is taken as a voltage, the amplifier is a transresistance amplifier, and delivers a voltage dependent on the load impedance; for example $V_o = I_i R_L$ for a resistor load R_L much smaller in value than the amplifier output resistance R_o. That is, the voltage gain in this case is:

$$V_o = I_o R_L \approx I_i R_L = \frac{V_{Src}}{R_{Src}} R_L$$

so that

$$A_V = \frac{V_o}{V_i} \approx \frac{R_L}{R_{Src}} \tag{6.28}$$

We can note that for source impedances such that $R_{Src} >> r_E$, the output impedance approaches $R_o = R_C \| [g_m (r_\pi \| R_{Src}) r_o]$.

For a very low impedance source, the common-base amplifier functions as a voltage amplifier. In this case, when $R_{Src} << r_E$ and $R_L << R_o$, the voltage gain becomes:

$$A_V = \frac{V_o}{V_{Src}} = \frac{R_L}{r_E} \approx g_m R_L \tag{6.29}$$

where $g_m = I_C / V_T$ is the transconductance. Notice that for low source impedance, $R_o = r_o \| R_C$.

The inclusion of r_o in the hybrid-π model predicts reverse transmission from the amplifiers output to its input, that is the amplifier is *bilateral* (as opposed to unilateral where signals flow only one way). One consequence of this is that the input/output impedance is affected by the load/source termination impedance, respectively. For example, the output resistance, R_o, may vary over the range

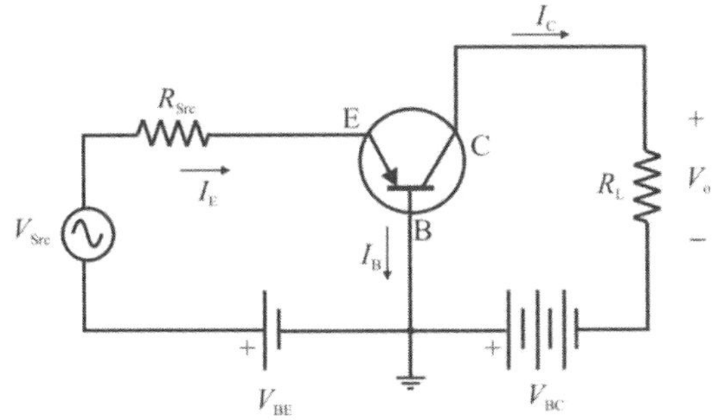

Figure 6.36 CB amplifier.

$r_o \| R_C \leq R_o \leq (\beta+1)r_o \| R_C$ depending on the source resistance, R_{Src}. The amplifier can be approximated as unilateral when neglect of r_o can be accommodated (valid for low gains and low to moderate load resistances), of course simplifying the analysis. This approximation often is used in discrete amplifier manifestations, but generally less accurate in RF circuits, and in integrated circuit designs where active loads are normally used.

6.5.5.1 Voltage Amplifier

The circuit is shown in Figure 6.36 for the case when the common-base circuit is used as a voltage amplifier. When the source impedance is low ($R_S << r_E$) the output resistance is large, at least $R_C \| r_o$. In a voltage amplifier a large output resistance is normally undesirable, as it leads to poor voltage division between the output resistance and the load at the output. Nonetheless, there is still considerable voltage gain even for small loads: according to (6.23), with $R_{Src} = r_E$ the gain is $A_V = g_m R_L / 2$. For larger source impedances, the gain is determined more by the resistor ratio R_L / R_{Src}, and less so by the transistor properties, which is a considerable advantage where insensitivity to temperature or transistor variations is important (which is most of the time).

The amplifier is transitional between voltage amplifier and current buffer when the R_{Src} value is in the vicinity of r_E. For $R_{Src} >> r_E$ the driver representation as a Thévenin source should be replaced by a Norton source. The common-base circuit stops behaving like a voltage amplifier and behaves like a current follower, as discussed next.

The input current flowing into the emitter is quite large as it's the sum of the base current and collector current

$$I_E = I_B + I_C$$

therefore, the collector current output is less than the emitter current input, resulting in a current gain for this type of circuit of unity or less. In other words, the common base configuration attenuates the input signal (see Figure 6.36).

This type of amplifier configuration is a noninverting voltage amplifier circuit, in that the signal voltages V_i and V_o are in-phase. This type of transistor arrangement is not very common due to its unusually high voltage gain characteristics. Its output characteristics represent that of a forward-biased diode while the input characteristics represent that of an illuminated photodiode.

Also, this type of BJT configuration has a high ratio of output to input resistance or more importantly load resistance (R_L) to input resistance (R_i) creating the notion of *resistance gain*. The voltage gain (A_v) for a common-base configuration is therefore given as

$$A_V = \frac{V_o}{V_i} = \frac{I_C R_L}{I_E R_i} = \alpha \frac{R_L}{R_i} \tag{6.30}$$

where: $\alpha = I_C / I_E$ is the current gain and R_L / R_i is the resistance gain.

The CB circuit is generally only used in single-stage amplifier circuits such as RF amplifiers due to its very good high-frequency response.

6.5.5.2 Current Follower

A low- to mid-frequency CB amplifier used as a current follower is shown in Figure 6.37. The input signal is provided by an AC Norton source (current I_{Src}, Norton resistance R_{Src}) at the input, and the circuit has a resistor load R_L at the output. We assume, which is normally the case, that the load is active with an infinite output resistance so R_C is omitted.

6.5.6 Common-Base Amplifier High-Frequency Response

The AC model of a BJT CB amplifier is shown in Figure 6.38. Since there are no capacitors between the input and output nodes, there is no Miller effect to take into consideration. Therefore, the poles are at very high frequencies since the values of the capacitors that are present are quite small.

We will assume that the base current is small so the voltage across r_b is also small. Then r_b can be ignored so that $V_E \approx -V_\pi$. The poles are located at

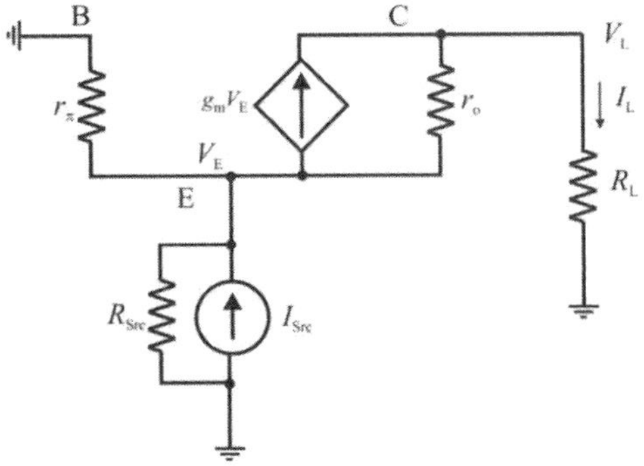

Figure 6.37 Common-base circuit with Norton driver; R_C is omitted because an active load is assumed with infinite small-signal output resistance.

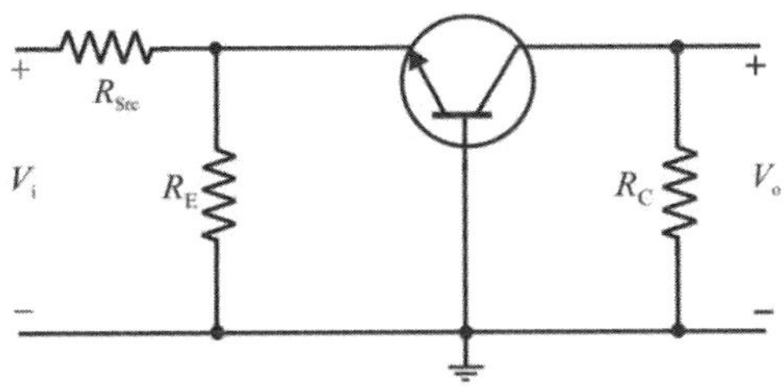

(a)

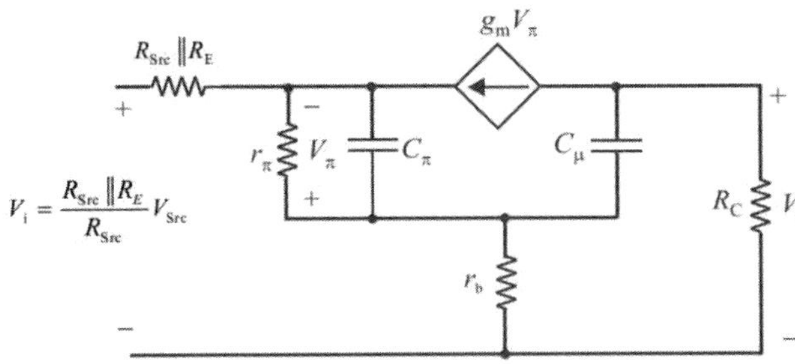

(b)

Figure 6.38 CB equivalent circuits: (a) midband AC equivalent, (b) high-frequency equivalent.

$$\omega_{p1} = \frac{r_\pi + (1 + g_m r_\pi)(R_{Src} \| R_E)}{C_\pi r_\pi (R_{Src} \| R_E)} \approx \frac{\beta_F}{C_\pi r_\pi} = \omega_t \quad \text{and} \quad \omega_{p2} = \frac{1}{C_\mu R_C} \tag{6.31}$$

Both of these poles are located at very high frequencies with the result that CB amplifiers are seldom the bandwidth-limiting elements in a multistage amplifier. They are normally used for impedance transformation for other kinds of RF amplifiers.

6.6 Cascode Amplifiers

6.6.1 BJT Cascode

As we discussed, FET amplifiers can provide extremely high input impedances. Another circuit configuration that provides high input impedances is the two (or more) BJT cascode configuration. The cascode amplifier we discuss here consists of a CE input stage with a CB output stage, illustrated in Figure 6.39. Of course, FETs can be used in lieu of BJTs. Together, the two transistors overcome some of the performance limitations of either the CE or CB configurations.

The voltage gain for this circuit is given by

$$A_V = \frac{V_o}{V_i} = -\alpha_1 \alpha_2 \frac{R_L}{R_B / (\beta_1 + 1) + r_{e1} + R_E} \tag{6.32}$$

where r_{e1} is the Q_1 dynamic emitter resistance. This gain is the same as a CE amplifier except for the additional α_2 loss of Q_2. The advantage of the cascode is that when the output resistance, r_o, of Q_2 higher than for the CE. For a BJT, this may be insignificant at low frequencies but is important for RF applications. The CB isolates the collector-base capacitance, C_{BC} (or C_μ of the hybrid-π BJT model), from the input by returning it to a dynamic ground at V_B. At the output, R_L is shunted by C_{BC} only, without a Miller-effect multiplier. The Q_1 collector voltage is also nearly constant and C_{BC} of Q_1 appears from the input with essentially no Miller effect. The C_{BC} of the CE has, in the cascade, been isolated from the output and the Miller effect eliminated. This is its primary advantage and is why it is used in fast amplifiers and RF stages.

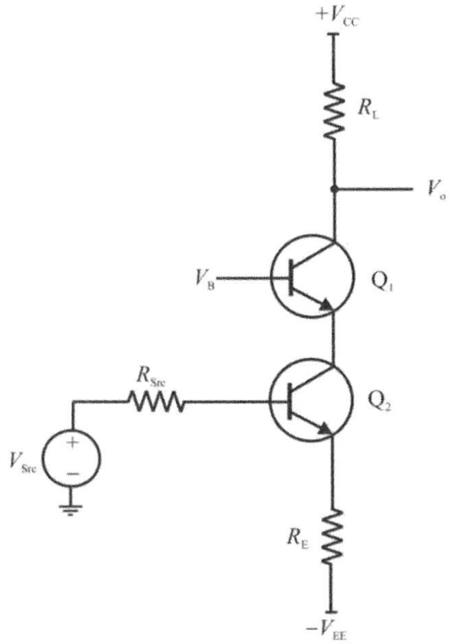

Figure 6.39 Cascode amplifier.

Small-signal analysis of the circuit shown in Figure 6.40 yields the following results:

$$R_S = r_{\pi 1} \| (r_{Bb'1} + R_{Src})$$

$$R_{\pi 1} = R_S$$

$$R_{\mu 1} = R_S(1 + g_{m1}R_{c1}) + R_{c1}$$

$$R_{c1} = r_{o1} \| \left[r_{e2} \left(\frac{r_{o2} + R_L}{r_{o2} + R_L / (\beta_2 + 1)} \right) \right]$$

$$\tau_H = C_{\pi 1}R_{\pi 1} + C_{\mu 1}R_{\mu 1} + (C_{s1} + C_{\pi 2})R_{c1} + (C_L + C_{cs2} + C_{\mu 2})(R_L \| R_o)$$

$$R_o \approx \beta_2 r_{o2}$$

$$f_H \approx \frac{1}{2\pi\tau_H}$$

$$A_M = -\frac{r_\pi}{r_\pi + r_{Bb'1} + R_{Src}} g_m (\beta r_o \| R_L)$$

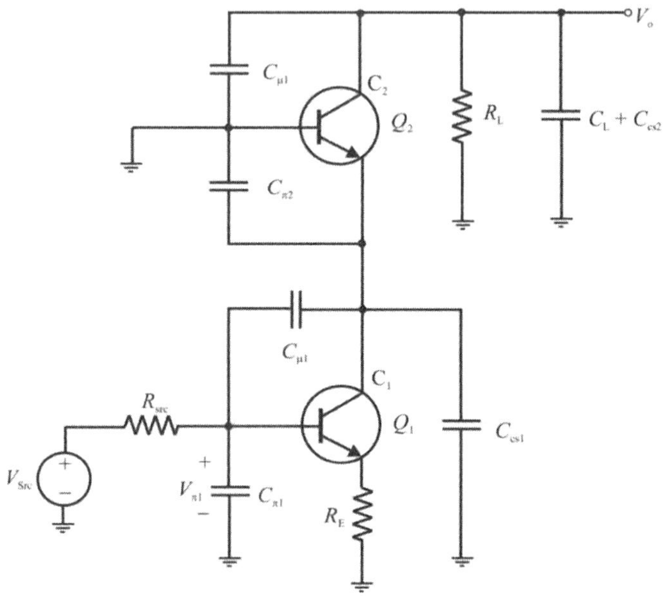

Figure 6.40 Cascode BJT amplifier.

Another advantage of the cascode over a CE is that the right-half plane zero that causes preshoot in a step response is also eliminated. In the CE configuration alone, C_{BC} provides a parallel, passive path from input to output. When a step is applied as V_i, it is coupled to the output node of the CE collector uninverted and precedes the amplified and inverted step as preshoot, but not for the cascode. Consequently, the step response of the cascode is not only faster, but cleaner than for the CE alone.

The cascode incremental output resistance is (with infinite r_o) simply R_L, and the incremental input resistance is

$$r_i = R_B + (\beta_1 + 1)(r_{e1} + R_E) \tag{6.33}$$

The resistance in the emitter branch of the input circuit is referenced to the base as $\beta + 1$ times the emitter side resistance and adds to the base resistance in series with it.

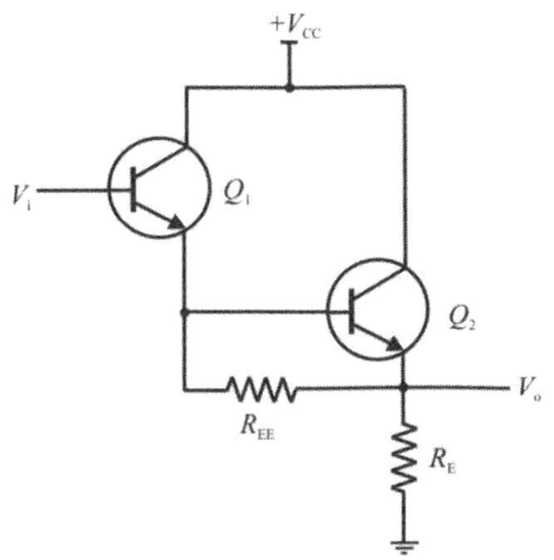

Figure 6.41 Darlington pair.

6.7 Darlington Pair

The Darlington configuration of an amplifier consists of two transistors where the second transistor is the emitter load of the first (see Figure 6.41 for the BJT NPN configuration). In essence the emitter current of the input transistor Q_1 becomes the base current of the output transistor, Q_2. The Darlington pair acts like one transistor with a beta substantially higher than either transistor alone, thereby forming a current amplifier. The overall beta is given by

$$\beta = \beta_1\beta_2 + \beta_1 + \beta_2 \tag{6.34}$$

They are used where high output currents are needed. In addition, the input impedance of the Darlington configuration is quite high.

The Darlington configuration has more phase shift at higher frequencies compared to a single transistor so therefore can more easily become unstable with negative feedback.

Switching of the second transistor is often slow, limiting the usefulness at RF. This is because the first transistor can't suddenly inhibit the base current of the second. To resolve this problem a resistor is sometimes connected between the two

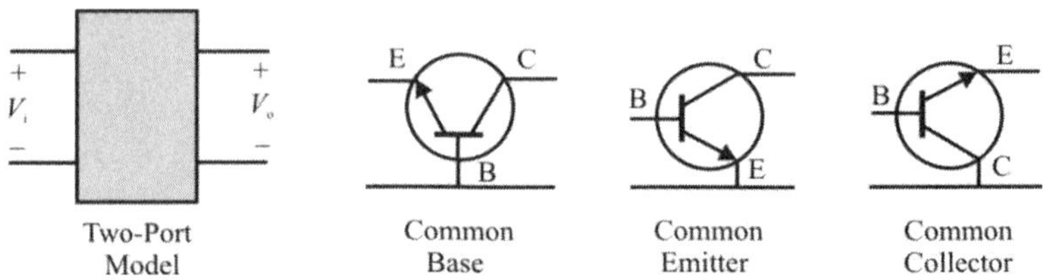

Figure 6.42 BJT summary.

emitters ((R_{EE} in Figure 6.41). This resistor acts as a low impedance discharge path for the charge accumulated on the base-emitter junction of Q_2. R_{EE} is not present when the slow switching speed is not an issue. Darlington pairs are available as single packages, usually with the resistor included.

One drawback of the Darlington pair is that the quiescent leakage current of Q_1 is amplified by Q_2. This causes significant steady-state leakage. As such, Darlington configurations rarely consist of more than two transistors.

6.8 Concluding Remarks

6.8.1 Bipolar Transistor Amplifier Summary

To summarize, the behavior of the bipolar transistor in each of the configurations shown in Figure 6.42 is very different and produces different circuit characteristics with regards to input impedance, output impedance, and gain whether this is voltage gain, current gain or power gain. These characteristics are summarized in Table 6.3.

Table 6.3 BJT Amplifier Configuration Characteristics

Characteristic	Common Base	Common Emitter	Common Collector
Input impedance	Low	Medium	High
Output impedance	Very high	High	Low
Phase shift	0°	180°	0°
Voltage gain	High	Medium	Low
Current gain	Low	Medium	High
Power gain	Low	Very Highh	Medium

References

[1] Gonzalez, G., *Microwave Transistor Amplifiers Analysis and Design,* 2nd Ed., Upper Saddle River, NJ: Prentice Hall, 1996.
[2] Pozar, D. M., *Microwave Engineering,* 3rd Ed., New York: Wiley, 2006.
[3] Razavi, B., *RF Microelectronics,* Upper Saddle River, NJ: Prentice Hall, 1998.
[4] Krauss, H. L., C. W. Bostian, F. H. Raab, *Solid State Radio Engineering*, New York: John Wiley & Sons, 1980.
[5] White, J. F., *High-frequency Techniques*, New York: John Wiley & Sons, 2004.

Chapter 7

MOSFET Amplifiers

7.1 Introduction

We examine MOSFET RF amplifiers in this chapter. We explore the basic characteristics of these amplifiers focusing mostly on the frequency response characteristics. We also discuss the impedance characteristics as well as the gain [1 – 3].

The structure of this chapter is as follows. First the RF models of MOSFETs are presented. These models are used to approximate the characteristics of these devices at high-frequency. We then examine the three circuit topologies in order: common source (CS), common drain (CD, and common gate (CG). At the end these topologies are compared.

7.2 Long- and Short-Channel MOSFETs

MOSFET devices are frequently categorized as short channel or long channel. These are depicted in Figure 7.1.

The long-channel MOSFET is defined as a device with width and length long enough so that edge effects from the four sides can be neglected. Channel length L must be much greater than the sum of the drain and source depletion widths.

A short-channel MOSFET is defined as a device with width and length short enough so that the edge effects cannot be ignored. The channel width L is comparable to the depletion widths associated with drain and source. Considerably more investigations have been explored for long-channel devices than short-channel so substantially more results are available.

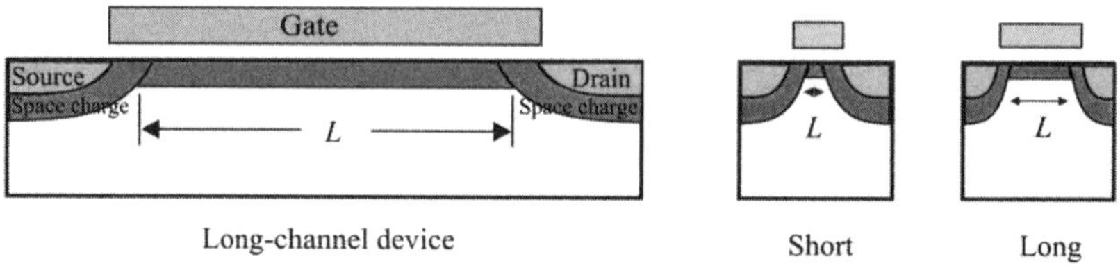

Figure 7.1 Short-channel MOSFET.

7.3 MOSFET High-Frequency Model

7.3.1 High-Frequency MOSFET Model

The MOSFET high-frequency hybrid-π model is shown in Figure 7.2. A simplified high-frequency model based on Figure 7.2 is shown in Figure 7.3. For this simplification, the source and body terminals are shorted, C_{GD} plays an important role in the amplifier response primarily because of the Miller effect, and C_{DB} is neglected. The circuit including the Miller effect capacitance is shown in Figure 7.4.

7.3.2 Unity-Gain Frequency

Smaller parasitic capacitances C_{gs} and C_{gd} are desirable for higher unity-gain frequency. For the MOSFET it is given by

$$f_t = \frac{1}{2\pi} \frac{g_m}{C_{gs} + C_{gd}}$$

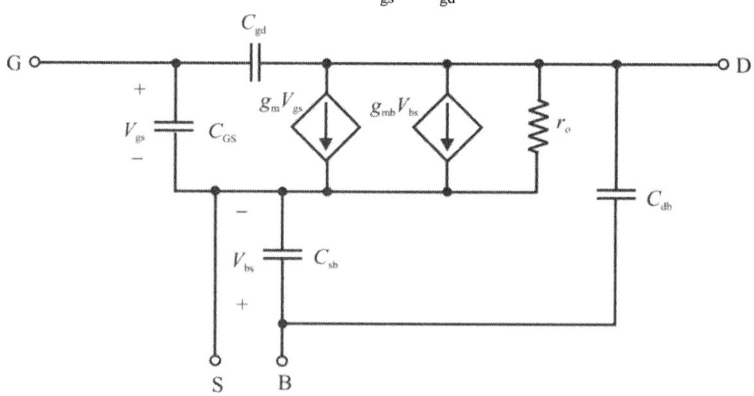

Figure 7.2 MOSFET high-frequency model.

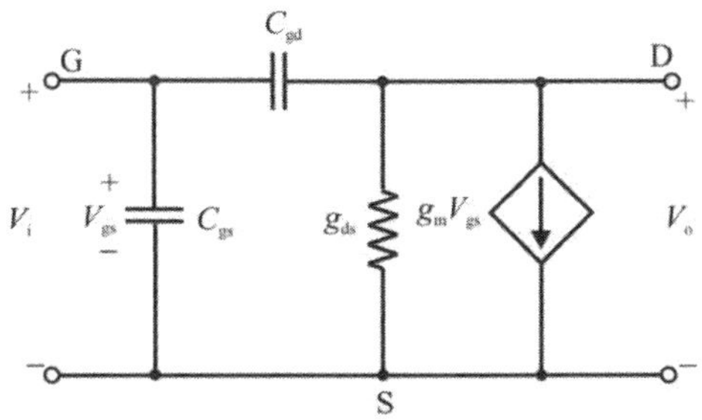

Figure 7.3 MOSFET simplified high-frequency model. Assumes $r_d = r_s = 0$.

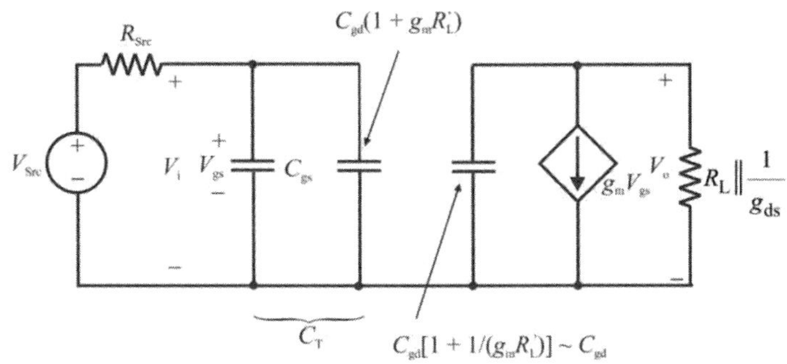

Figure 7.4 MOSFET high-frequency amplifier with Miller effect capacitance.

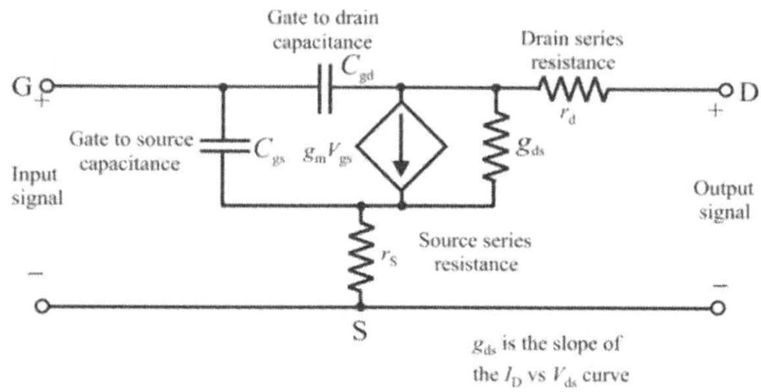

Figure 7.5 MOSFET complete high-frequency model. g_{ds} is given by the slope of the I_D versus V_{ds} curve at the quiescent point and $g_{ds} = 1/r_o$. The low- and mid-frequency model is obtained by removing the capacitors.

and $f_t \propto 1/L^2$. We can see that f_t is strongly influenced by the channel length, L, because of this relationship. A higher unity-gain frequency can be achieved for a given MOSFET by increasing the bias current or the overdrive voltage.

7.3.3 Common Source

At high frequencies, the MOSFET behaves as illustrated in Figure 7.5. There are two significant capacitances that are present that limit the high-frequency performance: (1) C_{gs}, the gate to source capacitance, and (2) C_{gd}, the gate to drain capacitance. The model shown in Figure 7.5 includes the drain and source resistances. If we assume that $r_d = r_s = 0$ then this model is that shown in Figure 7.4.

We assume a dominant pole introduced by C_{gd} in parallel with C_{gs} and get a pole in the amplifier transfer function as

$$\omega_H \approx \frac{1}{\left\{C_{gs} + C_{gd}[1 + g_m(R_L \| \frac{1}{g_{ds}})]\right\} R_S} = \frac{1}{C_T R_S}$$

We see that the Miller effect has a significant effect on the input side of the amplifier, whereas there is little impact on the output side.

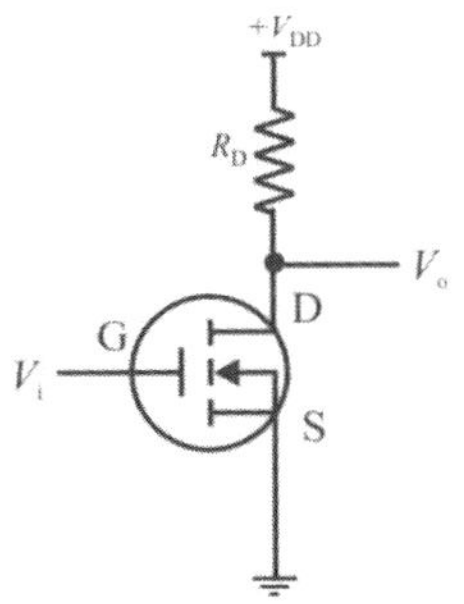

Figure 7.6 Basic N-channel MOSFET common source circuit (neglecting most biasing details).

7.4 Common Source

Like the common emitter BJT amplifier configuration, the common source MOSFET amplifier topology is perhaps the most prolific, for reasons discussed in this section.

In a common source MOSFET amplifier, the source is shared between the input and output circuits and it is held at zero (AC) potential. The output of the amplifier is taken from the drain of the MOSFET, which is connected to bias voltage through a drain resistor, R_D. The input is provided to the gate of the FET (see Figure 7.6). Such a configuration is typically used as a voltage amplifier. At low-frequency, the input impedance is very high due to the essentially zero gate current draw from the previous stage.

Adding the resistance of a source, a biasing resistor to the gate over which to develop the MOSFET input voltage (rather than let it float to whatever value it wants to), and a resistor to represent the load impedance at the drain produces the circuit shown in Figure 7.7.

The high-frequency model which adds the device input capacitance C_{gs} and the feedback capacitor C_{gd} is shown in Figure 7.8.

7.4.1 CS MOSFET Characteristics

The small-signal, low and midfrequency common source model of a MOSFET amplifier is shown in Figure 7.7 which is based on the simplified hybrid-π model shown in Figure 7.2. Using this model, the circuit is shown in Figure 7.8 with the characteristics are described in this section.

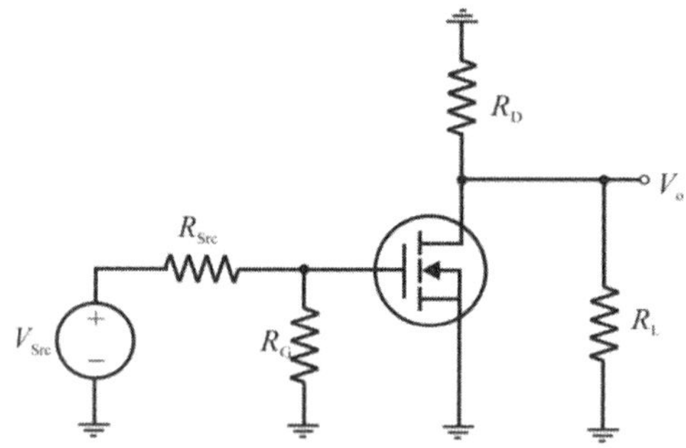

Figure 7.7 CS equivalent circuits with one form of biasing.

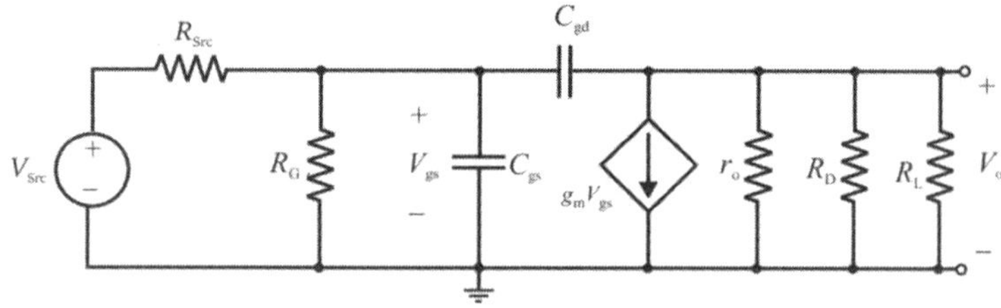

Figure 7.8 CS equivalent circuit of the circuit shown in Figure 7.7.

7.4.1.1 Midband Voltage Gain

The midband voltage gain is given by

$$A_{\mathrm{M}} = -\frac{R_{\mathrm{G}}}{R_{\mathrm{G}} + R_{\mathrm{Src}}} g_{\mathrm{m}} R'_{\mathrm{L}}$$

where

$$R'_{\mathrm{L}} = r_{\mathrm{o}} \| R_{\mathrm{D}} \| R_{\mathrm{L}}$$

and where its frequency response is given by

$$A_{\mathrm{V}} = \frac{V_{\mathrm{o}}}{V_{\mathrm{Src}}} = -\frac{A_{\mathrm{M}}\left(1 - s\frac{C_{\mathrm{gd}}}{g_{\mathrm{m}}}\right)}{1 + s\xi + s^2\zeta} \tag{7.1}$$

$$\xi = \left(\frac{R_{\mathrm{Src}} R_{\mathrm{G}}}{R_{\mathrm{Src}} + R_{\mathrm{G}}}\right)\left[\left(\frac{R'_{\mathrm{L}}}{R_{\mathrm{Src}}} + \frac{R'_{\mathrm{L}}}{R_{\mathrm{G}}} + 1 + g_{\mathrm{m}} R'_{\mathrm{L}}\right) C_{\mathrm{gd}} + C_{\mathrm{gs}}\right]$$

$$\zeta = \frac{R'_{\mathrm{L}} R_{\mathrm{Src}} R_{\mathrm{G}}}{R_{\mathrm{Src}} + R_{\mathrm{G}}} C_{\mathrm{gd}} C_{\mathrm{gs}}$$

The amplifier high-frequency cutoff frequency is given by

$$\omega_{\mathrm{H}} = \left\{\left[R_{\mathrm{Src}} \| R_{\mathrm{G}}\right]\left[\left(1 + g_{\mathrm{m}} R'_{\mathrm{L}}\right) C_{\mathrm{gd}} + C_{\mathrm{gs}}\right]\right\}^{-1}$$

The CS amplifier has one zero and two poles at higher frequencies. The amplifier gain falls off at frequencies beyond midband. The amplifier bandwidth is defined by the 3-dB frequency that is typically evaluated by the dominant pole (that is, the lowest frequency) in the transfer function. The dominant pole is normally determined by the Miller effect capacitance C_{μ} given by

$$C_{\mu} = C_{\mathrm{gd}}(1 + g_{\mathrm{m}} R_{\mathrm{L}})$$

7.4.1.2 Short-Circuit Current Gain

Without R_G, $I_i = 0$, and this is a meaningless parameter. With R_G present all of the input current flows through it, so $R_i = R_G$.

7.4.1.3 Open-Circuit Voltage Gain

The midband impedance

$$A_V = -\frac{g_m R'_L}{R_G + R_{Src}}$$

$$\approx -\frac{R_D}{R_{Src}} \qquad g_m R_{Src} >> 1 \tag{7.2}$$

7.4.1.4 Input Resistance

Without R_G, $I_i = 0$ and, the input resistance is given by

$$R_i = \infty \tag{7.3}$$

7.4.1.5 Output Resistance

The output resistance is given by

$$R_o = R_D \tag{7.4}$$

7.4.2 Source Degeneration.

Adding a source resistance increases the stability of the CS MOSFET amplifier. This topology is shown in Figure 7.9. We use the hybrid-π model to analyze this amplifier and ignore r_o of the FET. The model is shown in Figure 7.10. r_o is generally very much larger than the other resistors in the drain circuit so this is an acceptable assumption and leads to little error in most cases.

7.4.2.1 Current Gain

The AC current gain of the CS MOSFET amplifier with source degeneration is given by

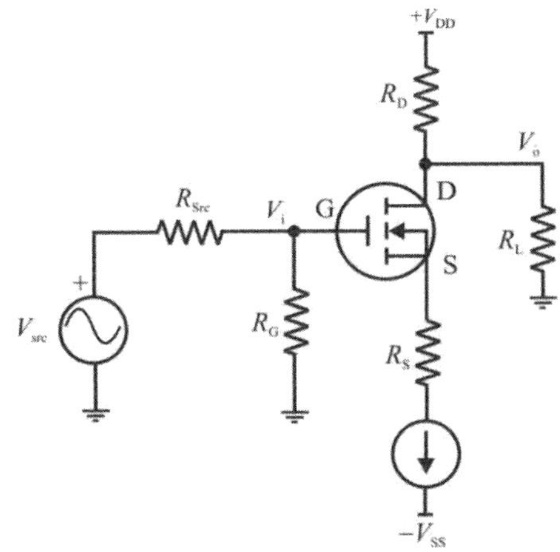

Figure 7.9 Basic N-channel MOSFET common source circuit with source degeneration.

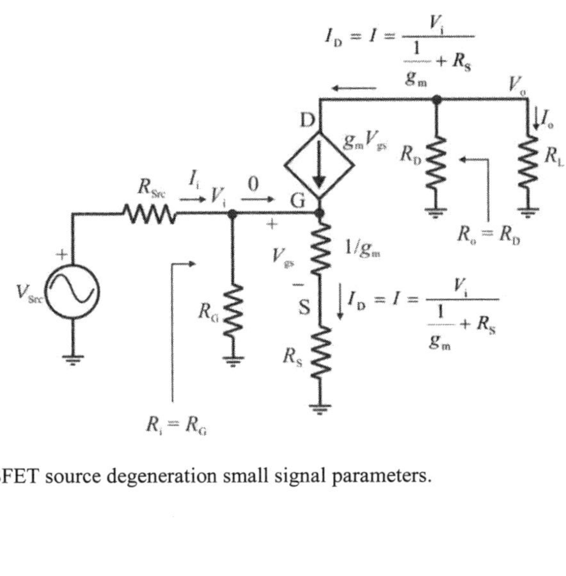

Figure 7.10 MOSFET source degeneration small signal parameters.

$$A_{\rm I} = \frac{I_{\rm o}}{I_{\rm i}} = -\frac{g_{\rm m} R_{\rm D}}{R_{\rm D} + R_{\rm L}} \frac{R_{\rm G}}{1 + g_{\rm m} R_{\rm S}} \tag{7.5}$$

7.4.2.2 Voltage Gain

The AC voltage gain of the CS MOSFET amplifier with source degradation included is given by

$$A_{\rm V} = -\frac{R_{\rm G}}{R_{\rm G} + R_{\rm Src}} \frac{g_{\rm m} \left(R_{\rm D} \| R_{\rm L}\right)}{1 + g_{\rm m} R_{\rm S}} \tag{7.6}$$

$$\approx -\frac{R_{\rm G}}{R_{\rm G} + R_{\rm Src}} \frac{\left(R_{\rm D} \| R_{\rm L}\right)}{R_{\rm S}} \qquad g_{\rm m} R_{\rm S} >> 1 \tag{7.7}$$

Thus, we see that when $g_{\rm m} R_{\rm S} >> 1$, the voltage gain becomes independent of the MOSFET characteristics. This is a major benefit of using source degeneration.

Conversely, without $R_{\rm S}$ ($R_{\rm S} = 0$), we see from (7.6) that $A_{\rm V}$ is proportional to $g_{\rm m}$ and is directly dependent on the physical properties of the transistor (and biasing) because

$$g_{\rm m} = \frac{I_{\rm d}}{V_{\rm gs}} = k'_{\rm n} \frac{W}{L} (V_{\rm gs} - V_{\rm t}) \tag{7.8}$$

in the case of an NMOS device.

The price we pay for this desirable behavior in (7.7), where $A_{\rm V}$ is not dependent on $g_{\rm m}$, is a reduced value for $A_{\rm V}$. This $A_{\rm V}$ is largest when $R_{\rm S} = 0$, as we can see from (7.6).

7.4.2.3 Input Resistance

Since $i_{\rm i} = 0$, then

$$R_{\rm i} = R_{\rm G} \tag{7.9}$$

7.4.2.4 Output Resistance

From Figure 7.10, with $V_{\rm Src} = 0$, then I must be zero, leading to

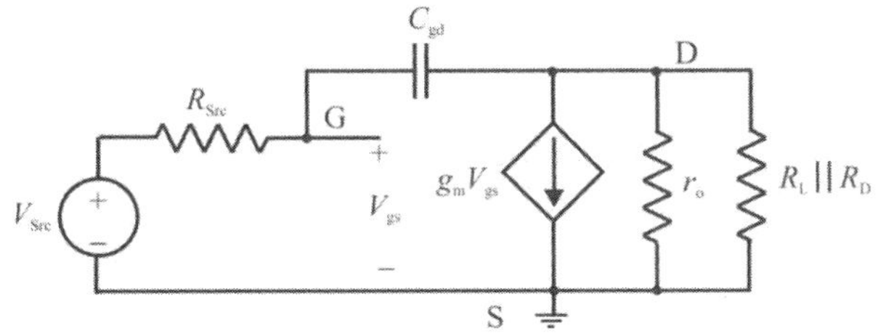

Figure 7.11 High-frequency, small-signal model for N-channel MOSFET common source amplifier.

$$R_o = R_D \tag{7.10}$$

7.4.3 CS MOSFET Amplifier Frequency Response

The high-frequency CS MOSFET amplifier topology is shown in Figure 7.11. This model includes the capacitance between the gate and drain, which forms a feedback path.

7.4.3.1 Bandwidth

At higher frequencies the internodal capacitances establish the frequency response. In particular, the Miller effect capacitance from the drain to the gate dominates the cutoff frequency. Other parasitic capacitances can frequently be neglected because they have only a minor effect on bandwidth in comparison to the Miller effect capacitance.

The bandwidth of the common source amplifier tends to be low, due to high equivalent input capacitance resulting from the Miller effect. The gate-drain capacitance is effectively multiplied by the factor $1+|A_v|$, thus increasing the total input capacitance and lowering the overall bandwidth. Figure 7.11shows the high-frequency small-signal model when a load resistor R_L is added at the output node and a driver of applied voltage V_S and series resistance R_S is added at the input. The limitation on bandwidth in this circuit stems from the coupling of parasitic transistor capacitance C_{GD} between gate and drain and the series resistance of the source R_S.

Incorporating the Miller capacitance, the circuit of Figure 7.11 is transformed to that of Figure 7.12, which shows the Miller capacitance C_μ on the input side of the circuit. C_μ is given by

$$C_\mu = C_{gd} \frac{V_{gd}}{V_{gs}} = C_{gd}\left(1 - \frac{V_D}{V_G}\right)$$

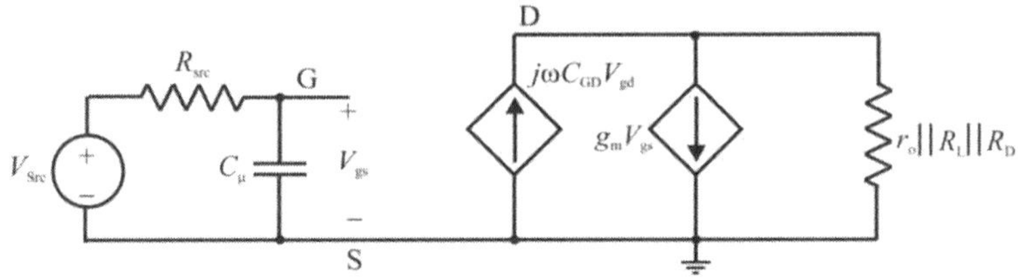

Figure 7.12 Small-signal model for N-channel MOSFET common source amplifier using Miller's theorem to introduce Miller capacitance C_μ .

where

$$\frac{V_D}{V_G} \approx -g_m (r_o \parallel R_L) \tag{7.11}$$

yielding

$$C_\mu = C_{gd}[1 + g_m (r_o \parallel R_L)] \tag{7.12}$$

The gain $g_m r_o \parallel R_L$ is large for large R_L, so even a small parasitic capacitance C_{gd} can have large influence in the frequency response of the amplifier. In addition, if the input capacitance, C_{gs}, is taken into effect it is simply added to the value of C_μ, further reducing the bandwidth.

From Figure 7.12, the gate voltage is related to the input signal level by

$$V_G = V_{Src} \frac{1}{1 + j\omega C_\mu R_{Src}} \tag{7.13}$$

By convention, the bandwidth (also called the 3-dB frequency) is the frequency where the signal drops to $1/\sqrt{2}$ of its low-frequency value. A reduction to $1/\sqrt{2}$ occurs when $\omega C_\mu R_S = 1$, making the voltage gain at ω_{3dB} $V_G / V_{Src} = 1/(1+j)$ where $|1+j| = \sqrt{2}$. As a result the 3-dB frequency, f_{3dB}, is given by (including C_{gs})

$$f_{3dB} = \frac{1}{2\pi R_{Src}\{C_{gs} + C_{gd}[1 + g_m (r_o \parallel R_L)]\}} \tag{7.14}$$

We can see that the bandwidth as indicated by f_{3dB} becomes larger if the source resistance R_{Src} is small, so the amplification of the capacitance caused by the Miller effect has less effect upon the bandwidth. Thus adding a voltage-follower (discussed later) between the driver and the common source stage reduces the bandwidth limiting effects of the Miller capacitance.

7.4.3.2 MOSFET Cutoff Frequency

The short-circuit current gain of the MOSFET high-frequency model is given by

$$A_I = \frac{-g_m}{j\omega(C_{gs} + C_{gd})} \tag{7.15}$$

The cutoff frequency of a MOSFET is where $A_I = 1$. Thus, by rearranging (7.15) we get

$$f_t = \frac{-g_m}{2\pi(C_{gs} + C_{gd})} \tag{7.16}$$

Realistic cutoff frequencies must also include the parasitic and fringing capacitances, C_p, which add to the gate capacitance so that

$$f_t = \frac{-g_m}{2\pi(C_{gs} + C_{gd} + C_p)} \tag{7.17}$$

7.4.3.3 Low-Frequency Response of the Common Source Amplifier

The common source amplifier topology and its equivalent circuit are shown in Figure 7.13. Using small-signal analysis, we get the following

$$V_g = V_{Src}\frac{R_G}{R_G + \frac{1}{sC_{C1}} + R_{Src}} = V_{Src}\frac{R_G}{R_G + R_{Src}}\frac{s}{s + \frac{1}{C_{C1}(R_G + R_{Src})}}$$

so

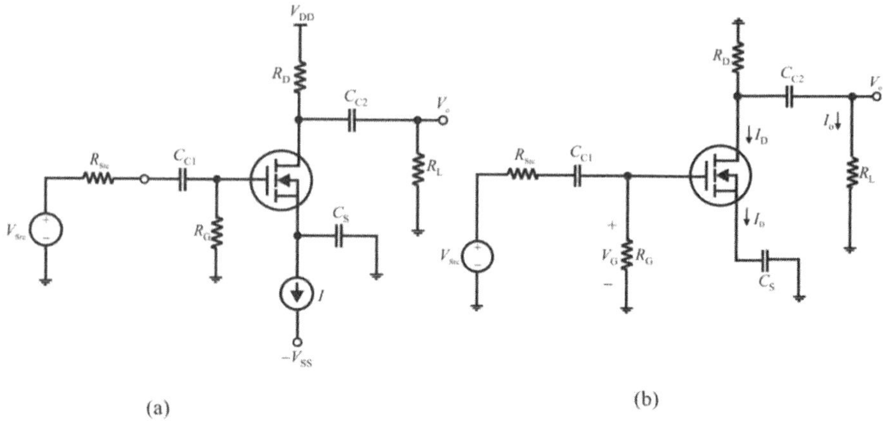

Figure 7.13 Common source: (a) circuit and (b) small signal equivalent.

$$\omega_{p1} = \frac{1}{C_{C1}(R_G + R_{Src})}$$

The drain current is given by

$$I_D = \frac{V_G}{\dfrac{1}{g_m} + \dfrac{1}{sC_C}} = g_m V_G \frac{s}{s + \dfrac{g_m}{C_S}}$$

so

$$\omega_{p2} = \frac{g_m}{C_S}$$

The output voltage is given by

$$V_o = I_o R_L = -I_D \frac{R_D}{R_D + \dfrac{1}{sC_{C2}} + R_L} = -I_D \frac{R_D R_L}{R_D + R_L} \frac{s}{s + \dfrac{1}{C_{C2}(R_D + R_L)}}$$

so

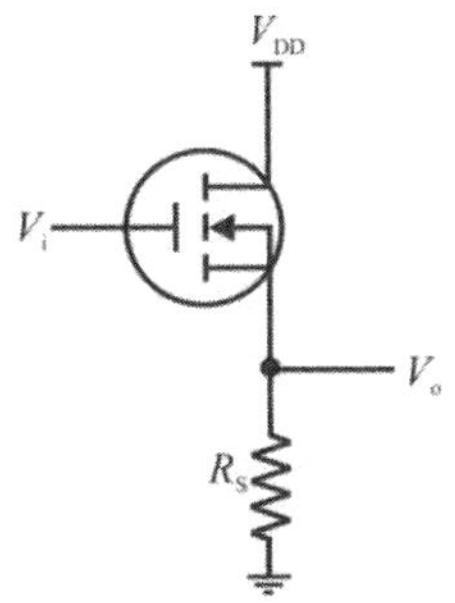

Figure 7.14 Basic enhancement-mode N-channel MOSFET source follower circuit (neglecting most biasing details).

$$\omega_{p3} = \frac{1}{C_{C2}(R_D + R_L)}$$

Thus, we have

$$\frac{V_o}{V_{Src}} = -\underbrace{\frac{g_m R_G (R_D \| R_L)}{R_G + R_{Src}}}_{\text{Midband gain}} \left(\frac{s}{s+\omega_{p1}}\right)\left(\frac{s}{s+\omega_{p2}}\right)\left(\frac{s}{s+\omega_{p3}}\right)$$

and

$$A_M = -\frac{g_m R_G (R_D \| R_L)}{R_G + R_{Src}}$$

7.5 Common Drain

The common drain (also known as the *source follower*) is architecturally similar to the BJT emitter follower. The input signal is applied to the gate and the output is taken from the source—that is, connected to ground through the source resistor (see Figure 7.14). The drain is common to both the input and output and is connected to V_{DD} (zero RF voltage). As with BJT emitter follower, this configuration is typically used as a voltage buffer and impedance transformation module.

The input impedance is that of the gate of an MOSFET, ideally infinite but at least very large (at low-frequency anyway). The source follower places very little load on the previous circuit due to this high impedance. On the other hand, the output impedance is low, making the source follower an excellent voltage source for the following stage. Low impedance sources for amplifiers is normally a desirable characteristic.

7.5.1 Characteristics of the Source Follower

At low frequencies, the source follower (see Figure 7.14) has the following small signal characteristics.

7.5.1.1 Short-Circuit Current Gain

Since $I_\mathrm{i} = 0$, this parameter makes no sense.

7.5.1.2 Open-Circuit Voltage Gain

The open-circuit voltage gain is given by

$$A_\mathrm{V} = \left.\frac{V_\mathrm{o}}{V_\mathrm{i}}\right|_{R_\mathrm{L}=\infty} = \frac{g_\mathrm{m} R_\mathrm{S}}{g_\mathrm{m} R_\mathrm{S} + 1}$$

$$\approx 1 \qquad g_\mathrm{m} R_\mathrm{S} >> 1 \tag{7.18}$$

7.5.1.3 Input Resistance

Since $I_\mathrm{i} = 0$,

$$R_\mathrm{i} = \infty \tag{7.19}$$

7.5.1.4 Output Resistance

The output resistance of the CD amplifier is given by

$$R_\mathrm{o} = R_\mathrm{Src} \| (1/g_\mathrm{m})$$

$$\approx \frac{1}{g_\mathrm{m}} \qquad g_\mathrm{m} R_\mathrm{Src} >> 1 \tag{7.20}$$

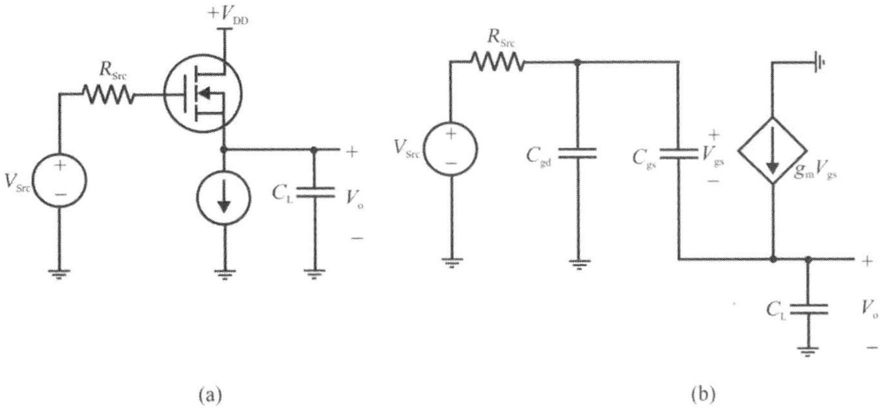

Figure 7.15 MOSFET source follower at high-frequency: (a) circuit and (b) equivalent circuit.

7.5.2 MOSFET CD High-Frequency Response

7.5.2.1 Voltage Gain

Consider the high-frequency small-signal model shown in Figure 7.15. Solving for the voltage gain $A_V(s) = V_o/V_i$ yields

$$V_{GS}sC_{gs} + g_m V_{gs} = V_o sC_L$$

$$V_{gs} = \frac{sC_L}{g_m + sC_{gs}} V_o$$

$$V_i = R_{Src}[V_i sC_{gsGS} + (V_i + V_{gs})sC_{gd}] + V_{gs} + V_o$$

$$A_V(s) = \frac{V_o(s)}{V_i(s)}$$
$$= \frac{g_m + sC_{gs}}{R_{Src}(C_{gs}C_L + C_{gs}C_{gd} + C_{gd}C_L)s^2 + (g_m R_{Src} C_{gd} + C_L + C_{gs})s + g_m} \tag{7.21}$$

The zero is due to C_{gs} that directly couples the input to the output. If the poles are far apart, then the s term in the denominator represents the dominant pole:

$$\omega_{p1} \approx \frac{g_m}{g_m R_S C_{gd} + C_L + C_{gs}}$$

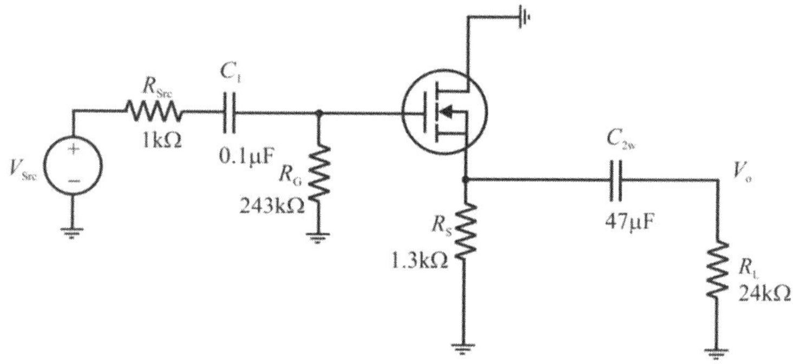

Figure 7.16 Common drain amplifier circuit.

7.5.2.2 Common-Drain Amplifier High Cutoff Frequency ω_H

In this section we determine an estimate of the high cutoff frequency of the CD amplifier with the topology shown in Figure 7.16.

Example 7.1: We will derive an expression for the high cutoff frequency and the bandwidth for the CD amplifier in Figure 7.16. We will assume that g_m = 1 mA/V. Note that since the drain is grounded C_{GD} can be redrawn as in Figure 7.17.

The small-signal circuit can be simplified in obvious ways as shown in Figure 7.18 yielding

$$V_{eq} = V_{Src} \frac{R_G}{R_{Src} + R_G}$$

and

$$R_{eq} = \frac{R_{Src} R_G}{R_{Src} + R_G}$$

From Figure 7.18 the resistance seen across the terminals of C_{gs} may be calculated using Figure 7.19 as

$$R_{gso} = \frac{V_x}{I_x}$$

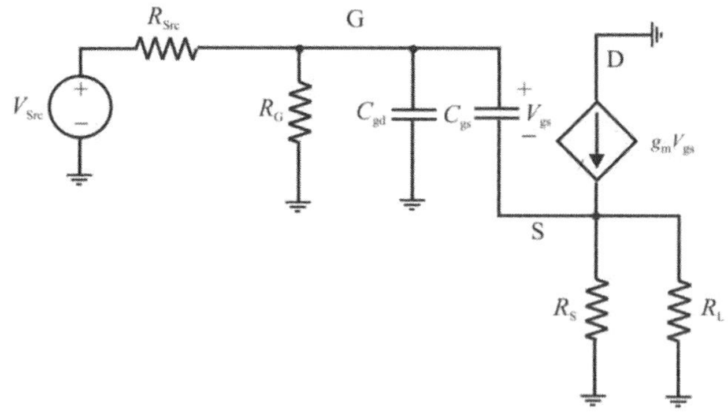

Figure 7.17 The CD amplifier in Figure 7.16 redrawn with the transistor replaced by its high-frequency small signal model.

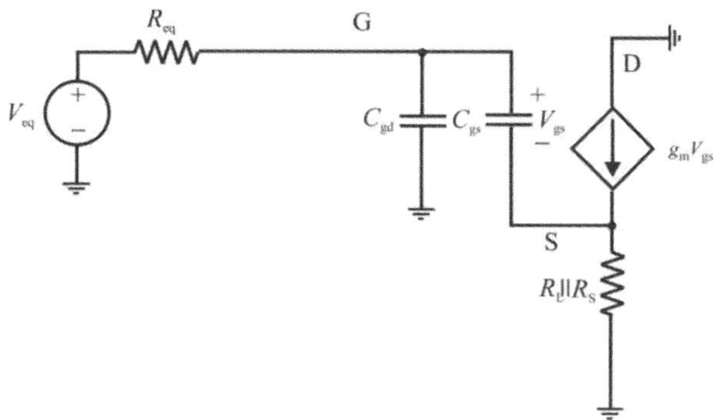

Figure 7.18 Simplified circuit to calculate the high-frequency response of the CD amplifier.

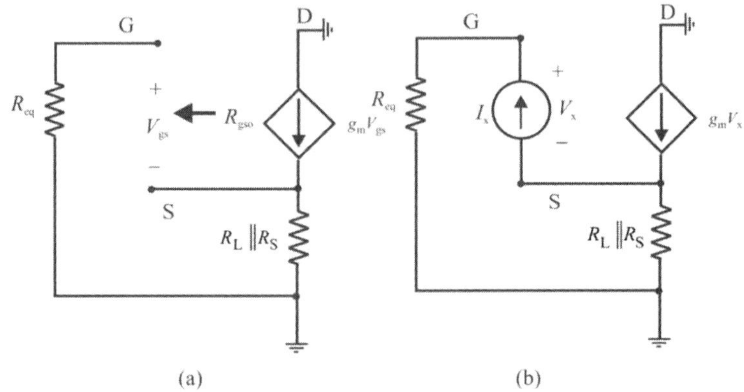

Figure 7.19 Circuit used to calculate the open-circuit time constant associated with C_{gs}.

$$V_x = V_g - V_S = I_x R_{eq} - (g_m V_x - I_x)(R_L \| R_S)$$
$$V_x[1 + g_m(R_L \| R_S)] = I_x[R_{eq} + (R_L \| R_S)]$$
$$R_{gso} = \frac{V_x}{I_x} = \frac{R_{eq} + (R_L \| R_S)}{1 + g_m(R_L \| R_S)}$$

so the time constant associated with C_{gs} is $C_{gs}R_{gso}$.

For C_{gd} the resistance seen across the capacitor terminals may be calculated from Figure 7.20. Once the current source between G and D is removed, R_i^{CD} becomes an open circuit, which yields

$$R_{gdo} = R_{eq} \| R_i^{CD} = R_{eq}$$

so the time constant associated with C_{gd} is $C_{gd}R_{eq}$.

The high cutoff frequency is

$$\omega_H = \frac{1}{\sum_{i=1}^{2} R_{io}C_i} = \frac{1}{R_{gso}C_{gs} + R_{gdo}C_{gd}} = \frac{1}{C_{gs}\left(\frac{R_{eq} + (R_L \| R_S)}{1 + g_m(R_L \| R_S)}\right) + C_{gd}R_{eq}}$$

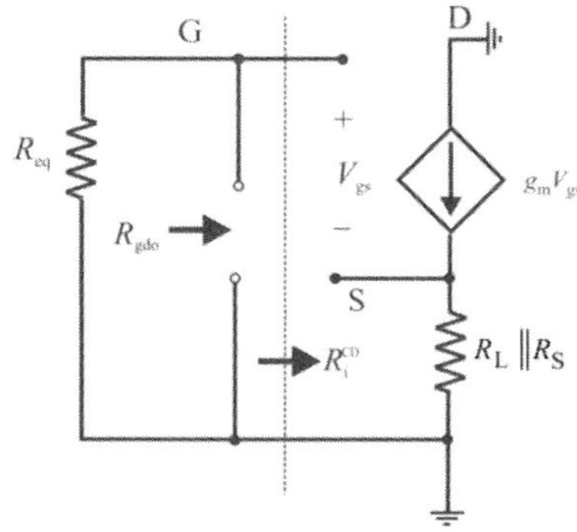

Figure 7.20 Circuit used to calculate the open-circuit time constant associated with C_{gd}.

Note that this expression may be obtained from that for the BJT common collector by setting r_π as infinite and r_x as zero in the final expression.

The bandwidth is approximated by assuming that ω_L is near zero so that

$$W = \frac{\omega_H - \omega_L}{2\pi} \approx \frac{\omega_H}{2\pi}$$

We can see that the bandwidth of the CG amplifier is much higher than that of the CS amplifier. This is due to the absence of the Miller's capacitance associated with the CS configuration.

7.5.2.3 CD High-Frequency Input and Output Impedances

Since source follower amplifiers are often used as voltage buffers, the input and output impedances are important considerations. We will calculate the input and output impedances in this section.

The input impedance is calculated by using the high-frequency small-signal model shown in Figure 7.3. This impedance is the impedance looking into the input port, which is comprised of just C_{gs}—see Figure 7.21. The result is

$$Z_i = \frac{1}{sC_{gs}} + \left(1 + \frac{g_m}{sC_{gs}}\right)\frac{1}{g_{mb} + sC_L} \tag{7.22}$$

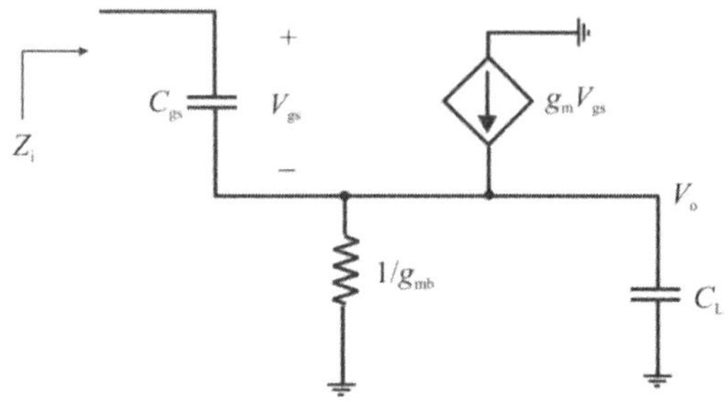

Figure 7.21 MOSFET CD Z_i.

The output impedance (ignoring g_{mb} for simplicity), Z_o, is shown in Figure 7.22 and we get

$$Z_o = \frac{R_{Src} sC_{gs} + 1}{g_m + sC_{gs}} \tag{7.23}$$

Thus, at low-frequency, $Z_o \approx 1/g_m$, and at high-frequency, $Z_o \approx R_{Src}$. The shape of the response depends on the relative size of R_{Src} and $1/g_m$. Two examples of Z_o response are shown in Figure 7.23. In addition, Z_o can look inductive or capacitive depending on $1/g_m$ and R_{Src}.

7.5.2.4 Common-Drain Amplifier Low Cutoff Frequency ω_L Estimation

Example 7.2: In this section the low cutoff frequency of the CD amplifier shown in Figure 7.16 will be estimated using the SCTC method. We will assume that the MOSFET is biased such that g_m = 1 mA/V.

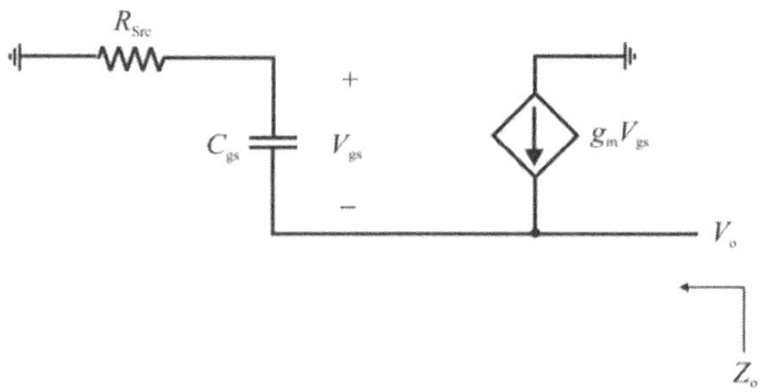

Figure 7.22 MOSFET CD Z_o.

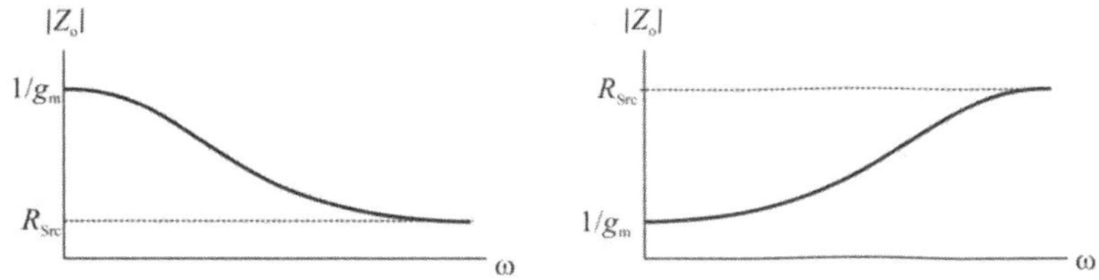

Figure 7.23 MOSFET CD Z_{out} plots.

For C_1 the resistance seen across the capacitor terminals may be calculated from Figure 7.24

$$R_{1S} = R_{Src} + (R_G \| R_i^{CD}) = R_{Src} + R_G$$

and the time constant associated with C_1 is $C_1 R_{1S}$.

The resistance seen across C_2 may be calculated from Figure 7.25 yielding

$$R_{2S} = R_L + (R_S \| R_o^{CD}) \approx R_L + \left[R_S \| (1/g_m) \right]$$

and the time constant associated with C_2 is $C_2 R_{2S}$.

The low cutoff frequency is calculated as:

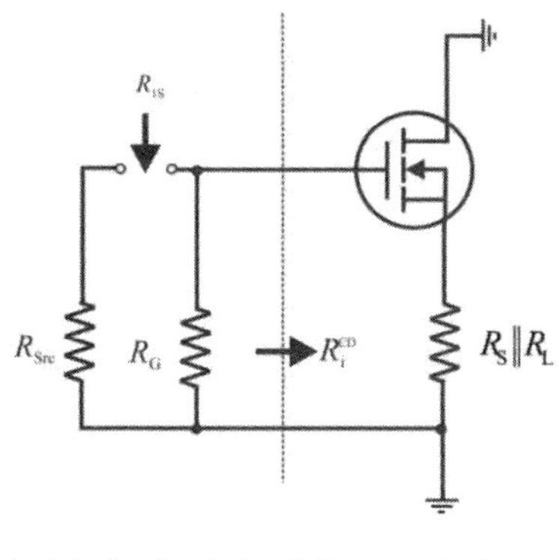

Figure 7.24 Circuit used to calculate the short-circuit time constant associated with C_1.

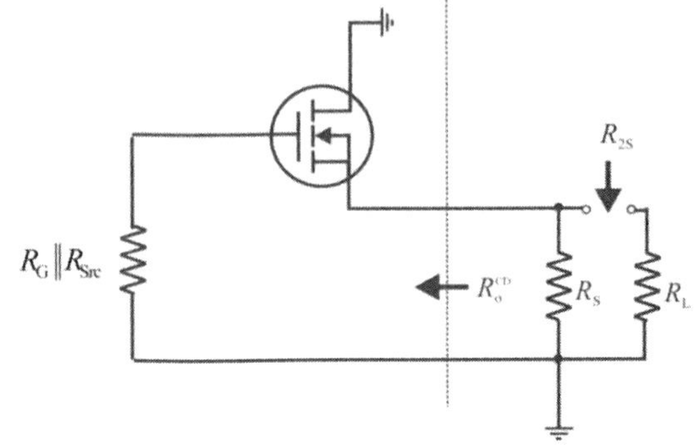

Figure 7.25 Circuit used to calculate the short-circuit time constant associated with C_2.

$$\omega_L = \sum_{i=1}^{2} \frac{1}{R_{iS} C_i} = \frac{1}{C_1 (R_{Src} + R_G)} + \frac{1}{C_2 \left[R_L + \left(R_S \| \frac{1}{g_m} \right) \right]}$$

$$= 41 + 0.9 = 41.9 \text{ rad/sec}$$

7.6 Common Gate

The common gate configuration shares the gate between the input and output loops. They are generally used as impedance transformation circuits.

The common gate amplifier configuration, shown in Figure 7.26, also serves as a voltage amplifier or a current buffer. The gate is shared with the input applied to the source node of the MOSFET and the output taken from the drain. In the topology shown in Figure 7.26, current source I_D represents an active load, as we discussed in Chapter 4. The signal to be amplified is applied at node V_i and the output is taken from node V_o. The output can be current or voltage.

This configuration is especially useful in implementation of CMOS RF applications when approaching the frequency limitations of the FETs because of the ease of impedance matching and potentially lower noise. However, it is used less often than the common source or source follower.

7.6.1 Low-Frequency Characteristics

Some of the ideal characteristics of the CG amplifier are given below, based on the model shown in Figure 7.27 where the hybrid-π model for the MOSFET has been

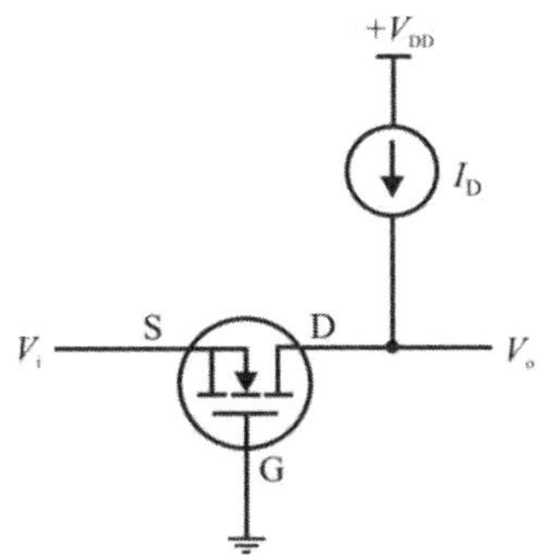

Figure 7.26 N-channel common gate circuit (neglecting most biasing details); with an active load.

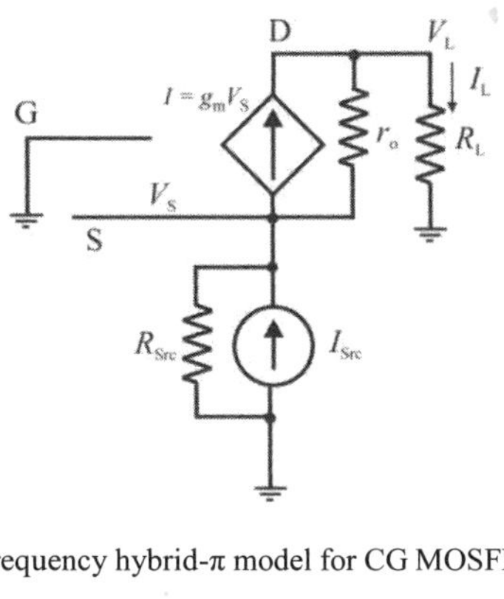

Figure 7.27 Small-signal low-frequency hybrid-π model for CG MOSFET amplifier driven by a Norton signal source.

employed. The approximate expressions are based on the (commonly true) assumptions that $r_o >> R_L$ and $g_m r_o >> 1$.

7.6.1.1 Current Gain

The current gain including the loading effects of R_D is given by

$$A_i = \frac{I_o}{I_i} = \frac{R_D}{R_D + R_L} \tag{7.24}$$

From (7.24) we get the short-circuit current gain by letting $R_L = 0$, which is the same as assuming an active load (rather than R_D). It is given by

$$A_I = \left. \frac{I_o}{I_{Src}} \right|_{R_L = 0} = 1 \tag{7.25}$$

7.6.1.2 Voltage Gain

The voltage gain including input and output loading provided by R_{Src}, R_D, and R_L is given by

$$A_v = \frac{V_o}{V_S} = \frac{g_m (R_D \| R_L)}{1 + g_m R_{Src}} \tag{7.26}$$

Because the input impedance of the common gate amplifier is very low, like the CB BJT amplifier, the cascode amplifier often is used instead. The cascode places a common source amplifier between the voltage driver and the common gate circuit to permit voltage amplification using a driver with $R_{Src} >> 1/g_m$. We discuss the cascode amplifier later.

The open- circuit voltage gain is when $R_L \to \infty$ in (7.26) and is given by (with R_D)

$$\begin{aligned} A_v = \left. \frac{V_o}{V_S} \right|_{R_L = \infty} &= \frac{g_m R_D}{1 + g_m R_{Src}} \\ &\approx \begin{cases} g_m R_D & g_m R_{Src} << 1 \\ R_D / R_{Src} & g_m R_{Src} >> 1 \end{cases} \end{aligned} \tag{7.27}$$

With an active load, R_D is replaced by $r_o \| r_{o,load}$.

7.6.1.3 Input Resistance

The input resistance for the MOSFET CG amplifier is given by

$$
\begin{aligned}
R_i &= \frac{V_S}{I_{Src}} = \frac{R_L + r_o}{g_m r_o + 1} \\
&\approx \frac{1}{g_m} \qquad r_o >> R_L \text{ and } g_m r_o >> 1
\end{aligned} \tag{7.28}
$$

7.6.1.4 Output Resistance

Since we are assuming an active load for the MOSFET, as we discussed above, the equivalent resistance of the active load is $r_{o,load}$. From the small-signal circuit above, with $V_S = 0$ we find that $I = 0$ since the gate is grounded. Consequently, the output resistance is given by

$$R_o = r_{o,load} \| r_o \tag{7.29}$$

If the load is R_D rather than an active load, then

$$R_o = R_D \tag{7.30}$$

In general the overall voltage/current gain may be substantially less than the open/short-circuit gains listed above (depending on the source and load resistances) due to the loading effect.

7.6.2 Common-Gate Amplifier Low Cutoff Frequency ω_L Estimation

In this section the low cutoff frequency of the CG amplifier shown in Figure 7.28 will be estimated. We do this by way of an example.

Example 7.3: We will assume that the MOSFET is biased where g_m = 1 mA/V. For C_1 the resistance seen across the capacitor terminals may be calculated from Figure 7.29.

$$R_{1S} = R_{Src} + \left(R_S \| R_i^{CG}\right) = R_{Src} + \left[R_S \| (1/g_m)\right]$$

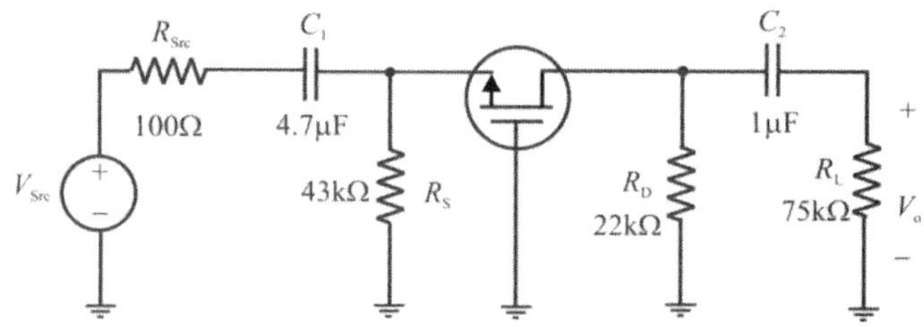

Figure 7.28 Common gate amplifier circuit.

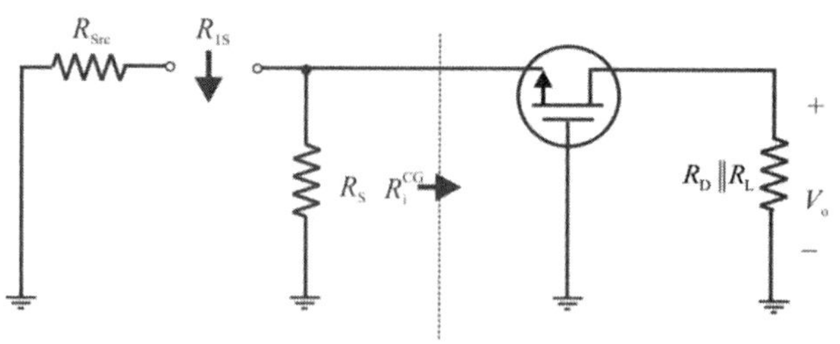

Figure 7.29 Circuit used to calculate the short-circuit time constant associated with C_1.

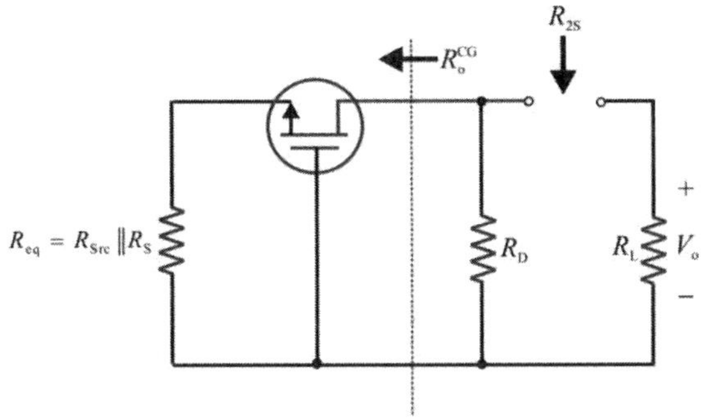

Figure 7.30 Circuit used to calculate the short-circuit time constant associated with C_2.

so the time constant associated with C_1 is $C_1\left[R_{Src} + R_S \| (1/g_m)\right]$.

For C_2 the resistance seen across the capacitor terminals may be calculated from Figure 7.30.

$$R_{2S} = R_L + (R_D \| R_o^{CG}) \approx R_L + (R_D \| r_o)$$
$$\approx R_L + R_D$$

using Miller's theorem and assuming high gain. So the time constant associated with C_2 is $C_2(R_L + R_D)$.

The low cutoff frequency is

$$\omega_L \approx \sum_{i=1}^{2} \frac{1}{R_{iS}C_i} = \frac{1}{C_1[R_{Src} + R_S \| (1/g_m)]} + \frac{1}{C_2(R_L + R_D)}$$
$$= 197.5 + 10 = 207.5 \text{ rad/sec}$$

Note that since the input resistance of the CG amplifier is higher than that of the CB amplifier (since the transconductance of the MOSFET is less than that of the BJT), the low cutoff frequency will be lower in the CG amplifier compared to the CB amplifier for the same capacitance and resistance values.

7.6.3 MOSFET CG High-Frequency Response

One way to avoid the frequency limitations of Miller multiplication of C_{GD} is to utilize a CG amplifier configuration as shown in Figure 7.31. We use the same basic

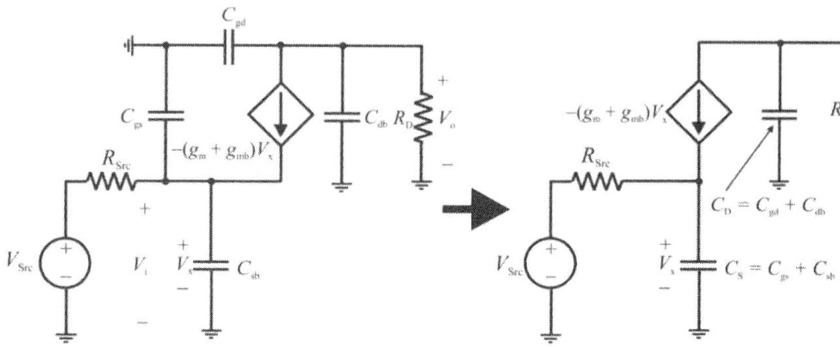

Figure 7.31 MOSFET common-gate amplifier high-frequency model.

model for the MOSFET but the input is changed to the source. This configuration has two time constants:

At the input (source node)

$$\tau_S = \frac{C_S}{1/R_{Src} + g_m + g_{mb}}$$

At the output (drain node)

$$\tau_D = C_D R_D$$

The output drives additional load capacitance such that the output pole is dominant. The frequency response of CG amplifiers is then combined with a CS stage to form a cascode amplifier discussed later.

7.6.3.1 Common-Gate Amplifier High Cutoff Frequency ω_H Estimation

In this section we will develop the estimated high cutoff frequency expression using the OCTC method for the CG amplifier in Example 7.3.

> **Example 7.4:** The CG amplifier is redrawn for small-signal analysis as shown in Figure 7.32 by replacing the transistor in Figure 7.28 by its high-frequency small-signal model. All the coupling and bypass capacitors are considered short circuits due to the high-frequency as well.

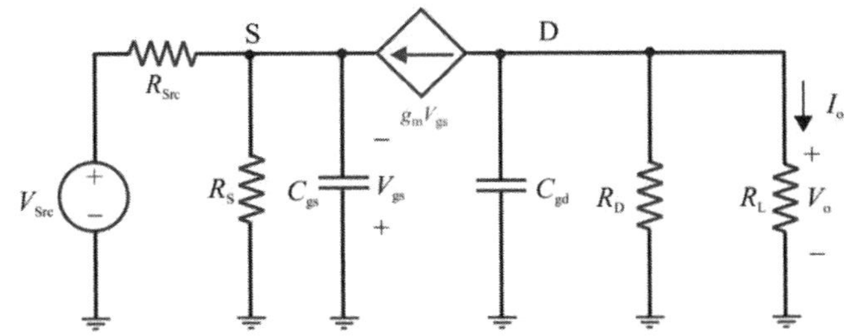

Figure 7.32 The CG amplifier in Figure 7.28 redrawn with the transistor replaced by its high-frequency small-signal model

The small-signal circuit can be simplified as shown in Figure 7.33.

$$V_{eq} = V_{Src}\frac{R_S}{R_{Src}+R_S}$$

$$R_{eq} = \frac{R_{Src}R_S}{R_{Src}+R_S}$$

From Figure 7.33, the resistance seen across C_{gs} may be calculated using Figure 7.34. At the source node in Figure 7.34(b)

$$I_x = g_mV_x + \frac{V_x}{R_{eq}}$$

so

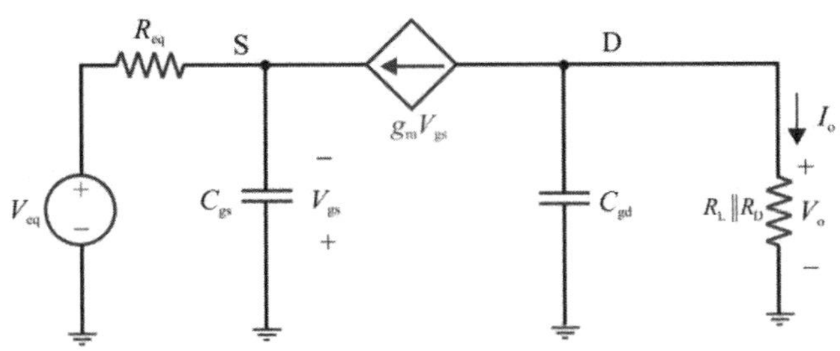

Figure 7.33 Simplified circuit to calculate the high-frequency response of the CG amplifier.

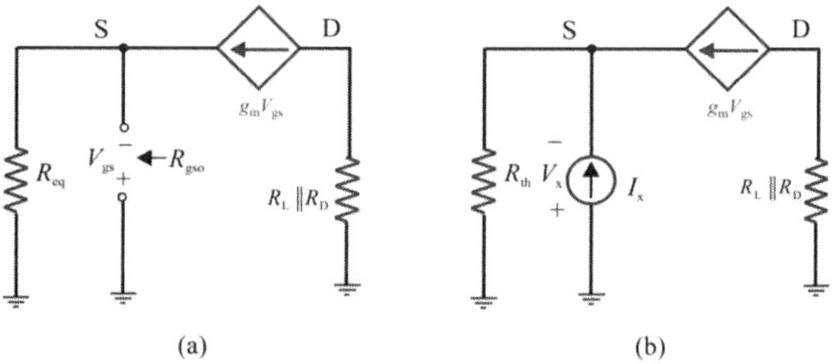

Figure 7.34 Circuit used to calculate the open-circuit time constant associated with C_{gs}.

$$R_{gso} = \frac{V_x}{I_x} = \frac{R_{eq}}{1 + g_m R_{eq}}$$

and the time constant associated with C_{gs} is $C_{gs} R_{gso}$.

For C_{gd} the resistance seen across the capacitor terminals may be calculated from Figure 7.35. Applying KVL

$$-V_{gs} = g_m V_{gs} R_{eq}$$
$$V_{gs}(1 + g_m R_{eq}) = 0$$

therefore

$$V_{gs} = 0$$

and

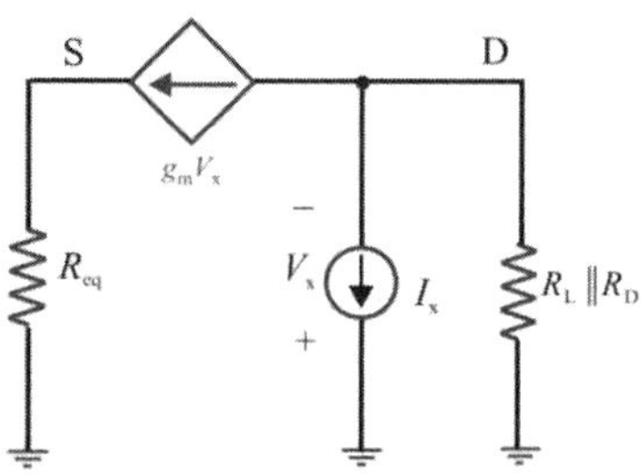

Figure 7.35 Circuit used to calculate the open-circuit time constant associated with C_{gd}.

$$R_{\text{gdo}} = \frac{V_x}{I_x} = R_L \| R_D$$

So, the time constant associated with C_{gd} is $C_{gd} R_L \| R_D$.

The high cutoff frequency is

$$\omega_H = \frac{1}{\sum_{i=1}^{2} R_{io} C_i} = \frac{1}{R_{\text{gso}} C_{gs} + R_{\text{Do}} C_{gd}} = \frac{1}{C_{gs}\left(\frac{R_{eq}}{1 + g_m R_{eq}}\right) + C_{gd} R_L \| R_D}$$

and

$$W = \frac{\omega_H - \omega_L}{2\pi} \approx \frac{\omega_H}{2\pi}$$

We can see that, like the CD topology, the bandwidth of the CG amplifier compared to the CS amplifier is again much higher. Again, this is because of the absence of the Miller's capacitance associated with the common source configuration between the output and input nodes.

7.7 MOSFET Cascode

We will use the OCTC method to estimate the high-frequency response of the MOSFET cascode amplifier. A MOSFET cascode amplifier topology is illustrated in Figure 7.36, where the device capacitances have been made explicit.

- Capacitance C_{gs1} sees a resistance R_{Src}
- Capacitance C_{gd1} sees a resistance R_{gd1} given by

$$R_{gd1} = (1 + g_{m1} R_{d1}) R_{Src} + R_{d1}$$

- Capacitance ($C_{db1} + C_{gs2}$) sees a resistance R_{d1}

$$R_{d1} = r_{o1} \| R_{i2} = r_{o1} \| [(r_{o2} + R_L) / g_{m2} r_{o2}]$$

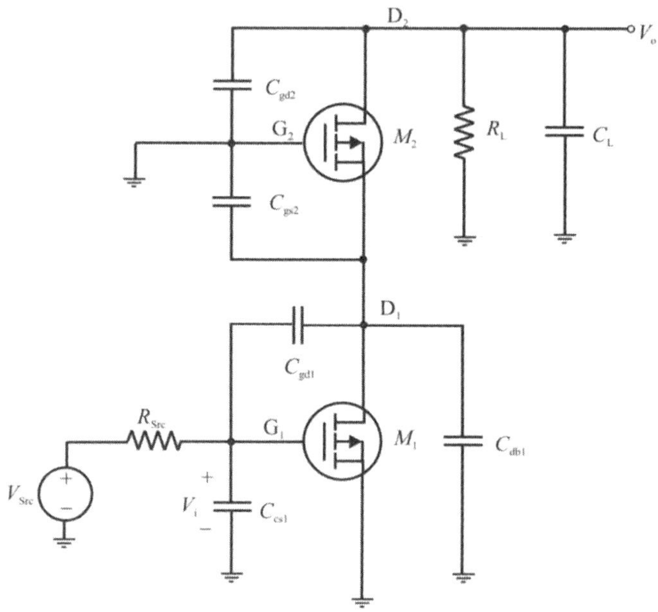

Figure 7.36 Cascode amplifier.

- Capacitance ($C_L + C_{gd2}$) sees a resistance $R_L \| R_o$

$$R_o = r_{o2} + r_{o1} + g_{m2} r_{o2} r_{o1}$$

- Effective time constant

$$\tau_H = C_{gs1} R_{Src} + C_{gd1}[(1 + g_{m1} R_{d1}) R_{Src} + R_{d1}] \\ + (C_{db1} + C_{gs2}) R_{d1} + (C_L + C_{gd2})(R_L \| R_o) \tag{7.31}$$

- The upper 3-dB frequency

$$f_H = \frac{1}{2\pi\tau_H}$$

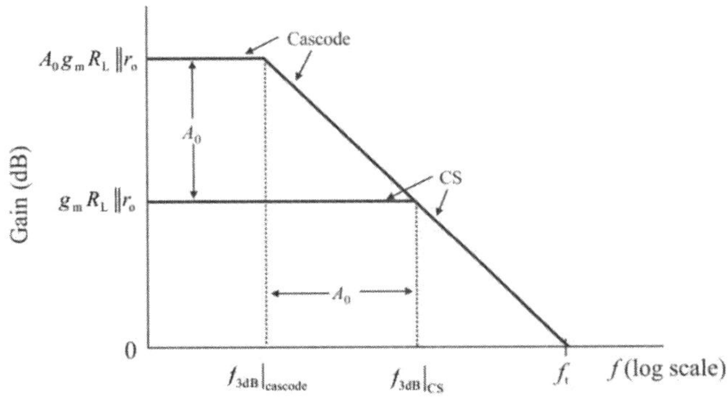

Figure 7.37 MOSFET Cascode CS comparison.

In the case of a large R_{Src} the first term in (7.31) dominates, especially if the Miller multiplier is large (typically with large R_{D1} and R_L). A small R_L (on the order of r_o) is needed for extended bandwidth. The midband gain drops as the value of R_L decreases—thus we see a trade-off between gain and bandwidth.

If R_{Src} is small the first term in (7.31) becomes negligible and the third term dominates. A large R_L (on the order of $A_0 r_o$) can be used to boost the amplifier gain. In the case of zero R_{Src} we have

$$f_H \approx \frac{1}{2\pi} \frac{1}{(R_L \| R_o)(C_L + C_{gd2})}$$

$$f_t \approx \frac{1}{2\pi} \frac{g_m}{C_L + C_{gd2}}$$

A comparison of the cascode MOSFET topology with the MOSFET CS configuration is shown in Figure 7.37. We can see that the low-frequency gain of the cascode topology is higher than that of the CS. On the other hand, the 3-dB bandwidth of the CS is larger.

7.8 Concluding Remarks

MOSFET Transistor Amplifier Summary

The summary of the MOSFET amplifier configurations shown in Figure 7.38 is contained in Table 7.1.

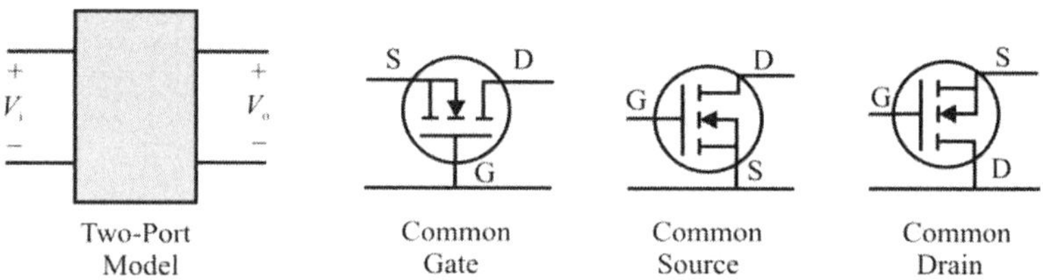

Figure 7.38 MOSFET summary.

Table 7.1 MOSFET Amplifier Configuration Characteristics

Characteristic	Common Gate	Common Source	Common Drain
Input impedance	Low/moderate	Very high	Very high
Output impedance	Medium/high	Medium/high	Low
Phase shift	0°	180°	0°
Voltage gain	High	Medium/high	Low
Current gain	Low	Very high	Very high
Power gain	Low	Very high	Very high

References

[1] Gonzalez, G., *Microwave Transistor Amplifiers Analysis and Design,* 2nd Ed., Upper Saddle River, NJ: Prentice Hall, 1997.
[2] Pozar, D. M., *Microwave Engineering,* 3rd Ed., New York: Wiley, 2005.
[3] Razavi, B., *RF Microeclectronics,* Upper Saddle River, NJ: Prentice Hall, 1997.

Chapter 8

Active Device Biasing

8.1 Introduction

As mentioned previously, biasing is an integral part of amplifier (any electronic circuit really) design. Biasing a circuit places the quiescent operating points of active devices where they need to be in their characteristic curves to obtain the desired operating feature. We address an introduction to biasing techniques for BJT and MOSFET amplifiers in this chapter; however, just as in any design effort, the coverage is not comprehensive. For those who wish to pursue more details on semiconductor biasing methods [1 – 3] are recommended.

This chapter is structured as follows. We begin the discussion of bias approaches considering the BJT RF amplifier. We put forth five methods of biasing and discuss the merits of each. We then cover biasing MOSFET devices for the RF amplifier function. Due to their much higher input impedance, bias circuit design for MOSFETs is much simpler than for BJT devices. Next, we discuss biasing both circuit types with passive components, followed with a discussion on biasing with active circuits. The chapter concludes with a brief introduction to biasing MMIC amplifiers.

8.2 BJT Amplifier Biasing

The quiescent bias points for these classes of BJT amplifier are depicted in Figure 4.2, repeated here as Figure 8.1.

The most predominant biasing schemes used to obtain both temperature stabilization and single-supply operations are:

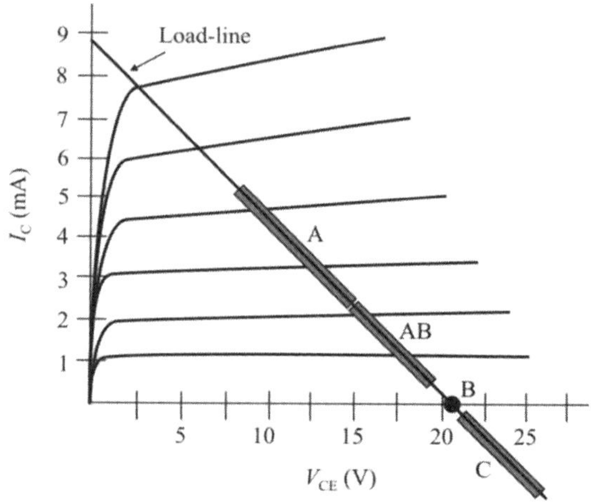

Figure 8.1 Quiescent points.

- Base-biased emitter feedback;
- Voltage-divider emitter feedback;
- Collector-feedback;
- Diode feedback;
- Active-feedback bias.

All five are found in Class A and AB operation, while Class B and C amplifiers can implement other methods. We discuss these methods below.

8.2.1 Base-Biased Emitter Feedback

Base-biased emitter feedback works in the following way (Figure 8.2): The base resistor (R_B), the 0.7 V base-to emitter voltage drop (V_{BE}), and the emitter resistor (R_E), are all in series. As the collector current (I_C) increases due to a rise in the transistor's temperature, the current through the emitter resistor will also increase, which increases the voltage dropped across R_E. This lowers the voltage that would normally be dropped across the base resistor and, since the voltage sum around a closed loop must always equal zero, this reduction in voltage across R_B decreases the base current, which then lowers the collector current.

The emitter capacitor (C_E) across R_E bypasses the RF signal around the emitter resistor to eliminate RF gain degeneration by R_E. The higher the voltage across R_E (V_E), the more temperature stable the amplifier, but the more power will be wasted

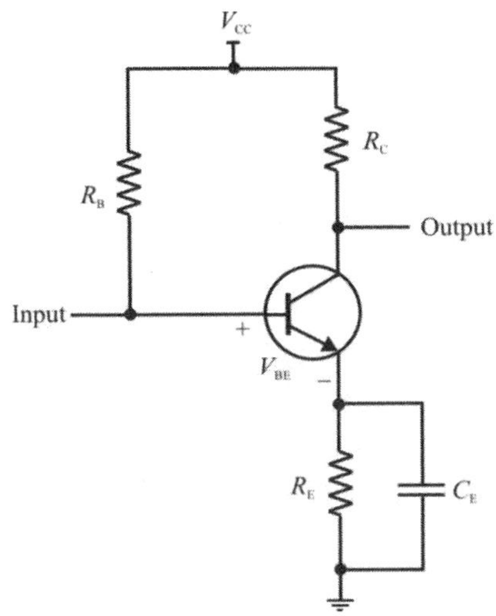

Figure 8.2 Base-biased emitter feedback.

in R_E due to V_E^2 / R_E, as well as the decreased AC signal gain if R_E is not bypassed by a low reactance capacitor.

Typical values of V_E for most HF amplifiers are between 2 to 4V to stabilize ΔV_{BE}. UHF and higher frequency amplifiers normally avoid these emitter resistors completely.

8.2.2 Voltage Divider Emitter Feedback

The *voltage divider emitter feedback* configuration is shown in Figure 8.3. The roles of R_C, R_E, and C_E are the same. R_1 and R_2 form the necessary base bias current. Some low-frequency high gain *intermediate frequency* (IF) amplifiers will split the single emitter feedback resistor (R_E) into two separate emitter resistors, with only one of these resistors having an AC capacitor bypass, while the other one is providing constant degenerative feedback to enhance amplifier stability reducing the chance of oscillations. This topology allows solid setting of the gain of the amplifier by just changing the value of R_4. This topology is depicted in Figure 8.4.

8.2.3 Diode Temperature Compensation

For better temperature compensation the most common method is *diode temperature compensation*. Two diodes, D_1 and D_2, attached to the transistor's heatsink or to the device itself, will carefully track the transistor's internal

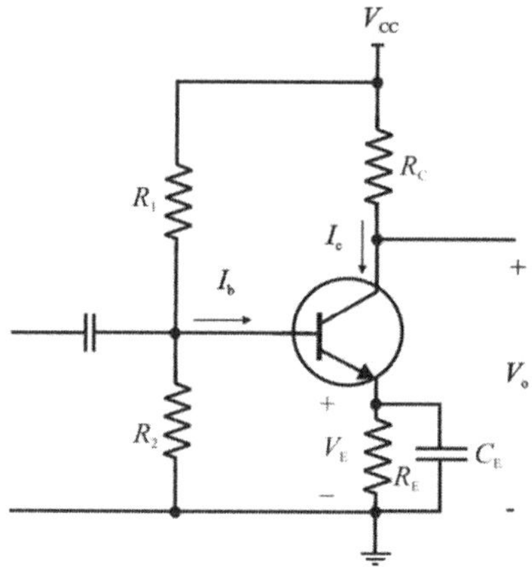

Figure 8.3 Voltage divider emitter feedback for a common emitter BJT amplifier.

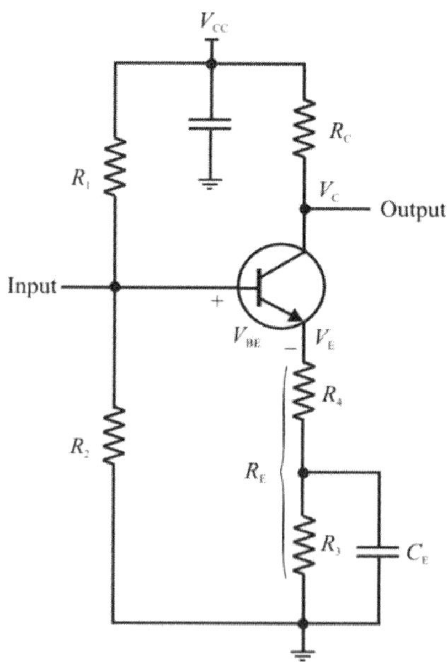

Figure 8.4 Stable IF amplifier.

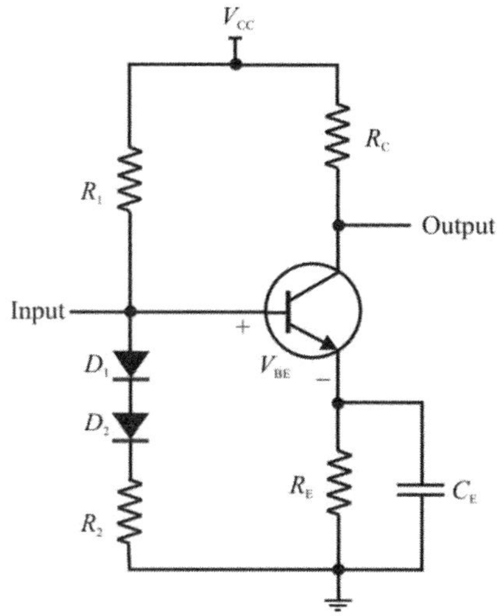

Figure 8.5 Diode temperature compensation bias.

temperature changes. This scheme is depicted in Figure 8.5. This is accomplished by the diode's decrease in its internal resistance with any increase in heat, which reduces the diode's forward voltage drop, thus lowering the transistor's base-emitter voltage, and diminishing any temperature-induced current increase in the BJT.

8.2.4 Collector Feedback

A very low-cost biasing scheme for RF and microwave circuits, but with less thermal stability than above, is called *collector feedback* bias as illustrated in Figure

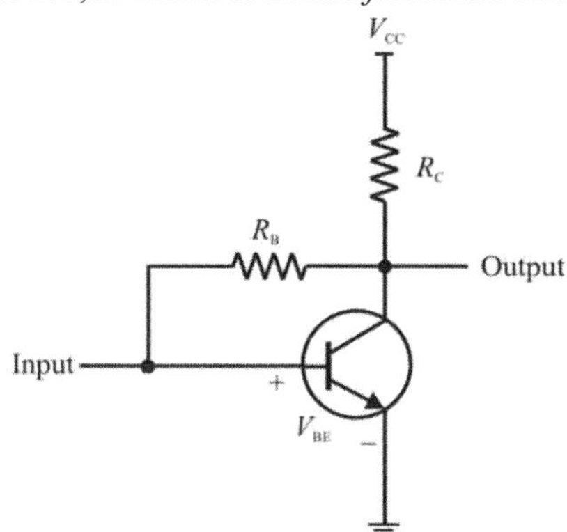

Figure 8.6 Collector feedback bias.

8.6. The circuit employs only two resistors, along with the active device, and has very little lead inductance due to the emitter's direct connection to ground.

Collector-feedback bias temperature stabilization functions so as the temperature increases, the transistor will start to conduct more current from the emitter to the collector. But the base resistor is directly connected to the transistor's collector, and not to the top of the collector resistor as in the above biasing techniques, so any rise in I_C causes more voltage to be dropped across the collector resistor. This forces less voltage to be dropped across the base resistor, which decreases the base current and, consequently, I_C.

8.2.5 Class C BJT Bias

For bipolar transistors, class-C amplifiers permit the use of three biasing techniques:

- Signal;
- External;
- Self-bias.

8.2.5.1 Signal Bias

Some class-C transistor amplifier are not given any bias at the base whatsoever, the so-called signal biased technique, but in order to lower the chances of any BJT

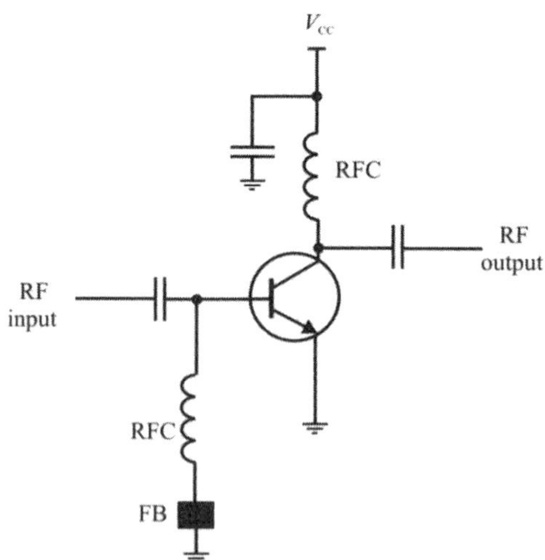

Figure 8.7 Class C signal bias.

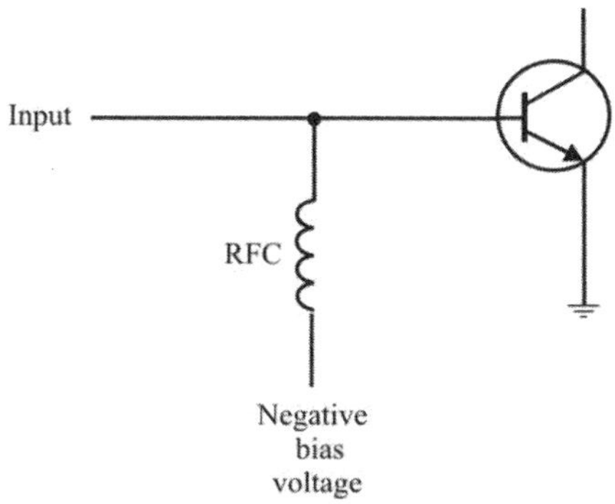

Figure 8.8 Class C external bias.

power device instability the base should be grounded through a low-Q choke, with a *ferrite bead* (FB) on the base lead's grounded end as illustrated in Figure 8.7.

These biasing techniques still require an RF signal with amplitude high enough to overcome the reverse (or complete lack of) bias at the class-C input.

8.2.5.2 External Bias

A less common method in class-C amplifiers is to use an external bias as illustrated in Figure 8.8. This circuit uses a negative bias supply to bias the base, and a standard positive supply for the collector circuit. The RFC acts as high impedance for the RF frequency itself so that it does not enter the bias supply.

8.2.5.3 Self-Bias

Self-bias of class C BJT amplifiers is the same as that shown in Figure 8.9, collector feedback bias. This bias technique utilizes the collector voltage to provide the necessary V_x and I_B. See Figure 8.9. One important characteristic of this approach is that the collector has a higher potential than the base, thus guaranteeing active-mode operation of the BJT regardless of the value of β as the base bias is derived from the collector voltage.

In this approach, since the base bias resistor, R_B is connected to the collector, instead of to V_{CC}, if the collector current increases, the collector voltage drops, reducing the base drive and thereby reducing the collector current. This is a form of negative feedback that tends to increase the stability of the amplifier. The opposite reaction occurs when the transistor's collector current decreases. This method of biasing is called *self-biasing*, producing good stability for most amplifier designs.

To ensure proper bias the following guidelines should be followed:

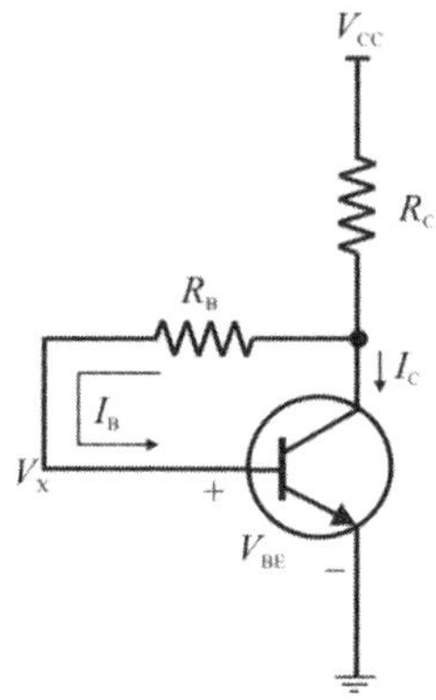

Figure 8.9 Self-bias.

- $R_C >> R_B / \beta$
- $\Delta V_{BE} << V_{CC} - V_{BE}$

Significant variation to transistor parameters is common even among transistors of the same variety. These variations arise due to production variables as well as environmental conditions. This bias approach provides good insensitivity to variations in β as well as insensitivity to variations in V_{BE}. It also produces decreased sensitivity to time and temperature variations as well.

8.2.6 Transformer Biasing

Transformers provide an alternative means of isolating the bias voltages between two amplifier stages, as opposed to using DC blocking capacitors. One example is illustrated in Figure 8.10.

This approach has the added advantage of providing wideband tuning of the amplifier stage. The transformers can be incorporated in the matching networks at both the input and output.

8.2.7 Highly Stable Active Bias for High-Frequency Amplifiers

The most common form of biasing RF circuits is the *current mirror*. This basic stage is used everywhere and it acts like a current source. It takes a current as an input and this current is usually generated, along with all other references, by a circuit called a *bandgap reference generator*. A bandgap reference generator is a temperature-independent bias generating circuit that balances V_{BE} dependence on temperature, to result in a voltage or current nearly independent of temperature. The most basic

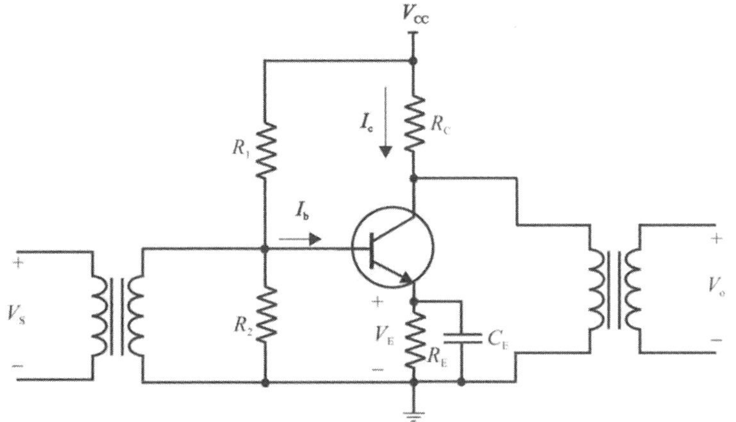

Figure 8.10 Transformer coupled CE BJT.

current mirror topologies are illustrated in Figure 8.11. In this mirror, the bandgap reference generator produces current I_{bias} and forces this current through Q_1. Scaling

the second transistor allows the current to be multiplied up and used to bias working transistors.

One major drawback to the circuit in Figure 8.11(a) is that it can inject a lot of noise at the output due primarily to the gain of the transistor Q_1 which acts like an amplifier for noise. A capacitor can be used to clean up the noise as in Figure

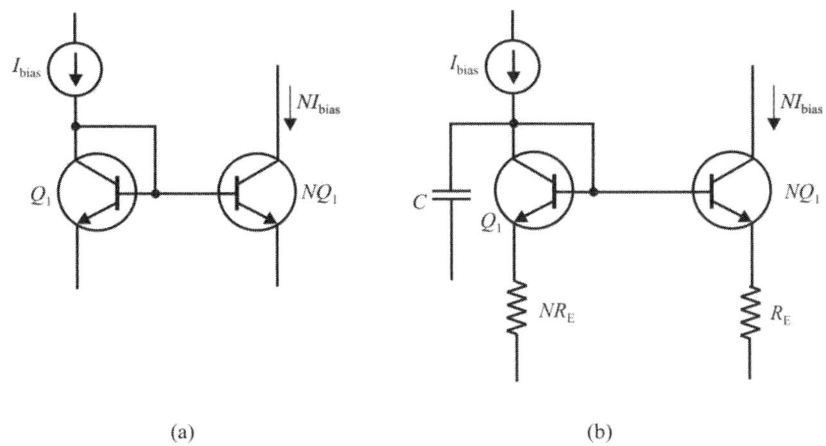

Figure 8.11 Current mirrors: (a) basic current mirror and (b) with de-noising capacitor C.

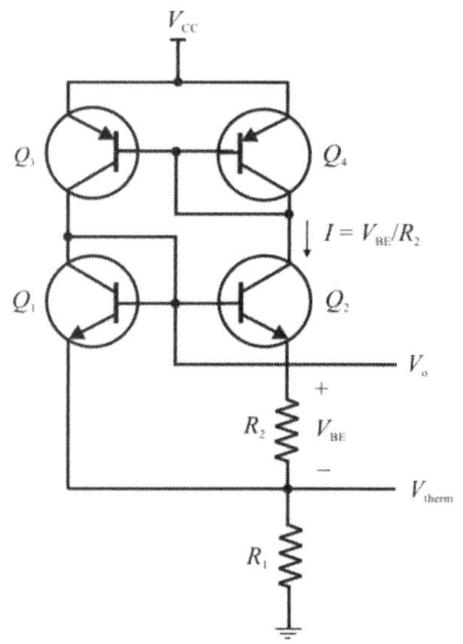

Figure 8.12 Temperature reference circuit.

8.11(b), and resistor degeneration can be put into the circuit to reduce the gain of the transistor, also shown in Figure 8.11(b).

With any of the mirror topologies, a voltage at the collector of NQ_1 must be maintained above a minimum level or else the transistor will go into saturation. Saturation will lead to bad matching and nonlinearity.

8.2.8 Temperature Reference Circuits

The general method for making temperature independent references is to add a voltage that goes up with temperature to one that goes down with temperature. If the two slopes cancel, the sum will be independent of temperature. Such a temperature reference circuit is depicted in Figure 8.12.

To realize this idea, bandgap voltage reference circuits are typically used. This produces an output voltage that is traceable to fundamental constants and therefore relatively insensitive to circuit variations, temperature, and supply. With proper choice of R_1 and R_2, the output voltage will have zero variation with temperature, providing at the same time (at the emitters of Q_1 and Q_2) a voltage proportional with temperature (thermometer output, V_{therm}).

Generally, the output impedance of the bias circuitry should be kept small in order to increase the linearity of the output bias stage. However, the output impedance is typically designed to have a large resistance in order to reduce the noise contribution from the bias circuitry, and to avoid significant loading on the RF input port. For example, using an inductor in the bias circuit to form low impedance near DC, and high impedance near the RF signal band. An RF choke (RFC) is sometimes used for this purpose; such devices are essentially short circuits at DC and open circuits at RF.

8.2.9 BJT LNA Biasing

It is difficult to compare different bipolar LNAs if the biasing arrangement is not mentioned. Using an external bias from an external power supply can increase the IIP_3 of an LNA compared to an LNA with on-chip bias.

8.2.9.1 Active Voltage-Feedback Bias

The biasing scheme shown in Figure 8.13 uses two transistors (PNP and NPN) in a voltage-feedback scheme from the collector to the base of Q_1. The voltage-feedback maintains the collector of the RF LNA Q_1 at one-half the supply voltage plus one V_{be} (about 0.4V). Transistor Q_2 measures the voltage difference between the collector of Q_1 and the center point of the two resistors R_1 and R_2, which is amplified and fed back to the base of Q_1. Q_1's collector voltage holds accurately at half the

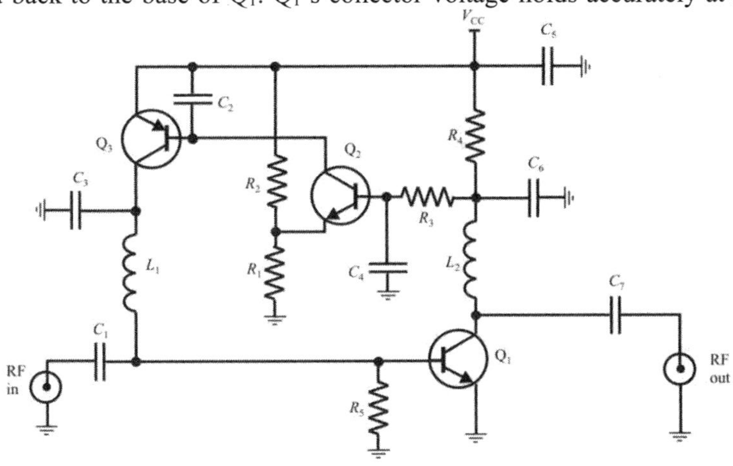

Figure 8.13 Voltage feedback bias.

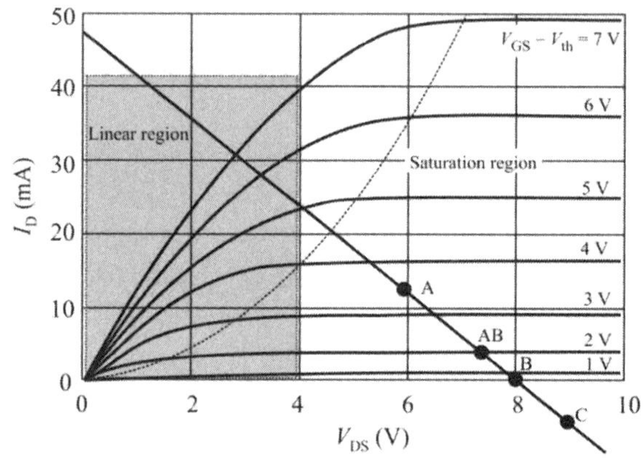

Figure 8.14 FET output characteristics.

supply voltage plus one V_{BE} because of the gain in the feedback loop. Gain changes with temperature in all of the transistors are corrected for, and no adjustable resistor is required. This bias circuit has to be carefully bypassed at both high and low frequencies as well, and for the same reasons.

8.3 MOSFET Bias

FET output operating characteristics are similar to those for the BJT. An example is shown in Figure 8.14 along with bias points. Bias networks are used to put an FET at the intended quiescent operating point. For example, we might want to operate an FET in a PA at 12V V_{DS} and at 50% of the saturated drain current ($I_{DSS}/2$). This is the quiescent point. The same as in the BJT case, there are at least three ways to bias an FET amplifier to get to the intended quiescent operating point.

One option is to have separate DC power supplies for the gate and drain connections, with the gate supply being adjustable, and ground the source. Grounding the source will provide the most gain and efficiency from the FET. In this case generally the gate bias supply is a fixed negative power supply around –5V, with an adjustable resistor-divider network being employed to supply the needed gate voltage.

Another method of biasing an FET is with an active bias network, generally designed in a feedback configuration to maintain constant quiescent current.

Common source FETs can utilize a common Class-A biasing technique called *source bias*, a form of self-bias, shown in Figure 8.15. With FETs, unlike BJTs, no gate current will flow with an input signal present, so the drain current will always

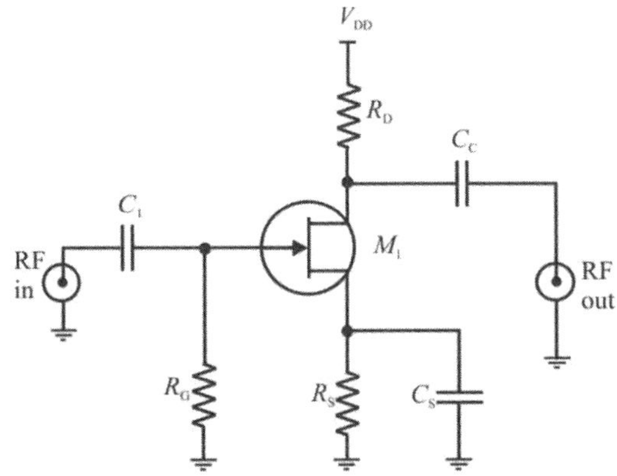

Figure 8.15 FET source bias; in this case a JFET.

be equal to the source current. However, source current does flow through the source resistor R_S, creating a positive voltage at the top of this resistor. Since the FET's source is shared by both the drain and the gate circuits, and the gate will always be at 0V with respect to ground (since no gate current equals no voltage drop across R_G), then the gate is now negative with respect to the common source. This allows the FET to be biased at its Class A, AB, or B quiescent currents I_{CQ}'s, depending on the value chosen for R_S, while a capacitor can be inserted across R_S in order to restrain the bias voltage to a steady DC value and as an RF bypass for R_S. The Source decoupling capacitor C_S is required to ensure that there is no RF power loss on R_S. A downside to using self-bias schemes is that amplifier efficiency is lost due to the voltage drop of the source resistor.

8.3.1 FET Active Feedback Bias Example

Consider the GaAs FET amplifier in Figure 8.16. With the components shown, this amplifier is designed for 423-MHz operation. When the DC current through diode D_1 increases (due to temperature variation or change in device), the voltage drop across R_3 increases which, in turn reduces V_{BE} at Q_1. The collector current of Q_1 drops, reducing the voltage drop across R_8. This reduces V_{GS} and I_D of Q_2 to correct for the initial change in current through D_1.

8.3.2 Biasing MOSFET Amplifiers

Since MOSFETs have gate threshold voltages up to 5 to 6 volts, they require some gate bias voltage in most applications. They can be operated in Class C (zero gate bias), but at a cost of low power gain. Zero bias is often used in amplifiers intended

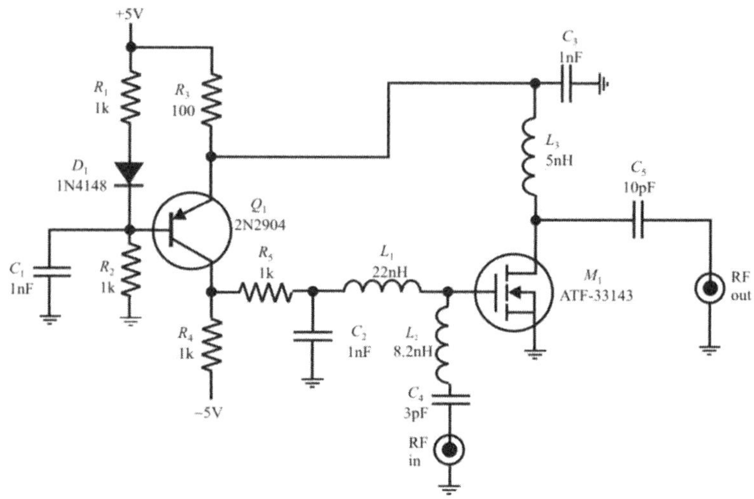

Figure 8.16 GaAs FET LNA. Operation at 423MHz.

for signals that do not need linear amplification (such as FM signals and some forms of CW signals), and efficiencies in excess of 80% are not uncommon. In Class B, the gate bias voltage is set just below the threshold, resulting in zero drain idle current flow. The power gain is higher than in Class C but the drain efficiency is 10 to 15% lower. Class B is also suitable only for nonlinear amplification. Between classes of operation, we must decide whether the system has power gain to spare and how important efficiency is. At higher frequencies, such as UHF, a good compromise may be Class B or even Class AB.

In Class AB, the gate bias voltage is somewhat higher than the device threshold, resulting in drain idle current flow. The idle current required to place the device in the linear mode of operation is usually given in a data sheet. In this respect, MOSFETs are much more sensitive to the level of idle current than are BJTs. They also require somewhat higher current levels compared to BJTs of similar electrical size.

The temperature compensation of MOSFETs can be most readily accomplished with networks consisting of thermistors and resistors as illustrated in Figure 8.17 (R_3 is the symbol for a thermistor). The resistor network values must be determined according to the thermistor characteristics and the g_{FS} of the FET. The changes in the gate threshold voltage are inversely proportional to temperature and amount to approximately 1 mV/°C. These changes have a larger effect on the I_{DQ} of an FET with a high g_{FS} than one with a low g_{FS}. Unfortunately, the situation is complicated by the fact that g_{FS} is also reduced at elevated temperatures, making the drain idle current dependent on two variables.

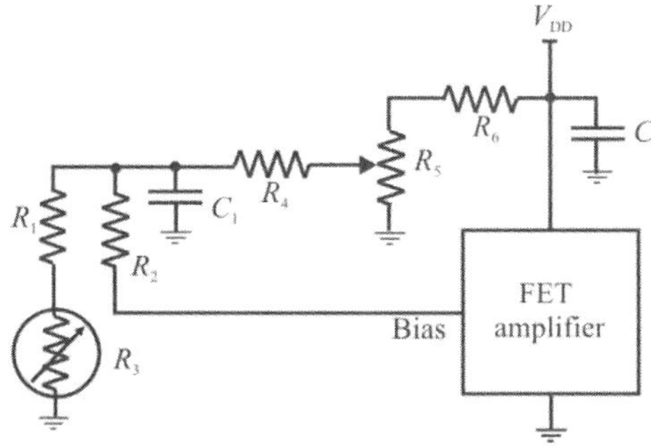

Figure 8.17 Thermistor biasing of an FET amplifier.

The thermistor is thermally connected to a convenient location in the heat source in a manner similar to that described for the compensating diodes with BJTs discussed earlier.

8.3.3 Cascode MOS LNA Bias

A cascode MOS LNA is shown in Figure 8.18. This amplifier is for 1.3-GHz operation with the component values given in the figure. Cascading transistor M_2 is used to reduce interaction of the tuned output with tuned input. Transistor M_3 essentially forms a current mirror with M_1. The current through M_3 is set by the supply voltage and R_2 in conjunction with V_{GS} of M_3. The bias resistor R_1 is chosen

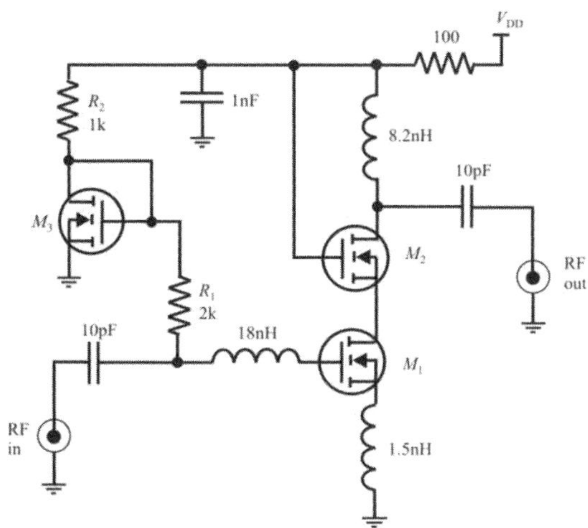

Figure 8.18 Cascode MOS LNA bias. Component values shown are for 1.3 GHz.

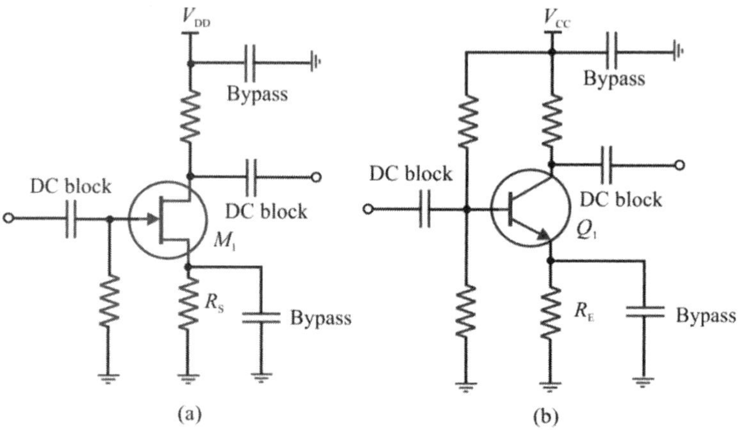

Figure 8.19 Passive bias circuits for (a) MESFET (depletion mode)/JFET and (b) BJT.

large enough that its equivalent noise current is small enough to be ignored. In a 50 Ω system, values of several hundred ohms to kiloohms or so are adequate.

8.4 Passive Bias Methods

In some cases, bias stabilization may be accomplished with only passive circuit elements. An emitter or source resistor provides negative feedback to stabilize bias current. For example, as seen in Figure 8.19, R_S is a self-bias resistor. $V_{gs} = -I_D R_S$ in order to provide a negative gate-source voltage. If I_D increases, V_{gs} decreases to compensate. But wiring inductance is introduced in the source circuit, even with a bypass capacitor across R_S. This will become significant when $\omega L \rightarrow 1/(10g_m)$ which is usually a very small inductance.

Referring again to Figure 8.19, the BJT circuit uses the conventional 4 resistor bias approach where the emitter resistor provides negative feedback stabilization against drift of the bias point with temperature or device parameter variation from batch to batch of devices.

But these circuits are not often used for RF applications because the biasing resistances load the circuit, reduce the gain, and introduce noise. Therefore, circuit techniques that permit use of a directly grounded source or emitter connection are preferred for high-frequency amplifiers when implemented using discrete components on PC boards.

The MESFET (or JFET or PHEMT) circuit uses two power supplies when the source is grounded on a PCB because the threshold voltage of a typical microwave FET is negative. Microwave FETs are always N-channel.

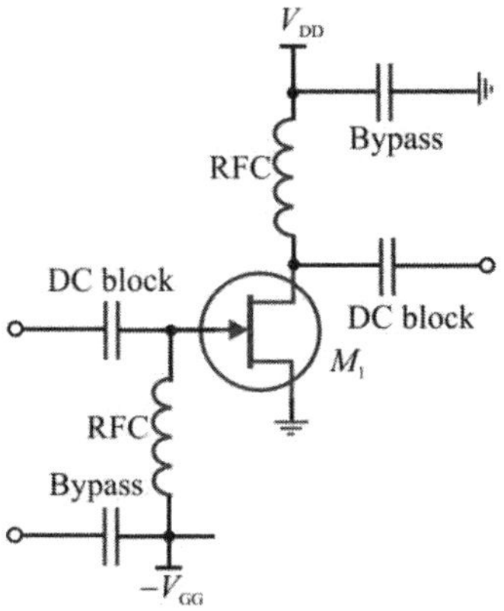

Figure 8.20 Example of GaAs or InP FET bias insertion using two power supplies and an RFC.

So, RFCs are used for biasing in some cases as shown in Figure 8.20. They provide low Z at DC and high Z at the design frequency and so gain is not usually sacrificed. The RF noise introduced is minimal. However, they always have resonances, so a complete RFC equivalent circuit model is essential if the circuit performance is to be accurately predicted.

When possible, a better approach for bias insertion is to use the matching network itself to insert bias voltages. This is possible when a shunt shorted stub or shunt inductor is connected to the input or output terminals of the device without any series DC blocking or impedance matching capacitor. The grounded end is bypassed with a grounded capacitor with low impedance over the full range of frequencies (see Figure 8.21).

Another bias insertion method that is widely used and more suitable for microwave circuits utilizes high impedance lines that are one-quarter wavelength at the design frequency. This is shown in Figure 8.22 as a bias feed for the drain. If a low impedance bypass capacitor is used to short the bias feed end, the $\lambda/4$ line transforms the short into an open at f_0. An added bonus comes from short circuiting the even harmonics because the line has lengths of $\lambda/2$ at $2f_0$, λ at $4f_0$, and so forth. This can improve efficiency of some power amplifiers or can be used with a resistor to force the stability factor K to increase at $2f_0$ by killing high frequency gain with lossy loading.

8.5 Active Bias Circuit

Active devices are useful for providing bias levels in power amplifiers. We will demonstrate that here with an example shown in Figure 8.23.

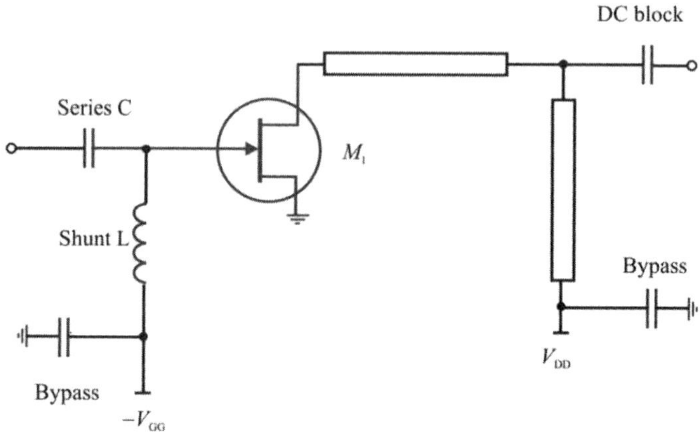

Figure 8.21 Biasing through matching networks. Examples of lumped (input) and distributed element (output) matching are shown.

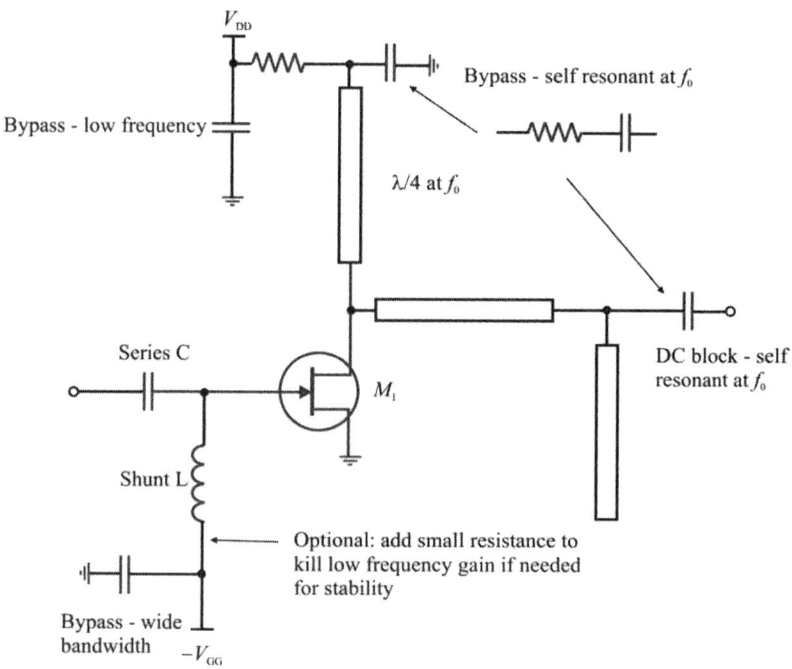

Figure 8.22 Bias insertion with quarter-wave high impedance line.

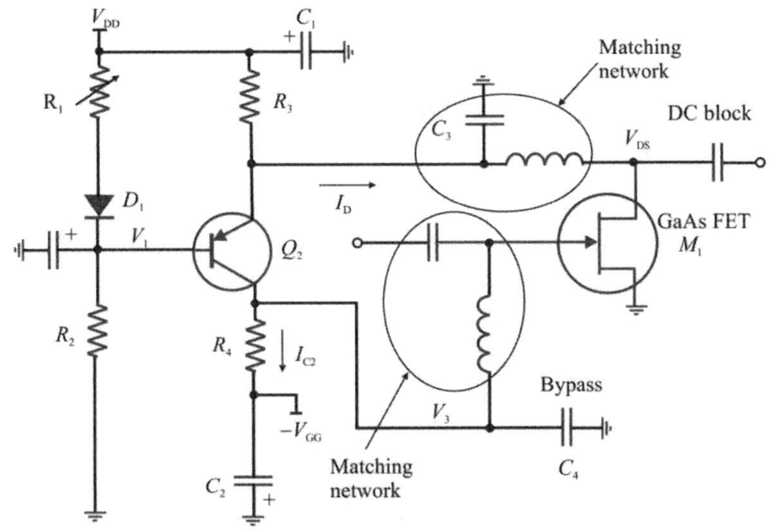

Figure 8.23 Active bias generator (Q_1 on left) provides drain and gate voltages to the GaAs FET M_1.

Example 8.1: We have shown in Figure 8.19 that passive biasing approaches using resistors can load an amplifier creating extra losses and add source or emitter inductance. From a performance point of view, the best practice is to directly ground the emitter or source for microwave amplifiers. But, grounding the emitter or source leaves the devices wide open to DC bias problems such as thermal runaway on the BJTs or temperature drift of the bias point for FETs. So, an active feedback circuit as shown in Figure 8.23 is popular for biasing. An approach using a PNP transistor as shown is frequently adequately efficient.

Q_2 can be nearly any lower frequency PNP with adequate current capacity. V_{DS} is set by the voltage divider as follows.

$$V_1 = V_{DD}\left\{\frac{R_2}{R_1 + R_2}\right\} - V_{D1}$$

and

$$V_{D1} = -V_{be,Q2} = 0.7\text{V}$$

So

$$V_{\text{DS}} = V_1 - V_{\text{be,Q2}} = V_{\text{DD}}\left(\frac{R_2}{R_1 + R_2}\right)$$

$$I_{\text{D}} + I_{\text{C2}} = \frac{V_{\text{DD}} - V_{\text{DS}}}{R_3}, \qquad I_{\text{C2}} << I_{\text{D}}$$

$$I_{\text{C2}} = \frac{V_3 + V_{\text{GG}}}{R_4}$$

D_1 provides for temperature compensation and can be a diode-connected PNP of the same type as Q_2.

The DC feedback can be seen as follows: Suppose that

1. I_{D} increases (maybe due to temperature variation or change in device).
2. Then, the voltage drop across R_3 will increase, reducing $V_{\text{be,Q2}}$.
3. I_{C2} drops, reducing the voltage drop across R_4. This reduces V_{gs} and I_{D} to correct for the change in step 1.

C_1 and C_2 are low-frequency bypass capacitors. These are typically polarized tantalum or electrolytic caps that are used for stability of the active bias circuit. C_3 and C_4 are high-frequency bypass capacitors. These must NOT be tantalum or electrolytic types, but must be high-quality ceramic chip caps with very-low series inductance. Any series L must be included as part of the matching network.

R_3C_3 and R_4C_4 add two low-frequency poles to the feedback circuit of the active bias feedback system. We know that a second order FB system can be underdamped (ringing) or oscillatory if the phase margin is incorrect. For stability, a dominant pole should be created with one of the two RC networks with a time constant at least 10 times larger than the other. This provides nearly a 90° phase margin.

While Example 8.1 implemented biasing for a MOSFET, a similar approach can be used for BJTs. Such an approach is illustrated in Figure 8.24.

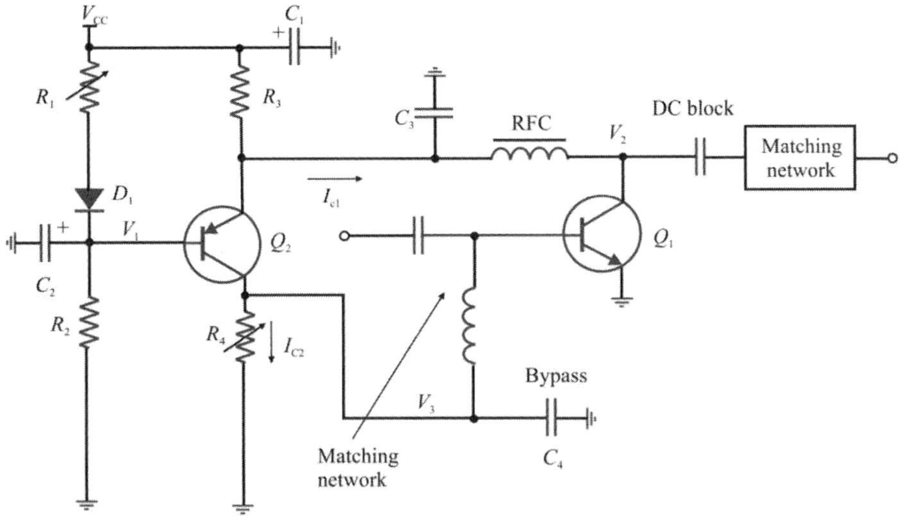

Figure 8.24 Active bias BJT. A similar active feedback approach can be used for biasing a BJT.

8.5.1 DC Biasing of RF Feedback Amplifiers

This approach provides good bias stability for microwave feedback amplifiers (see Figure 8.25).

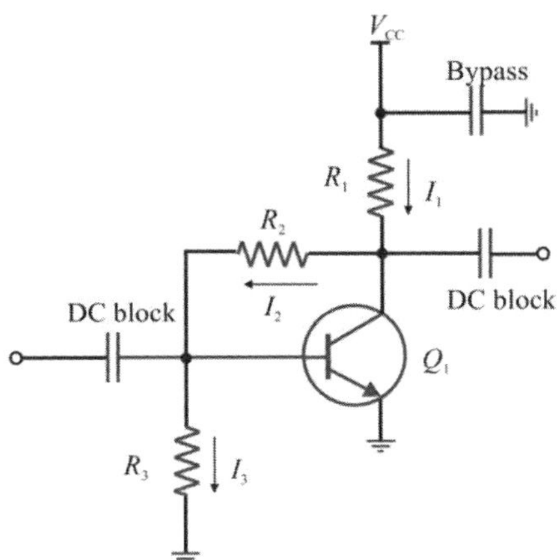

Figure 8.25 Shunt-shunt feedback DC biasing example.

$$I_3 = \frac{V_{\mathrm{BE}}}{R_3}$$
$$I_2 = I_1 \quad \text{if } I_{\mathrm{B}} \text{ is small}$$
$$V_{\mathrm{CE}} = V_{\mathrm{BE}} + I_2 R_2$$
$$I_1 = \frac{V_{\mathrm{CC}} - V_{\mathrm{CE}}}{R_1}$$
$$I_{\mathrm{C}} = I_1 - I_2$$

8.6 MMIC Amplifier Biasing

A typical MMIC biasing scheme is illustrated in Figure 8.26. The current-biased MMIC will attempt to draw more current as the temperature rises. The biasing is primarily determined by the current, where the voltage is relatively unimportant. The effectiveness of this temperature bias control is dependent on the voltage drop across R_{BIAS}, a value of up to 4V may be required for proper stabilization across a –25°C to +100°C temperature range.

If R_{BIAS} does not add up to 600 Ω or more, then the gain of the MMIC stage will suffer. If R_{BIAS} does not compute to be at or over 600 Ω, then an RFC should be added in series with R_{BIAS} to increase the output to this value, or approximately $R_{\mathrm{BIAS}} + X_{\mathrm{L}} > 600\,\Omega$.

These amplifiers are unconditionally stable at all frequencies, and they can be easily cascaded for higher gain. The inductor bias-feed circuit is preferable to obtain low-noise performance, high gain, and linearity. By using an inductor as the bias feed, the current becomes constant and can increase in the large-signal region to extend its output power.

8.7 Concluding Remarks

We have discussed biasing for transistor (BJT and MOSFET) amplifiers in this chapter. There is a plethora of methods available to accomplish this biasing. The methods we presented are only examples of the many approaches possible. These designs are only limited by to the imagination of the designer.

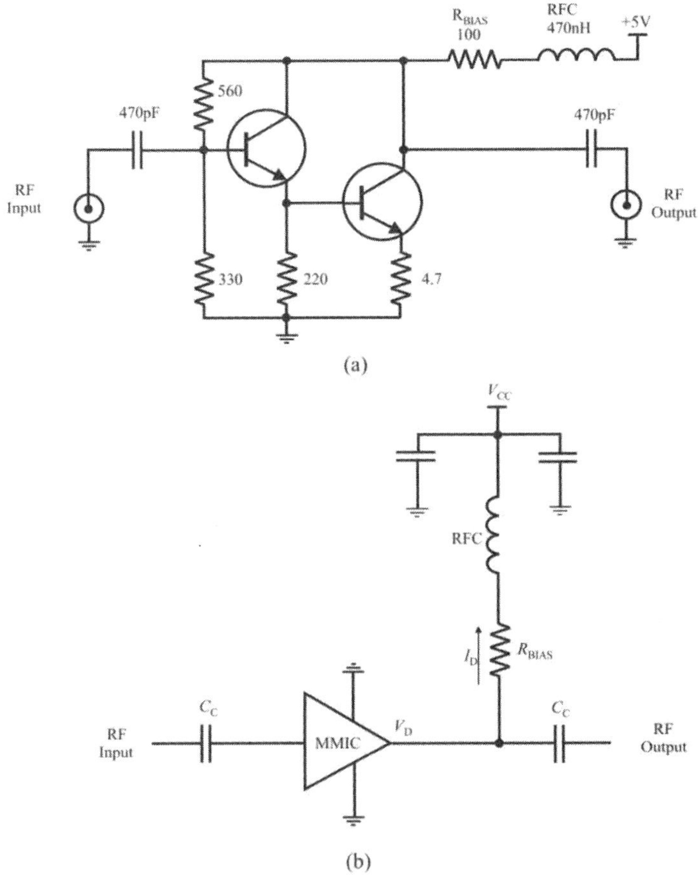

Figure 8.26 MMIC biasing (a) internal schematic and (b) actual components. DC blocking caps, R_{BIAS} and RFC are external components.

References

[1] Gonzalez, G., *Microwave Transistor Amplifiers Analysis and Design,* 2nd Ed., Upper Saddle River, NJ: Prentice Hall, 1997, Section 3.9.

[2] Millman, J, and C. C. Halkias, *Electronic Devices and Circuits*, New York: McGraw Hill, 1967, Chapter 10.

[3] *Transistor Manual*, *General Electric,* 7th Ed., Syracuse, NY: 1964, Ch 4.

Chapter 9

RF Power Amplifiers

9.1 Introduction

The power amplifier (PA) in an EW system is used to amplify signals generated in the exciter to significantly higher power levels for radiation via the antenna, eventually to find their way into the target receivers. Such signals disrupt the operation of the adversary's command and control communication system.

The signals generated by this subsystem are emitted by the system with the intent of interfering with the opponent's communications. As mentioned in Section 1.4, such interference is, of course, on a noncooperative basis. Therefore, most of the time it is necessary to inject more energy into the target receiver than the target transmitter does. Typically (although not always) the cooperating transmitter and receiver are closer together than the EW system and the receiver. These factors, then, dictate that the EW system must emit considerably more raw power than the cooperating transmitter.

This chapter introduces the characteristics of RF and microwave power amplifiers. The microwave band is rather loosely defined, but most definitions have it starting at 300 MHz, the low end of the UHF frequency band. While we cover the basic introduction to these amplifiers here, for those interested in a more thorough treatment, [1] is recommended.

Whereas noise performance is paramount in the low-level wideband amplifiers discussed in Chapter 4, the attention shifts to nonlinear performance and its associated intermodulation performance, efficiency, and input/output impedance matching for power amplifiers. Heat removal also is an issue with PAs. The power amplifier (PA) is usually the last amplification stage before the signals are emitted from the antenna. Therefore, maximum power transfer into and out of the PA is important.

There is a type of unwanted response in electronic systems that arises due to imperfections in the constituent components that make up EW systems. In particular, nonlinear responses of circuits generate unwanted spurious signals at the output. The sources of these spurious signals and the methods to reduce their impact are discussed in Chapter 11.

This chapter is organized as follows. After this introduction, some definitions and general characteristics of RF amplifiers as they are used in EW systems are presented. The two principal types of amplifiers used in EW systems are the low-noise amplifier (LNA) at the system input and the high-power RF amplifier used to transmit the jamming signal. Characteristics of these two types of amplifiers are discussed next. In general, EW systems must be broadband to cover a significant portion of the RF spectrum. The factors that limit the frequency response of amplifiers are then discussed. Lastly, some of the characteristics of RF and microwave amplifiers that are more or less unique to EW systems are presented.

9.2 Power Amplifiers

Consistent with discussions in Chapter 4, the common designations of RF PAs are classes A, B, AB, C, D, E, F, F^{-1}, and S [2–5]. These classes define differences in the methods of operation, amplifier efficiency, linearity, and power-output capability. Classes A, B, AB, and C were discussed in Chapter 6, as they are classes that are not specifically focused on PA applications. The remainder of the classes are focused on PA applications and therefore are discussed in this chapter. The RF PA is one of the most critical building blocks in EW/RF transmitters and is also the most power-hungry component in the RF front end.

An RF PA is a circuit that converts DC power into RF/microwave output power. The active components in RF PAs are FETs (of several varieties), BJTs (also of several varieties), and vacuum tubes. When originally discovered, RF power was generated by sparks and arcs. The triode vacuum tube (the triode is the simplest kind of vacuum tube that can amplify) was invented in 1906. Even though the transistor was invented in 1948, the vacuum tube was the main active component in RF PAs from about 1920 until the 1970s when the transistor PA became feasible. Vacuum-tube PAs still dominate some applications such as high-power generation at the higher frequencies. Discrete solid-state RF-power devices appeared at the end of 1960s with the introduction of silicon bipolar transistors. New kinds of transistors for PA application as well as for lower power uses are continuing to be discovered.

Typically, one of the key characteristics of power amplifiers in communication EW systems is that they must operate over a wide frequency range. Communication EW systems must be designed to address a wide range of targets—the target array is typically not completely known ahead of time. A ground-based communication

EW system, for example, might be designed to cover the 20–150 MHz frequency range (higher than this limit becomes impractical for ground-based systems since RF signals do not propagate very well close to the surface of the Earth). Airborne EW systems, on the other hand, can have a substantially broader frequency range than this since they are not necessarily limited by the signal propagation characteristics close to the Earth. This broad frequency range makes the design of such amplifiers somewhat different from those found in the commercial marketplace.

The basic topology of a single stage RF/microwave PA utilizing an N-channel MOSFET is illustrated in Figure 9.1. The amplifying device (an FET shown in this case) is placed between an input matching circuit and an output matching circuit. DC power, which provides bias voltages and currents, and which gets converted into the output RF power, is supplied though an RF choke, L_2, whose purpose is to keep the RF signal within the PA from returning to the power supply–it is a low-pass filter in this function. There are two bias supplies shown in Figure 9.1. V_{DD} supplies the voltage for the drain and V_{GG} supplies the bias for the gate. V_{GG} also is filtered by an RF choke (L_1) whose purpose is the same as L_2 for V_{DD}. C_G and C_D are coupling capacitors to keep the bias voltages and currents isolated to this stage while the C_b capacitors provide for high-frequency filtering for this stage close to the components making up this stage.

Sometimes there is more than one stage required to amplify weak RF signals to the power levels adequate for EW purposes. When this occurs matching between stages may or may not be used. However, it is highly desirable that a single power amplification stage be used when possible. This may be accomplished, for example,

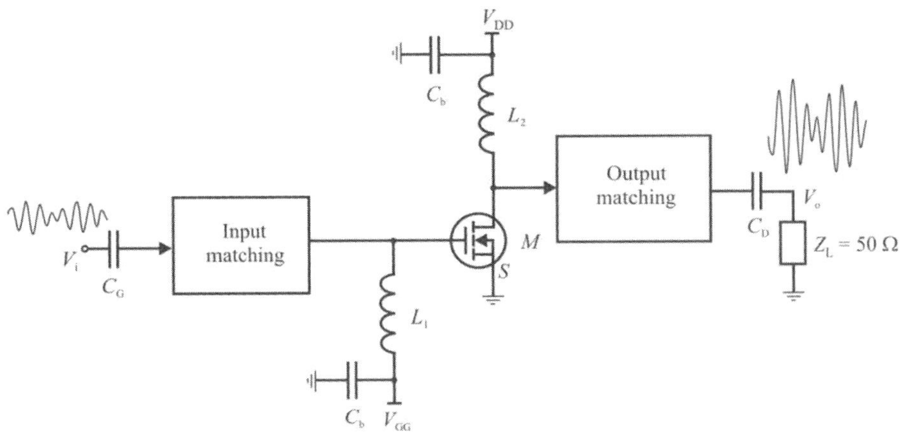

Figure 9.1 Basic topology of a PA (in this case using a MOSFET).

by using multiple power amplification stages in parallel, splitting the input signal for these stages, and combining their outputs.

The matching circuits are necessary for proper termination and interfacing between stages or between the last stage and the output load. The impedances of active devices used in RF/microwave applications are in general different on the inputs and outputs. The input matching circuit matches the impedance level of the previous stage to the PA stage. This is usually to accomplish maximum power transfer. The load, which for an EW system PA is usually an antenna, is connected at the output of the output matching circuit. This matching circuit matches the load to the output of the active device, which are usually at different impedance levels. Again, this circuit is usually intended to accomplish maximum power transfer by providing a conjugate matching impedance to the load.

9.2.1 Typical Solid-State Power Amplifiers

For many years, solid-state amplifier designs were based solely on power BJT technology; however, power FET amplifying devices are also now used. Amplifiers using BJT power transistors are still used, however. Many of these component devices were designed to operate directly from 12.6 VDC (nominal) vehicular power sources, whereas, power FET's classed at 25W or above usually require and allow higher operating voltages, complicating power supply requirements, particularly in mobile EW applications. For fixed installations or other tactical configurations where adequate space is available (i.e., shipborne) a power supply limitation is not necessarily as critical.

RF power BJTs are available that produce from under 1 to 60 or more and power FETs can generate up to the 250Ws output and more. It is typical in BJT PA designs to use a single stage driver feeding hybrid splitters at the input to two or four "push-pull, parallel" devices and recombining the outputs using hybrid devices.

Figure 9.2 shows a flow diagram of the typical arrangement of components in solid state amplifiers. The active devices are usually operated in class B mode, drawing very small amounts of current without a driving signal. Their operation is determined by drive level so the greater the amount of drive power, the greater is the DC current drawn from the power source.

90° hybrid couplers can be used, or as shown in Figure 9.2, a Wilkinson design, using stripline elements. Two half wavelength 70 Ω lines provide 50 Ω source and output impedances over reasonable bandwidths. The load terminations are 100 Ω resistors, usually mounted on a heat sink that also is the sink for the active devices. The resistors dissipate little (theoretically no) power as long as the energy at their opposing ends are 180° out of phase and equal in voltage amplitude as long as circuit balance is maintained.

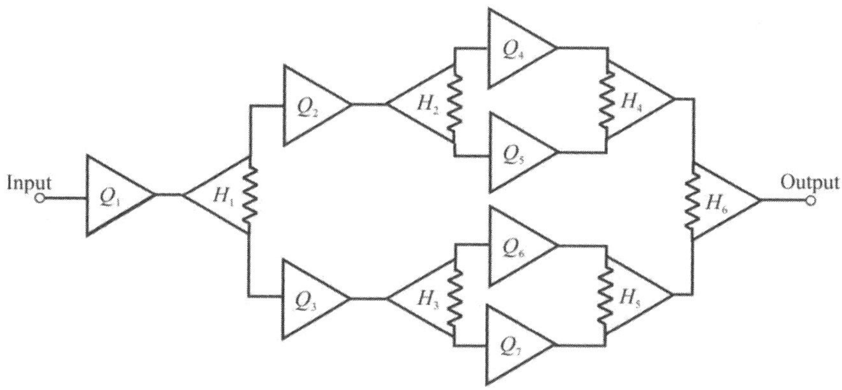

Figure 9.2 Typical topology of a solid-state PA showing a single stage driving a push-pull stage which, in turn, is driving a quad stage arrangement of components. Q_1–Q_7 are power BJTs or MOSFETs while H_1–H_6 are hybrid couplers. H_1–H_3 function as splitters while H_4–H_6 function as couplers/combiners.

The dynamic impedance at both the input and output of these stage(s) will change as applied driving power changes. In an amplifier there is only a relatively narrow range of output levels for a particular dynamic operating impedance. Thus, class B and class C amplifiers must be designed to meet particular target output power levels, and can be expected to perform at their highest efficiency within a relatively narrow range of output power levels.

Power FET amplifiers have the same considerations. Their input impedance is much higher than BJT amplifiers and correct input matching is even more critical if stable, low IMD performance is to result. If any amplifier is significantly overdriven, device failure often results. Overexcitation will result if excessive current is drawn by the BJTs or FETs, causing component failures and/or damage to circuit boards or other components due to the generation of excessive heat.

Most power transistors employ multiple junctions on a single substrate. Partial component failure such as the failure of some of the internal junctions in a power transistor in a single-ended stage can lead to unpredictable performance and/or spurious signal generation. If the stage employs multiple transistors (i.e., a push-pull stage) the failure of one device will result in significant circuit unbalance, normally causing total failure of the remaining device(s) and/or the generation of unwanted, spurious signals before complete failure occurs.

9.2.2 Basic Power Modules

High-gain amplifier stages tend to become unstable and oscillate with age and temperature variations. These oscillations produce unwanted spurious output

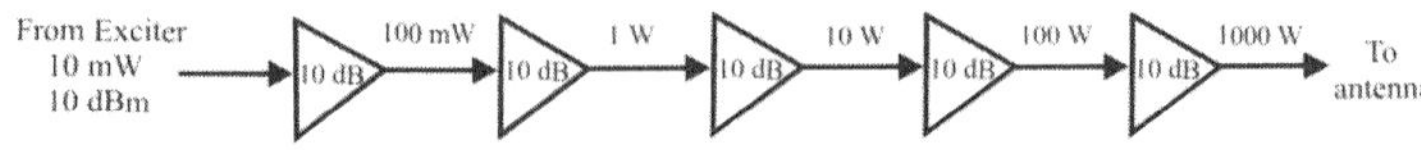

Figure 9.3 Amplification chain using basic power modules. Each stage provides modest gain, while boosting the exciter signal to a level adequate for EA operations.

signals. Therefore, it is sometimes necessary to gang these circuits in series to achieve the necessary power output (see Figure 9.3). The individual circuits that form one element of this gang are called basic power modules. Forming the high-power signal is done in stages with each stage contributing only a moderate amount of gain. Moderate gain amplifiers such as these are easier to design for stability and reliability than high gain stages.

A parallel configuration for a high-power amplifier using relatively low-gain basic power modules and combiners for RF EW applications is shown in Figure 9.4. The input signal is routed through a splitter that could just be wires (but not necessarily). The splitter is often included for impedance matching. Once the signal is routed to the amplifiers, the outputs are combined to generate the high-power signal.

Sometimes switches are added to this tree of amplifiers for more flexibility. If two or more signals at different frequencies are to be amplified, the amplifier tree can be split in half or some other combination in order to accomplish this.

It is not at all uncommon to power share the power amplifier in an EA system. Power sharing is splitting the available power from the amplifier between two or more signals at the same time. This allows a single amplifier to jam more than one signal simultaneously. Such sharing must be carefully managed, because the output

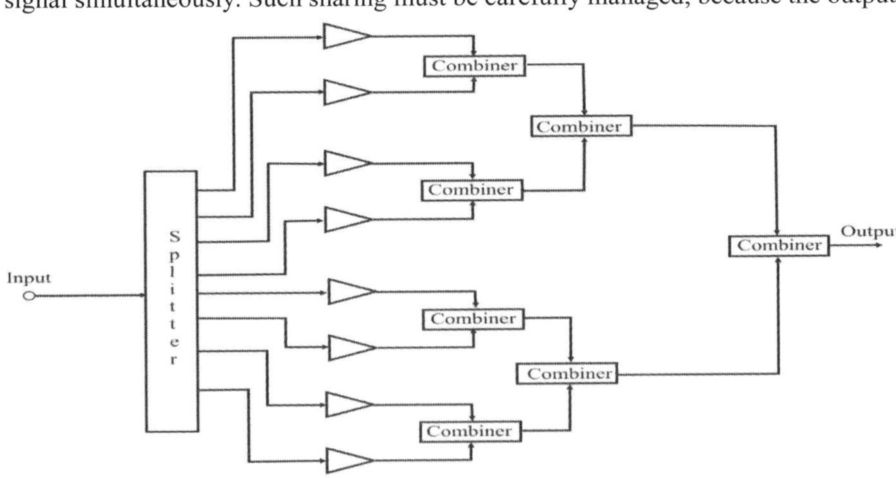

Figure 9.4 Typical configuration of a high-power amplifier made up of basic power modules.

signal power per frequency decreases approximately as the square of the number of signals sharing the available power [6].

Power sharing places significant linearity constraints on the amplifiers in the chain. Nonlinear amplifiers generate frequency cross products that can occur at frequencies that interfere with friendly communications. Time-sharing of the power amplifier is also possible. Time-sharing implies that the frequency of the signal source can be changed very rapidly.

9.2.3 Broadband Power Amplifier Considerations

It was noted in Chapter 4 that achieving a wide bandwidth in an amplifier is difficult to do. One approach to do so, however, is to use balanced amplifiers where two distinct amplification paths are implemented. The one amplifier is fed by the no shift output of a 90° hybrid while the second amplifier is fed by the output of the hybrid that has been shifted. Likewise, the outputs of the amplifiers are fed to a 90° hybrid that effectively removes the input phase shift. The advantage of doing this is to provide good impedance matching to the input and output and wide bandwidths are possible.

9.3 Power Amplifier Parameters

In this section we discuss some of the major parameters associated with PAs.

9.3.1 Crest Factor

The crest factor and peak-to-mean ratio are measures of the statistics and variability of a signal $s(t)$. The *crest factor* (CF) is the ratio of the peak value of $s(t)$ to its RMS value. Amplitude-symmetric square waves, or DC levels, have a crest factor of one since their peak value is equal to the RMS value. Other waveforms with more complex structures have higher crest factors. A sine wave has a crest factor of $\sqrt{2}$ since $s_{\text{peak}} = \sqrt{2}s_{\text{RMS}}$.

Consideration of the CF in PAs is important for efficiency and compatibility issues. A PA is most efficient (highest level of conversion of primary power to output power) when it is operated just under its 1-dB compression point. This is the point of average power amplification. However, as the signal rises above that point, the PA becomes nonlinear in its amplification performance and this places unwanted energy into adjacent channels, thereby causing interference to other communication services that may be using that channel. This causes the amplifier to be biased further down its characteristic curve, reducing efficiency while also reducing interference.

The *peak-to-mean ratio* (PMR) is the ratio of the peak power in $s(t)$ to its RMS power. Typical values of PMR are in the range of 6-dB to 13-dB.

9.3.2 Peak Envelope Power

The *peak envelope power* (of a radio transmitter) is the average power supplied to the antenna transmission line by a transmitter during one radio frequency cycle at the crest of the modulation envelope taken under normal operating conditions.

9.3.3 Stage Matching

Whereas careful design of the input matching network is required to realize the desired noise characteristics of LNAs, in power amplifiers more emphasis is placed on optimizing the output matching network. This is because the signal levels at the input to PAs are usually of sufficient amplitude relative to the noise at that point. The low input impedance typical of bipolar power devices, however, creates special problems that require special treatment in the input matching network for wideband amplifiers.

A key issue for multistage amplifiers is the ability to cascade individually designed stages without a requirement for retuning or redesign to account for the characteristics of the driving or following stages. In many cases, the use of balanced amplifiers permits the benefit of 3-dB coupler interstages, which direct reflected power to the isolated port rather than the driving stage. As we will see later, there are special problems of nonlinear oscillations arising from interaction between signal harmonics and modes of the output matching structure.

9.3.4 Bias Circuits and Bias Circuit Instabilities

Once the amplifier is designed, it remains to provide the dc bias voltages and currents required for the active device. This is no simple problem, as the arrangements to introduce the biases can disturb the functioning of the circuit. Generally, high impedance microstrip traces can be used as decoupling inductors, but caution must be exercised not to create a low-frequency oscillator circuit in the bias network.

A common cause of trouble is the use of an inductor with a large bypass capacitor, which can create a resonator in the megahertz region that can support oscillation of the active element, which will have very high gain at lower frequencies. Bias-circuit instabilities are a common source of problems in amplifiers and other active circuits. These generally result from the use of inductors and capacitors in the bias circuit without regard to resonances or situations where 180° phase shift can occur.

Two examples of circuit configurations that can promote bias oscillations at low frequencies are (see Figure 9.5):

- The use of an inductor with bypass capacitors on both sides as shown in Figure 9.5(a) to filter the DC supply to earlier stages of an amplifier; this can have 180° phase shift at a lower frequency where the active element has substantially more gain than at microwave frequencies. To avoid oscillations lossy inductors should be used, either because of resistive loading or by use of lossy ferrite cores to provide RF isolation.
- The use of an inductor with a bypass capacitor to isolate dc supply input to the base or gate of the active element; this can form a resonator for lower frequency oscillations. The circuit in Figure 9.5(b) contains a resonant circuit consisting of the series inductor and capacitor that can result in parasitic oscillation; in this case, the small current required for base or gate bias permits the use of resistance in series with the inductor. A wire-wound resistor can be used to accomplish both functions at the same time.

Even though we are addressing microwave amplifiers in this section, it is generally useful to sketch the low-frequency equivalent circuit to be sure that no instabilities can exist. Lossy elements are required to guarantee stability in the bias circuit. In the past, for economic and reliability reasons, emphasis was placed on using the minimum number of components. Integrated circuits have superseded this concept, and active feedback bias circuits are generally used to insure the stability of device operating conditions.

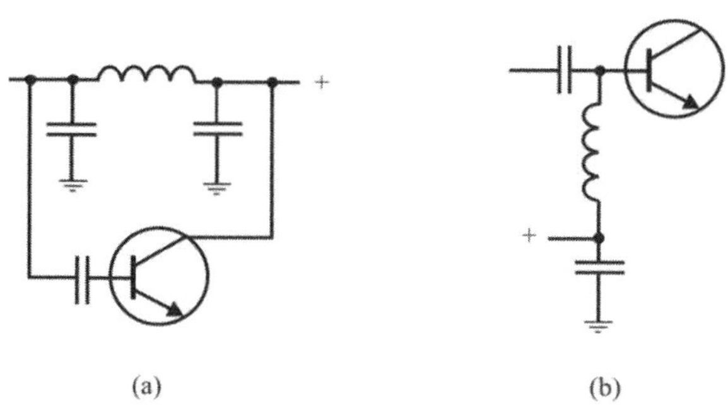

Figure 9.5 (a, b) Parasitic bias.

9.3.5 Efficiency

Efficiency is a measure of how well a device, component, or system converts energy from one form to another. The efficiency of a power amplifier in an RF EW system reflects its ability to convert prime power, such as from a generator, into power out of the amplifier that can be radiated. Typically, such efficiencies are not very high. Achieving a system efficiency of 50% would be quite remarkable; values of 20% to 30% or less are more common. What doesn't get converted to RF energy goes into heat, which must be dissipated by some means, and in EW applications is universally a bad byproduct of energy conversion.

The efficiency of an RF device is a function of the operating parameters such as frequency, temperature, input drive level, load impedance, bias point, device geometry, and intrinsic device characteristics. The maximum efficiency under different conditions can be determined using load pull.

Characteristics of RF and microwave devices are usually specified at a single set of conditions (such as characteristic impedance of 50 Ω). Characteristics at different conditions from the nominal are typically determined by a process called *load pull*. Load pull is configuring the device whose characteristics are to be determined in a test jig whose parameters, such as load resistance, can be varied. The parameters of interest are then adjusted and the device characteristics under different than nominal conditions determined.

The lower-case Greek letter eta (η) is often used to denote efficiency with a subscript frequently attached to specify a particular type of efficiency.

Efficiency is not a first-order concern with EW PAs, as there is usually sufficient source power available. It becomes a concern, however, as the PA power increases because the power that is not radiated is converted into heat, and that heat must be dissipated somehow. Such heat decreases the reliability of the PA. We investigate heat removal in more depth in Chapter 12.

9.3.5.1 Measures of Efficiency

In RF power amplifiers the goal is to convert power supply power (normally DC power) to the RF signal output power. Amplifier efficiency is normally specified in one or more of three ways: drain efficiency, power added efficiency, and total efficiency.

Drain Efficiency

Drain efficiency is the ratio of output RF power to input DC power:

$$\eta_D = \frac{P_{RFo}}{P_{DC}} = \frac{P_{RFo}}{V_{DC} I_{DC}} \tag{9.1}$$

Drain efficiency is a measure of how much DC power is converted to RF power. The drain efficiency does not consider the incident RF power that goes into a device so is not a true measure of the efficiency of the device itself. In the case of a single-stage RF PA, the RF input power can represent a substantial portion of the power because the gain is normally low.

Power-Added Efficiency

The input power is taken into consideration when the *power added efficiency* (PAE), denoted here by η_{PA}, is specified. It is given by

$$\eta_{PA} = \frac{P_{RFo} - P_{RFi}}{P_{DC}} \tag{9.2}$$

For a real amplifier, η_{PA} will always be less than η_D, but once we get to 30-dB gain or so [it can be argued that it is even less (10–20 dB)], the two quantities become very close in value because $P_{RFo} = 1{,}000 \times P_{RFi}$ and the second term in the numerator of (9.2) can be ignored. η_{PA} gives a reasonable indication of PA performance when gain is high; however, it can become negative for low gains. η_{PA} is the most popular efficiency measure of the RF/microwave technical cognoscenti.

Total Efficiency

Total efficiency, sometimes called overall efficiency, gives a complete picture of the ratio of output power to both types of input power (DC and RF):

$$\eta_{total} = \frac{P_{RFo}}{P_{DC} + P_{RFi}} \tag{9.3}$$

Average Efficiency

The above definitions imply an instantaneous efficiency that is the efficiency at one specific output level. For most PAs, the instantaneous efficiency is highest at the peak output power (PEP) and decreases as the output decreases. Signals with time-varying amplitudes such as those with amplitude modulation, or pulses, therefore produce time-varying efficiencies. The average efficiency then becomes a more useful measure of performance. The average efficiency [7] is defined as the ratio of the average output power to the average DC input power

$$\eta_{avg} = \frac{P_{RFo,avg}}{P_{RFi,avg}} \tag{9.4}$$

which can be applied to any of the definitions of efficiency given in (9.1)–(9.3).

Bandwidth Effects on Efficiency

Generally, as the bandwidth of the PA is broadened the efficiency will decrease. High efficiency in PA designs is achieved by matching components at specific frequencies. At different frequencies those matches will not apply. Usually approximately optimized efficiency is only achieved over a limited bandwidth. For example, a solid-state amplifier with a bandwidth from 2 to18 GHz will have $\eta_{PA} <$ 0.10 (10%).

Thermal Management

The power represented by the inefficiency of an amplifier is manifest in heat generated. If an amplifier were perfectly efficient, then all of the prime power would be converted into signals to be output. Since there is no such amplifier as yet, some mechanism must be provided to remove the heat generated in the process. If there is a 0.5V drop across the junctions in a transistor, which is passing a current of 100A, then the transistor must dissipate 50W of power. This power is converted into heat and is removed by properly configured heat sinks. Diamond has been found to be very good at heat removal.

Sometimes it is necessary to remove this heat with a cooling solution. Water, suitably conditioned for cold weather, has been used for this function. Refrigerants such as Freon have also been used.

We include much more on heat removal in Chapter 12.

9.4 Push-Pull Architectures

A type of arrangement for power amplifiers known as a *push-pull configuration* is shown in Figure 9.6. As the input signal oscillates between positive and negative voltages, the output transistors conduct alternately. It has the advantage that the

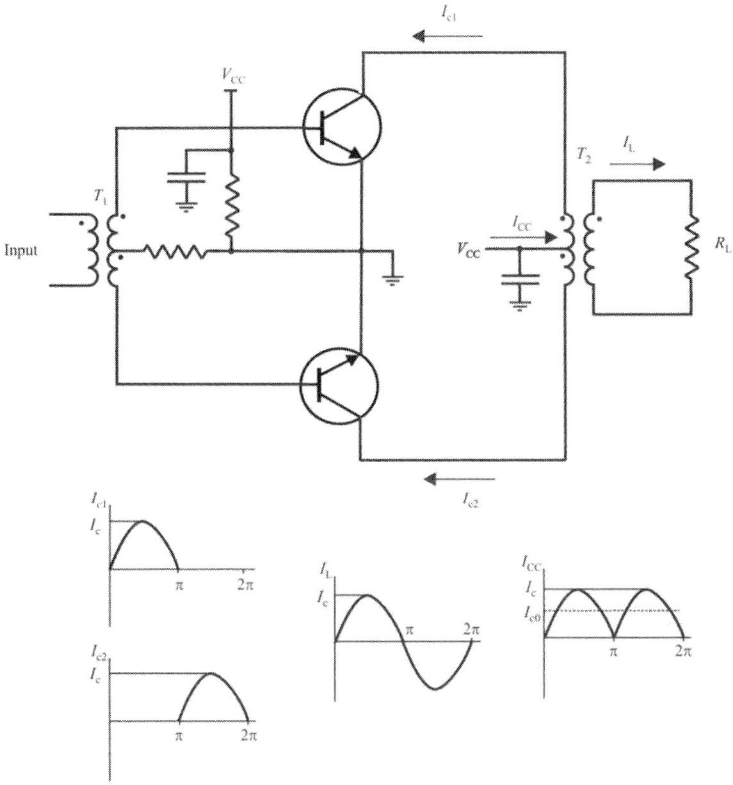

Figure 9.6 Push-pull architecture.

configuration of the circuit facilitates total suppression of the even harmonics in the output (assuming ideal components). Thus, as the transistors are pushed to their linear limits and beyond, the only significant harmonic in the output is the third, which can frequently be removed by simple low-pass filters. The suppression of even order harmonics is due to the way the output of the two channels combine in the output transformer.

The input and output impedances of push-pull PAs are higher than other types of amplifiers. In addition, even order harmonics are suppressed. For ideal class B the duty cycle of each device is 50%. With a sinusoidal voltage as input, the first transistor collector current is

$$I_{c1} = \begin{cases} +I_c \sin\theta, & 0 \le \theta < \pi \\ 0, & \pi \le \theta < 2\pi \end{cases} \tag{9.5}$$

while the second transistor collector current is

$$I_{c2} = \begin{cases} 0, & 0 \le \theta < \pi \\ -I_c \sin\theta, & \pi \le \theta < 2\pi \end{cases} \tag{9.6}$$

At the output of transformer T_2, the current through the load is

$$I_L(\theta) = I_{c1}(\theta) - I_{c2}(\theta) = I_c \sin(\theta) \tag{9.7}$$

The current flowing in the center tap of the primary winding of T_2 is

$$I_{CC}(\theta) = I_{c1}(\theta) + I_{c2}(\theta) = I_c \left| \sin(\theta) \right| \tag{9.8}$$

The total DC collector current is

$$I_{co} = \frac{1}{2\pi} \int_0^{2\pi} I_{CC}(\theta) d\theta = \frac{2}{\pi} I_c \tag{9.9}$$

Assuming zero saturation resistance when the collector voltage amplitude $V_C = V_{CC}$ and equal winding turns at the collectors so that $V_L = V_C$, DC and fundamental output powers are

$$P_0 = \frac{2}{\pi} I_c V_{CC} \tag{9.10}$$

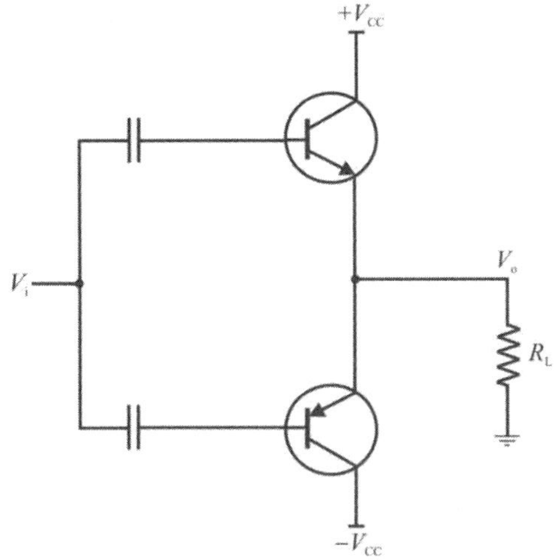

Figure 9.7 Push-pull amplifier configuration without an output transformer, also referred to as *complementary push-pull.*

$$P_o = \frac{1}{2} I_c V_{CC} \tag{9.11}$$

In theory, that is, with perfectly matched components (not ideal components, if the transistors were ideal, they would produce no harmonics at all), even-order harmonics are canceled because they are 180° out of phase and combined in the center tap of the primary winding of T_2. In reality, some will be present, of course.

It is also possible to construct push-pull amplifiers without using transformers. The advantage of such a configuration is that transformers frequently have limited frequency ranges. Figure 9.7 shows one such configuration. The disadvantage here is that two power supplies are needed. The bias circuitry is not shown in Figure 9.7 for clarity. Another disadvantage is this circuit requires both NPN and PNP transistors. Because of the limited mobility (speed) of charges in P-doped material, these amplifiers are limited to the lower frequency ranges and cannot be used, for example, in the UHF and above ranges.

9.4.1 Class E Push-Pull Architecture

The high efficiency of class E amplifier configurations can be exploited in push-pull architectures. One such configuration is illustrated in Figure 9.8. The transistors are driven 180° out of phase so when one is off the other is on, providing a current path from the DC power supply through the transformer to ground. The voltage waveforms at the collectors of the transistors are shown in Figure 9.8, as well as the voltage waveform across the load. The impedance seen from the transistor looking into the transformer is given by

$$Z = \frac{1}{2}\frac{m^2}{n^2} Z_0 \tag{9.12}$$

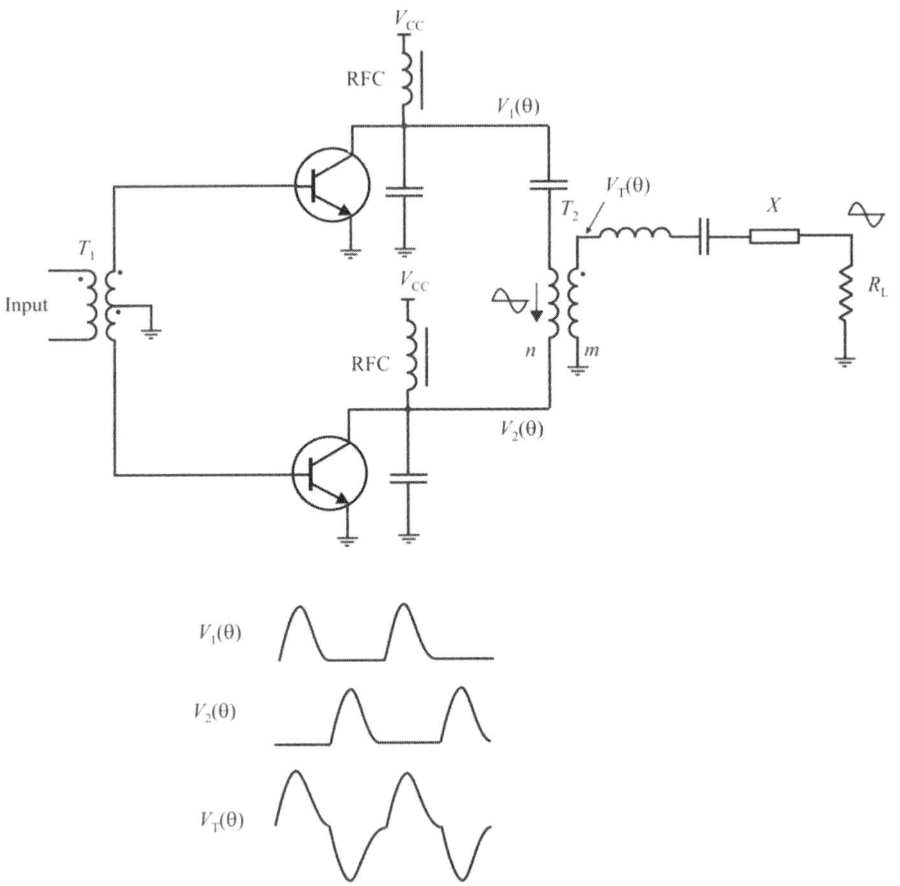

Figure 9.8 Push-pull class E architecture.

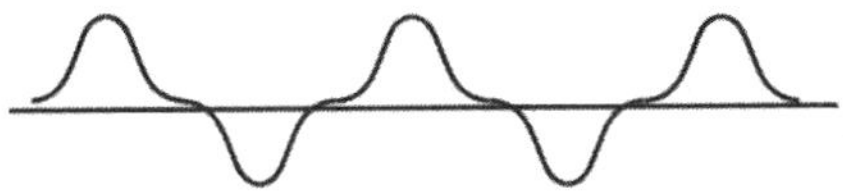

Figure 9.9 Crossover distortion in class B amplifiers (distortion is exaggerated).

9.4.2 Crossover Distortion

When class B amplifiers are used in the push-pull arrangement discussed above, there is a short time when both transistors are turned off. This produces what is called crossover distortion. One transistor is on during half of the input cycle and the other is on for the other half. The point where one is almost off and the other is almost on is called the crossover point. Crossover distortion is illustrated in Figure 9.9.

9.5 Doherty Amplifier

The Doherty amplifier is a type of PA that specifically addresses the nonlinearity that occurs when amplifiers are driven into saturation. We discuss this PA topology in Chapter 11, where linearity is addressed.

9.6 Safety Considerations

EW systems emit significant amounts of energy frequently in the microwave frequency ranges. Humans can only tolerate a certain amount of energy at these frequency ranges. Microwave ovens, for example, cook food at a power level of around 1 kW at a frequency around 1 GHz. Safety limits on radiation are provided at [7].

9.7 PA Matching

9.7.1 Frequency Response of Series/Parallel RLC Networks

The frequency response of series and parallel RLC networks is illustrated in Figure 9.10. The parallel network impedance peaks at the resonant frequency to a value or

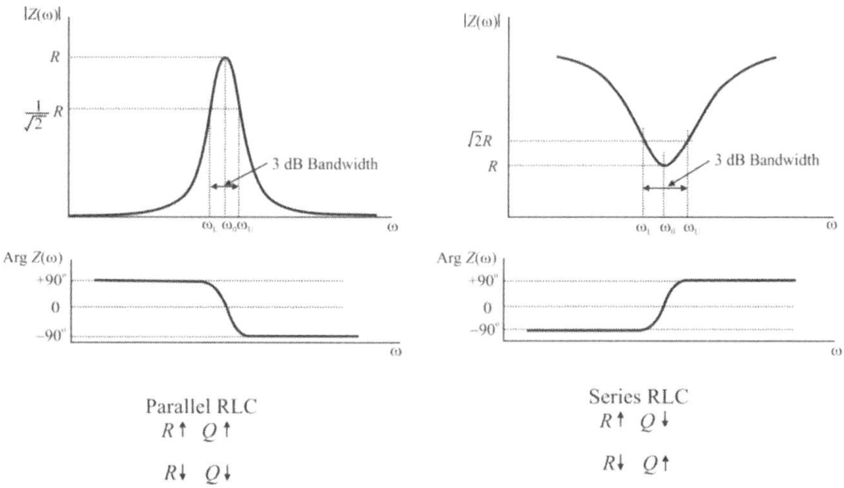

Figure 9.10 Frequency response of RLC networks.

R. The reactive impedances cancel. The series impedance reaches a minimum, equivalent to *R*, at the resonant frequency, due to the reactance cancelation.

9.7.2 Adding L and C Elements to Matching Networks

When adding capacitors and inductors in series or shunt topologies for matching purposes, the resulting reactances and susceptances change, but the resistances and conductances do not. Therefore, the resulting values can be determined. Referring to Figure 9.11, we add a series inductor to the impedance. The effect on the Smith chart is shown in the figure. Assuming we start at point A, we follow the constant resistance circle (in this example it is at 2 Ω) and wind up at point B. The reactance was changed from $-j1$ Ω (capacitive) to $+j0.1$ Ω (inductive) in the process.

Likewise, if we add a capacitor in series with the impedance as illustrated in Figure 9.12, the effect on the impedance on the Smith chart is as shown in Figure 9.12 as we move from point A to point B.

In a similar fashion, inserting a capacitor in shunt with an admittance load as in Figure 9.13, has the effect on the admittance on the Smith chart as shown. At point A, the admittance is $y_{\text{in}} = jy$ (inductive in this case) while at point B it becomes $y_{\text{in}} = -jy_{\text{L}} + jy$ (capacitive).

Finally, adding an inductor in shunt as in Figure 9.14 moves the admittance value as shown from A (capacitive) to B (inductive).

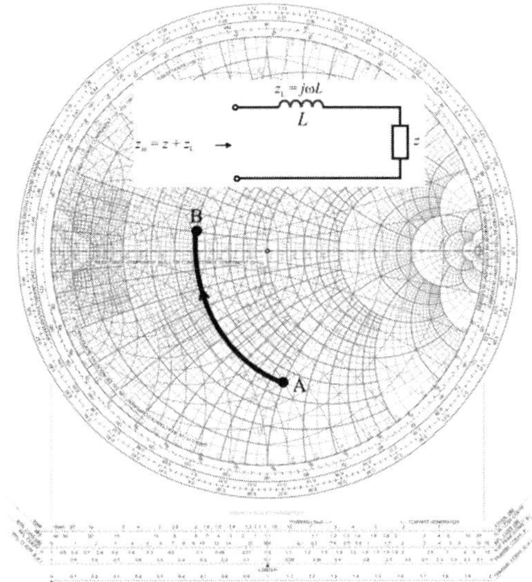

Figure 9.11 Effects of adding a series inductor.

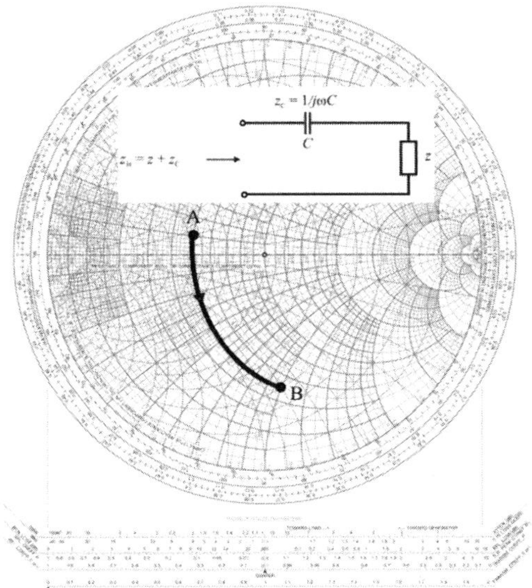

Figure 9.12 Effects of adding a series capacitor.

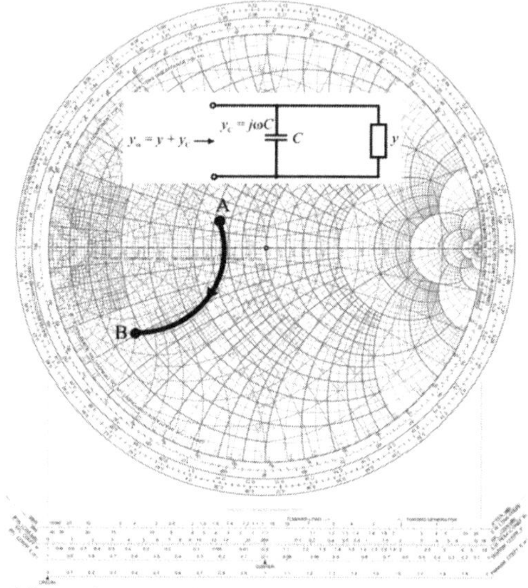

Figure 9.13 Effects of adding a shunt capacitor.

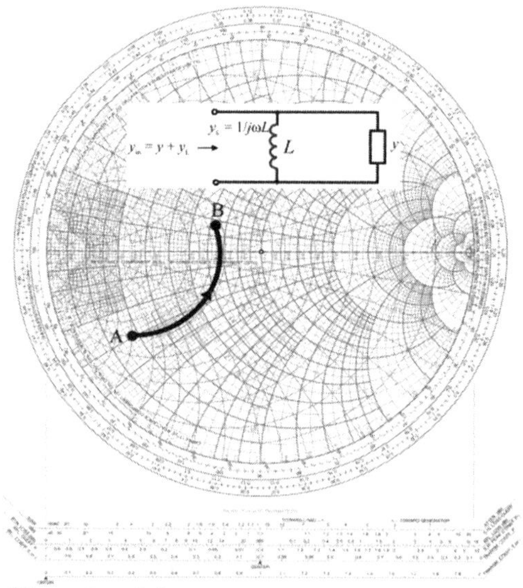

Figure 9.14 Effects of adding a shunt inductor.

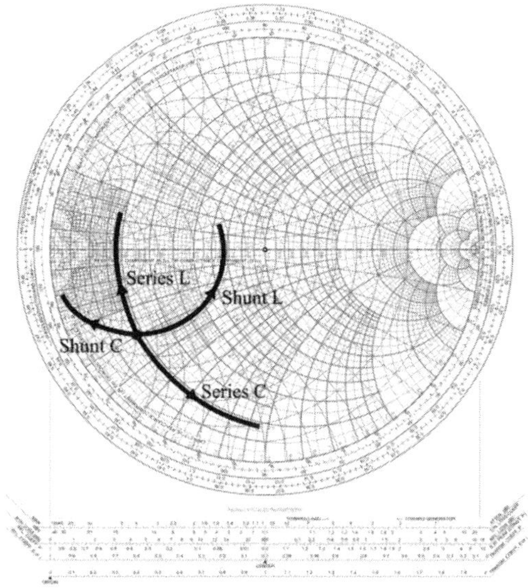

Figure 9.15 Movement summary.

These processes are summarized in Figure 9.15 showing which way the impedances and admittances move when the various elements are added. There are four basic L-network configurations using two elements that are widely used[4] [8]. These are shown in Figure 9.16. Which one is used depends of whether $R_g > R_L$ or $R_g < R_L$ and other factors (such as ease of implementation, and so forth).

With adding only two elements as we are here there are only two possibilities to take into consideration as we will see in the examples to follow. Furthermore, the goal is to match the source and load impedance/admittance with those of the active device. We will be trying to match to s_{11}^* and s_{22}^*. Therefore the end of the second move of the resistance or admittance in these processes will be at those points on the Smith chart. The movement at the input will start at the characteristic impedance and move towards s_{11}^* and at the output will start at the load and move toward s_{22}^*. Depending on whether these points are in the inductive region or the capacitive region will determine the direction the second move must be, which, in turn, dictates the direction of the first move.

[4] There are more than four configurations available as indicated in [8]. The most popular are shown here. If more than two elements are used, there are considerably more topologies possible.

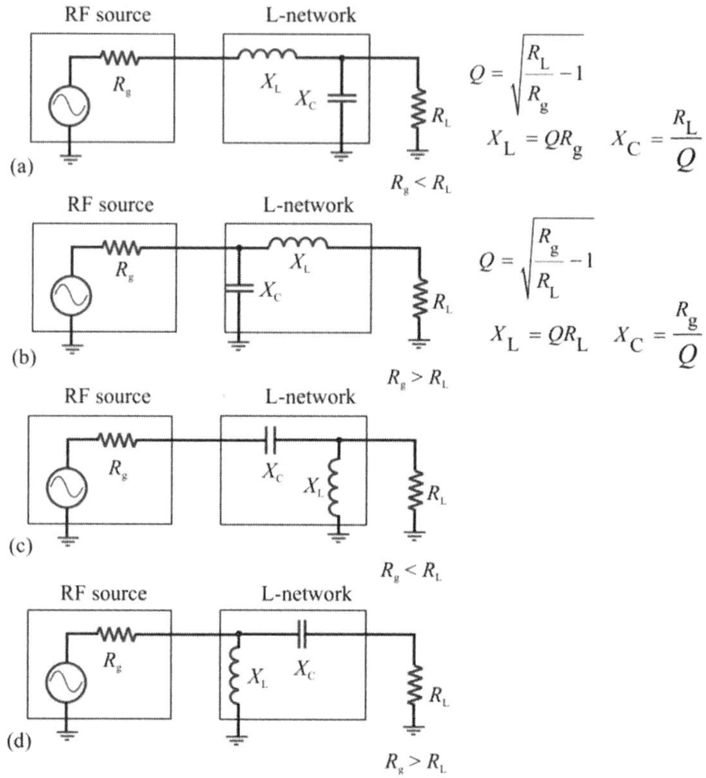

Figure 9.16 There are four basic L-network configurations. The network to be used depends in the relationship of the generator and load impedance values. Those in (a) and (b) are low-pass circuits and those in (c) and (d) are high-pass versions.

Example 9.1: We match the output impedance of a low-power RF transistor amplifier to a 50 Ω output load—50 Ω is a universal standard for most receiver, transmitter, and RF circuits. On the other hand, most power amplifiers have a low output impedance, typically less than 50 Ω.

Figure 9.17 shows the desired circuit. Assume an amplifier output (generator) impedance of 10 Ω at a frequency of 76 MHz. The inductor and capacitor values are computed as:

$$Q = \sqrt{\frac{R_L}{R_g} - 1} = \sqrt{\frac{50}{10} - 1} = \sqrt{4} = 2$$

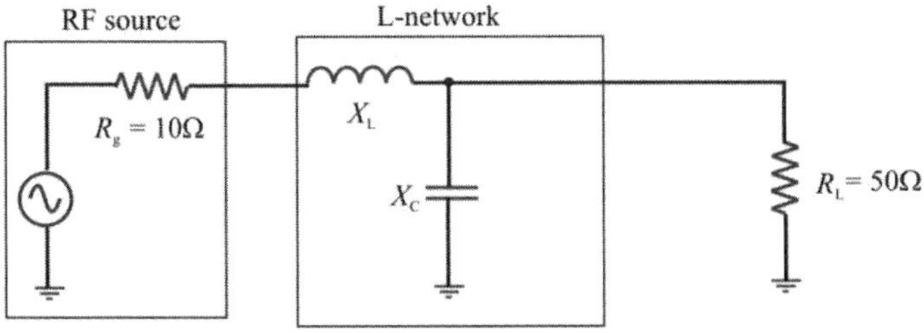

Figure 9.17 The RF source is a transistor amplifier with an output impedance if 10 Ω that is to be matched to a 50 Ω output impedance load. The L-network with a parallel output capacitor is used since $R_g < R_L$.

$$X_L = QR_g$$

$$= 2 \times 10 = 20\Omega$$

$$L = \frac{X_L}{2\pi f}$$

$$= \frac{20}{2\pi \times 76 \times 10^6} = 42\text{nH}$$

$$X_C = \frac{R_L}{Q}$$

$$= \frac{50}{2} = 25\Omega$$

$$C = \frac{1}{2\pi f X_C}$$

$$= \frac{1}{2\pi \times 76 \times 10^6 \times 25} = 83.8\text{pF}$$

These values ignore any output impedance reactance such as transistor amplifier output capacitance or inductance and any load reactance that could be shunt capacitance or series inductance. When these factors are known, the computed values can be adjusted by incorporating their values into the components found here.

The bandwidth (W) of the circuit is relatively wide given the low Q of 2:

$$W = \frac{f}{Q} = \frac{76 \times 10^6}{2} = 38\text{MHz}$$

The performance characteristic of this network is shown in Figure 9.18. We can see the wide bandwidth. In fact, the lower leg of this curve actually never reaches –3 dB from the peak—at least not after 10 MHz anyway.

9.7.3 Parallel and Series Circuit Equivalents

Sometimes it's necessary to convert a series RC or RL circuit into an equivalent parallel RC or RL circuit or vice versa. Such conversions are useful in RLC circuit analysis and design. Figure 9.19 shows the network topologies for these conversions. These equivalents sometimes help us understand how the L-networks and other impedance matching circuits work. The designations are:

R_S = series resistance
R_P = parallel resistance
X_S = series reactance
X_P = parallel reactance

The conversion formulas are given as follows:

$$R_S = \frac{R_P}{Q^2 + 1}$$

$$X_S = \frac{X_P}{(Q^2 + 1) / Q^2}$$

$$R_P = R_S(Q^2 + 1)$$

$$X_P = X_S[Q^2 + 1) / Q^2]$$

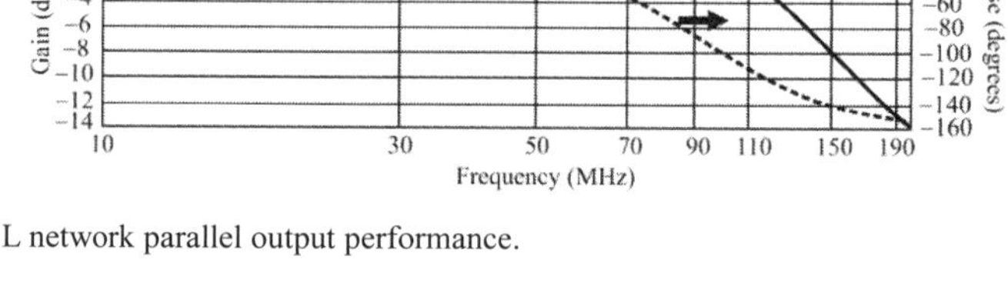

Figure 9.18 L network parallel output performance.

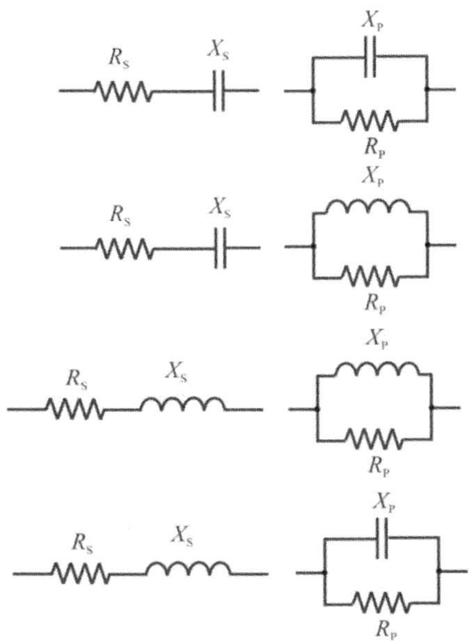

Figure 9.19 RC and RL equivalents. The text shows equations for converting among these values.

$$Q = \sqrt{R_p / R_s - 1}$$

$$Q = X_L / R_S$$

$$Q = R_P / X_C$$

Example 9.2: We can see how the matching network in Example 9.1 functions by converting the parallel combination of the 50 Ω resistive load and the 25 Ω capacitive reactance in Example 9.1 into its series equivalent (see Figure 9.20):

$$R_S = \frac{R_P}{Q^2 + 1} = \frac{50}{2^2 + 1} = 10\Omega$$

$$X_S = \frac{X_P}{(Q^2 + 1) / Q^2} = \frac{25}{1.25} = 20\Omega$$

Note how the series equivalent capacitive reactance equals and cancels the series inductive reactance. Also, the series equivalent

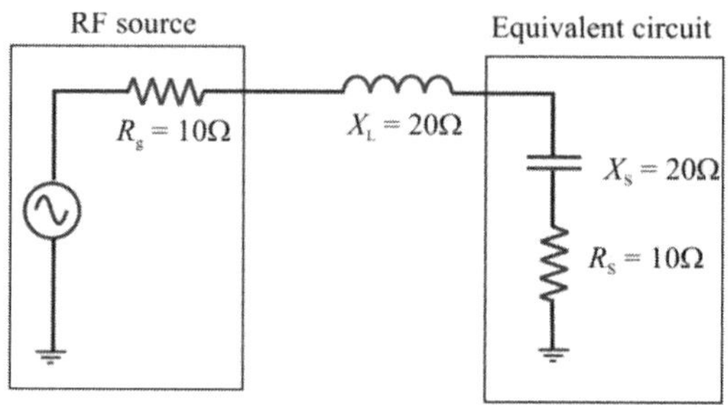

Figure 9.20 The series equivalent circuit of the network in Figure 9.17 is a simple RLC network where the reactances cancel and the source and load impedances match.

load of 10 Ω matches the generator resistance for maximum power transfer.

This circuit is shown in Figure 9.20.

Example 9.3: Match the output impedance of 50 Ω transmitter at 76 MHz to a 5 Ω loop antenna impedance (Figure 9.21).

$$Q = \sqrt{\frac{R_g}{R_L} - 1}$$

$$= \sqrt{\frac{50}{5} - 1} = \sqrt{9} = 3$$

$$X_L = QR_L$$

$$= 3 \times 5 = 15\Omega$$

$$L = X_L / 2\pi f$$

$$= \frac{15}{2\pi \times 76 \times 10^6} = 31\text{nH}$$

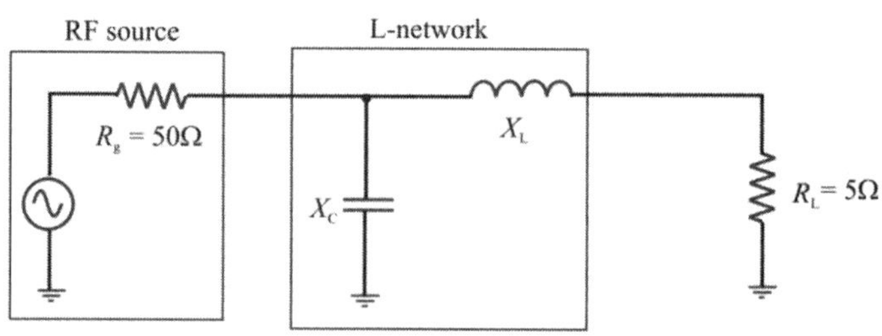

Figure 9.21 L-network match to loop. The RF source is a transmitter at 76 MHz with an output impedance of 50 Ω. The load is a loop antenna with an impedance of 5 Ω. This topology is selected since $R_g > R_L$.

$$X_C = \frac{R_g}{Q}$$

$$= \frac{50}{3} = 16.17\Omega$$

$$C = \frac{1}{2\pi f X_C}$$

$$= \frac{1}{2\pi \times 76 \times 10^6 \times 16.67} = 126\text{pF}$$

In this example, the capacitor, inductor, and load resistance form a parallel resonant circuit (see Figure 9.22).

Recall that a parallel resonant circuit acts like an equivalent resistance. That resonance equivalent resistance (R_R) of a parallel circuit can be calculated as:

$$R_R = \frac{L}{CR}$$

or

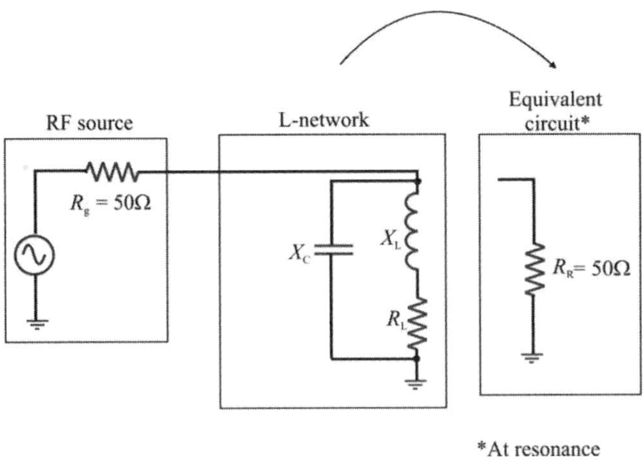

Figure 9.22 The equivalent circuit of the L-network and load as a parallel resonant circuit. At resonance, the parallel circuit has an equivalent resistance equal to the generator resistance of 50 Ω for a match.

$$R_R = R(Q^2 + 1)$$

$$R_R = \frac{L}{CR} = \frac{31\times10^{-9}}{126\times10^{-12}\times5} = 49.2\Omega$$

$$R_R = R(Q^2 + 1) = 5(3^2 + 1) = 50\Omega$$

In both cases the parallel resonant load equivalent resistance is 50 Ω and equal to the generator resistance allowing maximum power transfer. As above, any load reactive component must be taken into account. The equivalent high-pass networks could also be used. One benefit of doing so is that the series capacitor can block DC if required.

The simulated performance of this matching network is shown in Figure 9.23.

9.7.4 Impedance Matching with Microstrip Lines

Simple lumped element networks, or with the equivalent short lengths of transmission line, can be used for input and output impedance matching. The element values can be determined by use of the Smith chart or by calculations [9]. We know that for a short length of transmission line of characteristic impedance Z_0 and velocity v, the inductance L per unit length (same units as v) is given by

$$\frac{L}{d} = \frac{Z_0}{v}$$

where

$$v = \frac{c}{\sqrt{\varepsilon_{\text{eff}}}}$$

For a capacitive transmission line (short, open line), C per unit length d is

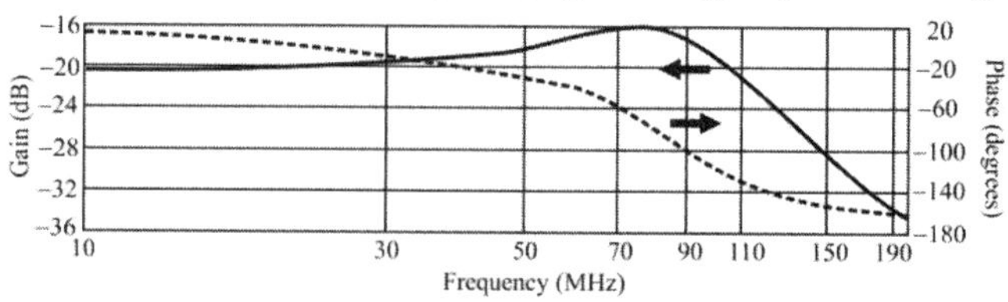

Figure 9.23 L network feeding loop performance.

$$\frac{C}{d} = \frac{1}{Z_0 v}$$

The fringing capacitance must be taken into consideration by subtraction from the capacitance value desired from the transmission line [10].

One of the major causes of reduced gain or instability in microwave amplifiers arises from incorrect treatment of common lead grounding (see Figure 9.24). Frequently, plated-through vias are used to carry the ground plane of the microstrip up to the common leads, but this results in common-lead inductance unless the microstrip is quite thin (< 1 mm). Other methods include mounting the active device in a hole, with the common leads soldered to the ground side of the board.

9.8 Diode and Transistor Bias for Linear Power Amplifiers

The use of class A and class AB amplifiers for linear PAs requires the use of a steady bias current, applied to the base (in case of a BJT) in order to bias the RF device into partial or full conduction with no input signal present (see Figure 8.1). This bias current must remain constant, despite the varying amplitude of the input signal to the amplifier.

Simple resistive biasing techniques are often used when considering low power RF amplification. Such techniques are not appropriate for medium- or high-power amplifiers, however, due to their requirements for relatively high current and voltage sources, which in turn, implies use of low impedances.

Diode biasing, illustrated in Figure 9.25 (which could be provided with a transistor as well), provides a low impedance voltage source. This topology provides

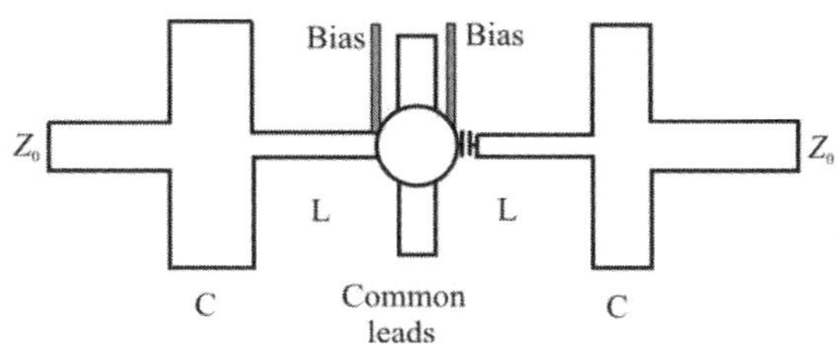

Figure 9.24 Treatment of common lead grounding.

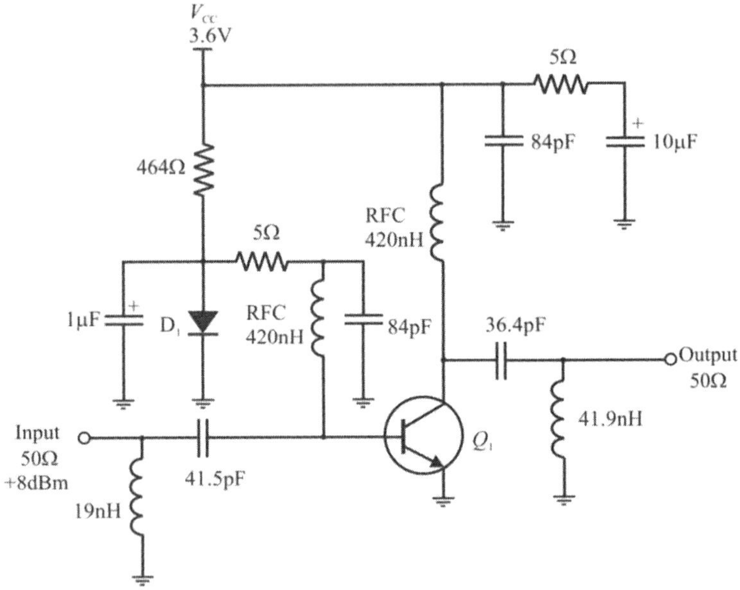

Figure 9.25 Class AB power amplifier. Operating frequency is about 23.5 MHz with the component values indicated.

a higher current to flow through it than the bias current required by the transistor. The bias components must be thermally connected to the RF transistor itself. This allows the bias components to maintaining the PA's collector current at a steady DC level by tracking the PA's temperature variations. The 0.7V across Q_1's base increases/ decreases in value as the temperature varies because as the temperatures rises, a silicon semiconductor junction's voltage decreases from its room temperature value of nominally 0.7V.

The circuit in Figure 9.25 is a band-pass amplifier biased for class AB operation at a frequency of about 23.5 MHz with the component values shown. The frequency response is shown in Figure 9.26.

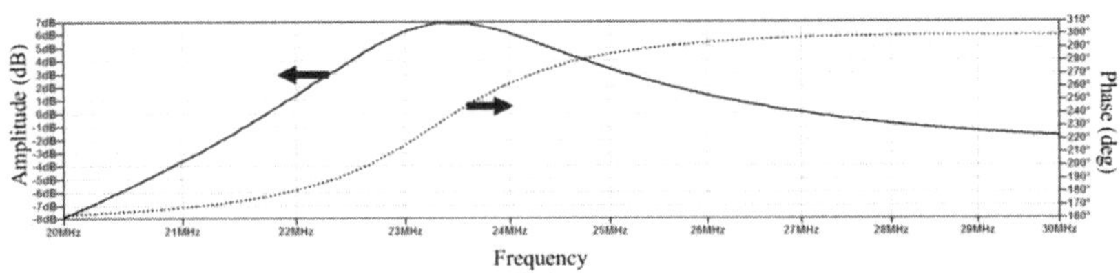

Figure 9.26 Class AB frequency response. $f_T = 400$ MHz.

In addition, in order to provide for PA bias stability no matter what the input and output RF power levels, the standing current through the diode or transistor bias components must be sufficient to permit a steady voltage to be maintained across the PA's base. We discuss stability considerations in more depth in Chapter 10.

An emitter follower that amplifies the diode current, illustrated in Figure 9.27, improves the bias performance of clamping diode circuit presented in Figure 9.25.

The current required through the diode is reduced by a factor of the current gain of Q_2. This arrangement employs two diodes in series—one is required to compensate for the base emitter voltage drop of the transistor.

9.8.1 Power MOSFET Bias

Power FET device biasing is typically simpler than biasing BJT devices. In most cases it is sufficient to provide a bias voltage directly on the gate through suitable impedance. These impedances are typically resistors and RF chokes; the latter minimizes the amount of RF feeding back into the bias supply, a function not provided by resistors alone. Capacitors can also be used to provide this functionality.

Although many variations are possible, a typical bias network is shown in Figure 9.28, consisting of a voltage supply (a variable voltage regulator illustrated in this case) fed via a bias resistor. The primary function of the voltage regulator is to aid the decoupling of the gate from the bias supply. A quarter-wave transmission line is used in this case, which provides the RF isolation between gate and the bias supply.

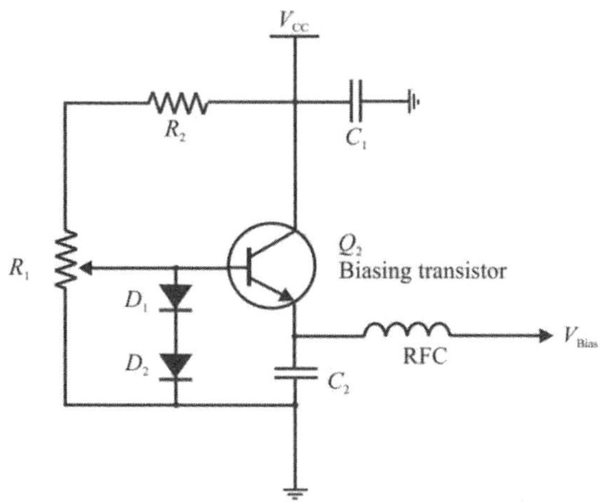

Figure 9.27 Amplified diode bias.

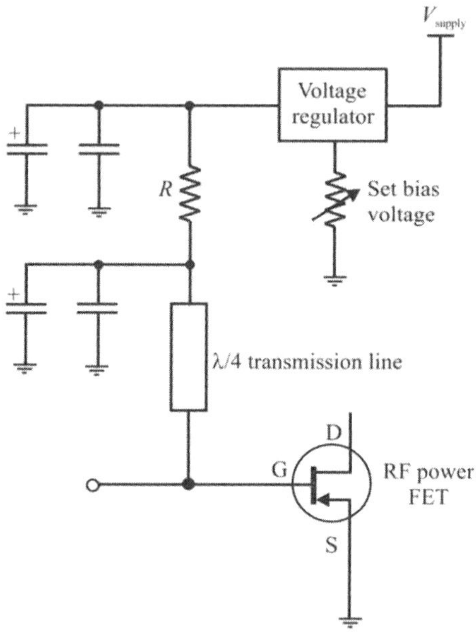

Figure 9.28 Power MOSFET bias.

9.8.2 Power LDMOS Bias

LDMOS transistors, as discussed in Chapter 7, are CMOS devices designed for high-frequency and high-voltage usage. They are specifically designed for HPA applications.

LDMOS transistors exhibit I_{DQ} *drift* which is a detrimental characteristic. All LDMOS transistors exhibit a hot carrier injection effect to varying degrees. The gate field changes due to charge buildup in the gate-drain region because of this hot carrier injection results. This causes a change in the quiescent current (I_{DQ}) with a fixed gate voltage. Thus, adaptive bias circuits are required that adjust V_{GS} periodically to maintain a constant I_{DQ}.

A simplified circuit of an LDMOS amplifier bias circuit is shown in Figure 9.29. As with all MOS devices, the DC bias is established by applying a DC voltage to the gate (V_{gs}). The drain current (I_D) is continually monitored. Ideally, this I_{DD} is constant over temperature, but since the V_{gs} of LDMOS amplifier devices varies with temperature due to the I_{DQ} drift phenomena, temperature compensation is required.

A popular way to bias an LDMOS amplifier is by digitally converting temperature information and adjusting the DC bias using a *look-up table* (LUT).

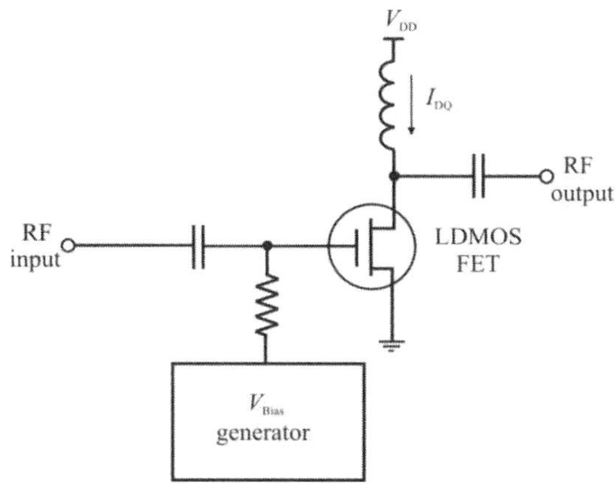

Figure 9.29 Power LDMOS FET biasing.

The LUT is programmed at final test using measured parameters from the amplifier. DC bias performance can thus be optimized over the required temperature range.

LDMOS amplifiers also have a characteristic I_D drift over time (drain current reduces for a given V_{gs}), as well as temperature. A LUT correction can be used to address this issue as well, so that over time the I_D will drift closer to the target bias value, not further away.

Example 9.5: LDMOS PA Bias

An LDMOS HPA topology with active bias is shown in Figure 9.30. The gate-source bias voltage is supplied through a voltage divider set by adjusting the variable resistor P_1 to control the optimum M_1 drain current I_{DQ}. M_1 is an LDMOS power transistor originally developed for PCS cell base stations operating in the 1800–2000 MHz range. With power handling capability in the 100W range, it makes an excellent choice for EW/EA applications requiring about this power in this frequency range. A UAS could use this type of application.

Optimum performance from this amplifier is achieved with I_{DQ} set to 750mA. Deviations from this optimal bias point result in suboptimal trade-offs of performance parameters such as gain compression and efficiency. The device is more linear, but less efficient at higher drain voltages. If the efficiency requirements are more severe, 26V may be more suitable on the drain. If

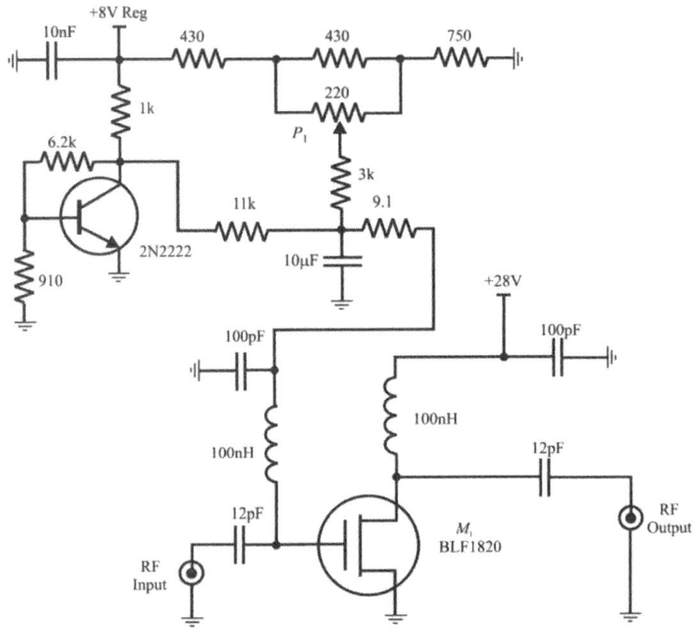

Figure 9.30 LDMOS PA with operating frequency in the 1800-2000 MHz range.

linearity is a driving parameter, 30V on the drain may be more appropriate.

9.9 Switched Power Amplifiers

9.9.1 Introduction

There are some additional amplifier classes that are used as RF amplifiers and that rely on the ability of active devices to rapidly switch ON and OFF. These classes are called switching architectures; they were introduced in Chapter 4 and are investigated in more depth here.

9.9.2 Architectures

This is a class of amplifier architectures that switch rapidly to perform the amplification. Discussed here are classes D, E, F, and S architectures.

The most significant advantage of these architectures is their efficiency; generally, they are much more efficient than the analog architectures discussed previously.

9.9.2.1 Class D

Class D topologies can be used in low RF applications—useful class D amplifiers have been reported that operate to 1 GHz. A block diagram of a class D amplifier is illustrated in Figure 9.31. In a class D amplifier, the active devices are driven to produce an output square-wave that switches between a high and low level at a frequency outside the range of human hearing. Instead of modulating the amplitude, internal circuitry modulates the duty cycle of the square-wave at a rate corresponding to the level of the input signal when the output is filtered down to audio band.

Like classes E and F, class D's active devices are driven as switches rather than in linear mode. A square-wave with a frequency that's significantly higher than the highest frequency component of the input waveform drives the class D amp's two devices between saturation and cutoff.

The square-wave's pulse width or pulse density is variable, and the input signal controls one or the other. At the amplifier output, a low-pass filter attenuates the switching frequency and its harmonics, leaving only the amplified version of the input waveform.

With the FETs operating in either cutoff or saturation, losses come primarily from the transistors' forward-voltage drops. Class D amps can achieve efficiencies as high as 90%, with distortion levels approaching class AB. A drawback of class D concerns the challenging task of suppressing radiated and conducted interference from the switching circuitry.

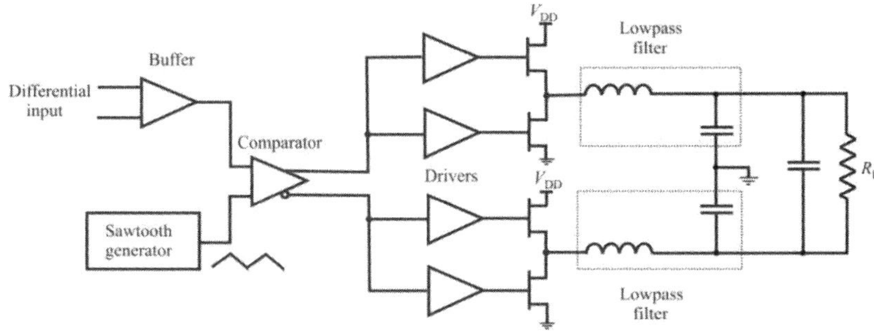

Figure 9.31 Block diagram of a class D amplifier. Switching amplifiers sequentially saturate and turn off power transistors in the output stage using pulse width modulation.

A class D amplifier does not operate linearly at all. They are sometimes referred to as a switching amplifier because the active devices operate as electronic switches. The signal to be amplified is a train of constant amplitude pulses, so the active devices switch rapidly back and forth between a fully conductive and nonconductive state. The analog signal to be amplified is converted to a series of pulses by pulse width modulation, pulse density modulation or other A/D method before being applied to the amplifier. After amplification, the output pulse train normally is converted back to an analog signal with a passive low-pass filter.

A significant advantage of a class D amplifier is that it can be more efficient than analog amplifiers, with less power dissipated as heat in the active devices.

A class D amplifier is substantially different from those of the other classes mentioned previously. An amplifier operating in class D rapidly turns the power transistors on and off in the output stage, producing a digital output of 1s and 0s. With *pulse width modulation* (PWM), the longer the transistor is ON, the greater the average output power.

In the block diagram of a class D amplifier shown in Figure 9.31, the differential input of the signal to be amplified is input to one side of a comparator. A sawtooth waveform is applied to the other input as shown. When the amplitude of the input signal exceeds that of the sawtooth, the output of the comparator changes state to a logical one in this case. As seen in Figure 9.32, which displays the time waveforms involved, the lower the amplitude of the input signal, the more of the sawtooth is included in that which is less than the input signal, causing the output to remain at

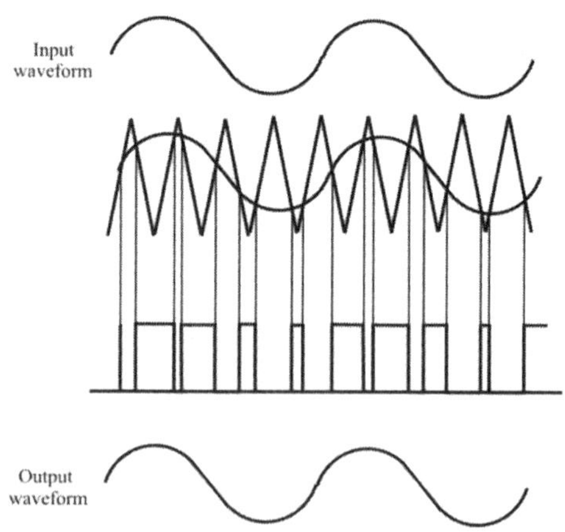

Figure 9.32 Waveforms within a class D amplifier. Every time the sawtooth waveform goes above the input waveform, the output of the comparator goes to a high state and vice versa. The comparator output, through drivers, turns the output transistors ON and OFF.

a high state longer. In this way the circuit adjusts the width of the output of the comparator as a function of the amplitude of the input signal. This PWM waveform then drives the high-power output stage that switches on and off according to this signal. The wider the pulse, the longer the power stage stays on and the more of the prime power is fed to the output low-pass filter. The output, being a square wave, would generate spurious signals if transmitted in that form. This output is lowpass filtered prior to transmission to extract only the fundamental frequency.

The primary advantage of class D amplifiers is their efficiency compared to the other classes. Efficiencies greater than 80% are possible. However, because the transistors must be switched much faster (say, by a factor of 10) than the highest frequency in the signal, the use of class D amplifiers is limited to low frequencies with modern transistors.

9.9.2.2 Classes E, F, and F^{-1}

Classes E and F are subsets of class C. Classes E and F, much like class C, feature RF amplifier topologies that use LC tank circuits. Where class C amplifiers are widely used below 100 MHz, class E amplifiers tend to fall into the VHF and microwave frequency ranges. The difference between class E and class C amps is the active device becoming a switch, rather than operating in the linear portion of its transfer characteristic. They have the potential to be much more efficient than class C topologies.

Class F amplifiers resemble class E amplifiers, but use a more complex load network. In part, this network improves the impedance match between the load and the switch. Moreover, it's designed to eliminate the input signal's even harmonics so the switching signal is more nearly a square-wave. It improves efficiency because the switch runs at saturation or cutoff for a longer period.

Class E Amplifiers

The design of class E amplifiers is based on using a series or parallel resonant load network. The current and voltage time-waveforms at the active device output terminal are optimized in such a way as to minimize the DC power dissipation. The active device acts as a switch, driven by the RF input signal to ON and OFF conditions.

The ideal AC load lines for switching transistors (class D, E, F) are shown in Figure 9.33. We can see that the operating point moves along the V_{DS} and I_D axes; that is, the device is either OFF (in the saturated region) or ON (in the linear region). Under this ideal switching, the output voltage and current waveforms at the device

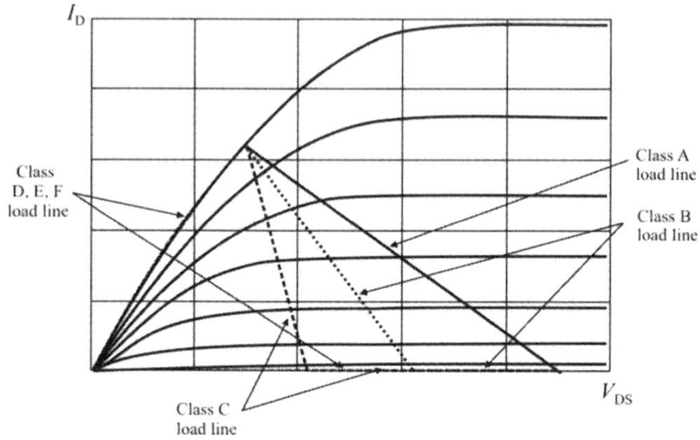

Figure 9.33 Ideal transistor load lines for several classes of operation.

output terminal do not simultaneously exist and, therefore, the dissipated energy within the device is zero, leading to 100% theoretical power conversion efficiency.

With the advances in active-device performance, advent of nonlinear modeling and monolithic circuit technology in the last several years, significant progress has been made toward the development of high-efficiency RF and microwave components. For class E high-efficiency power amplifiers, the useful operating frequency of these circuits have been increased to ever-higher frequencies [11 – 13].

Design Methodology

We can obtain an approximate maximum frequency (f_{max}) for class E operation knowing the device drain to source capacitance (C_{DS}) and the drain voltage (V_{DS}). Similarly, assuming a load resistance of 50 Ω, approximate values for the circuit

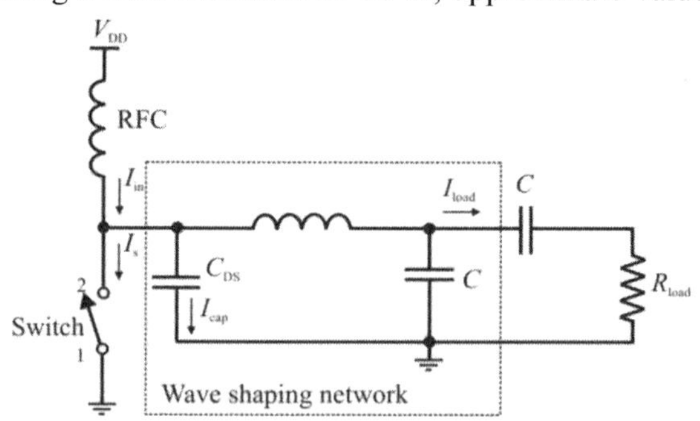

Figure 9.34 Class E topology.

elements (L and C) of the shunt load network shown in Figure 9.34 can be obtained by using the following expressions:

$$f_{\max} = \frac{I_{\max}}{56.5 C_{DS} V_{DS}} \tag{9.13}$$

$$L = \frac{0.28}{\omega^2 C_{DS}} \left(\sin\theta + \cos\theta_0 \sqrt{\frac{\omega_s C_{DS} R}{k_0 \cos\theta}} - 1 \right) \tag{9.14}$$

$$C = \frac{1}{\omega_s R} \sqrt{\frac{\omega_s C_{DS} R}{k_0 \cos\theta_0 - 1}} \tag{9.15}$$

where

$k_0 = 0.28$
$\theta_0 = 49.05^\circ$
$\omega = 2\pi f_{\max}$

Having obtained the starting values for the load network provided by (9.13)–(9.15) a time-domain simulation can be performed to optimize the current and voltage waveforms at appropriate terminals of the ideal class E amplifier shown in Figure 9.34. Figure 9.35 (same as Figure 4.25 but with specific values) shows simulation results for the circuit after optimization of the load network. The voltage waveform across the switch rises slowly at switch-off and falls to zero at the end of the half-cycle. As we discussed in Chapter 4, it also has a zero rate of change at the

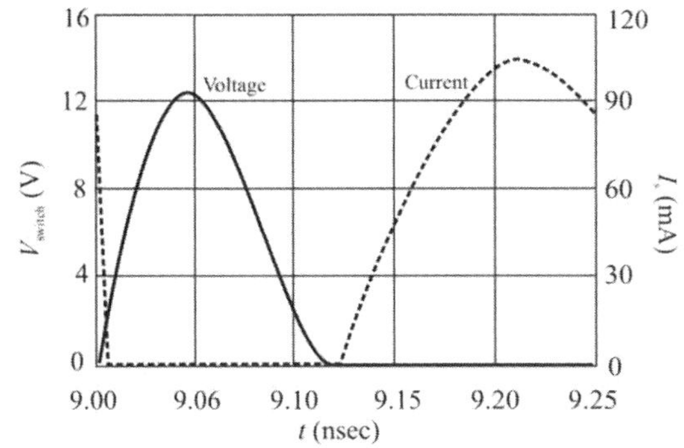

Figure 9.35 Class E voltage and current waveforms.

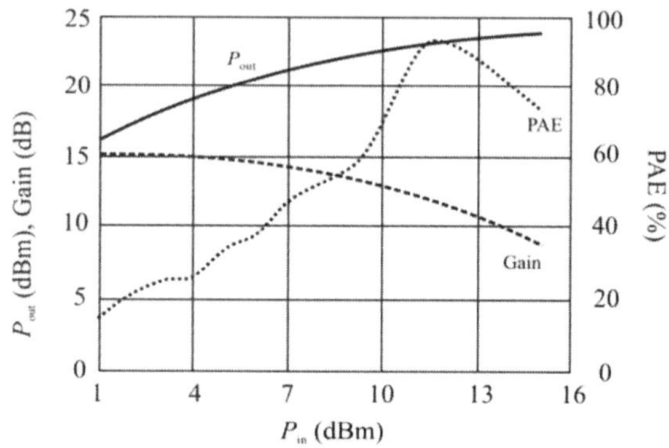

Figure 9.36 Class E performance shown in Figure 9.34 at 3.25 GHz.

end of half-cycle, the so-called soft-switching condition as expressed by (4.44) and (4.45) in Section 4.7.3, thereby ensuring a soft turn-on condition. The voltage across the switch when it is OFF is defined by the integral of the current flowing through C_{ds}. The phase shift introduced by the LC circuit adjusts the point at which the current is diverted from the switch to the capacitor C_{ds}. Therefore, to ensure class E operation, it is essential that the integral of capacitor current over the half-cycle is zero and that the capacitance current has dropped to zero by the end of the half-cycle.

Figure 9.36 shows the measured performance of a prototype class E amplifier that has been optimized at 3.25 GHz using the procedure described. The peak in PAE is above 90% while P_{out} reaches 24 dBm.

Figure 9.37 shows that the optimized current and voltage waveforms comply with the aforementioned criteria for the class E amplifiers.

Class E amplifier configurations have limited utility as the power amplifier in EW systems because they are obviously tuned amplifiers. EW systems typically require amplifiers that can quickly (instantaneously) cover a broad frequency band. A class E amplifier would need its output circuit changed quickly to function as a broadband amplifier. In those cases when broadband is not required, the Class E is an excellent choice because of its efficiency.

Class F and F^{-1} Power Amplifiers

This section begins with a description of the principles of operation of Class F RF PAs including their theory of operation and design techniques. Various possible approaches to the design of Class F PAs are explained, and some examples of practical designs are illustrated [14].

Due to their relative square waveforms at the output, class F and F^{-1} are popular choices for driver topologies for the other types of switching configurations.

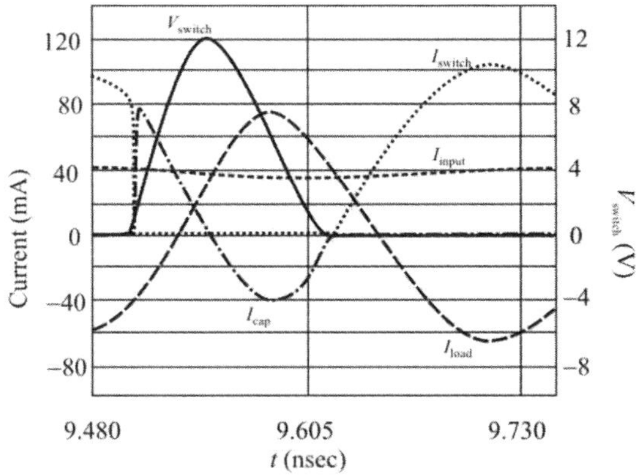

Figure 9.37 Switch voltage and current waveforms for the circuit in Figure 9.34.

Fundamentals of an Ideal Class-F PA

A Class-F PA boosts both efficiency and output power by using harmonic resonators in the output network to shape the drain waveforms such that the load appears to be a short at even harmonics and an open at odd harmonics. The drain voltage waveform includes one or more odd harmonics and approximates a square wave, while the drain current waveform includes even harmonics and approximates a half sine wave (see Figure 9.38) (same as 4.27). Because there is no overlap between the drain voltage and current, an ideal efficiency of 100% is achieved. In the case of its dual, i.e., Class F^{-1}, the shapes of drain voltage and current are swapped (i.e., the drain voltage waveform is close to a half sine wave while the drain current is close to a square wave [15]). The characteristics of class F and class F^{-1} are illustrated in Figure 9.39.

The output network of a classical Class-F PA is shown in Figure 9.40, which consists of a quarter-wavelength transmission line and a parallel-tuned inductance/capacitance (*LC*) output network [16]. The output tank is tuned to resonance at the fundamental frequency f_0. At f_0, the drain sees a pure resistive load of R_L, since the output tank is an open circuit, and the quarter-wave transformer transforms the 50 Ω load to R_L at the drain. The load resistance is chosen as the optimum load impedance for the PA, which can lead to maximum excursions of drain current and voltages. This load condition can be determined from load-pull

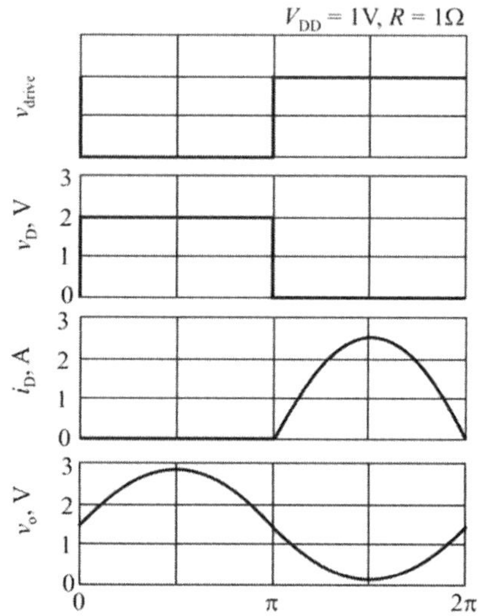

Figure 9.38 Class F waveforms.

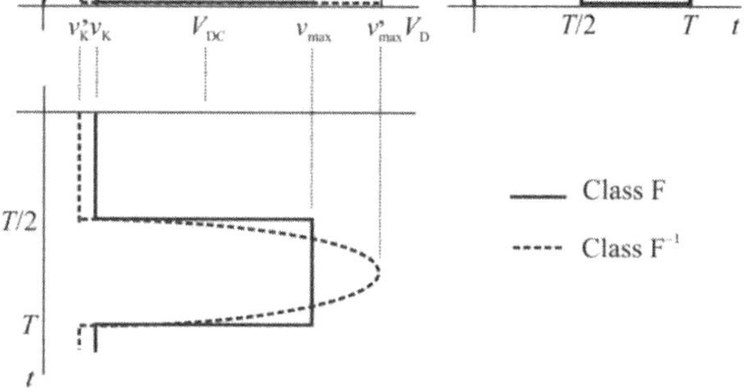

Figure 9.39 Class F and F^{-1} conduction characteristics.

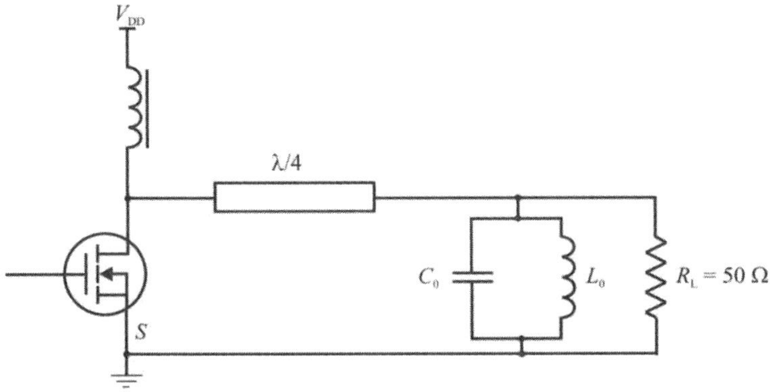

Figure 9.40 Output network of a classical class-F PA.

measurement or simulations. At even harmonics $2nf_0$ ($n = 1, 2, \cdots$), the output tank becomes a short, and the transmission line appears to be a length of $n\lambda_g/2$ to the drain, where λ_g is the guided wavelength of the transmission line. Therefore, the drain sees a short at all even harmonics, which would result in a half-rectified sinusoid current output as desired. Conversely, at odd harmonics $(2n + 1)f_0$ ($n = 1, 2, \cdots$), the output tank also becomes a short, and the transmission line appears to be a length of $[(2n+1)\lambda_g]/4$ to the drain. Due to the quarter-wavelength TL, the drain sees an open circuit at odd harmonics. The drain voltage contains only the odd harmonics, while the drain current includes only the even harmonics. Thus, the harmonic impedance is zero at even harmonics and infinite at odd harmonics.

Theoretical Analysis and Design

The ideal Class F PA assumes the inclusion of an infinite number of harmonics, which is, of course, unrealistic. For example, the drain-source capacitance C_{DS} will short circuit most high-order harmonics at microwave frequencies. It is nearly impossible to make even harmonics short-circuited and odd harmonics open-circuited due to the parasitic capacitance and inductance at the drain output.

Most real-world designs of Class F RF PAs consider only a few harmonics, mostly the second and third harmonics, which have the most effects on the efficiency and output power. Raab investigated the effects of finite harmonics on the output power capability and efficiency [17 – 19]. The maximum efficiency for a given set of harmonics is shown in Table 9.1 [18]. As we can see, the maximum efficiency of an ideal PA is 50% for Class A, where only the fundamental frequency components are considered (i.e., n = 1 for both drain voltage and current). The

Table 9.1 Maximum Efficiency of Class-F PAs.

	$n = 1$	$n = 2$	$n = 5$	$n = \infty$
$m = 1$	50%	57.7%	60.3%	63.7%
$m = 2$	70.7%	81.7%	85.3%	90.0%
$m = 4$	75%	86.6%	90.5%	95.7%
$m = \infty$	79.5%	90.7%	94.8%	100%

m, *n* denote maximum order of harmonics in drain current and voltage, respectively.
Source: [18].

maximum efficiency increases from 50% (Class A) toward unity (e.g., 70.7, 81.7, 89.6, and 90.5%, for second, third, fourth, and fifth harmonics, respectively), and the utilization factor increases from 0.125 to $1/2\pi$.

A frequency-domain analysis is also given for the ideal Class C, E, and F PAs in [19], which are based on a finite number of harmonics. It is shown that the number of harmonics determines the maximum attainable efficiency. The harmonic impedances determine the power-output capabilities, which is the highest for ideal Class F operations.

In the case of considering only the second and third harmonics (called *third-harmonic peaking*), the maximum efficiency reaches 81.7%. An output network that includes a third-harmonic peaking circuit is shown in Figure 9.41. The quarter-wavelength TL in Figure 9.40 is removed here and, instead, a parallel resonant tank at $3f_0$ is added in series with the drain output, which provides a short at $2f_0$ and an

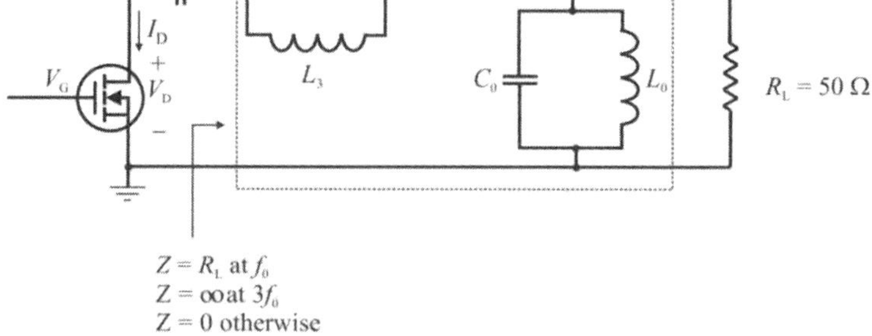

Figure 9.41 Output network including third-harmonic peaking circuit.

open at $3f_0$. Another parallel resonant tank at f_0 is added in parallel with the load resistance, the same as in Figure 9.40, which ensures optimum transistor load at f_0. The parallel resonant circuit transforms R_L to the optimum load required by the drain port.

Due to the bandwidth requirements of EW PAs, several matching networks would be needed, being switched in and out with pin-diode switches (or other mechanisms). Therefore, if the output tuning can provide the required filtering, the input matching network shown in Figure 9.1 would probably not be considered in the tuning process as it would double the number of networks required.

Theoretical designs of third-harmonic peaking load networks for Class F PAs are presented in [20] and [21], where the component values in the load network are derived. It is important to take into account the active device output capacitance C_{ds}, which can be a part of the output loading network. Figure 9.42 shows two possible circuit configurations with additional parallel resonant circuit (a) or series resonant circuit (b) and derived values of their parameters [20]. Figure 9.42(c) gives an equivalent microstrip impedance-peaking circuit with derived values of its elements, which could provide short-circuit termination for all even harmonics and open-circuit termination for the third harmonic [20].

The practical designs of Class F PAs are further complicated by parasitic reactance of the device packages, the nonlinearities of drain current I_D and nonlinearities of C_{gs} and C_{ds}. However, the equations in (9.42) can serve as a good starting point for practical Class F RF PA designs.

Class F Amplifier Design Approaches

There are several different ways of structuring Class F PAs; four are summarized in Figure 9.43. Figure 9.43(a) shows the conventional approach to Class F PA design, which uses output harmonic tuning. The output network provides a short at even harmonics and an open at odd harmonics, while the input network provides a conjugate match to the previous stage at the gate input. An optimum load at the fundamental frequency is obtained at the drain output by using the output network. Some work on the impact of input driving signals or input harmonic tuning on Class F PA performances [22 – 24] has been performed. RF transistors usually have a severe nonlinear variation of C_{gs} with respect to applied input voltage. This becomes the main cause of signal distortion at the gate terminal, especially when the bias is set around the pinch-off voltage and the input voltage variation is large, as in PA applications. In [17], this nonlinearity is compensated by putting an extra diode in parallel with the gate. In [16] and [18], improvements in gain and efficiency are reported by using short-circuit harmonic terminations at the input of the device,

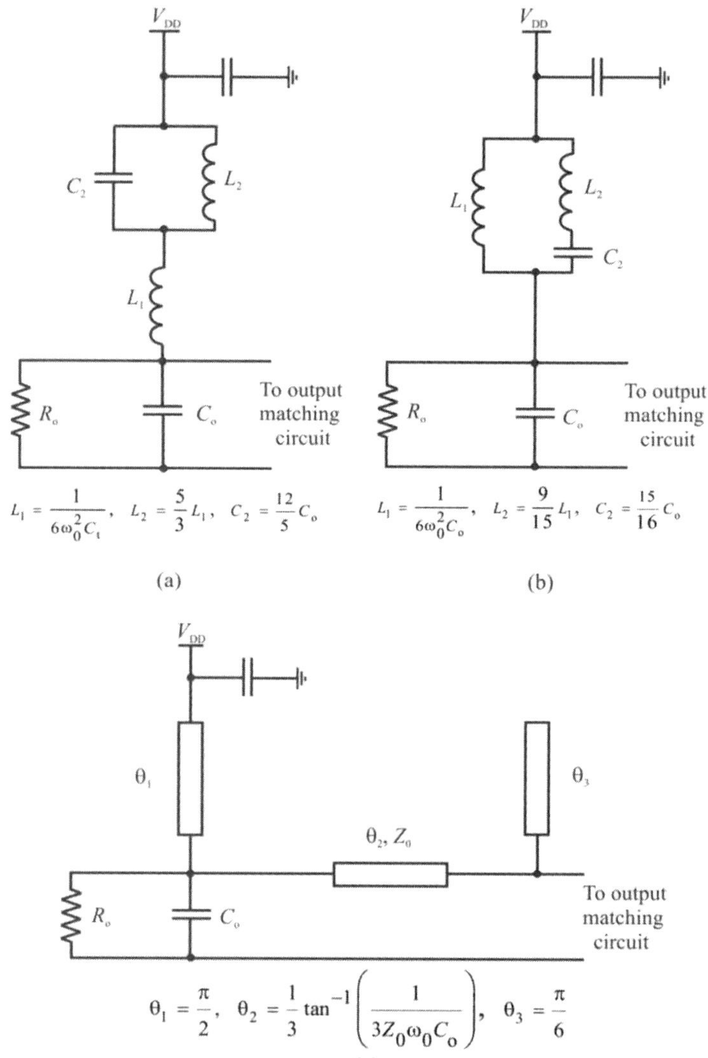

Figure 9.42 Third harmonic peaking circuit of ideal class F amplifiers with derived values of their parameters: (a) parallel resonant circuit and (b) series resonant circuit [20]. Figure (c) gives an equivalent microstrip impedance-peaking circuit with derived values of its elements, which could provide short-circuit termination for all even harmonics and open-circuit termination for the third harmonic [20].

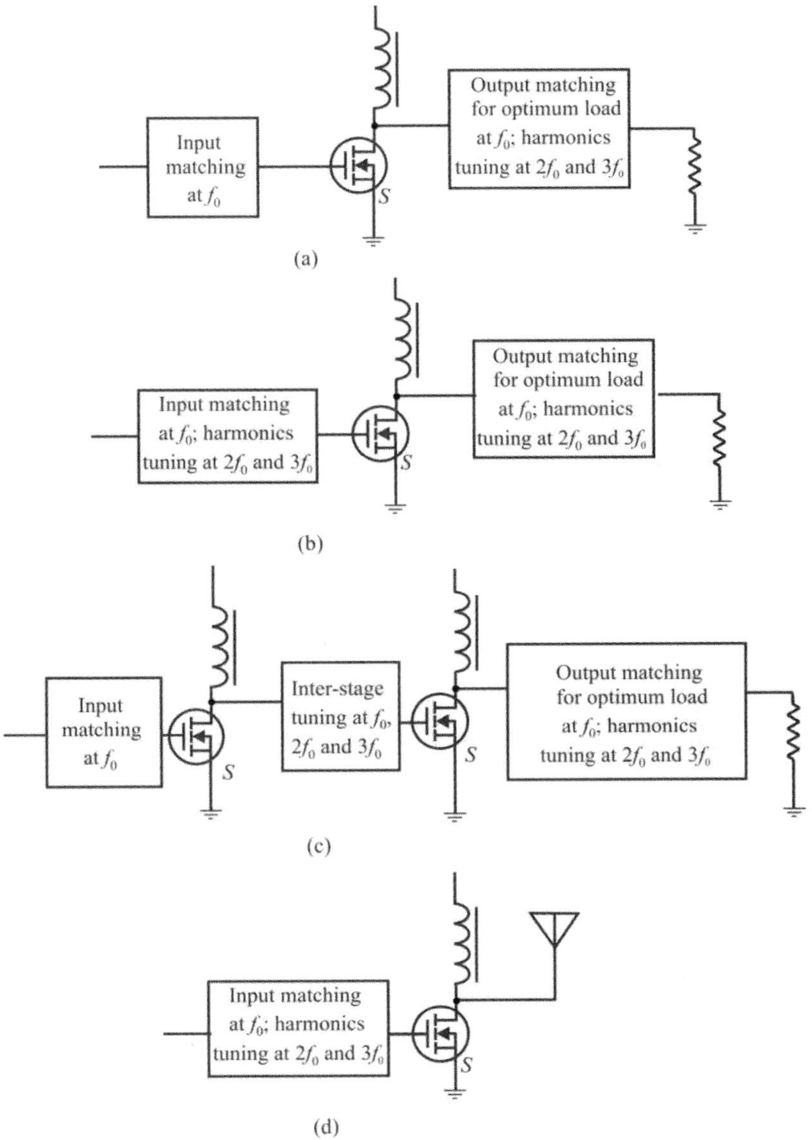

Figure 9.43 (a) Output harmonics tuning strategy. (b) Both input and output harmonics tuning. (c) Multistage class F PAs. (d) Active-antenna approach for class F PAs.

which eliminates or at least reduces the distortions in the input voltage and resultant output current waveform.

It is commonly accepted that Class F PA designs should consider both input- and output-harmonics tuning, as shown in Figure 9.43(b).

Figure 9.43(c) shows a two-stage amplifier, which includes a driver/waveform-shaping stage and a final stage. The first stage will serve two functions: one is to provide enough power at the input of the final stage and the other is to shape the input voltage waveform for the final amplifier stage, so that the input voltage waveform is a half sinusoid at the input of the final stage. The waveform shaping could be achieved by using appropriate proportions of the fundamental component and the second harmonic, as in [25]. This is also called *the harmonic-controlled amplifier* (HCA) [25], which has the advantages of the high gain of class A and the high efficiency of class F, due to the elimination of the negative V_{gs} at the input of the final stage. The problem is the complexity of its design and circuit layout, which may also introduce some losses. An alternative way of realizing the driver/waveform shaping is to use the class F^{-1} PA at the first stage and the class F PA at the final stage. The class-F^{-1} PA produces the output voltage as a half sinusoid, which feeds directly into the input of the final stage through a capacitance as DC block.

Figure 9.43(d) shows an active antenna approach for class F PA designs [26]. The antenna, the active device, and the input network are considered a single entity. In addition to its function as a radiator, the antenna achieves both the harmonics tuning and the filtering without using an additional output-matching network between the antenna and the drain output. This has the obvious advantages of compact design, eliminating the cable and matching network loss, thus improving the overall system efficiency and output power. For additional information on active antennas see [27].

9.9.2.3 Class S Amplifiers

Class S amplifiers operate similarly to class D described earlier in that the signal to be amplified is initially in digital form and is converted to analog. In its digital state it drives a delta-sigma modulator.

Delta-sigma modulators were discovered in the field of A/D and D/A converter design. One of the difficulties with D/A converters is the precision required in some of the components to convert a digital signal to an analog form. The goal was to remove all analog components from the D/A design and the delta-sigma modulator was discovered. A simplified block diagram of perhaps the simplest of these devices is shown in Figure 9.44. The current state of the modulator, B, is subtracted from the level of the digital input signal forming $A - B$. B is a function of the output of the 1-bit quantizer. If the quantizer output is 0, then $B = 0$. If the quantizer output is 1, then $B = 2^M - 1$, where M is the number of bits in the modulator. $A - B$ is called

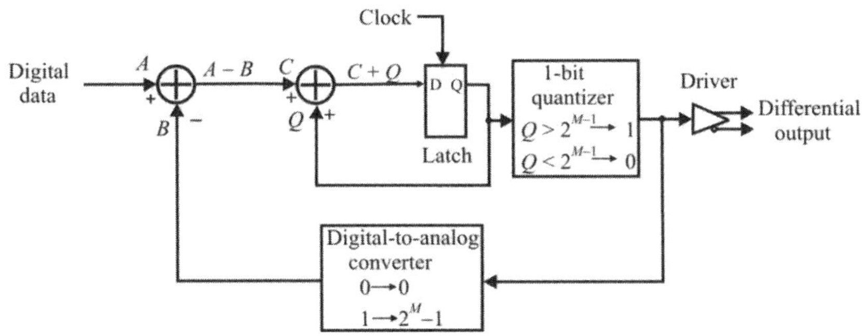

Figure 9.44 Block diagram of a sigma-delta modulator.

the error signal. The second adder adds this error to the accumulated error as stored in the latch. On the next clock cycle, this updated accumulated error is clocked into the latch, forming the next state of the modulator. The 1-bit quantizer outputs a zero if the accumulated error is less than one-half the maximum value allowed ($= 2^M / 2 = 2^{M-1}$), and a 1 if it is greater than half the maximum.

The net effect of the modulator is shown in Figure 9.45. When the input is largely positive, the output of the quantizer is mostly 1, whereas when the input amplitude is small, this output is mostly 0. Another way of saying this is the output remains in a 1-state, on average, and longer when the input signal level is greater. The output of the quantizer is periodic.

Although overall operation produces results that appear to be PWM, as with class D amplifiers, in fact, a delta-sigma modulator produces *pulse density modulation* (PDM). The output width of the pulses is not changed per se according to the amplitude of the input signal, but the number of pulses in the periodic cycle is. The differential outputs are then used to drive the output circuit as shown above

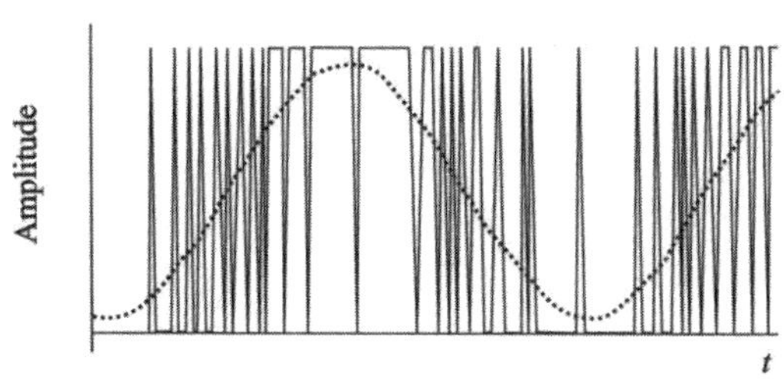

Figure 9.45 Input/output of the sigma-delta modulator driving a class S output stage.

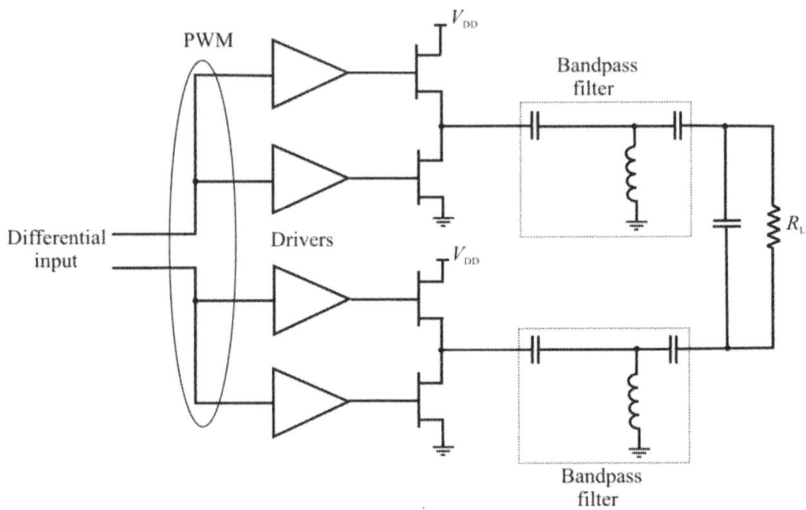

Figure 9.46 Output stage of a class S amplifier. The input takes the output of the sigma-delta modulator and turns the output power transistors ON and OFF using PWM. The output filters can be band-pass, as shown here, or low-pass for broadband operation.

for the class D amplifier. This is shown in Figure 9.46. In this case there is no need for the sawtooth generator or the comparator, and a band-pass filter has replaced the low-pass filter.

It is desirable to keep the differences between one sample and the next relatively small. In order to do this, the sample clock frequency must be much higher than the highest input frequency. This is called oversampling. Nevertheless, much higher frequencies of operation are possible than class D, achieving cutoff frequencies in the gigahertz range. In addition, there are other forms of delta-sigma modulators. In particular, there are higher-order modulators. The order of a delta-sigma modulator is the number of delay stages it has. They tend to calculate the error better and therefore replicate the input signal better. The first-order modulator shown in Figure 9.44 is one of the simplest to illustrate their basic operation.

9.9.3 Load Lines Revisited

Each of the amplifier classes has its own characteristic load line. These are illustrated in Figure 9.47. (Although the load lines in Figure 9.47 do not extend to the ordinate, in reality they do.)

The class A load line extends to $2V_{DD}$ on the abscissa (V_{DS}) axis. Thus, the linear output range extends to twice the supply voltage level. On the other hand, the linear portion of the class B load line terminates on the abscissa at V_{DD}, and is zero after

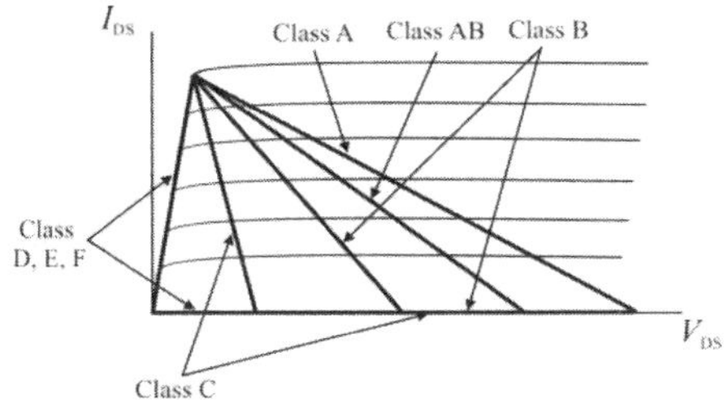

Figure 9.47 Load lines of the amplifier classes.

that. Current only flows for half of the input cycle. The class AB load lines are between these two. The linear part of the class C load lines ends at a voltage less than V_{DD}, and therefore output current conduction only occurs for less than half of the input cycle.

The load lines for the switching classes (D, E, and F) are different from those for the analog classes. Since the active device is either ON or OFF for these classes, the load lines are either represented by an open circuit (zero current, any voltage) which corresponds to the abscissa axis or a short circuit (ideally) when the switch is on. This latter condition corresponds to the ordinate axis (ideally) or the linear region (nonsaturated) curve (practically) shown in Figure 9.47. This limit is due to the fact that the on resistance of a practical device is not zero, but corresponds to the value of the slope of the linear region curve.

9.10 Power Gate and Base Drivers

A power-MOSFET gate driver or a power-BJT base driver is a PA driven by a low-power input and produces the appropriate high-current gate or base drive. Such a driver is used when a PWM controller cannot provide the necessary output current required to drive the gate capacitance of the associated MOSFET [28] or base current of a BJT.

It is important that the gate driver provide sufficient drive current to quickly pass through the Miller plateau region of the power-MOSFET's switching transition (see Figure 5.27). This interval occurs when the transistor is being driven both ON or OFF, and the voltage across its gate-to-drain parasitic capacitor (C_{gd}) is being charged or discharged by the gate driver.

Some have the view that the design of a driver stage is much less important than the design of the output stage, because the power level at the driver stage is much lower than that at the output stage, by a factor equal to the power gain of the output stage, typically a factor of about 10 to 100. That simplistic view is not correct, because the output transistor will not operate as intended if its input is not driven properly. If the output transistor does not operate as intended, the output stage will not operate as intended, either. The output-stage transistor will operate properly as a switch, as intended, if its input port (gate-source of an FET or base-emitter of a BJT) is driven properly by the output of its driver stage. The driver stage must provide the output specified below. We show FETs as the active device, however, BJTs work as well.

1. Enough OFF bias during the OFF interval to maintain the drain or collector current at an acceptably small value. If you are willing to tolerate a power loss of x fraction of the normal DC input power due to nonzero OFF-state current, the drain or collector current during the OFF interval can be up to

$$I_{\mathrm{D}}(\mathrm{OFF}) = xI_{\mathrm{DD}}\frac{1}{1-D} \tag{9.16}$$

where I_{DD} is the DC current drawn from the V_{DD} DC drain-voltage supply, and D is the output-transistor's ON duty ratio (usually 0.50, but it can be any value you choose and provide for in the choice of R, L, and C values in the load network).

Example 9.6: Power Stage Driver 1

If 1% additional power consumption from the V_{DD} voltage supply caused by the nonzero OFF-state current is tolerable, if I_{DD} is 5A, and if D is the usual value of 0.50, you can tolerate an OFF-state drain current of 0.01 [5A] [1/(1 - 0.50)] = 0.1 A = 100 mA.

2. Enough ON drive during the latter 3/4 of the ON interval to maintain a low-enough R_{ON}. What is low-enough depends on the application. Why the latter 3/4 of the ON interval?: The current $i(t)$ during the first 1/4 of the ON interval is small enough that $[i(t)]^2R_{ON}(t)$ can be acceptably small for a fairly high $R_{ON}(t)$ because the small $i(t)$ during the first 1/4 of the ON interval causes an even smaller $[i(t)]^2$.

3. Enough turn-OFF drive to turn-OFF the drain or collector current from 100% to 0% in a fall-time t_f fast enough to make the turn-OFF power dissipation an acceptably small fraction of the output power. That fraction is

$$\frac{(2\pi A)^2}{12} \tag{9.17}$$

where

$$A = \left(1 + \frac{0.82}{Q_L}\right)\frac{t_f}{T} \tag{9.18}$$

and $T = 1/f$ is the period of the operating frequency f. Choose the acceptable fraction of the output power to be dissipated during the non-zero turn-OFF switching time. Then calculate the required drain or collector-fall time t_f that must result from the enough turn-OFF drive. Then provide sufficient turn-OFF drive to accomplish the required objective, according to the characteristics of the chosen output transistor.

Example 9.7: Power Stage Driver 2

If the required objective is to have the turn-OFF power dissipation ($P_{diss, turn-OFF}$) be 6% of the output power, and if $Q_L = 3$, the allowable value for

$$\frac{t_f}{T} = \frac{\sqrt{12P_{diss,turnOFF} / P}}{2\pi(1 + 0.82/Q_L)}$$

is

$$\frac{\sqrt{12(0.06)}}{2\pi(1 + 0.82/3)} = 0.106$$

i.e., t_f can be as large as 10.6% of the period.

4. Enough turn-ON drive to turn-ON the output transistor fast enough to make an acceptably small power dissipation during the turn-ON switching. Most driver circuits turn the transistor ON and OFF with about the same switching times. If the more-important turn-OFF switching time is fast enough, the accompanying turn-ON switching time will be more than fast enough. The input-port characteristics of BJTs, MOSFETs, and MESFETs are different enough that different types of driver circuits should be used to drive those three different types of transistors.

A brief summary of driving a MOSFET or a MESFET follows. The polarity descriptions assume N-channel or NPN; reverse the polarity descriptions for P-channel or PNP.

The best gate-voltage drive is a trapezoid waveform, with the falling transition occupying 30% or less of the period. (Trade-off: The shorter the turn-OFF transition time, the smaller will be the power dissipation in the output transistor during turn-OFF switching, but the larger will be the power consumption of the driver stage. For both MOSFETs and MESFETs, the optimum drive minimizes *the sum* of the output-stage power dissipation and the driver-stage power consumption.) The upper level of the drive waveform should be safely below the MOSFET's gate-source maximum voltage rating, or the MESFET's gate-source voltage at which the gate-source diode conducts enough current to cause either of two undesired effects: (a) metal migration of the gate metalization at an undesirably rapid rate (making the transistor operating lifetime shorter than desired) or (b) enough power dissipation to reduce the overall efficiency more than the efficiency is increased by the lower dissipation in the lower $R_{DS}(on)$ that results from a higher upper level of the drive waveform. The lower level of the trapezoid should be low enough to result in a satisfactorily small current during the transistor's OFF state, discussed in requirement 1 above.

A sine-wave is a usable (but not optimum) approximation to the trapezoid waveform described above. To obtain an output-transistor ON duty ratio of 50% (usually the best choice, but a larger or smaller duty ratio can be used if appropriate component values are used in the load network), the zero-level of the sine wave should be positioned slightly above the FET's turn-on threshold voltage.

A better approach is to remove the part of the sine wave that goes below the V_{GS} value that ensures fully OFF operation, replacing it with a constant voltage at that V_{GS} value. This reduces the input drive power by slightly less than 50%, almost doubling the power gain of the output stage.

9.11 Microwave Monolithic Power Amplifiers

Microwave monolithic integrated circuits (MMICs) are used in the vast majority of commercial and military microwave system. Monolithic implementations of most receiver and transmitter building blocks are readily available for many applications and frequency bands. These functions include mixers, modulators, switches, digital/analog attenuators, digital/analog phase shifters, LNAs, driver amplifiers, and PAs. In the past these MMICs were based primarily on GaAs or InP. However, many new monolithic circuit functions are being realized with the silicon-based technologies discussed in Chapter 5. Examples include SiGe, metal-oxide based devices CMOS, LDMOS, and combined technologies, an example of which is SiGe-

BiCMOS [31]). GaN has found application for microwave high-power applications, which may be epitaxially deposited on SiC or silicon substrates [32]. These devices have simultaneous high-voltage and high-frequency operation, and they are well suited for high-performance EW PAs.

The PA MMIC is quite frequently the most expensive component in a system. It is also common that it is the component that dominates system reliability. For EW applications, these devices are often run at near-saturated output power levels. For analog and digital communication systems, linear operation is required, and the PA may be run at high bias current, high supply voltage, and/or output power levels where the efficiency is low. In either case, the PA MMICs drive thermal management requirements for the entire system because they are normally the highest power dissipation and hottest components in the subsystem. Clearly, the PA MMICs have a significant influence on the operation of the overall system due to their performance and reliability requirements.

Monolithic PAs have some significant advantages over designs using discrete components.

- MMIC PAs are usually much smaller and lighter than discrete circuits.
- There is less performance variation unit to unit for the as-built circuits or, equivalently, less postfabrication tuning and adjustment of the amplifiers required.
- Higher levels of integration and complexity for MMIC PAs are possible because other circuit functions are available in monolithic format. Entire transmit/receive systems have been integrated on to a single die [33].
- Additional possibilities are available to the designer. For example, the transistor cells used in the circuit can be designed for optimum electrical and thermal performance. Restriction to a particular set of available transistor cell sizes, as is the case with a discrete amplifier, is not required.
- For MMIC PAs that incorporate multiple gain stages (frequently the case in EW system applications), the impedance levels between the stages are arbitrary and can be optimized for bandwidth, gain, and efficiency. Discrete PA circuits tend to be designed for fixed real impedance, typically 50 Ω, so that they can be cascaded.
- As the operating frequency increases, discrete circuit architectures become more difficult to realize. Above 10 GHz, the selection of high-quality surface mount components used for quasi-lumped element matching becomes limited and distributed circuit approaches are more

common [7]. Above 18 GHz, monolithic implementations for the PA begin to become the only feasible option.

Monolithic circuits do have some significant drawbacks, however.

- The smaller PA footprint means that more heat must be removed from the MMIC, which impacts the cost and complexity of the thermal management system. Thermal management issues at the next level assembly typically get past back to the MMIC manufacturer as a higher temperature requirement for which the device must maintain reliable operation. Reliable operation generally means that transistor channel temperature must not exceed some maximum value [35].
- MMIC components are typically more costly than discrete circuits. The most direct way to reduce MMIC cost is to make the die as small as possible. However, in the case of PAs, die size compaction while maintaining low channel temperature are in direct opposition.
- Amplifiers constructed with discrete components are relatively easy to tune by element value adjustment, metal traces can be cut or widened, and so forth. Some limited circuit tuning is possible for MMIC PAs. However, adjustments are limited and require specialized equipment and highly skilled personnel, thereby contributing to increased costs. Generally speaking, the design of MMIC components must be right the first time.

The general subject of microwave power amplifier design has been covered extensively in the published literature. Excellent references on PA circuits, architecture, applications, analysis, and modes of operation are readily available. This section provided only a brief introduction to the topic. For more information for the interested reader a few excellent, concise publications [36 – 38] are recommended.

9.12 Power Amplifier Reference Design

In this section we present a reference design for an RF PA and summarize its performance. This design would be applicable for a thin jammer [39] targeting low powered communication targets such as IEDs. Its output power is relatively low, making it unacceptable for stand-off EW applications.

A schematic of a complete PA is shown in Figure 9.48. This circuit, with Polyfet nomenclature TB263, is based on the LS2641 dual LDMOS transistor from

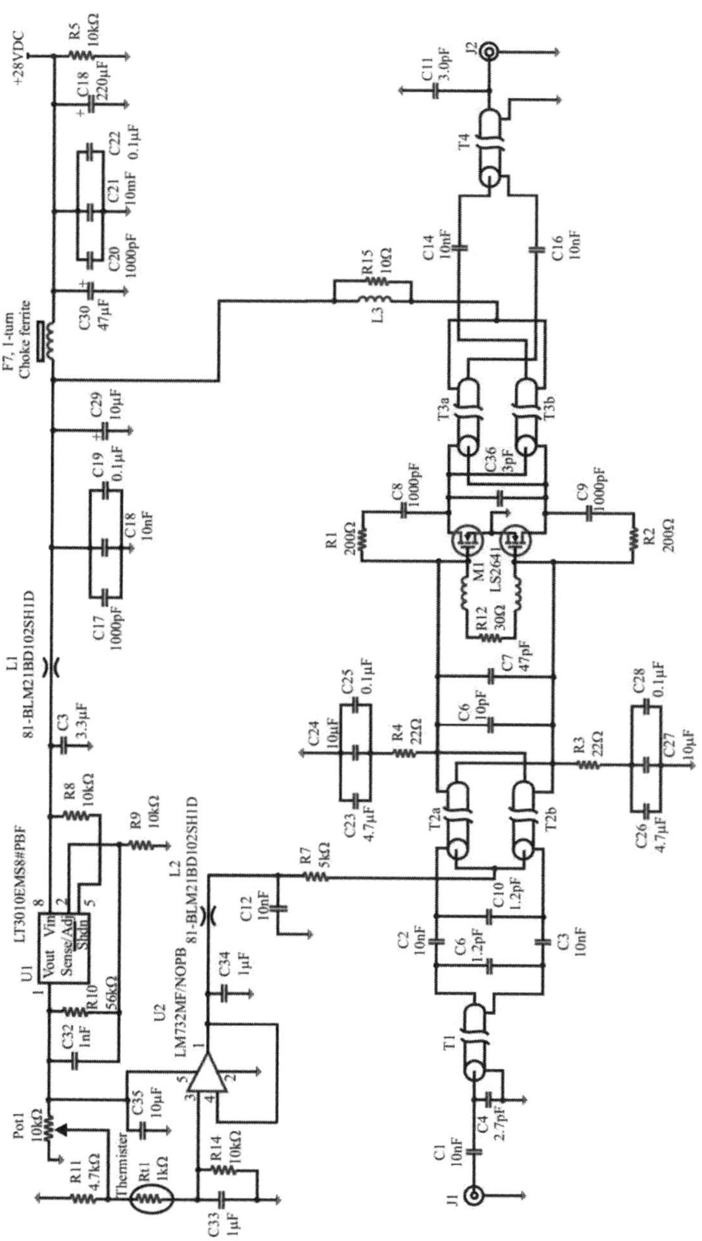

Figure 9.48 TB263 Schematic. (Source: [40]. © Polyfet. All rights reserved. Reprinted with permission.)

Figure 9.49 PA prototype layout. (Source: [40]. © Polyfet. All rights reserved. Reprinted with permission.)

Polyfet identified as M1 in Figure 9.48. These MOSFETs are configured in a push-pull arrangement and the drivers for the power transistors are implemented on the same chip carrier as the power transistors.

The prototype amplifier is pictured in Figure 9.49. It measures 3.8" × 3.2" so is quite suitable for a thin jammer application [39]. The overall frequency range of this amplifier is 30–512 MHz with a minimum output power of 180W.

Broadband *transmission line transformers* (TLTs) are utilized to achieve the bandwidth. These devices are inherently broadband. We will digress and discuss the characteristics of TLTs shortly.

The upper half of the schematic shown in Figure 9.48 provides the bias voltages for the LS2641. The drain bias is passive and is shown on the upper right. The gate bias circuitry is active, and is shown in the upper left.

The physical layout of the PA is shown in Figure 9.50.

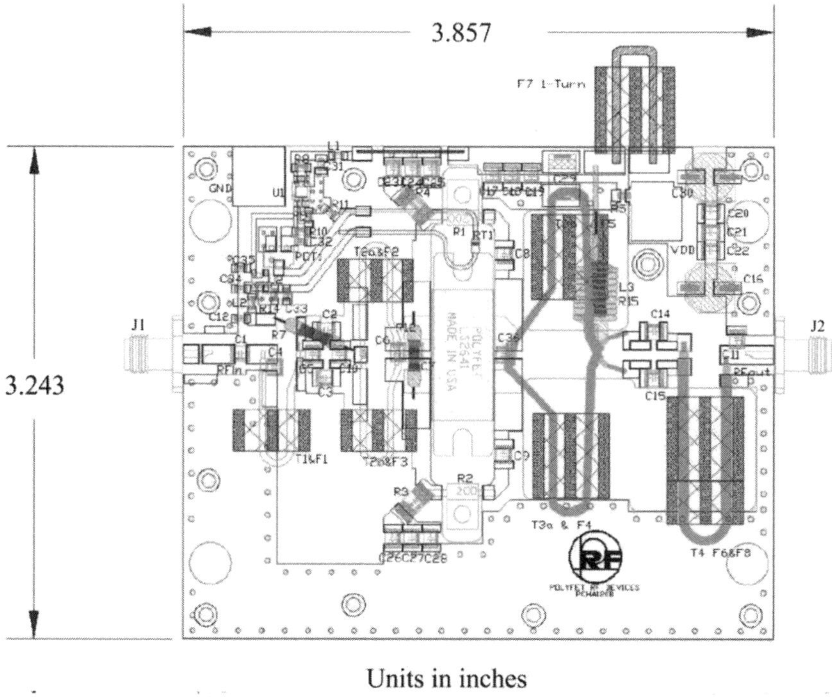

Figure 9.50 TB263 layout. (Source: [40]. © Polyfet. All rights reserved. Reprinted with permission.)

9.12.1 LS2641 General Description

The LS2641 Si enhancement mode RF power LDMOS transistor works from a 28VDC (or less) power source. The device has low values of capacitance which makes it ideal for broad band applications such as communication EW purposes. Employing back-to-back gate diodes for enhanced ESD protection and having a high drain breakdown voltage makes this device exceptionally rugged. The overall frequency range is 1–1300 MHz. The circuit consists of a pair of MOSFETs on a single chip, Polyfet part number SQ202. They are used as drivers for another pair of MOSFETs, also on the same chip, Polyfet part number LY402. They both use coaxial transformers (one instantiation of transmission line transformers) for coupling, which are very broadband and low loss.

The parts list for the PA is provided at Figure 9.51.

NOMENCLATURE	DESCRIPTION	VENDER	VENDER PART #
C4	2.7pF +/-2% Ultra-low ESR, Microwave Cap., 1111N	Passive Plus, Inc	1111N1R0BW501
C5, C10	1.2pF +/-2% Ultra-low ESR, Microwave Cap., 1111N	Passive Plus, Inc	1111N3R0BW501
C6	10pF +/-2% Ultra-low ESR, Microwave Cap., 1111N	Passive Plus, Inc	1111N9R1BW501
C7	47pF +/-2% Ultra-low ESR, Microwave Cap., 1111N	Passive Plus, Inc	1111N430BW501
C11, C36	3.0pF +/-2% Ultra-low ESR, Microwave Cap., 1111N	Passive Plus, Inc	1111N3R3BW501
C16	Aluminum Electrolytic Capacitors - SMD 35Volts 220uF 20%	United Chemi-Con	EMZJ350ADA221MHA0G
C19,C22,C25,C28	0.1uF +/-10% RF By-Pass Capacitors,1111X	Passive Plus, Inc	1111X104KW500
C24,C27	10uF TANCERAM HIGH CAP. SMD 1206	Passive Plus, Inc	PPI-GMC31X5R106K10
C23,C26	4.7uF TANCERAM HIGH CAP. SMD 1206	Passive Plus, Inc	PPI-GMC31X7R475K25NT
C8,C9,C17,C20	1000pF +/-2% Ultra-low ESR, Microwave Cap., 1111N	Passive Plus, Inc	1111N102GW500
C1,C2,C3,C12,C14,C15, C18,C21	10,000pF +/-10% RF By-Pass Capacitors,1111X	Passive Plus, Inc	111X103KW500
C29	10uF 63V 20% AVX Tantalum Capacitors	AVX	TCJE106M063R0100
C30	Aluminum Electrolytic Capacitors - SMD 47UF 35V TK SMD	Panasonic	EE-TK1V470P
C31	3.3uF TANCERAM HIGH CAP. SMD 1206	Johanson Dielectrics	100r18x335KV4E
C32	1nF 10% Hi-Q/Low ESR	Passive Plus, Inc	0505x102KW500
C33,C34	1uF 10%TAMCERAM HIGH CAP. SMD 603	Johanson Dielectrics	250r14x105kv4t
C35	10uF 10%TAMCERAM HIGH CAP. SMD 805	Johanson Dielectrics	100r15x106kv4e
R1,R2	30W, 200OHM	Innovative Power Products-IPP	IPP-RB201-200
R3,R4	RES 22ohm 2W 1% 2512 SMD	Stackpole Electronics Inc	RHC2512FT22R0
R5	RES 10K OHM 1/4W 5% 1206 SMD	Rohm Semiconductor	MCR18EZPJ103
R7	RES 5K 1/2W 5% CARBON FILM	Stackpole Electronics Inc	CF12JT5K0
R8,R9,R14	10K 1/4W 0805 SMD	Rohm Semiconductor	MCR10EZHJ103
R10	56K 1/4W 0805 SMD	Rohm Semiconductor	MCR10EZPJ563
R11	4.7K 1/4W 0805 SMD	Rohm Semiconductor	MCR10EZPF4701
R12	RES 30 OHM 1W 5% METAL OXIDE	Xicon	281-30-RC
R15	RES 10 OHM 1W 5% METAL OXIDE	Xicon	281-10-RC
RT1	1000 OHM TEMP SENSOR AXIAL	NXP Semiconductors	KTY83/110,113
POT1	Trimmer Resistors - 10K OHM 0.25W SMD	Murata Electronics North America	PVG5A103C03R00
L1,L2	0805 1kohms HiSpeed EMI Filter Beads,Chokes & Arrays	Murata Electronics North America	81-BLM21BD102SH1D
L3	20turns 18AWG I.D 195mils	www.electromechanicsonline.com	MW280SPNRD
WIRE	18 AWG Hook-up Wire UL	Belden	83029 0061000
U1	IC REG LDO ADJ 50mA 8-MSOP	Linear Technology	LT3010EMS8E#PBF
U2	IC OP AMP R-R I/O SOT23-5	National Semiconductor	LM7321MF/NOPB
T1	Semi-ridge Coax ,1.8 inch, 50ohm	MICROWAVE	UT-34-M17
T2a,T2b	Semi-ridge Coax ,1.8 inch, 25ohm	MICROWAVE	UT-34-25
T3a,T3b	Semi-ridge Coax ,2.5 inch, 10ohm	MICROWAVE	UT-70-10
T4	Semi-ridge Coax , 3 inch, 50ohm	MICROWAVE	UT-85-50
F1,F2,F3	950uF BINOCULAR FERRITE	Ceramic Magnetic, Inc	003212-02
F4,F5,F6/ F8 Half	125uF BINOCULAR FERRITE/ BINOCULAR HALF	Ferronic/ Amidon	12-365-K/ 2861001502
F7	Material 4B1 250ui,EMI Supressor	Ferroxcube	MHB2-14/8.5/14-4B1
J1,J2	PE4000-SF SMA Female, 4 Hole Panel Mount	Pasternack	PE4000-SF
PCB	material RO4350B 60mils(1.524mm) 2oz/2oz (70um/70um)	Rogers. Corp	RO4350B
M1	LDMOS	Polyfet RF Devices	LS2641

Figure 9.51 TB263 parts list. (Source: [40]. © Polyfet. All rights reserved. Reprinted with permission.)

9.12.2 Transmission Line Transformers

Extremely low losses (a fraction of a dB) are possible over bandwidths of two or more decades are facilitated by using transmission line transformers (TLTs). Other transformer types can match the low loss over a narrow bandwidth, or they may exhibit wide bandwidth, but have 1-dB or more loss. The low loss of TLTs makes them especially useful in high power circuits typical of PA applications in EW systems.

The TLT operates by transmitting energy via the transverse transmission line mode. It does not rely on the coupling of magnetic flux lines like a conventional transformer. Figure 9.52 shows the simplest illustration of this difference. In the TLT, the purpose of the windings is to eliminate (that is, reduce as much as possible) common-mode currents from the input to the output. With input and output isolated in this manner, the voltages at the input and output can be treated as being independent of one another.

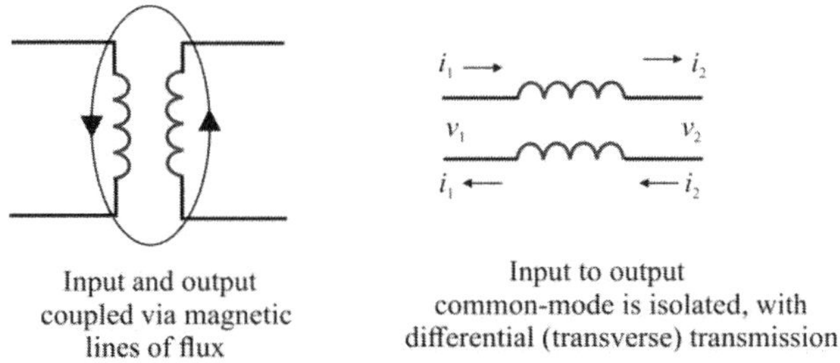

Figure 9.52 Transformers.

One of the primary functions of transformers in electronic circuits is impedance transformation. Figure 9.53 shows a simple transformer that has an impedance transformation ratio of 4:1. This is a Ruthroff design [41], one of the classic transmission line transformer types. In this circuit, two voltages are summed at the output by the two transformer leads connected in series to get twice the voltage. Since we must have conservation of energy, R_L must be four time the value of R_g in order for the voltage ratio to be two.

As an example of the utility of the simple 4:1 impedance transformer, the 4:1 ratio matches a 50 Ω source to a 12.5 Ω load. The former is a typical value for the

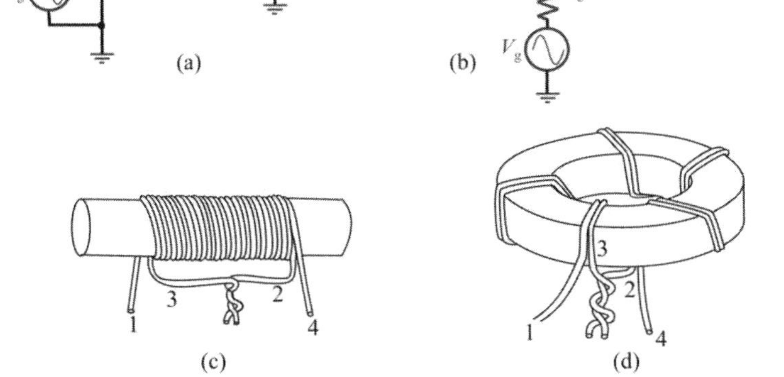

Figure 9.53 Ruthroff transformers.

output impedance of a PA while the latter is a typical value for the radiating impedance of several types of antennas.

Also, the phase difference of the voltage fed from terminal 2 to terminal 3 must not differ from the incident voltage. Thus, the physical length of the transmission line must be only a fraction of the wavelength. We also need to maintain the required isolation from input to output in order to add the delayed voltage at 2 back on itself at 3. This isolation is obtained by forming the transmission line (coaxial cable or closely spaced parallel wires) into an inductance. Inductance is maximum with a magnetic core of iron powder or ferrite material, as shown in Figure 9.53(c) and (d). [42, 43]

In this PA implementation T1–T4 are TLTs implemented with ferrite cores as illustrated in Figures 9.50 and 9.53. There are many other forms of TLTs that provide a wide variety of impedance transforming functions. The salient characteristics of these transformers are their broad bandwidth and low loss.

The concept of TLTs has been known since at least 1944, when Guanella presented an impedance transformer that consisted of a pair of interconnected transmission lines [44]. TLTs have been found to possess far wider bandwidth and much greater transmission efficiencies by arranging the windings of the TLT to have uniform transmission line properties. These devices are widely used for matching networks for antennas and amplifiers in the HF through UHF bands, up to 2 GHz and higher, and their low losses (a fraction of a dB) make them especially useful in high power circuits such as EW PAs [45].

Typical structures for TLTs consist of parallel wires [42], coaxial cable, or bifilar twisted wire pairs [46, 47], with the last being popular because the characteristic impedance can readily be determined by the wire diameter, the insulation thickness, and the twisting pitch [47, 48]. In the case of a coaxial cable transmission line with the correct characteristic impedance for the TLT, the high frequency bandwidth limit is reached when the cable length becomes on the order of a half wavelength ($\lambda/2$), with the overall achievable bandwidth being about a decade [46]. We will discuss the frequency limits shortly.

When using magnetic materials such as powdered iron or ferrite [49] to the TLT as core materials, both the low frequency limit and the high frequency limit are improved [46], and when low loss, high permeability ferrites are used along with semi-rigid coaxial cable, bandwidths of four decades or more are achievable [50].

9.12.2.1 TEM Transmission Line Mode

Transmission line transformers function by transmitting energy by way of the transverse electro-magnetic (TEM) transmission line mode, rather than by the more familiar coupling of magnetic flux as with a conventional transformer [4] (see

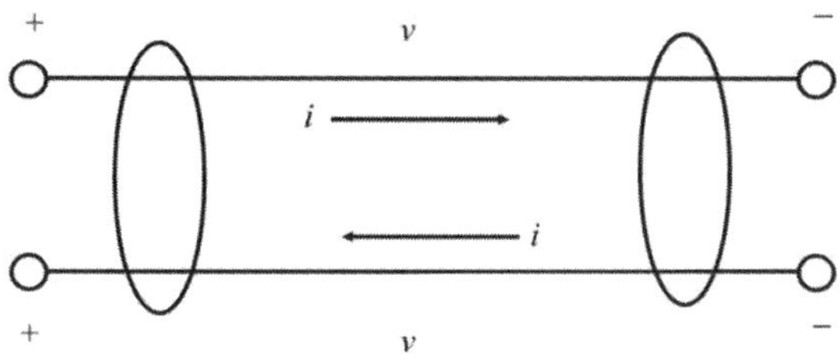

Figure 9.54 Transformer TEM mode.

Figure 9.52). Figure 9.54 illustrates this concept, where the two lines represent the two conductors of a transmission line, irrespective of whether it is made of parallel wires, twisted wires, coaxial cable, or any other means. The currents in the two conductors are equal in magnitude and opposite in phase, while the voltages along the two conductors are equal in magnitude [51]. In the TLT, the windings serve to eliminate, or at least substantially reduce, common-mode currents from the input to the output [46].

12.2.2.2 TEM Transmission Lines

For all types of TEM mode transmission lines (coaxial, parallel wire, twisted wire, waveguide, etc.) the equations that govern the functioning have the same basic form. Referring to Figure 9.55, if R, L, G and C are the total series resistance, series inductance, shunt conductance, and shunt capacitance per unit length z, then the voltage on the transmission line may be expressed as [51]:

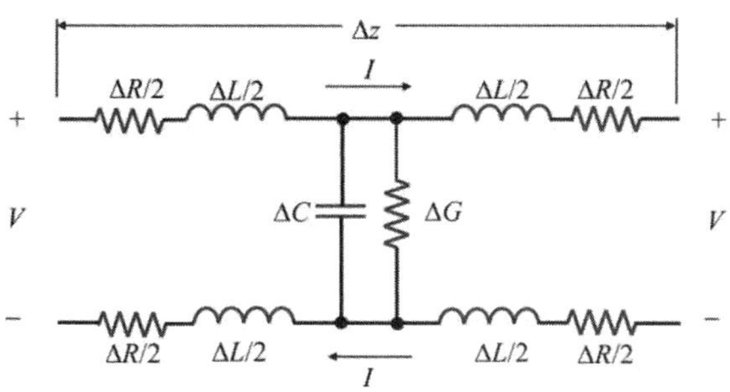

Figure 9.55 Transmission line equivalent circuit.

$$\frac{\partial V}{\partial z} = -(R + j\omega L)I \tag{9.19}$$

and the current as

$$\frac{\partial I}{\partial z} = -(G + j\omega C)V \tag{9.20}$$

which are derived from Maxwell's equations [51 – 53].

From (9.19) and (9.20) we can derive the characteristic impedance Z_0 of the transmission line as [51]:

$$Z_0 = \frac{\partial V}{\partial I} = \sqrt{\frac{R + j\omega L}{G + j\omega C}} \tag{9.21}$$

and the complex propagation constant γ, which can be approximated as [51]:

$$\begin{aligned} \gamma &= \alpha + j\beta \\ &= \sqrt{(R + j\omega L)(G + j\omega C)} \end{aligned} \tag{9.22}$$

where α is called the attenuation constant and β is called the phase constant.

For low-loss transmission line such as good quality coaxial cable [51]:

$$R << \omega L \tag{9.23}$$

and

$$G << \omega C \tag{9.24}$$

[Equations (9.23) and (9.24) are often referred to as the transmission line loss-free conditions.] These equations allow us to approximate the characteristic impedance Z_0 as [51] :

$$Z_0 = \sqrt{\frac{L}{C}} \tag{9.25}$$

and the complex propagation constant γ to be approximated as [51]:

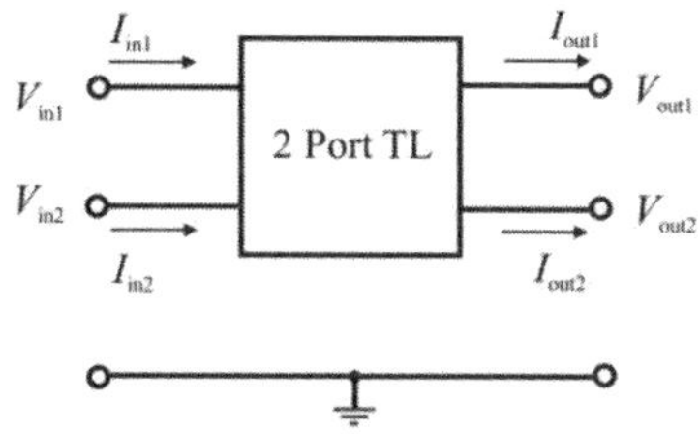

Figure 9.56 2-port TEM transmission line as 4-port ABCD element.

$$\gamma \approx j\omega\sqrt{LC} \tag{9.26}$$

The reciprocal of the square root of the product of L and C provides the phase velocity [51]:

$$v_p = \frac{\omega}{\beta} = \frac{1}{\sqrt{LC}} = \frac{1}{\sqrt{\mu\varepsilon}} \tag{9.27}$$

where μ is the *permeability* of the transmission line medium in Henries/meter (H/m) and ε is the *permittivity* of the transmission line medium in farads/meter (F/m) [51].

Considering the 2-port transmission line of Figure 9.54 as a 4-port, and pairing each port with ground, it may described with a 4-port ABCD matrix, as shown in Figure 9.56, and described analytically as:

$$I_{in2} = -I_{in1} = I_{in} \tag{9.28}$$

$$I_{out2} = -I_{out1} = I_{out} \tag{9.29}$$

$$V_{in1} = V_{out1} \cosh \gamma l + Z_0 I_{out} \sinh \gamma l \tag{9.30}$$

$$V_{in2} = V_{out2} \cosh \gamma l + Z_0 I_{out} \sinh \gamma l \tag{9.31}$$

$$\begin{aligned} I_{in} &= V_{out1} \sinh \gamma l - V_{out2} \sinh \gamma l + I_{out} \cosh \gamma l \\ &= (V_{out1} - V_{out2}) \sinh \gamma l + I_{out} \cosh \gamma l \end{aligned} \tag{9.32}$$

The second term of (9.30) and (9.31) represent the voltage along the length of the transmission line, and we can readily see that this voltage is equal in magnitude and phase for both conductors.

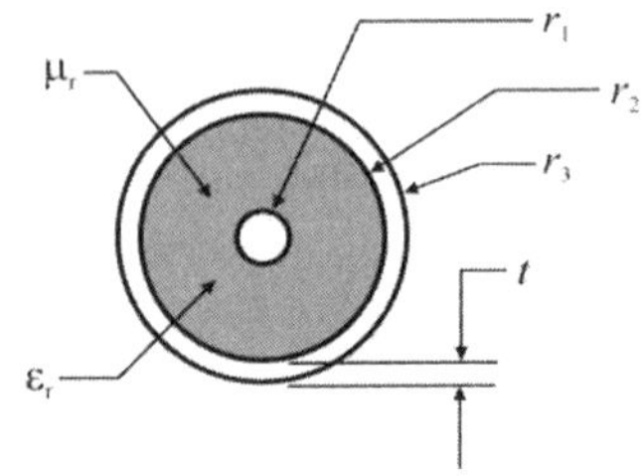

Figure 9.57 Coax transmission line cross section.

12.2.2.3 Coaxial Cable

Consider the cross section of a coaxial cable illustrated in Figure 9.57. The inner conductor has a radius of r_1 and the outer conductor has an inner radius of r_2, an outer radius of r_3, and a thickness of t. The space between the inner and outer conductors is filled with an insulating dielectric material such as PTFE[5] that has a dielectric constant ε_r and a relative permeability μ_r. If the conductors are considered lossless, the unit shunt capacitance C is [54]:

$$C = \frac{2\pi\varepsilon}{\ln\left(\frac{r_2}{r_1}\right)} = \frac{55.6\varepsilon_r}{\ln\left(\frac{r_2}{r_1}\right)} \quad \text{pF/m} \tag{9.33}$$

and the unit series inductance L is [54]:

$$\begin{aligned} L &= \frac{\mu}{2\pi}\ln\left(\frac{r_2}{r_1}\right) \\ &= 0.2\mu_r \ln\left(\frac{r_2}{r_1}\right) \quad \mu\text{H/m} \end{aligned} \tag{9.34}$$

The characteristic impedance of a coaxial TLT is given by:

$$Z_0 = 60\sqrt{\frac{\mu_r}{\varepsilon_r}}\ln\frac{r_2}{r_1} \tag{9.35}$$

---

[5] Wikipedia: *Polytetrafluoroethylene* (PTFE) is a synthetic fluoropolymer of tetrafluoroethylene that has numerous applications. The best-known brand name of PTFE-based formulas is Teflon by Chemours Chemours is a spin-off of DuPont, which originally discovered the compound in 1938.

The unit series resistance R depends on a phenomenon known as the *skin effect,* and is frequency dependent. A quantity known as the $1/e$ depth of penetration [54] is defined as:

$$\delta = \sqrt{\frac{2}{\omega \mu_v \mu_r \sigma}} \tag{9.36}$$

where μ_v is the permeability of free space ($4\pi \times 10^7$ H/m) and σ is the conductivity of the material. The current in a conductor will always concentrate on the surface that is nearest the source that creates the current [54]. This can be imagined by assuming that the conductor is plane sheet and the E-field is running parallel to it aligned with the axis. The current with density J induced into the conductor will concentrate in the conductor where that E-field is closest—that is, just underneath the E-field point under consideration. In the case of coaxial cable this is the electromagnetic field that is present between the inner conductor and the inner surface of the outer conductor. At high frequencies the skin effect causes the current to flow only on the outer surface of the inner conductor and the inner surface of the outer conductor [54], and this condition persists as long as the thickness t of the outer conductor (see Figure 9.57) is appreciably greater than the depth of penetration.

From (9.36) we can now define the surface resistance of the conducting material as [51]:

$$R_s = \frac{1}{2\pi\sigma\delta_s r_2}\left(\frac{1}{r_1} + \frac{1}{r_2}\right) \quad \Omega \tag{9.37}$$

where

$$\delta_s = \frac{1}{\sqrt{f}\sqrt{\pi \mu_r \mu_0 \sigma}} \quad \text{m} \tag{9.38}$$

In most cases it is sufficient to use the loss tangent data provided by the manufacturer of the cable we are using and from that derive the surface resistance R_S.

The unit shunt conductance is related to the resistivity of the dielectric insulating material, due to a phenomenon known as dielectric hysteresis, which is analogous to the magnetic hysteresis in magnetic materials. It is convenient to describe the total losses of the transmission line as the equivalent conductivity [55]:

$$\sigma' = \sigma + \omega\varepsilon'' \tag{9.39}$$

from which we can derive the *loss tangent* of the transmission line, which is [55]:

$$\tan\delta = \frac{\sigma'}{\omega\varepsilon'} \tag{9.40}$$

where ε′ and ε′′ are often referred to as *dielectric dispersion formulas* [55] and are complicated expressions using the parameters of the transmission line being considered. Using (9.40) we can calculate the shunt conductance as

$$G = \frac{2\pi\varepsilon_0\varepsilon_r}{\ln\frac{r_2}{r_1}}\tan\delta \qquad \text{S/m} \tag{9.41}$$

In most cases of interest to our applications, G is quite small compared with ωC.

As examples, for Teflon, $\tan\delta = 0.00015$ and for ¼” semirigid cable, $Z_0 = 50$ Ω, $r_2 = 7\times10^{-3}$ m, and $\varepsilon_r = 2.1$. For R214 with polyethylene, $Z_0 = 50$ Ω, $r_2 = 1\times10^{-2}$ m, $\varepsilon_r = 2.3$, and $\tan\delta = 0.00031$.

Low Frequency Bandwidth Limit

The low frequency bandwidth limit of a coaxial cable TLT is determined by way of the magnetizing inductance of the outer surface of the outer conductor, which results in the low frequency model illustrated in Figure 9.58. Here, the transmission line proper is represented by the ideal 1:1 transformer. The resistance R_0 represents the losses of the transmission line, and the inductance L_{ac} represents the magnetizing inductance of the outer surface of the outer conductor.

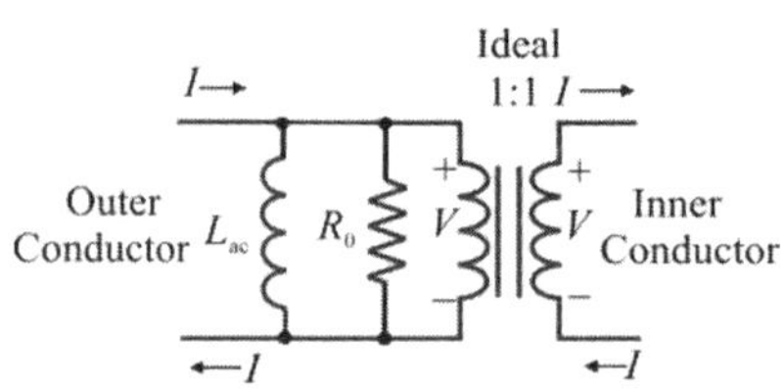

Figure 9.58 Low frequency model of 1:1 coaxial cable transmission line transformer.

An approximation to the magnetizing inductance can be made by considering the outer surface of the coaxial cable to be the same as that of a straight wire (or linear conductor) which, at higher frequencies where the skin effect would cause the current to be concentrated on the outer surface of the outer conductor, would have the self-inductance of: [56]

$$L_{ac} = 2l\left[\ln\left(\frac{2l}{r_3}\right) - 1\right] \qquad \text{nH} \tag{9.42}$$

where l is the length of the coaxial cable in cm and r_3 is the radius of the outer surface of the outer conductor in cm, as shown in Figure 9.57. As stated, this inductance is for a straight conductor, and will generally increase when the coaxial cable is formed into various shapes such as a helix, making the inductance L_{ac} of (9.42) a lower limit in the design process.

The length of the transmission line is kept to no more than an eighth of a wavelength ($\lambda/8$) at the highest frequency for the application. The characteristic impedance of the transmission lines used in the TLT should be equal to the geometric mean of the input and output impedances [56]:

$$Z_{TL} = \sqrt{Z_{in} \times Z_{out}} \tag{9.43}$$

although suboptimal transmission line characteristic impedances may be used provided that the increased losses, degraded return loss, and reduced bandwidth are acceptable, which is sometimes a necessary tradeoff when using commercial coaxial transmission line.

From (9.42) and (9.43) we can now estimate the low frequency bandwidth limit of a coaxial cable TLT as:

$$f_L = \frac{Z_{TL}}{2\pi L_{ac}} \tag{9.44}$$

Referring back to the low frequency model in Figure 9.58, we note that the voltage for the low frequency model is described along the length of the conductors rather than across the end terminals, which is represented by the second term of equations (9.30) and (9.31). This voltage orientation is commonly used in the design of TLTs.

9.12.3 Amplifier Performance

The prototype VHF/low UHF amplifier has been assembled and tested. Test results are illustrated in Figures 9.59–9.64.

Performance for this PA is shown in Figure 9.59 at 30 MHz, Figure 9.60 for 250 MHz, and Figure 9.61 for 512 MHz. In all these cases, P_{out} increases monotonically as the P_{in} is increased, as does the efficiency.

Gain versus frequency at the rated power output of 180W is displayed in Figure 9.62. The gain remains reasonably flat as P_{in} is increases and is above 17 dB across the frequency range, while the efficiency stays around 60%.

As we see from Figure 9.63, the 1-dB output compression point is 180W minimum, which, as we discuss later in Chapter 11, is typically used as a measure of the maximum linear output power available from an RF amplifier. For adjacent channel interference avoidance, an amplifier is not normally operated at the 1-dB compression point. Maximum input power is normally backed off from this point to keep the amplifier in its linear range.

The intermodulation performance, as given by IM3 and IM5 is depicted in Figure 9.64. At 180W PEP with a frequency separation of 100kHz, IM3 is below –15 dBc over the frequency range 30–512 MHz. IM5 is below –20 dBc over the same frequency range.

9.13 Concluding Remarks

In this chapter we discussed topologies and characteristics of RF power amplifiers. Such amplifiers are usually the final stage in the RF signal path in EW systems, and are used to boost signal strength to levels suitable for EA applications.

For stand-off applications such signal strength is on the order of hundreds of watts to be fed to an antenna for broadcast. For stand-in applications (i.e. IED jammers, UAS configurations, hand-emplaced, and so forth) power levels in the tens of watts are frequently adequate. In most situations, wideband amplifiers are required as the target set to be addressed is often not a priori defined.

We presented a PA reference design based on LDMOS transistors and TLT broadband transformers. This is a complete amplifier (needs a power supply and heat sink, of course) but is currently available from Polyfet. It is a relatively low power design (180W PEP) and is suitable for low and medium power applications but has good performance over the frequency band 30–512 MHz.

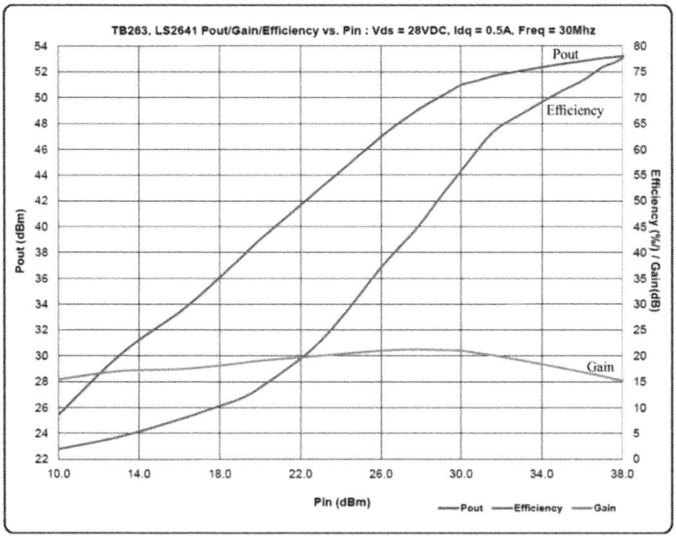

Figure 9.59 Performance at 30 MHz. (Source: [40]. © Polyfet. All rights reserved. Reprinted with permission.)

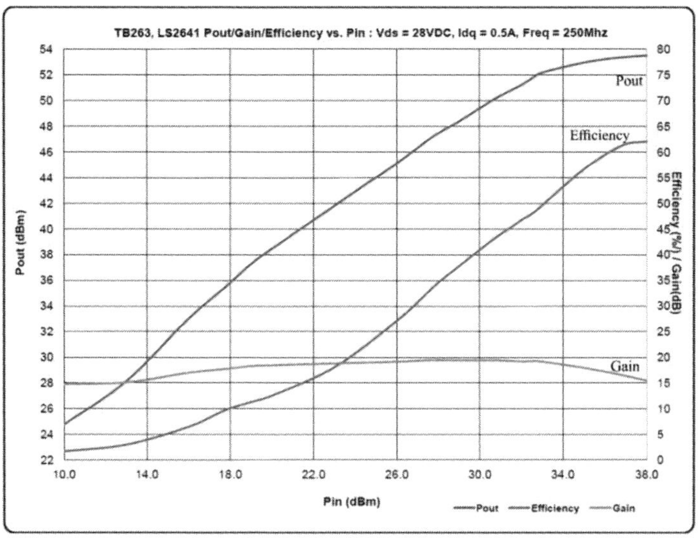

Figure 9.60 Performance at 250 MHz. (Source: [40]. © Polyfet. All rights reserved. Reprinted with permission.)

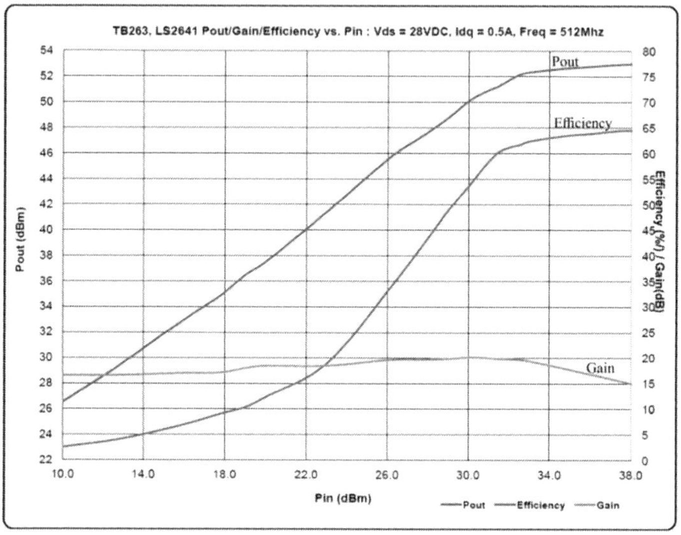

Figure 9.61 Performance at 512 MHz. (Source: [40]. © Polyfet. All rights reserved. Reprinted with permission.)

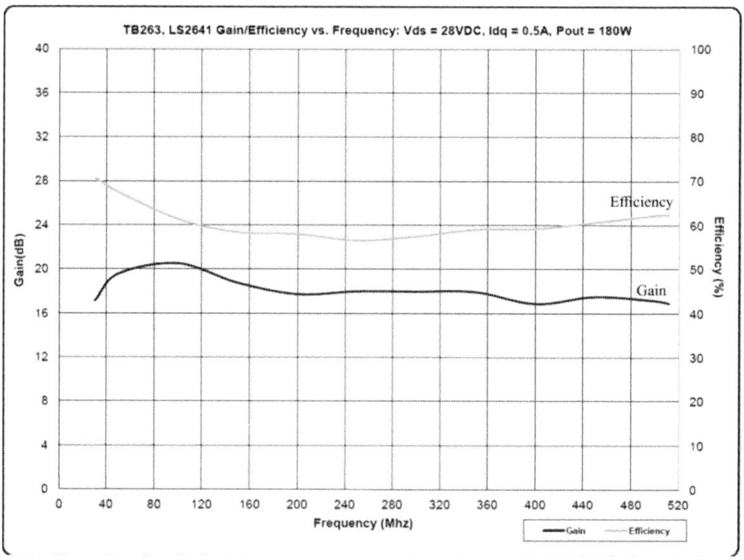

Figure 9.62 PA performance versus frequency. (Source: [40]. © Polyfet. All rights reserved. Reprinted with permission.)

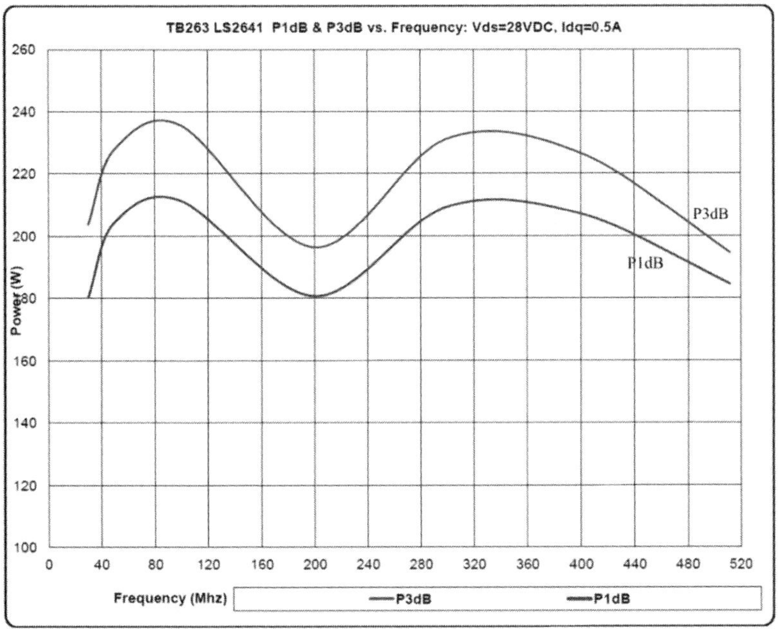

Figure 9.63 P3dB and P1dB performance versus frequency. (Source: [40]. © Polyfet. All rights reserved. Reprinted with permission.)

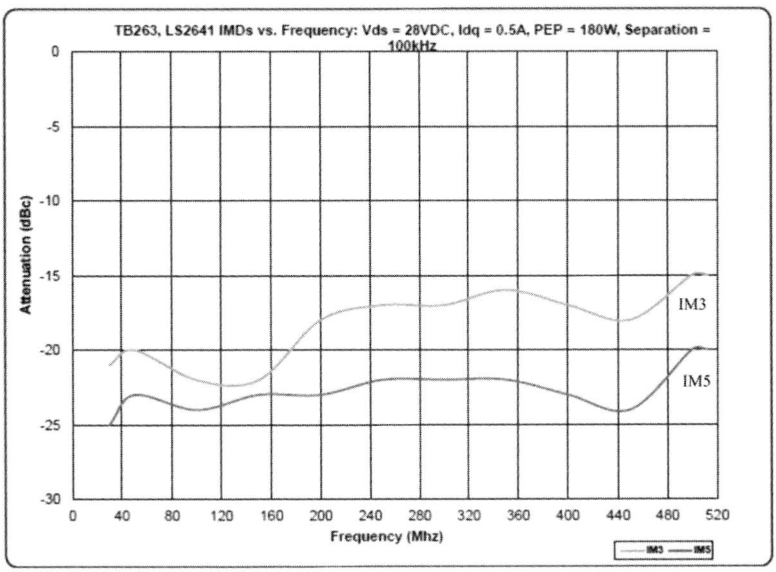

Figure 9.64 TB263 intermodulation performance versus frequency. (Source: [40]. © Polyfet. All rights reserved. Reprinted with permission.)

References

[1] Kenington, P. B., *High-Linearity RF Amplifier Design*, Norwood, MA: Artech House, 2000.
[2] Krauss, H. L., C. W. Bostian, and F. H. Raab, *Solid State Radio Engineering*, New York: Wiley, 1980.
[3] Cripps, S. C., *RF Power Amplifiers for Wireless Communication,* Norwood, MA: Artech House, 1999.
[4] Grebennekov, A., *RF and Microwave Power Amplifier Design*, New York: McGraw Hill, 2009.
[5] Raab, F. H., P. Asbeck, S. Cripps, P. Kenington, Z. Popovic, N. Pothecary, J. Sevic, and N. Sokal, "Power Amplifiers and Transmitters for RF and Microwave," *IEEE Transactions on Microwave Theory and Technology,* Vol. 50, No. 3, Mar. 2009, pp. 814–826.
[6] Poisel, R. A., *Introduction to Communication Electronic Warfare Systems*, Norwood, MA: Artech House, 2002.
[7] Poisel, R. A., *Introduction to Communication Electronic Warfare Systems,* 2nd Ed., Norwood, MA: Artech, 2008, p. 622.
[8] Poisel R. A., *Antenna Systems and Electronic Warfare Applications*, Norwood, MA: Artech House, 2012, Section 20.4.1.1.
[9] Poisel R. A., *Antenna Systems and Electronic Warfare Applications,* Norwood, MA: Artech House, 2012, Ch. 20.
[10] Silvester, S., and Benedek, "Equivalent Capacitances of Microstrip Open Circuits," *IEEE Transactions on Microwave Theory*, Vol. MTT-20, Aug. 1972, pp. 511–516.
[11] Mader, T. B., and Z. B. Popovic, "The Transmission- Line High-Efficiency Class-E Amplifier," 1999 IEEE M&GWL, Vol.5, No. 9, Sep. 1999, pp. 290–292.
[12] Tayrani, R., "A Monolithic X-band Class- E Amplifier," *IEEE GaAs IC Symposium Digest* 2001, pp. 205–209.
[13] Quach, T., et al., "Broadband Class-E Power Amplifier for Space Radar Application," *IEEE GaAs IC Symposium Digest 2001*, pp. 209–219.
[14] Gao, S, "High-Efficiency Class F RF/Microwave Power Amplifiers," *IEEE Microwave Magazine*, Feb. 2006, pp. 40–48.
[15] Inoue, A., A. Ohta, S. Goto, T. Ishikawa, and Y. Matsuda, "The Efficiency of Class F and Inverse Class F Amplifiers," *IEEE MTT-S International Symposium Digest*, June 6–11, 2004, Vol. 3, pp. 1947–1950.
[16] Raab, F. H., "Introduction to Class F Power Amplifiers," *RF Design*, Vol. 19, No. 5, May 1999, pp. 79–84.
[17] Raab, F. H., "Class F Power Amplifiers with Maximally Flat Waveforms," *IEEE Transactions on Microwave Theory and Technology*, Vol. 45, Nov. 1999, pp. 2007–2012.
[18] Raab, F. H., "Maximum Efficiency and Output of Class F Power Amplifiers," *IEEE Transactions on Microwave Theory and Technology*, Vol. 47, June 2001, pp. 1162–1166.
[19] Raab, F. H., "Class E, Class C, and Class F Power Amplifiers Based Upon a Finite Number of Harmonics," *IEEE Transactions on Microwave Theory and Technology*, Vol. 49, No. 8, pp. 1462–1468, Aug. 2001.
[20] Trask, C., "Class F Amplifier Loading Networks: a Unified Design Approach," *IEEE MTT-S International Symposium Digest*, Anaheim, Vol. 1, June 13–19, 1999, pp. 351–354.
[21] Grebennikov, A. V., "Circuit Design Technique for High Efficiency Class F Amplifiers," *IEEE MTT-S International Symposium Digest*, Boston, MA, June 11–16, 2000, Vol. 2, pp. 771–779.
[22] Maeda, M. and H. Takehara, "A High Power and High Efficiency Amplifier with Controlled Second-Harmonic Source Impedance," *IEEE MTT-S International Symposium Digest*, Orlando, FL, Apr. 1995, Vol. 2, pp. 579–589.
[23] Jeon, K., Y. Kwon, and S. Hong, "Input Harmonics Control Using Non-linear Capacitor in GaAs FET Power Amplifier," *IEEE MTT-S International Symposium Digest*, Denver, CO, Vol. 2, June 8–13, 1997, pp. 817–820.

[24] White, P., "Effect of Input Harmonic Terminations on High Efficiency Class B and Class F Operation of PHEMT Devices," *IEEE MTT-S International Symposium Digest*, Vol. 3, June 7–12, 1998, pp. 1611–1619.

[25] Ingruber, B., W. Pritzl, D. Smely, M. Wachutka, and G. Magerl, "Harmonic-Controlled Amplifier," *IEEE Transactions on Microwave Theory and Technology*, Vol. 46, No. 6, June 1999, pp. 857–863.

[26 Radisic, V., Y. Qian, and T. Itoh, "Class F Power Amplifier Integrated with Circular Sector Microstrip Antenna," *IEEE MTT-S International Symposium Digest*, Vol. 4, June 1997, pp. 687–690.

[27] Poisel R. A., *Antenna Systems and Electronic Warfare Applications*, Norwood, MA: Artech House, 2012, Ch 14.

[28] Davis, S., *Electronic Design*, Jul. 19, 2004, http//electronicdesign.com/power/power-mosfet-gate-drivers, accessed Aug. 2009. |

[29] Golio, J. M., *Microwave MESFETs & HEMTs*, Norwood, MA: Artech House, 1991.

[30] Vendelin, G. D., A. M. Pavio, and U. L. Rohde. *Microwave Circuit Design Using Linear and Nonlinear Techniques,* New York: John Wiley and Sons, 2005.

[31] Gray, P. R., P. J. Hurst, S. H. Lewis and R. G. Meyer, *Analysis and Design of Analog Integrated Circuits*, New York: John Wiley & Sons, 2009.

[32] Davis, R. F. and M. S. Shur, *GaN Based Materials and Devices—Growth, Fabrication, Characterization and Performance*, World Scientific: Singapore, 2004.

[33] Koch, S., I. Kallfass, R.Weber, A. Leuther, M. Schlechtweg, and S. Saito, "A Fully Integrated, Compound Transceiver MIMIC Utilizing Six Antenna Ports for 60 GHz Wireless Applications," *IEEE CSIC Symposium*, 2009, pp. 37–40.

[34] Campbell, C. F. and M. Poulton, "Compact Highly Integrated X-Band Power Amplifier Using Commercially Available Discrete GaN FETs," *Proceedings Asia-Pacific Microwave Conference*, 2011, pp. 243–246.

[35] Goyal. R., *Monolithic Microwave Integrated Circuits: Technology and Design*, Norwood, MA: Artech House, 1989.

[36] Gonzalez, G., *Microwave Transistor Amplifiers—Analysis and Design,* Upper Saddle River, NJ: Prentice-Hall Inc., 1997.

[37] Colantonio, P., F. Giannini and E. Limiti, *High Efficiency RF and Microwave Solid State Power Amplifiers*, New York: John Wiley and Sons, 2009.

[38] Raab, F. H., P. Asbeck, S. Cripps, P. B. Kenington, Z. B. Popovic', N. Pothecary, J. F. Sevic, and N. O. Sokol, "Power Amplifiers and Transmitters for RF and Microwave," *IEEE Transactions on Microwave Theory and Technology,* 2002, Vol. 50, No. 3, pp. 814–826.

[39] Poisel, R. A., *Introduction to Communication Electronic Warfare Systems,* 2nd Ed., Norwood, MA: Artech House, 2008, Ch. 15.

[40] www.Polyfet.com, Accessed May 2019.

[41] Ruthroff, C. L., "Some Broad-Band Transformers, *Proceedings IRE*, Vol. 47, Aug. 1959, pp. 1337–1342.

[42] Sevick, J., *Transmission Line Transformers*, *4th Ed*, Raleigh, NC: Scitech Publishing, 2001.

[43] Walder, J. L. B., et. al, *Classic Works in RF Engineering*, Norwood, MA: Artech House, 2006, Ch. 4.

[44] Guanella, G., "New Method of Impedance Matching in Radio-Frequency Circuits," *The Brown Boveri Review,* September 1944, pp. 327–329. Reprinted in Walker, J. L. B., et al., *Classic Works in RF Engineering,* Norwood, MA: Artech House, 2006, pp. 125–127.

[45] Breed, G.A., "Transmission Line Transformer Basics," *Applied Microwave and Wireless,* Vol. 10, No. 4, May 1998, p. 60.

[46] Horn, J. and G. Boeck, "Design and Modeling of Transmission Line Transformers," *IEEE 2003 International Symposium on Microwave Theory and Techniques,* pp. 421–424.

[47] Rotholz, E., "Transmission-Line Transformers," *IEEE Transactions on Microwave Theory and Techniques,* Vol. 29, No. 4, April 1981, pp. 327–331. Reprinted in Walker, J. L. B., et al., *Classic Works in RF Engineering,* Norwood, MA: Artech House, 2006, pp. 185–189.
[48] Lefferson, P., "Twisted Magnet Wire Transmission Line," *IEEE Transactions on Parts, Hybrids, and Packaging,* Vol. 7, No. 4, December 1971, pp. 148–154.
[49] Walker, J., F. Raab, D. Myer, and C. Trask, *Classic Works in RF Engineering: Combiners, Couplers, Transformers and Magnetic Materials,* Artech House, 2006.
[50] Belleman, J.M., "A Four-Decade Bandwidth Hybrid Coupler," European Organization for Nuclear Research, Geneva, Switzerland, Note 99-09, 20 May 1999.
[51] Jordan, E.C. and K.G. Balmain, *Electromagnetic Waves and Radiating Systems,* 2nd Ed., Prentice-Hall, 1968.
[52] Poisel, R. A., *Antenna Systems and Electronic Warfare Applications,* Norwood, MA: Artech House, 2012, Section 2.5.
[53] Shadowitz, A., *The Electromagnetic Field,* New York: Dover Publications, 1975, Ch. 11.
[54] Matick, R.E., *Transmission Lines for Digital and Communication Networks,* McGraw-Hill, 1969.
[55] Kraus, J.D. and K.R. Carver, *Electromagnetics,* 2nd ed., McGraw- Hill, 1973.
[56] Malinen, A., K. Stadius, and K. Halonen, "Characteristics and Modeling of a Broadband Transmission-Line Transformer," *IEEE 2004 International Symposium on Circuits and Systems,* Vol. 4, pp. 413–416.

Chapter 10

Amplifier Stability

10.1 Introduction

Many amplifier configurations can become unstable at some frequencies if care is not taken in their design. If an amplifier is unstable, it can start oscillating. This can occur at its input or output, or both.

We investigate the conditions for stability in this chapter. These conditions are conveniently displayed on the Smith chart, which renders them evident on visual inspection.

This chapter is structured as follows. We first cover the concept of negative resistance, particularly at the ports of two-port active devices. Stability circles are covered next as a graphical method for displaying stability. Methods for stabilizing potentially unstable active two-ports are discussed next.

10.2 Negative Resistance

Consider the generic linear, small-signal BJT amplifier in Figure 10.1. The transistor can be characterized by its S-matrix

$$\mathbf{S} = \begin{bmatrix} s_{11} & s_{12} \\ s_{21} & s_{22} \end{bmatrix} \tag{10.1}$$

Assume that the matching networks are passive so that

$$|\Gamma_{\rm S}| < 1 \quad \text{and} \qquad |\Gamma_{\rm L}| < 1 \tag{10.2}$$

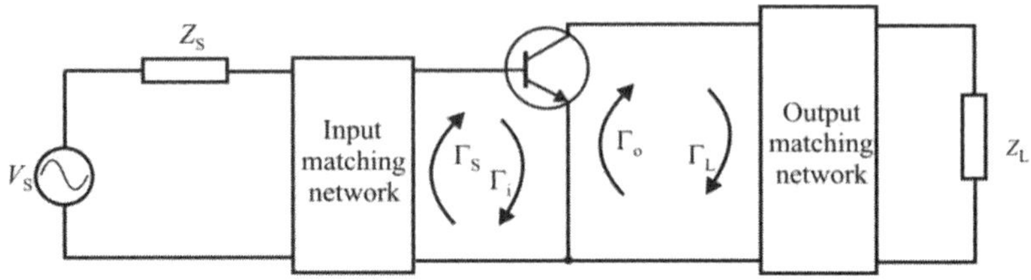

Figure 10.1 BJT amplifier.

Instability, that is oscillation, is possible if a signal incident on the input or output port of the transistor is reflected with a gain greater than 1. That is, if

$$|\Gamma_i| > 1 \quad \text{or} \quad |\Gamma_o| > 1 \tag{10.3}$$

This could occur when noise is incident on either port and is reflected with gain, is reflected again at the corresponding matching network, and so on. This noise could be amplified repeatedly to such a level that the device is forced into unstable operation. For $|\Gamma| > 1$, the real part of the impedance seen looking into the port must be negative. That is

$$|R_i| < 0 \quad \text{or} \quad |R_o| < 0 \tag{10.4}$$

for instability. This region lies outside the unit circle on a Smith chart, as we'll see shortly.

The input and output reflection coefficients are given by [1]

$$\Gamma_i = s_{11} + \frac{\Gamma_L s_{12} s_{21}}{1 - \Gamma_L s_{22}} \tag{10.5}$$

and

$$\Gamma_o = s_{22} + \frac{\Gamma_S s_{12} s_{21}}{1 - \Gamma_S s_{11}} \tag{10.6}$$

Using (10.5) and (10.6) in (10.2) and (10.3), for the circuit to be unconditionally stable:

$$|\Gamma_i| = \left| s_{11} + \frac{\Gamma_L s_{12} s_{21}}{1 - \Gamma_L s_{22}} \right| < 1 \tag{10.7}$$

and

$$|\Gamma_o| = \left| s_{22} + \frac{\Gamma_S s_{12} s_{21}}{1 - \Gamma_S s_{11}} \right| < 1 \tag{10.8}$$

Note that if $|\Gamma_S| > 1$ or $|\Gamma_L| > 1$ then (10.7) and (10.8) are requirements for conditional stability only.

10.3 Stability Circles

Stability circles are graphical depictions of the ranges of Γ_S and Γ_L values that may lead to instability. Because they are drawn on the Smith chart, the stability circles are particularly helpful because of the extra information provided with this depiction. Stability circles establish the boundary between stable and *potentially* unstable Γ_S or Γ_L.

To determine these boundaries, we set $|\Gamma_i| = 1$ (or $|\Gamma_o| = 1$) and draw these curves in the Γ_L (or Γ_S) plane.

On the output side of the network, for the load stability circle, from (10.5) we have

$$|\Gamma_i| = \left| s_{11} + \frac{\Gamma_L s_{12} s_{21}}{1 - \Gamma_L s_{22}} \right| = 1 \tag{10.9}$$

Equation (10.9) can be rearranged to

$$|\Gamma_L - C_L| = R_L \tag{10.10}$$

where

$$C_L = \frac{\left(s_{22} - s_{11}^* \Delta\right)^*}{|s_{22}|^2 - |\Delta|^2} \tag{10.11}$$

and

$$R_L = \left| \frac{s_{12}s_{21}}{|s_{22}|^2 - |\Delta|^2} \right| \tag{10.12}$$

Equation (10.10) is an equation for a circle in the complex Γ_L plane with center C_L and radius R_L. In (10.11) and (10.12)

$$\Delta = s_{11}s_{22} - s_{12}s_{21} \tag{10.13}$$

For a conditionally stable circuit, the load stability circle may look something like Figure 10.2 on the extended Smith chart. Again, this circle defines those Γ_L values where $|\Gamma_i| = 1$. The contour defines a boundary within which the circuit will be stable or unstable, but as yet we do not know which (remember that we assumed that $|\Gamma_L| \le 1$). Those Γ_L outside of the circle (and with $|\Gamma_L| \le 1$) define the opposite condition.

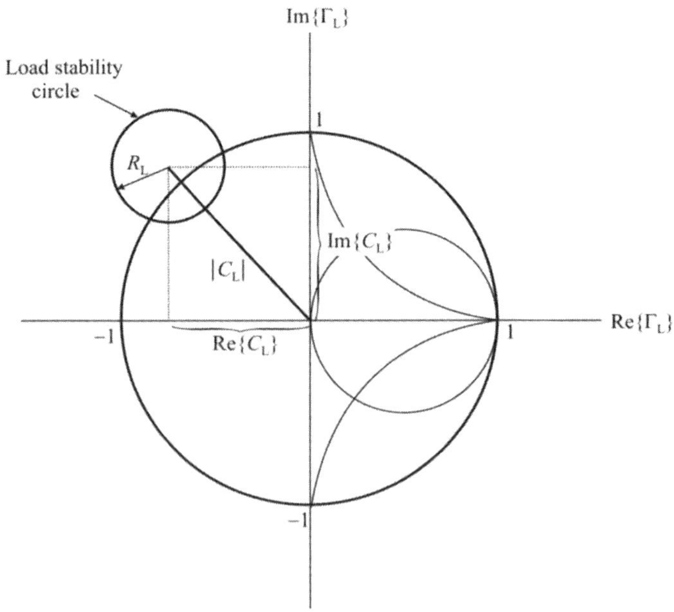

Figure 10.2 Extended Smith chart.

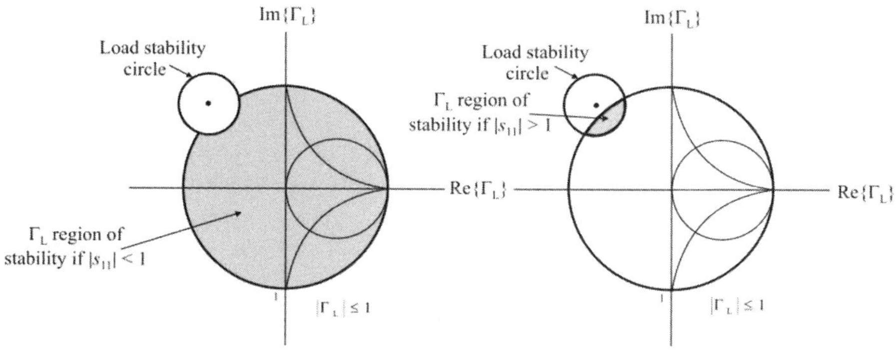

Figure 10.3 Two stability regions.

To determine the stable region we choose a load Z_L so that $Z_L = Z_0 \Rightarrow \Gamma_L = 0$, so that from (10.5)

$$|\Gamma_i| = |s_{11}| \tag{10.14}$$

Consequently, if $|s_{11}| < 1$, then the region containing the origin in the Γ_L plane is the stable region. Otherwise, if $|s_{11}| > 1$, then the region inside the stability circle plane is the stable region. Figure 10.3 summarizes these conditions.

In order for the circuit to be unconditionally sable, the stability circle must lie entirely outside the Smith chart.

Further, if $|s_{11}| > 1$ or $|s_{22}| > 1$ then it is not possible for the amplifier circuit to be unconditionally stable because we can always have a load (or source) impedance equal to Z_0. This implies Γ_L (or Γ_S) = 0. In this case from (10.5)

$$|\Gamma_i| = |s_{11}| > 1 \qquad \text{or} \qquad |\Gamma_o| = |s_{22}| > 1 \tag{10.15}$$

Both equations indicate the circuit is always potentially unstable.

On the other hand, we can show that the circuit is unconditionally stable if

$$\mu = \frac{1 - |s_{11}|^2}{|s_{22} - s_{11}^* \Delta| + |s_{12} s_{21}|} > 1 \tag{10.16}$$

This is the *mu stability factor* [2]. The larger the value of μ the greater the stability.

Table 10.1 BJT Parameters at $f = 2$ GHz

s_{11}	$0.177\angle -136^{\circ}$
s_{12}	$0.118\angle 44^{\circ}$
s_{21}	$1.70\angle 61^{\circ}$
s_{22}	$0.632\angle -78^{\circ}$

Beginning with (10.6), we analyzed the load stability circle; however the same analysis can be applied to the source stability circle. The region of stability for the source stability circle is now determined by s_{22}. If $|s_{22}| < 1$ the region of stability contains the origin in the complex Γ_S plane.

Example 10.1: The s-parameters of a microwave BJT are listed in Table 10.1 for a CE amplifier. These parameters are referenced to a 50 Ω system at 2 GHz. We will determine the stability of this amplifier assuming $|\Gamma_S| < 1$ and $|\Gamma_L| < 1$.

From (10.13)

$$\Delta = s_{11}s_{22} - s_{12}s_{21} = 0.260\angle 80.9^{\circ}$$

so that from (10.16)

$$\mu = \frac{0.969}{0.613} = 1.57 > 1$$

Since the μ stability factor is greater than unity the amplifier is unconditionally stable, at least at 2 GHz.

However, it is important to test for stability at all frequencies less than f_T of the transistor (that is, at all frequencies where gain > 1). For example, at 100 MHz, the s-parameters could be as shown in Table 10.2. Using these values in (10.16) we find that $\mu = 0.552 < 1$. Therefore, at 100 MHz the amplifier is only conditionally stable.

While 100 MHz may not be in the passband of the amplifier, it is possible that nonlinearities at 100 MHz could mix with incoming signals and produce output in the passband. Then the whole circuit would oscillate.

Table 10.2 BJT Parameters at $f = 0.1$ GHz

s_{11}	$0.574\angle -79^{\circ}$
s_{12}	$0.023\angle 67^{\circ}$
s_{21}	$11.65\angle 127^{\circ}$
s_{22}	$0.788\angle -56^{\circ}$

From (10.11) and (10.12), the load stability circles at 1 GHz are given by

$$C_L = \frac{(s_{22} - s_{11}^{*}\Delta)^{*}}{|s_{22}|^2 - |\Delta|^2}$$

$$= \frac{[0.632\angle -78^{\circ} - (0.177\angle 136^{\circ})(0.260\angle 80.9^{\circ})]^{*}}{0.632^2 - 0.260^2}$$

$$= 0.597\angle 80.9^{\circ}$$

and

$$R_L = \left| \frac{s_{12}s_{21}}{|s_{22}|^2 - |\Delta|^2} \right| = \left| \frac{(0.118\angle 44^{\circ})(1.70\angle 61^{\circ})}{0.632^2 - 0.260^2} \right| = 0.605$$

The load stability circle is shown in Figure 10.4. Since $|s_{11}| < 1$ at 2 GHz the region of stability is inside the circle. Therefore, the circuit is unconditionally stable.

We see from this example that if the circuit is unconditionally stable (as this one is at 2 GHz), there really isn't any need to plot the stability circle.

The stability circles at 100 MHz are shown in Figure 10.5. Since $|s_{11}| < 1$ the load stable region is inside the load circle. The same is true for the source stability circle since $|s_{22}| < 1$.

10.3.1 Rollett Stability Factor

Another stability condition that is often used can be derived from (10.7) and (10.8) is referred to as the *Rollett stability factor*, and is denoted by K_R. K_R is given by

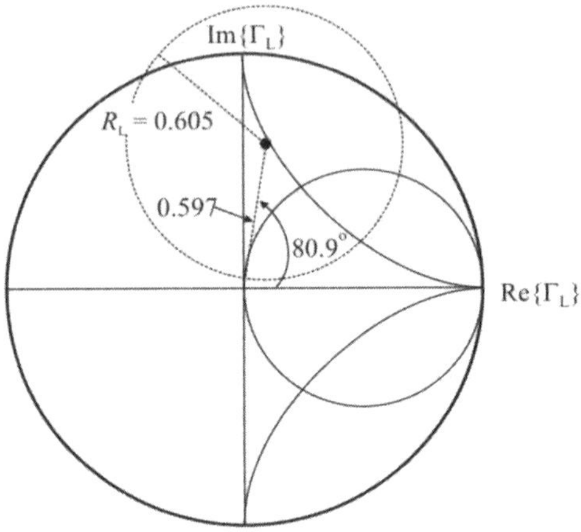

Figure 10.4 Example 10.1 load stability at 2 GHz.

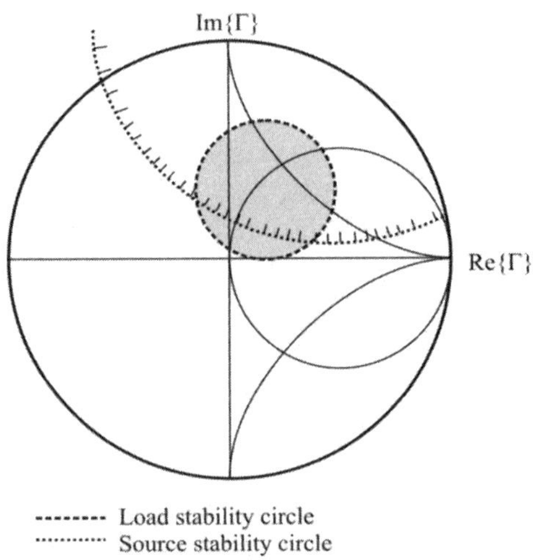

Figure 10.5 Example 10.1 stability circles at 100 MHz.

$$K_{\mathrm{R}} = \frac{1 - |s_{11}|^2 - |s_{22}|^2 + |\Delta|^2}{2|s_{12}s_{21}|} > 1 \qquad (10.17)$$

When this factor is greater than one, combined with the following two conditions,

$$|s_{12}s_{21}| < 1 - |s_{22}|^2 \qquad (10.18)$$

and

$$|s_{12}s_{21}| < 1 - |s_{11}|^2 \qquad (10.19)$$

then the circuit is absolutely stable [3]. For most devices $|\Delta| < 1$, so (10.17) – (10.19) are sufficient to guarantee absolute stability.

10.4 Stability of Active Two-Ports

Two-ports that map passive loads into input admittances with negative real parts or passive generator admittances into output admittances with negative real parts may be unstable, that is, sustain oscillations. This is an undesirable condition when the two-port in question is an amplifier. An example is shown in Figure 10.6, where a passive load y_{L} determined by

$$y_{\mathrm{i}} = g_{\mathrm{i}} + jb_{\mathrm{i}} = y_{11} - \frac{y_{21}y_{12}}{y_{22} + y_{\mathrm{L}}} \qquad (10.20)$$

is supposed to provide an input admittance with $g_{\mathrm{i}} < 0$ at frequency ω_0. An admittance Y_{G} of opposite sign is passive since $g_{\mathrm{G}} > 0$. Connected across the input terminals, Y_{G} gives a total parallel admittance of zero, which is a property of an undamped parallel circuit tuned to ω_0. Once the voltage amplitude of V_1 is reached, oscillations are sustained while power is delivered to the external conductance g_{G}.

A two-port through which passive load and generator admittances cannot transform to input and output impedances with negative real parts at a given frequency are said to be *absolutely stable* at that frequency. If the two-port is not absolutely stable, it called potentially unstable. We need to stress that absolute stability will not prevent oscillations if there is a feedback path between the ports outside the two-port.

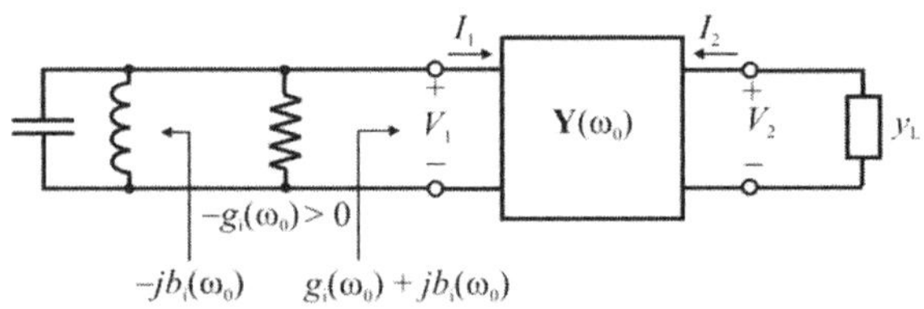

Figure 10.6 Example of an oscillatory circuit with a two-port that maps the passive admittance y_L into an input admittance having a negative real part, $g_i(\omega_0) < 0$.

On the other hand, *instability* means the possibility that the two-port will oscillate for some passive terminations at a specified frequency. Had we, for instance, in the example of Figure 10.6 used $g_L > -g_i(\omega_0)$ giving a total positive load conductance, any oscillation initiated would decay.

To determine the conditions for absolute stability we start by introducing the so-called *short circuit stability criteria* given by

$$g_{11} = \mathrm{Re}\{y_{11}\} > 0, \text{ and} \qquad g_{22} = \mathrm{Re}\{y_{22}\} > 0 \tag{10.21}$$

with unilateral two-ports where $y_{12} = 0$, (10.21) holds the only necessities for absolute stability. In the general case the effect of internal feedback (y_{12}) must be investigated.

Investigating for absolute stability may be based on the mismatch factor M_{mch} introduced in Chapter 5 with the short-circuit criteria from (10.21). If M_{mch} is positive, calculations of the port impedances and maximum gain follow directly using

$$y_{G,opt} = g_{G,opt} + jb_{G,opt}, \qquad \begin{cases} g_{G,opt} = g_{11} M_{mch} \\ b_{G,opt} = \dfrac{Q}{2g_{22}} - b_{11} = \dfrac{\mathrm{Im}\{y_{21}y_{12})}{2g_{22}} - b_{11} \end{cases} \tag{10.22}$$

$$y_{L,opt} = g_{L,opt} - jb_{L,opt}, \qquad \begin{cases} g_{L,opt} = g_{22} M_{mch} \\ b_{L,opt} = \dfrac{Q}{2g_{11}} - b_{22} = \dfrac{\mathrm{Im}\{y_{21}y_{12}\}}{2g_{11}} - b_{22} \end{cases} \tag{10.23}$$

$$G_{max} = \frac{|y_{21}|^2}{2g_{11}g_{22}(1 + M_{mch}) - \mathrm{Re}\{y_{21}y_{12}\}} \tag{10.24}$$

While the short-circuit conditions are common, criteria equivalent to M_{mch} are usually found in literature and data sheets.

Before considering constraints set by the lossless two-port, the power expressions

$$P_{\mathrm{i}} = |V_1| 2 g_{\mathrm{i}} \tag{10.25}$$

where g_{i} is the real part of the input admittance. The voltage gain of the port is

$$A_{\mathrm{v}} = \frac{V_2}{V_1} = -\frac{y_{21}}{y_{22} + y_{\mathrm{L}}} \tag{10.26}$$

So the power delivered to the load is given by

$$P_{\mathrm{L}} = |V_2|^2 g_{\mathrm{L}} = |V_1|^2 \left| \frac{y_{21}}{y_{22} + y_{\mathrm{L}}} \right|^2 g_{\mathrm{L}} \tag{10.27}$$

where g_{L} is the load conductance from $y_{\mathrm{L}} = g_{\mathrm{L}} + jb_{\mathrm{L}}$. Since the generator current I_{G} divides between the generator and the two-port input admittance the available generator power becomes

$$P_{\mathrm{av}} = \frac{|I_{\mathrm{G}}|^2}{4 g_{\mathrm{G}}} = \frac{|V_1|^2 |y_{\mathrm{G}} + y_{\mathrm{i}}|^2}{4 g_{\mathrm{G}}} \tag{10.28}$$

We express the available output power P_{o} using the output short-circuit current $I_2 \big|_{V_2=0}$ as

$$P_{\mathrm{o}} = \frac{\left| I_2 \big|_{V_2=0} \right|^2}{4 g_{\mathrm{o}}} = \frac{|y_{21}|^2}{4 g_{\mathrm{o}}} \left| V_{1|V_2=0} \right|^2 = \left| \frac{y_{21}}{y_{11} + y_{\mathrm{G}}} \right|^2 \frac{g_{\mathrm{G}}}{g_{\mathrm{o}}} P_{\mathrm{av}} \tag{10.29}$$

Equations (10.25)–(10.29) give the matching factors directly in terms of two connected admittances by

$$M_{\mathrm{mch,i}} = \frac{P_{\mathrm{i}}}{P_{\mathrm{av}}} = \frac{4 g_{\mathrm{G}} g_{\mathrm{i}}}{|y_{\mathrm{G}} + y_{\mathrm{i}}|^2}, \qquad M_{\mathrm{mch,o}} = \frac{P_{\mathrm{L}}}{P_{\mathrm{o}}} = \frac{4 g_{\mathrm{L}} g_{\mathrm{o}}}{|y_{\mathrm{L}} + y_{\mathrm{o}}|^2} \tag{10.30}$$

(see Figure 10.7).

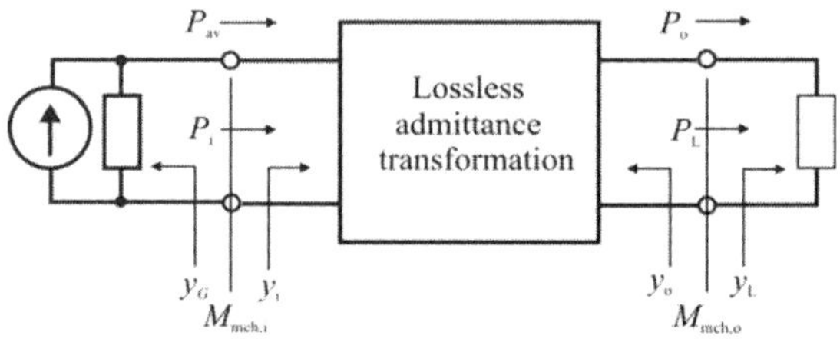

Figure 10.7 Power flow through a lossless two-port. P_{av} and P_o are available.

The other condition for absolute stability expressed in terms of the y-parameters is given by

$$2g_{11}g_{22} - \mathrm{Re}\{y_{21}y_{12}\} > |y_{21}y_{12}| \tag{10.31}$$

Together with the short-circuit stability requirements, the last condition is called *Llewellyn's absolute stability conditions*. A simple reorganization brings it to the Rollett's stability condition, equivalent to (10.17),

$$K_R = \frac{2g_{11}g_{22} - \mathrm{Re}\{y_{21}y_{12}\}}{|y_{21}y_{12}|} > 1 \tag{10.32}$$

Other obvious equivalent conditions that can be derived from (10.31) are:

Linvill's stability condition

$$C = \frac{|y_{21}y_{12}|}{2g_{11}g_{22} - \mathrm{Re}\{y_{21}y_{12}\}}, \quad 0 < C < 1 \tag{10.33}$$

Stern's stability condition

$$K_{\mathrm{stern}} = \frac{2g_{11}g_{22}}{\mathrm{Re}\{y_{21}y_{12}\} + |y_{21}y_{12}|}, \qquad K_{\mathrm{stern}} > 1 \tag{10.34}$$

10.4.1 Stabilizing Active Two-Ports

RF transistors are usually potentially unstable in a large part of the frequency range where they provide power gain. Absolute stability means that the two-port remains stable with any passive terminations and with predictable maximum power gain. By contrast, any power gain may be supported, if the two-port is allowed to be potentially unstable. The potentially unstable two-ports may still be useful, if it can be embedded to provide the required gain without oscillating. There are two main approaches to reach that goal. We may either reduce gain by adjusting matching conditions, possibly inserting additional lossy components, or we may add components to reduce feedback. In this section we initially consider the first method, where we examine the maximum stable gain. By the second method, called neutralization, we may reach maximum gain at the expense of more complex circuits.

The maximum stable gain, denoted by G_{ms} is the maximum gain of a non-unilateral two-port at the boundary of absolute stability, where $K = 1$. If the two-port is potentially unstable, G_{ms} is the maximum gain that may be obtained if the two-port is augmented with components placed across its ports as shown in Figure 10.8. G_{ms} is given by

$$G_{\mathrm{ms}} = |y_{21} / y_{12}| \tag{10.35}$$

The generator and load conductances presented at the ports should be chosen to fulfill the stability conditions from (10.21) and (10.32) with the resulting y-parameters for the augmented two-port given by

$$\hat{y}_{11} = (g_{11} + g_{\mathrm{G}}) + jb_{11},\ \hat{y}_{22} = (g_{22} + g_{\mathrm{L}}) + jb_{22},\ \hat{y}_{21} = y_{21},\ \hat{y}_{12} = y_{12} \tag{10.36}$$

Although we show one possible configuration of resistances at the two ports in Figure 10.8 to improve the stability of the two-port, there are several others possible in different configurations [4]. Some of them affect the gain while others do not. Usually, adding resistances to networks increases the noise, however.

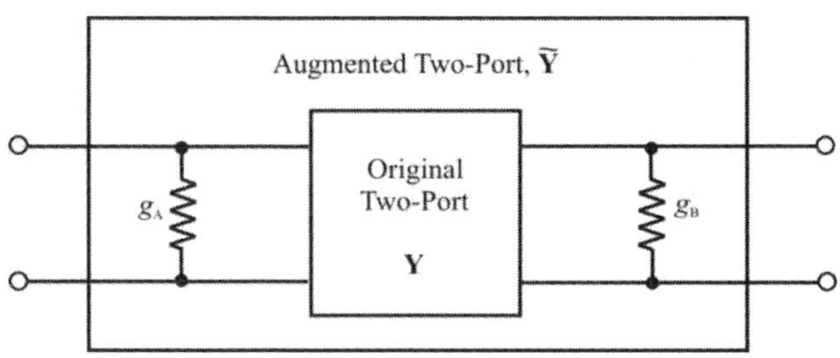

Figure 10.8 Augmented two-port. Conductances g_A, g_B control the stability factor K_R but keep the maximum stable gain G_{ms} unaffected.

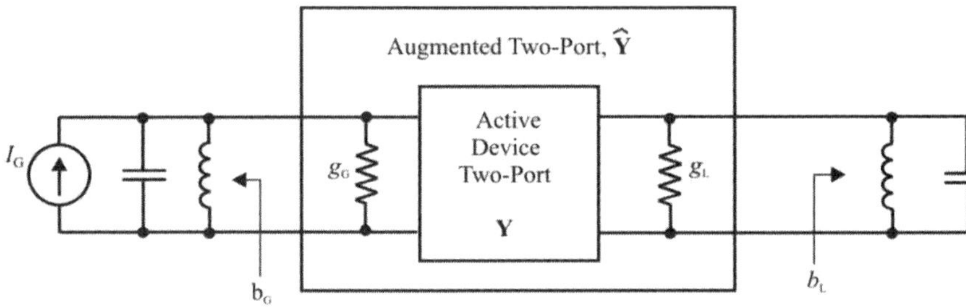

Figure 10.9 Augmentation of a two-port by adding generator and load conductances g_G and g_L for stability considerations. As long as $K > 1$, determined by proper selection of g_G and g_L, this configuration will remain stable for any values of the susceptances b_G and b_L.

One way of exerting control over the stability properties of a two-port is shown in Figure 10.9. The real parts of the generator and load admittances can be included into a hypothetical augmented two-port for stability computations. We can ensure that the terminated two-port stays stable with any generator and load susceptances using this technique. To find the pertinent power gains, however, it is not the parameters of the augmented two-port but the parameters of the original one that must be used together with y_G and y_L in the definition of gains in Chapter 4 for finding the type of gain that is appropriate. The latter question concerns the operational and the available power gains, where it may happen that the terminated two-port gets input or output conductances that are zero or negative, so these gain functions become meaningless.

While the partitioning in Figure 10.9 was dictated by our stability discussion, practical single stage RF amplifier design commonly gets a structure more like Figure 10.10. Lossless matching networks are inserted at either side of the active two-port to transform given generator and load admittances to the admittances that are required for the active device. If the device is absolutely stable, simultaneous conjugated matching at the device port provides maximum gain. Since it also implies transfer of available power (i.e., maximum power) to and from the device,

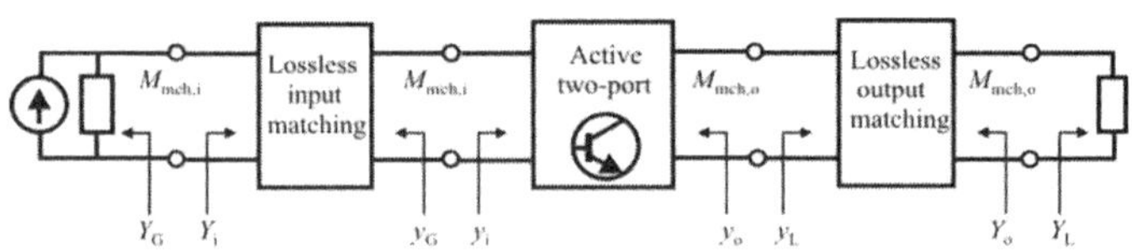

Figure 10.10 RF-amplifier structure. Matching ratios of input over available powers M_{mch} are the same on either side of lossless networks and equals one with conjugated matching to an absolutely stable device.

and since the matching networks cannot absorb power, transfers from the generator and to the load are maximum too, so the outer ports are also conjugate matched.

A potentially unstable device cannot be simultaneously matched at both ports. A K_R-factor greater than one, which has been obtained from augmentation estimates with a potentially unstable device, implies admittance mismatch at one or both device ports. If the generator and load admittances presented at the active two-port terminals, y_G and y_L respectively, are still transformed through lossless networks from the outer generator and load admittances, Y_G and Y_L, mismatch pertain to the outer connection ports too. We introduce a measure of matching, M_{mch}, as the ratio of power delivered to a load over the available power—sometimes called the mismatch factor that was introduced in Chapter 5. This ratio remains the same at either side of a lossless admittance transforming network.

There are situations where matching is required—there are no alternatives. The technique in Figure 10.8 can be used to adjust the real parts of y_{11} or y_{22} or both in the active two-port by parallel connecting resistors. There are several other possible concerns over and above stability that must be considered, however. Note that an additional resistive loss at the input may decrease the noise performance of the amplifier. An additional output conductance, on the other hand, may limit the power output capability of the active device.

Even if an active device is absolutely stable, either inherently or by resistive extensions, it is still useful to estimate the stability factor after augmentation encompassing the load and generator admittances. If the K_R-factor is close to one there is a potential that parameter variations with time, temperature, and/or device substitution could produce instabilities. The factor of the augmented network may indicate how sensitive the complete circuit is with respect to tuning and the parameter variations. The K_R-factor in the augmented estimation may be used as a degree-of-stability indicator like phase or gain margins in low-frequency designs. This property is discussed in the following example, and shows why a K_R-factor of five or more is a preferable choice in design.

Example 10.2: Degree of Stability

The K_R-factor is an indicator of the degree of stability as we will show in this example. Consider the amplifier illustrated in Figure 10.11. An FET with the high-frequency equivalent circuit in Figure 10.11(b) is matched at both ports with tuned narrowband resonant circuits. The input and output capacitors C_1 and C_2 compensate for the transistor input and output capacitances to give a total capacitance of C_p in both circuits; that is,

$$C_p = C_1 + C_\pi = C_2 + C_o \tag{10.37}$$

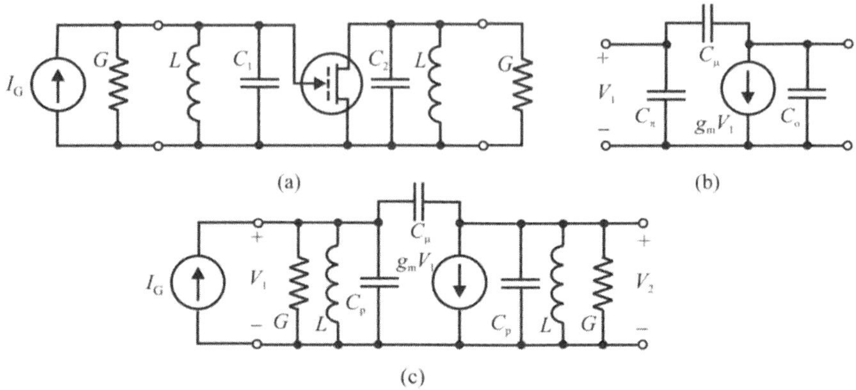

Figure 10.11 Simple, symmetric narrowband FET amplifier (a) with transistor equivalent circuit in (b). Stability conditions are calculated by the two-port in (c).

The two-port we use for the stability calculations is shown by Figure 10.11(c). The y-parameter matrix is expressed through $y_p(s)$, which represents the admittance function of the matched networks at either side of the transistor

$$\mathbf{Y} = \begin{bmatrix} y_p(\omega) & -j\omega C_\mu \\ g_m - j\omega C_\mu & y_p(\omega) \end{bmatrix} \tag{10.38}$$

where

$$y_p(\omega) = j\omega(C_p + C_\mu) + G + \frac{1}{j\omega L}$$

The transfer impedance of the amplifier is expressed as,

$$\frac{V_2}{I_G} = z_{21} = \frac{-y_{21}}{y_{11}y_{22} - y_{21}y_{12}} = \frac{-y_{21}}{y_p^2 - y_{21}y_{12}} = \frac{-y_{21}}{(y_p + \delta)(y_p - \delta)} \tag{10.39}$$

where

$$\delta = \sqrt{y_{21}y_{12}} \tag{10.40}$$

Around the frequency f_0, where the circuits are tuned, y_p can be approximated with the narrowband approximation

$$y_p(\omega) = 2C(j\omega - s_{p0}) \tag{10.41}$$

where

$$s_{p0} \approx -\frac{\omega_0}{2Q} + j\omega_0$$
$$\omega_0 = \frac{1}{\sqrt{LC}}$$
$$Q = \frac{\omega_0}{W_{3dB}} = \frac{C\omega_0}{G}$$

In (10.41), s_{p0} is the upper half-plane zero of $y_p(s)$. Had the transistor been unilateral, that is, $y_{12} = 0$, there would have been a double pole at s_{p0} in the amplifier transfer impedance (10.39). The feedback in the transistor changes that. The admittance of C_μ is much smaller than the transconductance g_m at the center frequency, so we approximate,

$$\begin{aligned} y_{21} &= g_m - j\omega C_\mu \approx g_m \\ y_{12} &= -j\omega C_\mu \approx -j\omega_0 C_\mu \end{aligned} \tag{10.42}$$

then, from (10.40)

$$\delta = \sqrt{-j}\sqrt{g_m C_\mu \omega_0} \tag{10.43}$$

y_{12} in (10.42) approximated as a constant is consistent with confining the frequency to a narrowband interval around f_0. Under the same assumptions, using (10.43), the stability factor of the two-port is related to Δ by

$$K_R = \frac{2\,\mathrm{Re}^2\{y_p(\omega_0)\}}{|\delta|^2} = \frac{2G^2}{g_m C_\mu \omega_0}$$

so that

$$\delta = \sqrt{-j}G\sqrt{\frac{2}{K_R}} \tag{10.44}$$

Using (10.44) in the transfer impedance in (10.39) shows the pole-splitting into s_{p1} and s_{p2},

$$z_{21}(\omega) \approx \frac{-g_m}{4C^2(j\omega - s_{p1})(j\omega - s_{p2})} \tag{10.45}$$

where

$$s_{p1} = j\omega_0 - \frac{\omega_0}{2Q}\left(1 - \sqrt{-j}\sqrt{\frac{2}{K_R}}\right)$$

and

$$s_{p2} = j\omega_0 - \frac{\omega_0}{2Q}\left(1 + \sqrt{-j}\sqrt{\frac{2}{K_R}}\right)$$

Because of the feedback through C_μ, the poles move along a line in directions of $j^{1/2}$ from the position of the unilateral double-pole at s_{p0}. The smaller K_R becomes, the greater the displacement from s_{p0}. If K_R becomes less than one, we see from the pole position sketch in Figure 10.12(a) that the lower pole moves into the right half of the s-plane and the amplifier becomes unstable. The frequency responses with various K_R-values are given in Figure 10.12(b), where we see that the larger K_R is, the closer the response is to the unilateral limit $K_R \rightarrow \infty$. We see that a K_R-value of more than five is preferable, if the design goal is to approximate an ideal unilateral characteristic, which it usually is.

10.5 Neutralization

The y-parameters for a two-port are the matrix relationship that expresses the port currents in terms of the port voltages and are given by (2.7)–(2.10), restated here as

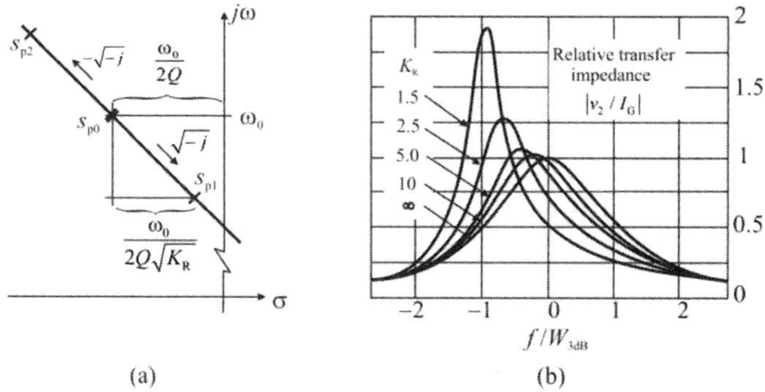

Figure 10.12 Pole positions, (a), and transfer characteristics, (b), for various K_R values corresponding to different C_μ's.

$$\vec{I} = \mathbf{Y}\vec{V} \tag{10.46}$$

$$\begin{bmatrix} I_1 \\ I_2 \end{bmatrix} = \begin{bmatrix} y_{11} & y_{12} \\ y_{21} & y_{22} \end{bmatrix} \begin{bmatrix} V_1 \\ V_2 \end{bmatrix} = \begin{bmatrix} g_{11} + jb_{11} & g_{12} + jb_{12} \\ g_{21} + jb_{21} & g_{22} + jb_{22} \end{bmatrix} \begin{bmatrix} V_1 \\ V_2 \end{bmatrix} \tag{10.47}$$

The y-parameters are often used for RF analysis of active circuits because linear small-signal RF models for many active devices are of π-shape for all but ultimate high frequency limits. We will use the y-parameters in this section.

The only requirement for absolute stability in *unilateral* two-ports, i.e. two-ports with $y_{12} = 0$, is the short-circuit conditions in (10.21). Adding external circuitry that counteracts the internal feedback is one means of keeping the gain high with a given device. Figure 10.13 illustrates this principal. Two two-ports connected in

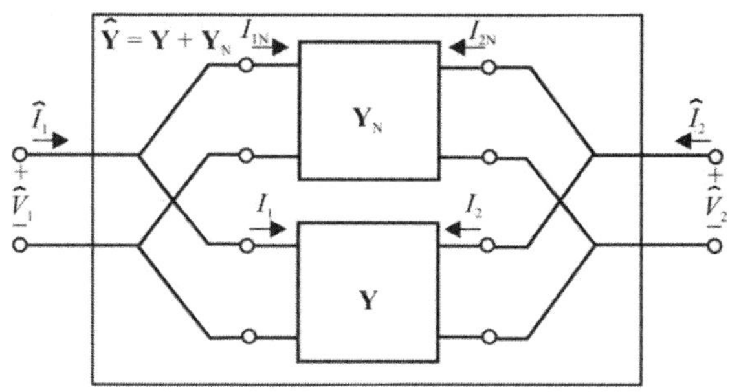

Figure 10.13 Parallel connection of two-ports. The resultant **Y**-matrix is the matrix sum of matrices for the paralleled two-ports.

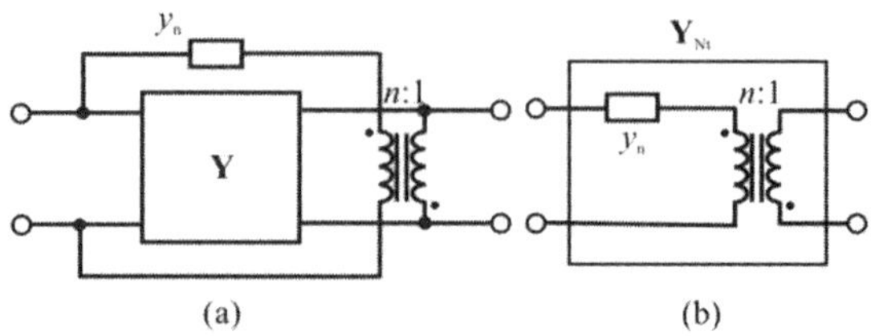

Figure 10.14 Broadband neutralization with transformer: (a) circuit configuration and (b) equivalent circuit. Neutralizing condition $y_{12} = -ny_n$.

parallel as shown get a total y-parameter matrix that is the sum of the y-parameter matrices for the individual two-ports. This is because the voltage sources, the independent variables, at the two ports are common to the other network so are equal, while the dependent variables, the currents, divide at the ports.

Consequently, corresponding y-matrix elements in such a parallel arrangement add to give the resultant element; that is,

$$\begin{aligned} \hat{y}_{11} &= y_{11} + y_{11,\mathrm{N}}, \quad \hat{y}_{12} = y_{12} + y_{12,\mathrm{N}} \\ \hat{y}_{21} &= y_{21} + y_{21,\mathrm{N}}, \quad \hat{y}_{22} = y_{22} + y_{22,\mathrm{N}} \end{aligned} \tag{10.48}$$

So the condition for neutralization is

$$\hat{y}_{12} = 0 \rightarrow y_{12,\mathrm{N}} = -y_{12} \tag{10.49}$$

To get optimum conductances and maximum gain in a unilateral two-port we must arrange conjugate matching to y_{11} and y_{22} at the input and the output ports, respectively. We get

$$y_{\mathrm{G,opt}} = -\hat{y}_{11}^{*}, \qquad y_{\mathrm{L,opt}} = -\hat{y}_{22}^{*}, \qquad G_{\max} = \frac{|\hat{y}_{21}|^2}{4\hat{g}_{11}\hat{g}_{22}} \tag{10.50}$$

Two common configurations used for neutralization are shown in Figures 10.14 and 10.15, where the y-parameters to be paralleled are

$$\mathbf{Y}_{\mathrm{Nt}} = \begin{bmatrix} y_{\mathrm{n}} & ny_{\mathrm{n}} \\ ny_n & n^2 y_{\mathrm{n}} \end{bmatrix}, \qquad \mathbf{Y}_{\mathrm{Ns}} = \begin{bmatrix} y_{\mathrm{n}} & -y_{\mathrm{n}} \\ -y_n & y_{\mathrm{n}} \end{bmatrix} \tag{10.51}$$

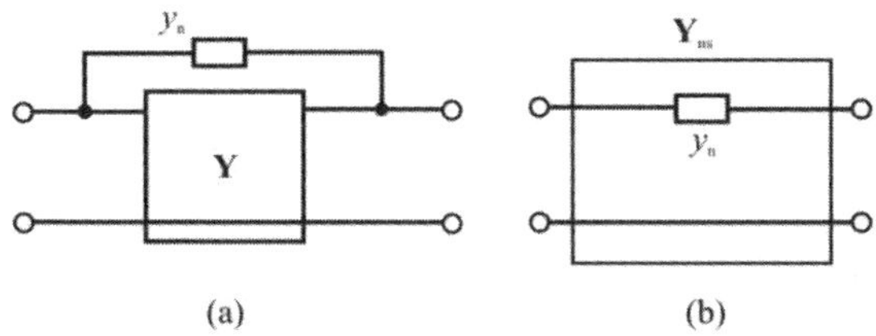

Figure 10.15 Narrowband neutralization of two-port: (a) circuit configuration and (b) equivalent circuit. Neutralizing condition $y_{12} = y_n$.

In Figure 10.15 the added two-port includes a common ground for the ports, a property that also must apply to the neutralized two-port. However, two-ports to be neutralized inherently have common ground due to active three-terminal devices.

Figure 10.16 illustrates neutralizations around a transistor using a π-type equivalent circuit. In case the two-port to be neutralized is of π-type, where $y_{12} = -y_\mu$, the transformer coupling in Figure 10.14 has the capability of covering a broad frequency range, an important consideration for EW applications. Here the counteracting network contains a scaled version of y_μ,

$$y_n = -y_{12} = \frac{1}{n} y_\mu \tag{10.52}$$

The neutralization also changes the other parameters as,

$$\hat{y}_{11} = y_{11} + \frac{y_\mu}{n}, \qquad \hat{y}_{21} = y_{21} + y_\mu, \qquad \hat{y}_{22} = y_{22} + n y_\mu \tag{10.53}$$

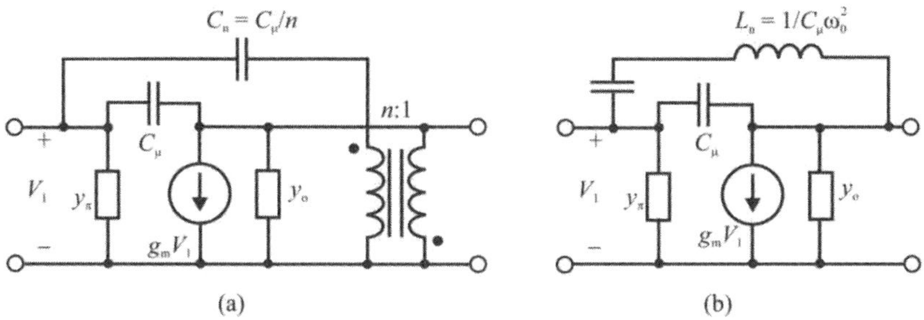

Figure 10.16 Neutralization of transistor feedback capacitance: (a) broadband with transformer and (b) narrowband with lumped element. The feedback capacitor in (b) is to remind us that bias separation between input and output is typically required and must be taken into consideration for frequency response effects.

The method in Figure 10.15 requires that y_n be the negative of y_{12}. Typically, an inductive y_n tunes out a capacitive feedback component in an amplifier.

Adjustable components may be required for total neutralization of active devices due to parameter spreading. Setting the resultant y_{12} exactly to zero is called *unilateralization* and is less demanding than the requirement to reduce the feedback sufficiently to guarantee stable operation.

Example 10.3: Stabilizing a Transistor by Neutralization

Suppose the transistor, at some frequency of interest, has the following scattering parameters:

$$\mathbf{S} = \begin{bmatrix} 0.73\angle -102^\circ & 0.10\angle 48^\circ \\ 2.21\angle 104^\circ & 0.47\angle -48^\circ \end{bmatrix} \tag{10.54}$$

The stability factor is calculated from (10.17) to be

$$K_R = 0.752 \tag{10.55}$$

and

$$|\Delta| = 0.294$$

So the circuit is potentially unstable.

One way to solve this potentially unstable condition is to use a shunt feedback to neutralize the internal feedback. The neutralization procedure is illustrated in Figure 10.17. The overall $\mathbf{Y}_T$ is

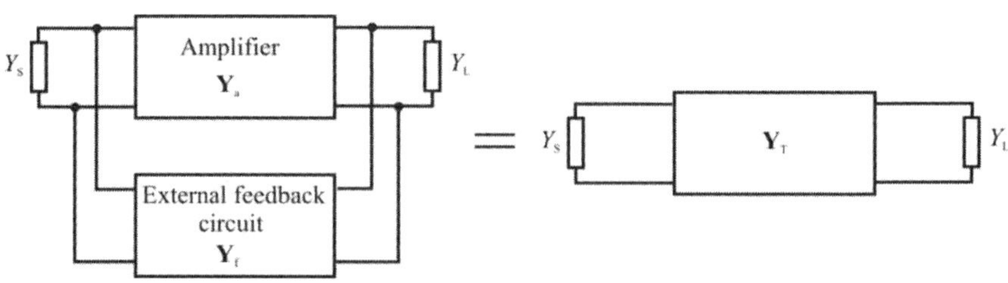

Figure 10.17 Neutralization.

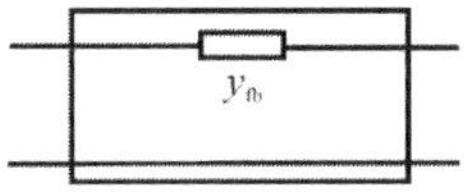

Figure 10.18 Simple feedback.

$$\mathbf{Y}_{\mathrm{T}} = \mathbf{Y}_{\mathrm{a}} + \mathbf{Y}_{\mathrm{f}} \tag{10.56}$$

We illustrate the simple narrowband feedback circuit as illustrated in Figure 10.18. In this case

$$\mathbf{Y}_{\mathrm{f}} = \begin{bmatrix} y_{\mathrm{fb}} & -y_{\mathrm{fb}} \\ -y_{\mathrm{fb}} & y_{\mathrm{fb}} \end{bmatrix} \tag{10.57}$$

We can find by transforming from the *s*-parameters resulting in (all *y*-parameters have the units of ℧)

$$\mathbf{Y}_{\mathrm{a}} = \begin{bmatrix} 5.5307\times10^{-3} + j1.9049\times10^{-2} & 3.9086\times10^{-4} - j2.3092\times10^{-3} \\ 4.7114\times10^{-2} - j2.1376\times10^{-2} & 5.4445\times10^{-3} + j5.1841\times10^{-3} \end{bmatrix}$$

Thus, the overall *y*-parameters are

$$\mathbf{Y}_{\mathrm{T}} = \begin{bmatrix} 5.5307\times10^{-3} + j1.9049\times10^{-2} + y_{\mathrm{fb}} & 3.9086\times10^{-4} - j2.3092\times10^{-3} - y_{\mathrm{fb}} \\ 4.7114\times10^{-2} - j2.1376\times10^{-2} - y_{\mathrm{fb}} & 5.4445\times10^{-3} + j5.1841\times10^{-3} + y_{\mathrm{fb}} \end{bmatrix}$$

Obviously if we set $y_{\mathrm{fb}} = -j2.3092\times10^{-3}$ then the internal feedback is cancelled, and we get

$$\mathbf{Y}_{\mathrm{T}} = \begin{bmatrix} 5.5307\times10^{-3} + j1.6739\times10^{-2} & 3.9086\times10^{-4} - j0 \\ 4.7114\times10^{-2} - j1.0967\times10^{-2} & 5.4445\times10^{-3} + j2.8750\times10^{-3} \end{bmatrix}$$

We can calculate

$$K_{\mathrm{R}} = 2.067$$

and

$$|\Delta| = 0.4037$$

These satisfy the Rollett criteria, the amplifier has been neutralized, and it is unconditionally stable (at the specified frequency).

Note that $y_{fb} = -j2.3092 \times 10^{-3}$ corresponds to an inductor of value 0.69 μH at 100 MHz.

10.5.1 Summary

A potentially unstable amplifier can be made unconditionally stable by neutralization as we demonstrated in this section. It is important to note that the neutralization process outlined in the example can be used up to the VHF range and applies to narrowband transmitters only.

There are other forms of neutralization; adding series resistances on the input and output for example [5]. We have shown one possible method. In addition, more complex feedback circuits could be used.

Example 10.4: VHF Amplifier Design

In this example we will design a 100 MHz single-stage amplifier with a BJT having the y-parameters in (10.59)

$$Y_{\text{trans}} = \begin{bmatrix} g_{11} + jb_{11} & g_{12} + jb_{12} \\ g_{21} + jb_{21} & g_{22} + jb_{22} \end{bmatrix} \tag{10.58}$$

$$= \begin{bmatrix} 43.7 + j10.6\text{mS} & -1.00 - j4.29\text{mS} \\ -13.9 - j331.0\text{mS} & 1.46 + j15.8\text{mS} \end{bmatrix} \tag{10.59}$$

The other relevant specifications of the amplifier are as follows:

- gain at 100 MHz: 15-dB
- 50 Ω match at both ports
- stable without load or/and generator connected
- 3-dB bandwidth, $W_{3\text{dB}}$, of 10 MHz with two-stage synchronous tuning

The topology of this amplifier is as illustrated in Figure 10.11.

We notice that the transistor data imply from (10.32)

$$y_{21}y_{12} = (1.00 + j4.29)(13.9 + j333) \tag{10.60}$$
$$= -1415 + j392.6(\text{mS})^2 = 1468\angle 164.4(\text{mS})^2 \tag{10.61}$$

so

$$K_{\text{trans}} = \frac{2g_{11}g_{22} - \text{Re}\{y_{21}y_{12}\}}{|y_{21}y_{12}|} = \frac{2 \times 43.7 \times 1.46 + 1415}{1468} = 1.051 \tag{10.62}$$

Although being absolutely stable the transistor is close to the stability bound at 1 so the requirement of stability without load and source connections will probably be highly sensitive to parameter variations. However, the transistor has a maximum stable gain of

$$G_{\text{ms}} = \left|\frac{y_{21}}{y_{12}}\right| = \sqrt{\frac{g_{21}^2 + b_{21}^2}{g_{12}^2 + b_{12}^2}} \tag{10.63}$$
$$= \sqrt{\frac{13.9^2 + 331^2}{1.00^2 + 4.21^2}} = 76.56 \quad (18.84\text{dB}) \tag{10.64}$$

Compared to the gain requirement (15 dB), there is room for enlarging the K_R-factor by resistive extension (there is inevitably gain loss and noise degradation when inserting resistance into amplifier stages). Writing the maximum gain as

$$G_{\max} = G_{\text{ms}} \frac{1}{K_R + \sqrt{K_R^2 - 1}} = G_{\text{ms}}(K_R - \sqrt{K_R^2 - 1}) \tag{10.65}$$

where G_{ms} is given by (10.63), we can solve for the ratio of $G_{\max}$ to G_{ms} and get the K_R-value for the extended two-port as

$$a = \frac{G_{\max}}{G_{\text{ms}}}, \qquad a = K_{\text{ext}} - \sqrt{K_{\text{ext}}^2 - 1}$$

so

$$K_{\text{ext}} = \frac{1}{2}\left(a + \frac{1}{a}\right) \tag{10.66}$$

Inserting the values, we have

$$G_{\max} = 15(\text{dB}) - 31.62 \Rightarrow a = \frac{31.62}{76.56} = 0.4130$$

so

$$K_{\text{ext}} = \frac{1}{2}(0.4130 + 2.421) = 1.417 \tag{10.67}$$

In this design we extend the transistor two-port by adding parallel conductance g_{B} across the collector port as indicated in Figure 10.8. If no other admittances are connected across the device ports, the K_{ext} value above places us further on the safe side for stability. The output conductance of the extended two-port is called $g_{22,\text{ext}}$, and it may be determined through the K_{R}-factor expression from (10.32). We get

$$\begin{aligned} g_{22,\text{ext}} = g_{22} + g_{\text{B}} &= \frac{K_{\text{ext}}\left|y_{12}y_{21}\right| + \text{Re}\{y_{12}y_{21}\}}{2g_{11}} \\ &= \frac{1.417 \times 1468 - 1415}{2 \times 43.7} = 7.610 \qquad \text{mS} \end{aligned} \tag{10.68}$$

$$g_{\text{B}} = g_{22,\text{ext}} - g_{22} = 7.61 - 1.46 = 6.15 \qquad \text{mS} \tag{10.69}$$

$$r_{\text{B}} = \frac{1}{g_{\text{B}}} = 163\Omega \tag{10.70}$$

Output conductance $g_{22,\text{ext}}$ is the only replacement that is required to find the y-parameter matrix of the extended two-port compared to the original one in (10.59). To find the generator and load admittances that imply simultaneous matching to the extended two-port, we start by calculating M given by

$$M = \pm\frac{\left|y_{21}y_{12}\right|}{2g_{11}g_{22,\text{ext}}}\sqrt{K_{\text{ext}}^2 - 1} \tag{10.71}$$

as

$$M = \frac{1468\sqrt{1.417^2 - 1}}{2 \times 43.7 \times 7.61} = 2.216 \tag{10.72}$$

Now the optimal generator and load admittances may be found using (10.22) and (10.23)

$$\begin{aligned} g_G &= g_{11}M = 43.7 \times 2.216 = 96.84 \quad \text{mS} \\ b_G &= \frac{\text{Im}\{y_{21}y_{12}\}}{2g_{22,\text{ext}}} - b_{11} \\ &= \frac{392.6}{2 \times 7.61} - 10.6 = 15.20\text{mS} \end{aligned} \tag{10.73}$$

$$g_L = g_{22,\text{ext}}M = 7.61 \times 2.216 = 16.86\text{mS}$$

$$b_L = \frac{\text{Im}\{y_{21}y_{12}\}}{2g_{11}} - b_{22} = \frac{392.6}{2 \times 43.7} - 15.8 = -11.31\text{mS} \tag{10.74}$$

Augmenting by the generator and load conductances, the stability factor of the amplifier under normal operation becomes

$$K_{\text{aug}} = \frac{2(g_{11} + g_G)(g_{22,\text{ext}} + g_L) - \text{Re}\{y_{21}y_{12}\}}{|y_{21}y_{12}|}$$

$$= \frac{2(43.7 + 96.84)(7.61 + 16.86) + 1415}{1468} = \frac{6878 + 1415}{1468} = 5.649 \tag{10.75}$$

As discussed in the previous example, a total stability factor of five or more is a practical design criterion that makes the amplifier relatively insensitive to parameter perturbations. Furthermore, it allows us to disregard internal feedback effects and design the two matching and tuning networks independent of each other.

The results of the design are shown in Figure 10.19.

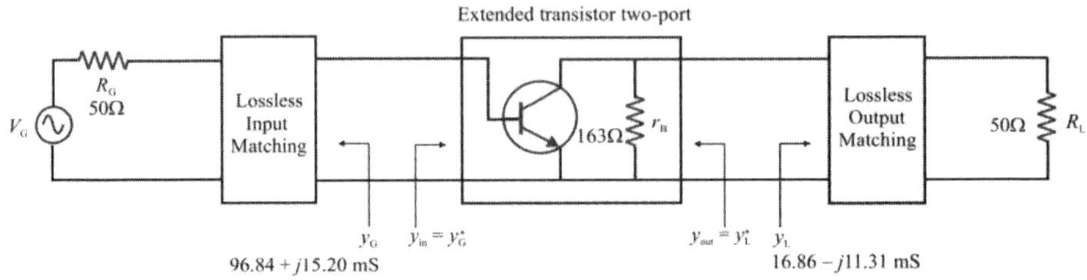

Figure 10.19 Amplifier principle including the data that are determined from gain and stability requirements.

10.6 Concluding Remarks

We discussed the requirements for an active two-port, active electronic circuit to remain stable in this chapter. Simply put, there is the possibility of oscillation at the input or output if a signal incident on the input or output port of the transistor is reflected with a gain greater than 1. That is, if

$$|\Gamma_i| > 1 \quad \text{or} \qquad |\Gamma_o| > 1 \tag{10.76}$$

We also covered some factors or network conditions that can be easily calculated based on *s*-parameters that must be met to ensure that (10.76) is true.

Last, we covered how networks can be augmented so that the conditions indicated in (10.76) can be met.

References

[1] Rizzi, P. A., *Microwave Engineering Passive Circuits,* Englewood Cliffs, NJ: Prentice Hall, 1988, p. 170.
[2] Pozar, D. M., *Microwave Engineering,* 3rd Ed., New York: Wiley, 2005, p. 545.
[3] Gonzalez, G., *Microwave Transistor Amplifiers,* 2nd Ed., Upper Saddle River, NJ: Prentice Hall, 1984, Appendix B.
[4] Poisel, R. A., *Electronic Warfare Receivers and Receiving Systems*, Norwood, MA: Artech House, 2014, Section 3.4.10.
[5] Poisel, R. A., *Electronic Warfare Receivers and Receiving Systems*, Norwood, MA: Artech House, 2014, Section 3.4.7.

Chapter 11

Power Amplifier Linearity

11.1 Introduction

The key design objectives for an RF power amplifier in an EW system are to boost the output signal level into the antenna with minimum size, distortion, power dissipation, and spectral pollution. Simultaneously optimizing all of these parameters is a major design challenge and one that becomes increasingly difficult as the complexity of the modulation format is raised.

At some level of input power, a power amplifier enters saturation, as indicated when the output power is no longer linearly related to the input power. (In fact, it can be argued that the output is *never* truly linearly related to the input.) It is the point on the transfer characteristic shown in Figure 11.1 when the transfer curve noticeably begins to curve. The 1-dB compression point (input or output) is the usual specification at that point.

When considering FET amplifiers, another way to view the linearity of the amplifier is illustrated in Figure 11.2. In this figure, the drain current is shown as a function of the gate voltage. The strongly nonlinear curve is ideal in the region before saturation begins. This can be approximated by a second, third, or higher order curve—a third-order power series is illustrated in Figure 11.2.

The effect of this nonlinear characteristic is to generate harmonics in the amplifier output that are not present in the input signal as illustrated in Figure 11.3. An odd-order nonlinearity model generates odd harmonics and an even order nonlinearity model generates even-order harmonics. The third order indicated in Figure 11.3 causes the output waveform to depart from being a larger equivalent of the input, as indicated by the flattened output voltage shown in Figure 11.3 in response to a sine wave input.

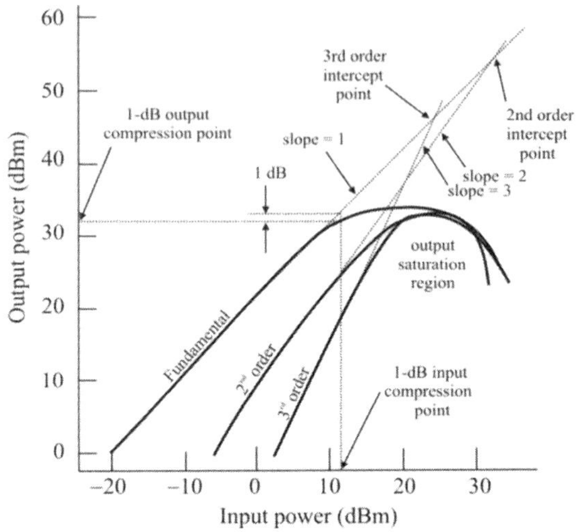

Figure 11.1 Amplifier power transfer characteristic with the second- and third-order harmonic responses.

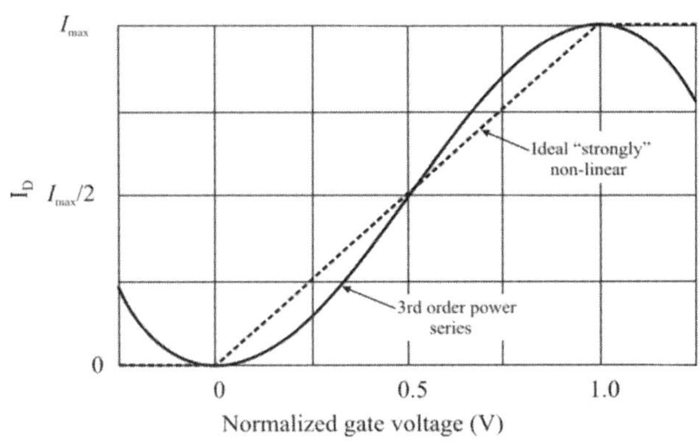

Figure 11.2 Third-order nonlinearity response at the output of a PA compare to a (limited range) linear response.

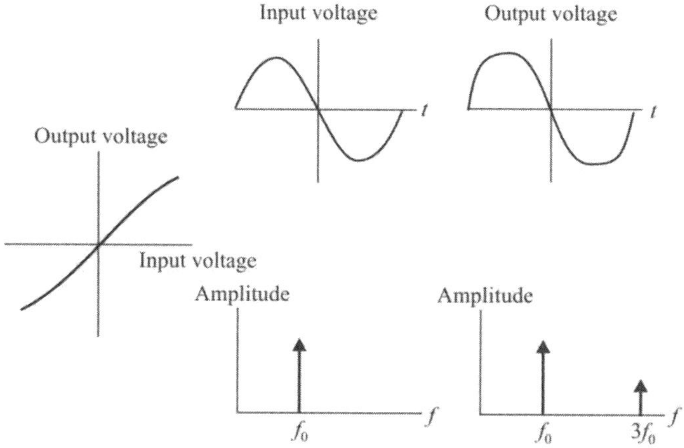

Figure 11.3 Amplifier nonlinearity generates harmonics and IMD in the output from a PA.

Nonlinearities lead to the aforementioned generation of harmonics, intermodulation distortion/spectral regrowth, SNR degradation, and constellation deformation in digital signals.

The *intermodulation products* (IMPs) generated by nonlinearities are illustrated in Figure 11.4 [1].

In order to improve amplifier efficiency and reduce cost while still achieving the required level of linearity, a range of amplifier linearization techniques have been developed, some more practical than others. In this chapter we look at some of the most common ones including the Cartesian loop, which is quite popular.

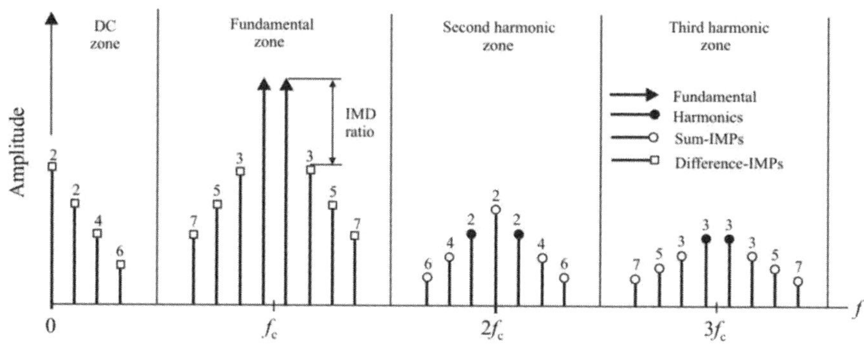

Figure 11.4 Intermodulation and harmonics. Source: [1].

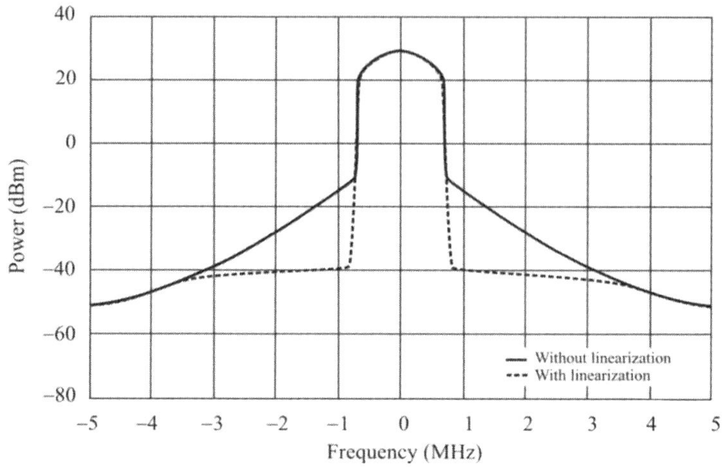

Figure 11.5 Predistortion, and linearity techniques in general, reduce the adjacent and other close-in skirts on the power spectrum from a PA.

The (ideal) net result of linearization techniques is illustrated in Figure 11.5. The goal of linearization is to minimize the energy in close-in adjacent channels, which, in this case, is approximately 20 dB or more as shown.

We provide a brief introduction to linearization of PAs in this chapter. For a more in-depth presentation and considerable expansion of the material presented here, see [2].

This chapter is structured as follows. After this introduction, which provides the motivation for addressing linearity in power amplifiers, an introduction to gain compression is provided. That is followed by a section on the several quantitative measures of linearity. PA linearization techniques are presented next followed by a discussion on feedforward and feedback spurious reduction methods. A section on some other linearization techniques is included next. This section includes cartesian feedback, predistortion methods, Doherty amplifier architectures, envelope elimination and restoral, and LINC. The chapter concludes with a discussion of load-pull measurement techniques for characterization of nonlinear PAs.

11.2 Gain Compression

All active devices (BJTs, MOSFETs, etc.) are nonlinear devices. That means that there is only a limited range of parameters over which they operate without becoming nonlinear. As such, some parameters are discussed here that address the linear range of amplifiers.

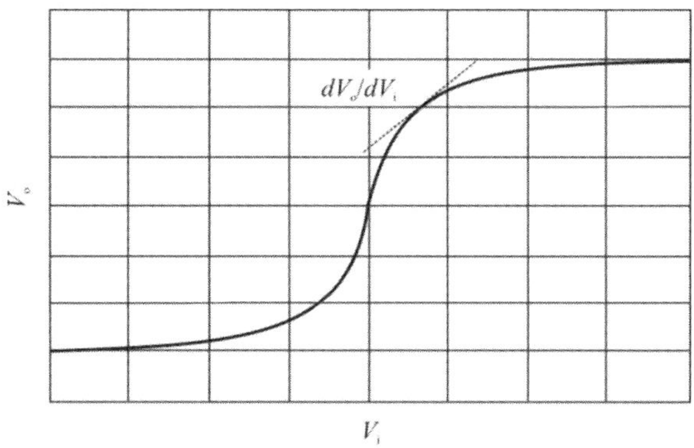

Figure 11.6 Gain compression in a nonlinear PA.

The large signal input/output relation can display gain compression or expansion as illustrated in Figures 11.6 and 11.7. Physically, most amplifier experience gain compression for large signals. The small-signal gain at a point on these curves is given by the slope of the curves at that point. For the graph in Figure 11.6, the gain decreases for increasing amplitude past the midpoint of the curve. This is known as *compression*. In Figure 11.7, the gain increases pat the midpoint. This is known as gain *expansion*.

11.2.1 Apparent Gain

Around a small deviation, the large signal curve can be described as a power series as given by the polynomial

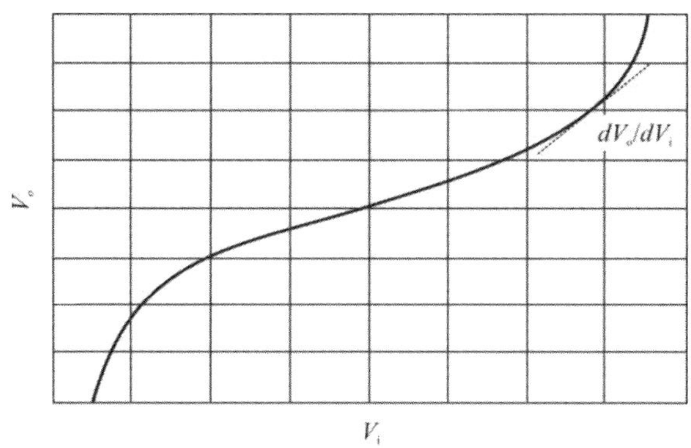

Figure 11.7 Gain expansion in a nonlinear PA. This is rarer than gain compression effects.

$$s_o = a_1 s_i + a_2 s_i^2 + a_3 s_i^3 + \cdots \tag{11.1}$$

For an input $s_i = S_1 \cos(\omega_1 t)$, the cubic term generates

$$\begin{aligned} S_1^3 \cos^3(\omega_1 t) &= S_1^3 \cos(\omega_1 t)\frac{1}{2}[1+\cos(2\omega_1 t)] \\ &= S_1^3 \left[\frac{1}{2}\cos(\omega_1 t) + \frac{2}{4}\cos(\omega_1 t)\cos(2\omega_1 t)\right] \end{aligned} \tag{11.2}$$

Recall that $2\cos a \cos b = \cos(a+b) + \cos(a-b)$ so that (11.2) becomes

$$S_1^3 \cos^3(\omega_1 t) = S_1^3 \left[\frac{1}{2}\cos(\omega_1 t) + \frac{1}{4}[\cos(\omega_1 t) + \cos(3\omega_1 t)]\right]$$

Collecting terms

$$S_1^3 \cos^3(\omega_1 t) = S_1^3 \left[\frac{3}{4}\cos(\omega_1 t) + \frac{1}{4}\cos(3\omega_1 t)\right]$$

The apparent gain of the system is therefore

$$\begin{aligned} G &= \frac{S_{o,\omega_1}}{S_{i,\omega_1}} = \frac{a_1 S_1 + \frac{3}{4} a_3 S_1^3}{S_1} \\ &= a_1 + \frac{3}{4} a_3 S_1^2 = a_1 \left(1 + \frac{3}{4}\frac{a_3}{a_1} S_1^2\right) = G(S_1) \end{aligned}$$

If $a_3/a_1 < 0$, the gain compresses with increasing amplitude.

Over the linear range of operation, the output signal power has a linear relationship with the input signal. This relationship is referred to as the *fundamental* in Figure 11.1, where the output power is plotted versus the input power. The second-order harmonics due to nonlinear operation have a slope on this plot of two, where the input signal produces a 6-dB increase in output power for every 3-dB increase in input power. Likewise, the third IM performance produces a 9-dB increase in output power for every 3-dB increase in input power. The amplifier reaches saturation at some point where these curves no longer increase in this fashion. Some of the parameters shown in Figure 11.1 are as follows.

1-dB output compression point: This is defined as the output power at which the fundamental gain has dropped 1 dB from its low-power value. Note that the slope of the output versus input power curve is 1 dB/dB. We often denote this point as P_{1dB}.

1-dB input compression point: This is defined as the input power at which the gain has dropped 1-dB from its low-power value.

Second order intercept point, (SOI, SOIP, IP_2): Since the second-order IM distortion products increase as s_i^2, we expect that at some power level the distortion products will overtake the fundamental signal. The extrapolated point where the curves of the fundamental signal and second-order distortion product signal meet is the *second-order intercept point* (IP_2). At this point, then, by definition IM_2 = 0-dBc is the IP_2 (see Figure 11.8). The second-order intercept point in Figure 11.8 is defined by two points, one on the abscissa called the *input second-order intercept point* and denoted by IIP_2, and one on the ordinate called *the output second-order intercept point* and denoted by OIP_2. Once the IP_2 point is known, the IM_2 at any other power level can be calculated. Note that for a dB backoff from the IP_2 point, the IM_2 improves dB for dB (see Figure 11.8).

Third order intercept point (TOI, TOIP, IP_3): Since the third-order IM distortion products increase as s_i^3, we expect that at some power level the distortion products will overtake the fundamental signal. The extrapolated point where the curves of the fundamental signal and third-order distortion product signal intersect is the *third-order intercept point* (IP_3). At this point, then, by definition IM_3 = 0-dBc. It also is defined by two points—one on the abscissa called the *input third-order intercept point* and denoted by IIP_3, and one on the ordinate called *the output third-order intercept point*, OIP_{11}. Once the IP_3 point is known, the IM_3 at any other power level can be calculated. Note that for a 10-dB backoff from the IP_3 point, the IM_3 improves 20 dB.

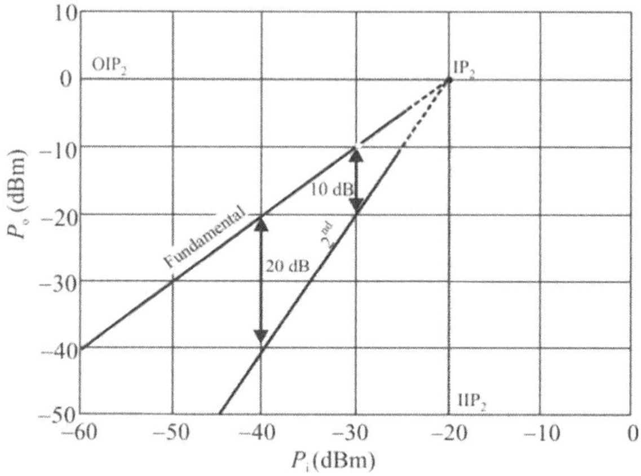

Figure 11.8 Second-order intercept point.

Example 11.1: Intercept Point

Suppose that the third-order characteristics of an amplifier have the form shown in Figure 11.9. We see that the amplifier has an $IIP_3 = -10$-dBm. What's the IM_3 for an input power of $P_i = -20$ dBm? Since the IM_3 improves by 20-dB for every 10-dB backoff, it's clear that $IM_3 = 20$ dBc. What's the IM_3 for an input power of $P_i = -110$ dBm? Since the IM_3 improves by 20 dB for every 10-dB backoff, it's clear that $IM_3 = 200$ dBc.

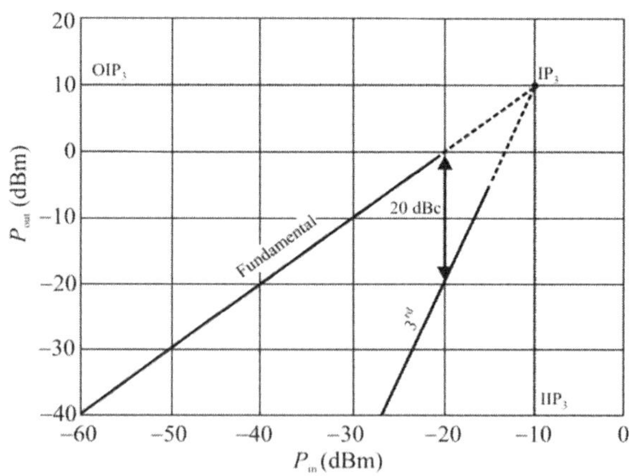

Figure 11.9. Third-order intercept point.

11.2.2 Calculated IIP_2/IIP_3

The IIP points can be calculated directly from the power series expansion (11.1). By definition, the IIP_2 point occurs when

$$IM_2 = 1 = \frac{a_2}{a_1} S_i$$

obviously

$$IIP_2 = S_i = \frac{a_1}{a_2}$$

Similarly, we can calculate IIP_3

$$IM_3 = 1 = \frac{3}{4}\frac{a_3}{a_1} S_i^2 \qquad\qquad IIP_3 = S_i = \sqrt{\frac{4}{3}\left|\frac{a_1}{a_3}\right|}$$

11.2.3 Dynamic Range

Consider a two-tone test where the input signal is

$$V(t) = A\cos 2\pi f_1 t + A\cos 2\pi f_2 t \tag{11.3}$$

The output frequencies will be of the form

$$f_o = mf_1 + nf_2 \tag{11.4}$$

where m and n are integers. The order of the intermodulation product (IP) is given by $|m| + |n|$. Note that $2f_1 - f_2$ and $2f_2 - f_1$ will be close to the original frequencies and filters may not be able to eliminate them. The third order intercept point P_{IP} is defined as the output power at which the third-order IP power intersects the linear power (assuming no gain compression/saturation occurs). The slope of the third-order intermodulation product output power versus input power is 3 dB/dB.

Dynamic range: The most common measure for the dynamic range of an amplifier is the two-tone spurious free dynamic range. To find this value, two equal valued tones are injected at the amplifier input and the harmonic responses are measured at

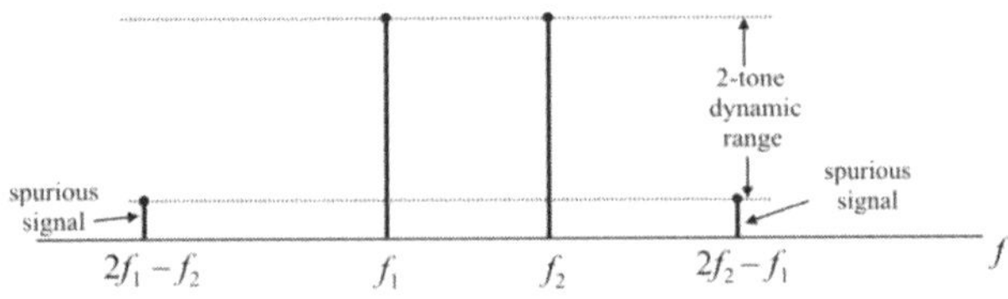

Figure 11.10. Two-tone dynamic range.

the output. The difference in the output response at the frequencies at the input and the largest spur is the two-tone spur free dynamic range (see Figure 11.10).

Another way of expressing this dynamic range is illustrated in Figure 11.11. The fundamental response as well as the third order IMD characteristic reach the output noise floor at some point as shown. (This noise floor is stochastic so will be described by a mean value, which is the point of concern.) The difference between these values is the SFDR.

11.2.4 Cascade

When we cascade two nonlinear systems, as shown in Figure 11.12, we have the nonlinear relationships between the variables

$$y = f(x) = a_1 x + a_2 x^2 + a_3 x^3 + \cdots$$
$$z = g(y) = b_1 y + b_2 y^2 + b_3 y^3 + \cdots$$

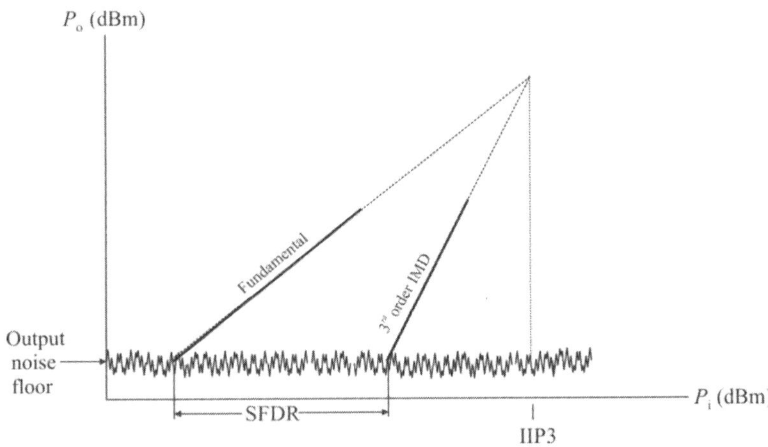

Figure 11.11 SFDR.

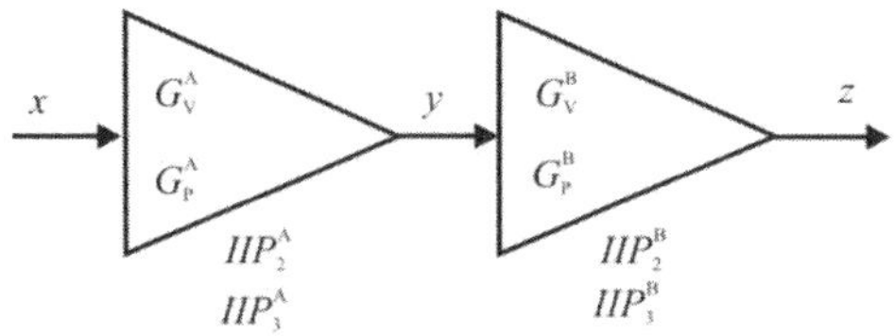

Figure 11.12 Cascade.

The task is to find the overall relation

$$z = c_1 x + c_2 x^2 + c_3 x^3 + \cdots$$

To find c_1, c_2, …, we simply substitute one power series into the other and collect like powers.

The linear term, as expected, is given by

$$c_1 = b_1 a_1$$

The square terms are given by

$$c_2 = b_1 a_2 + b_2 a_1^2$$

The first term is simply the second-order distortion produced by the first amplifier and amplified by the second amplifier linear term. The second term is the generation of second order by the second amplifier.

Finally, the cubic terms are given by

$$c_3 = b_1 a_3 + b_2 2a_1 a_2 + b_3 a_1^3$$

The first and last term have a very clear origin. The middle terms, though, are more interesting. They arise due to second harmonic interaction. The second-order distortion of the first amplifier can interact with the linear term through the second-order nonlinearity to produce cubic distortion. Even if both amplifiers have negligible cubic in and of themselves, that is, $a_3 = b_3 = 0$, we see the overall amplifier can generate a cubic response through this mechanism.

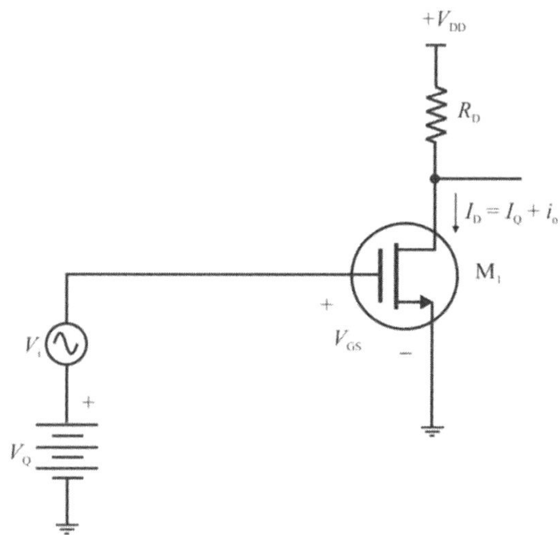

Figure 11.13 Long channel MOSFET amplifier.

Example 11.2: Distortion in Long-Channel MOSFET Amp

Consider the simple single-stage amplifier shown in Figure 11.13. Ignoring the output impedance (it can be combined with R_D) for the long-channel MOSFET in saturation we know that [3, 4]

$$I_D = \frac{1}{2}\mu C_{ox}\frac{W}{L}(V_{gs} - V_T)^2$$

$$I_o + I_Q = \frac{1}{2}\mu C_{ox}\frac{W}{L}(V_Q + V_i - V_T)^2$$

$$= \frac{1}{2}\mu C_{ox}\frac{W}{L}[(V_Q - V_T)^2 + V_i^2 + 2V_i(V_Q - V_T)]$$

$$= \underset{\text{DC}}{I_Q} + \underbrace{\mu C_{ox}\frac{W}{L}(V_Q - V_T)V_i}_{\text{Linear}} + \underbrace{\frac{1}{2}\mu C_{ox}\frac{W}{L}V_i^2}_{\text{Quadratic}}$$

So the output current has a DC component, a linear component (the desired one), and a quadratic (nonlinear) component. The last of these produces distortion in the output signal.

11.3 Measures of Linearity

Linearity is characterized, measured, and specified by various techniques depending upon the specific signal and application. The linearity of RF PAs is typically characterized by C/I, NPR, ACPR, and EVM (defined below).

11.3.1 Carrier-to-Intermodulation Ratio

A common measure of linearity is the *carrier-to-intermodulation* (C/I) ratio. The PA is driven with two or more tones of equal amplitudes. Nonlinearities cause the production of IMD products at frequencies corresponding to sums and differences of multiples of these tone frequencies. The amplitude of the third-order (usually) or maximum IMD product (other times) is compared to that of the tones to obtain the C/I (See Figure 11.4). A typical linear PA has a C/I of 30-dB or better.

11.3.2 Noise Power Ratio

The reduction in the SNR caused by nonlinearities in an amplifier is typically indicated by the *noise power ratio* (NPR). The concept is illustrated in Figure 11.14. An input signal is input to the amplifier which has the spectrum to the left in Figure 11.14. A notch is placed at the center of the spectrum as indicated. The nonlinearities in the amplifier cause noise in the output which fills in that notch. The noise level in the notch at the output is compared with the level of the rest of the spectrum, and the ratio is the NPR.

11.3.3 Adjacent Channel Power Ratio

The adjacent *channel power ratio* (ACPR) characterizes how nonlinearity effects cause energy to spill over from one channel to adjacent channels and is widely used with modern shaped pulse digital signals. ACPR is the ratio of the power in a specified band outside the signal bandwidth (channel) to the rms power in the signal

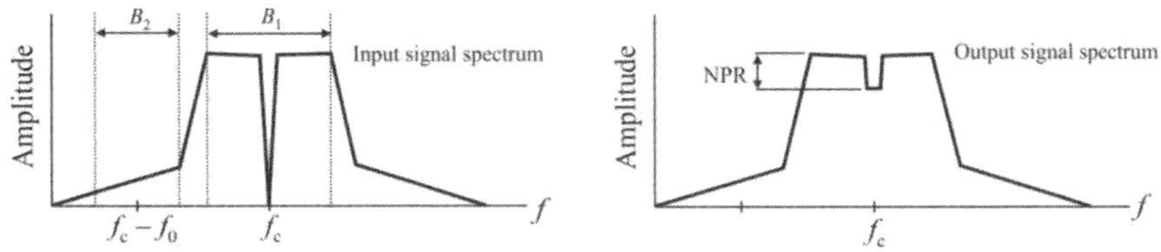

Figure 11.14 Reduced NPR.

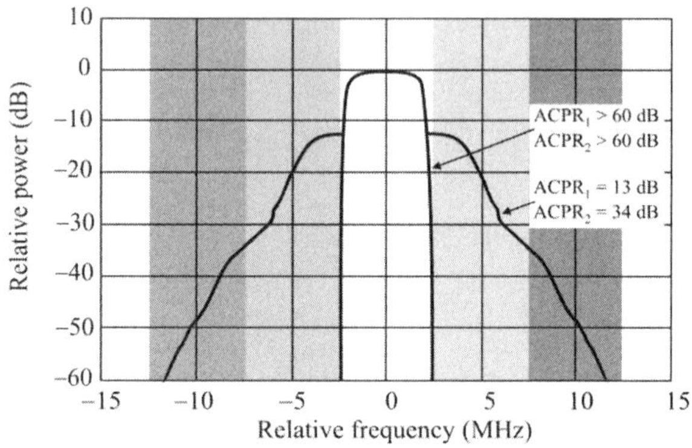

Figure 11.15 Spectral regrowth. The adjacent channels are shown shaded. ACPR.

(Figure 11.15). In some cases, the actual power spectrum $S(f)$ is weighted by the frequency response $H(f)$ of the pulse-shaping filter. The weighting, frequency offsets, and required ACPRs are determined by the applications. The ACPR can be specified for either upper or lower sideband, or both. In many cases, two different ACPRs for two different frequency offsets are specified. $ACPR_2$, based upon the outer band, is sometimes called *alternate channel power ratio*.

Spectral regrowth is when some of the energy in the signal in one channel falls over the channel edges into adjacent channels. This is illustrated in Figure 11.15. The *adjacent channel leakage power ratio* ($ACPR_i$) is the ratio of the power in the signal in the intended channel to the power from that same signal in a channel that is i channels away from the intended channel. Thus, for the broad spectrum shown in Figure 11.15, $ACPR_1 = 13$-dB and $ACPR_2 = 34$-dB. The narrowband spectrum in Figure 11.15 has $ACPR_1 > 60$-dB and $ACPR_2 > 60$-dB.

11.3.4 Error Vector Magnitude

A convenient measure of how nonlinearity interferes with detection is known as the *error vector magnitude* (EVM). EVM is the distance between the desired and actual signal vectors, normalized to a fraction of the signal amplitude (see Figure 11.16).

11.3.5 Deformed Constellations

Another effect of nonlinearity is a deformed constellation when the PA is amplifying digital signals. This is illustrated in Figure 11.17. The 16-ary PSK input constellation is shown on the left in Figure 11.17 and the output constellation is

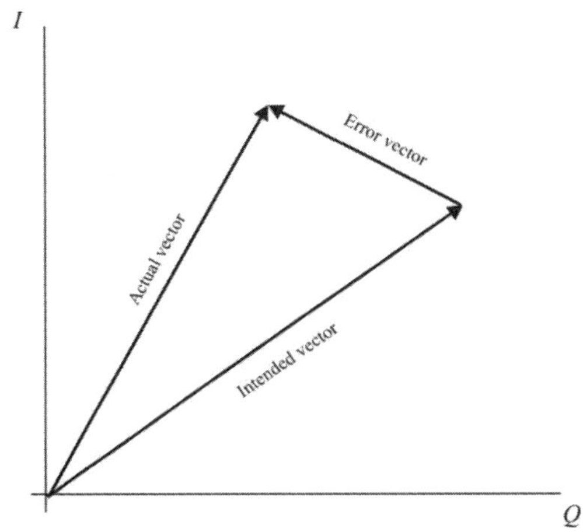

Figure 11.16 EVM.

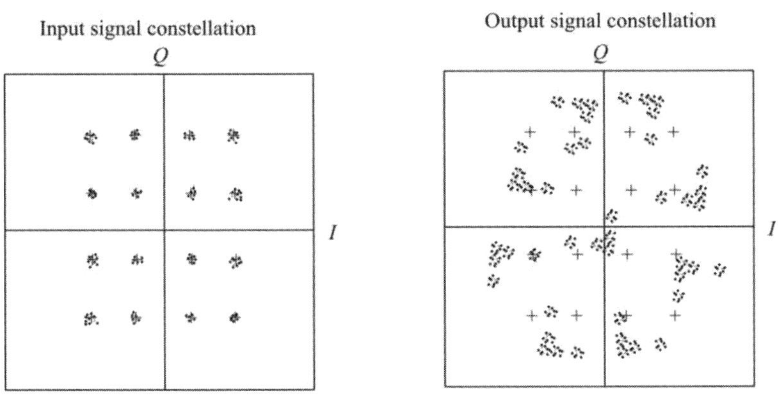

Figure 11.17 Constellation deformations.

shown in the right. It is very difficult to track which output symbol belongs to which constellation point in this case.

11.3.6 AM/AM and AM/PM Conversion

The transfer characteristic shown in Figure 11.1 for a PA illustrates how the output becomes nonlinearly related to the input when the input power reaches a certain level. This nonlinear relationship of the amplitude of the output relative to that of the input is sometimes referred to as *AM/AM conversion*.

There is a similar characteristic of amplifiers that when the input reaches a certain level the phase characteristics of the input cause an effect on the amplitude of the output power. This characteristic is sometimes referred to as *AM/PM conversion* of power amplifiers. Such nonlinearities in amplifiers cause the output to be noisier than the input signal, causing a reduced SNR than otherwise. A similar phenomenon occurs to the phase of the output relative to the input. In the linear region of the transfer characteristic shown in Figure 11.1 the output phase is, within reasonable bounds, the same as the input. However, as the saturation region is approached the phase of the output starts to deviate from that of the input. This is illustrated in Figures 11.18 and 11.19. The curves shown in Figure 11.18 are typical of GaAs amplifiers while those shown in Figure 11.19 are typical of LDMOS PAs.

11.4 Power Amplifier Linearization Techniques

11.4.1 Introduction

As just indicated, nonlinearities in power amplifiers can cause serious detrimental problems for signals. It has therefore been desirable to develop methods to extend the linearity in power amplifiers (it is impossible to eliminate the nonlinearity all together). This linearization should also simultaneously enhance or at least preserve the efficiency.

Linearity is particularly important in high-power EW applications. Nonlinearities cause unwanted energy to appear in nearby channels that are likely employed by friendly forces, and these unwanted energy sources are a cause of co-channel interference. PAs in EW systems are typically much more powerful than tactical combat net radios, and as such can easily cause interference where it isn't wanted.

The simplest way to preserve linearity is to back off the operating point so that the PA does not enter saturation for any input level. Other techniques for linearity improvement consist of distortion feedback, feed-forward distortion compensation,

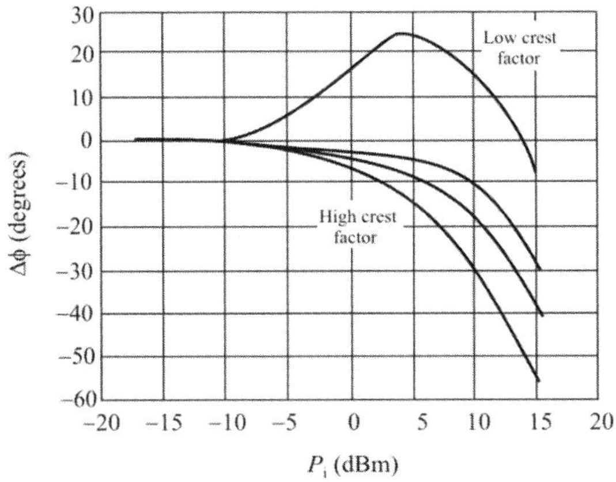

Figure 11.18 AM/PM conversion. The characteristic shown here is typical of GaAs power amplifiers.

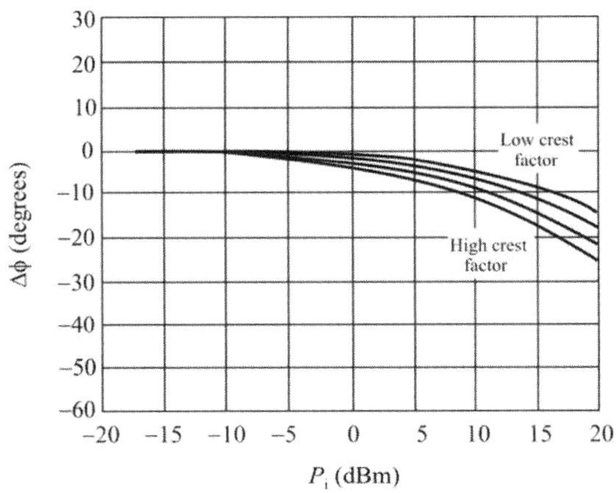

Figure 11.19 Typical AM/PM conversion characteristics for LDMOS power amplifiers.

Cartesian feedback, *linear amplification by nonlinear components* (LINC), and analog and digital predistortion. Techniques for preserving efficiency include utilizing Doherty architectures, envelope elimination, and restoral and adaptive bias. These techniques are presented in this section.

There is usually a trade-off between linearity and efficiency in RF amplifiers. With power efficiency comes battery life when that matters and, probably more important in general, easier thermal management. On the other hand, with linearity with sophisticated modulation techniques comes spectral efficiency.

For power efficiency the PA architectures of choice are the switching architectures, D, E, and F. For linearity, and therefore spectral efficiency, the architectures of choice are A, AB, B, and C.

An important measure of the strength of a given linearization technique is its robustness to poor characterization of the PA. A good method is to force the PA to behave as a benign nonlinear system, that is, look for ways to preserve time invariance.

11.4.2 Power Back-Off

One technique for linearizing an amplifier is referred to as *power back-off*. The theory is that if the output is given by $v_o(x)$ then

$$v_o(x) = a_0 + \underbrace{a_1 x}_{\substack{\text{dominant for} \\ x<<1}} + a_2 x^2 + \cdots = \sum_{k=0}^{\infty} a_k x^k \tag{11.5}$$

and the second, linear term dominates the response. While this technique does work, the problem is that $i_L << I_{bias}$, and poor efficiency results. This is not a very satisfactory answer for EW PAs, which must operate at maximum power most of the time.

The characteristic shown in Figure 11.20 illustrates the fundamental power amplification characteristic along with the representative higher order IM output characteristic. The essence of power backoff is to operate the amplifier well below the intersection of these curves so that the output power never reaches the intersection point.

11.4.3 Distortion Feedback

A flow diagram implementing distortion feedback is shown in Figure 11.21. The output signal is sampled and fed back to be summed with a delayed version of the input signal. The signal fed back is adjusted in gain and phase such that remnants of

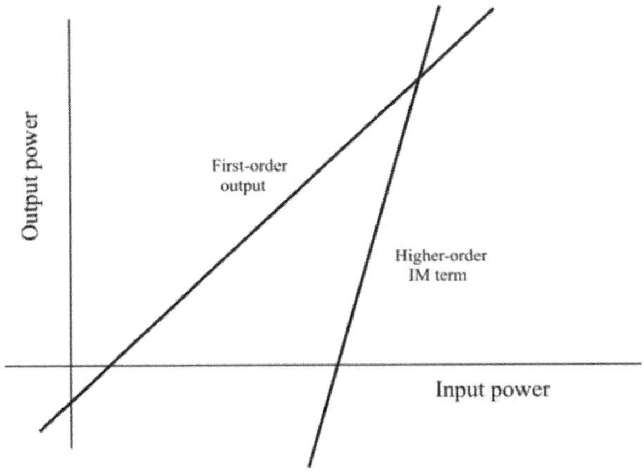

Figure 11.20 Power backoff.

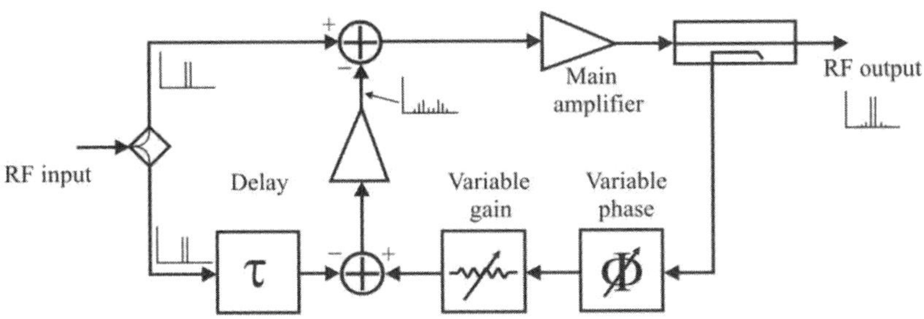

Figure 11.21 Distortion feedback.

signals that should not be there are removed or diminished, and that is the signal that is subtracted from the main channel with an adder.

Corrective Distortion

Corrective distortion (either predistortion and/or postdistortion) is illustrated in Figure 11.22. In general, the design issue is to choose $F_1(\ )$ and $F_2(\ )$ such that $F_2\{H[F_1(x)]\}$ is a linear function of x. Some design issues for a predistorter are:

- For analog predistorers, how are they realized;
- Initial calibration and training;
- Sensitivity to drift.

The advantages of adaptive predistortion techniques include:

- Does not suffer the bandwidth limitation incurred in continuous feedback techniques;
- They solve the problem of drift sensitivity.

There are some design issues with adaptive predistortion including:

- New, discrete-time feedback stability problem associated with model estimation;
- Depends on having a good power amplifier model.

11.5 Feedback and Feedforward Spurious Reduction Techniques in Power Amplifiers

11.5.1 Introduction

Phase noise (PM) and amplitude noise (AM) are of critical concern in EW systems. In many cases, the noise levels of components are above system demands and explicit noise reduction methods must be applied. Feedback and feedforward are

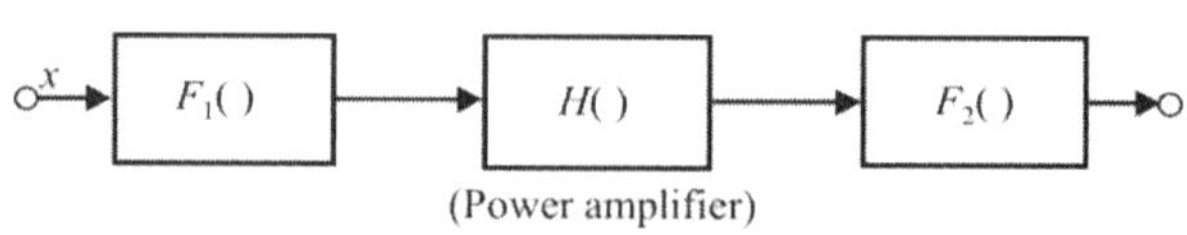

Figure 11.22 Predistortion.

two general noise reduction techniques that have been employed with amplifiers and oscillators. In this section, we compare feedback, feedforward, and other related noise reduction methods [5]. Key comparison criteria include the achievable noise performance, whether both PM and AM noise are reduced, system complexity, noise reduction bandwidth, operating frequency range, and stability.

Feedforward and feedback techniques have been around for about a century. Harold Black developed feedforward techniques in 1924 as a means to compensate for the noise and distortion in repeaters for long-distance telephony [6, 7]. He invented feedback three years later [8, 9]. From that time, feedback flourished as techniques for using and implementing feedback were developed. Feedforward was largely ignored until the late 1960s, when Seidel showed how to use feedforward to reduce *intermodulation distortion* (IMD) in TWT amplifiers [10, 11].

Since the 1970s, feedforward has been widely used for IMD reduction, while feedback has been used extensively for close-to-the-carrier phase and amplitude noise reduction. Due to the immense communications market and with limited spectrum availability, feedforward amplifiers are commercially available with excellent IMD performance, but these products are not usually optimized for low close-to-the-carrier noise performance (for example, net amplifier gain must be traded-off against noise). Feedback techniques have been applied to amplifier noise reduction, both at the active device level, around complete (e.g., packaged) devices, and with *systems on a chip* (SoC). Feedback has also been used to reduce phase noise of oscillators, with many different topologies being used with success (see Chapter 3). Current state-of-the-art oscillators are based on advanced feedback techniques [12].

11.5.2 Feedforward Distortion Reduction

The topology shown in Figure 11.23 is a feedforward amplifier structured for low-noise. The principles of feedforward have been described by many authors [13–15],

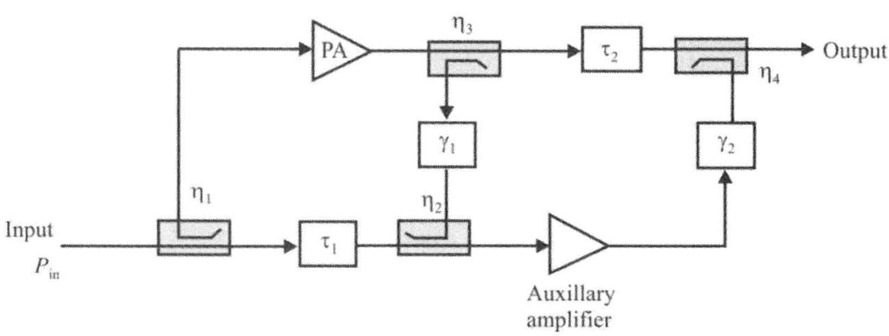

Figure 11.23 Feedforward amplifier flow diagram.

but mostly with IMD the major concern. This scheme exhibits excellent phase and amplitude noise performance.

The operation of the feedforward amplifier is as follows (see Figure 11.23). Noise and IMD generated by the main amplifier (PA) is sampled and compared to the input signal (which is assumed perfect for our purposes) at a subtracting junction (γ_1). This produces a signal at the input to an auxiliary amplifier that, since the carrier signal is subtracted, is devoid of carrier signal, but rich in the noise and IMD generated in the main amplifier. This information is enhanced by the auxiliary amplifier to a level comparable with the main amplifier output. Then, with appropriate phase and amplitude match at the output coupler, the noise and IMD can be removed by vector subtraction.

In Figure 11.23, τ_1 and τ_2 are delay elements to compensate for the delay of the main and auxiliary amplifiers, γ_1 and γ_2 are phase and amplitude adjustment means for balancing the two loops, and η_1-η_2 are the voltage coupling coefficients.

Effective noise reduction requires careful balancing of two loops, normally referred to as the carrier cancellation loop and the noise cancellation loop. Carrier suppression is performed by an interferometer with a combining junction at γ_1. In order to significantly cancel the carrier signals, the two signals should be almost equal amplitude and opposite phase.

Suppression of the carrier in the carrier cancellation loop is important for phase and amplitude noise reduction, since the auxiliary amplifier is a nonideal device. Generally, amplifiers exhibit multiplicative phase noise with a $1/f$ (flicker) signature, as well as AWGN (as represented by the amplifier's noise figure). It is imperative to keep the carrier power input to the auxiliary amplifier low so that its flicker noise contribution is small (also to reduce IMD in the auxiliary amplifier).

11.5.2.1 Distortion Performance of Feedforward Amplifier

The phase noise performance of the *feedforward amplifier* (FFA), S_ϕ^{FFA}, is determined by the input power to the system, the phase noise characteristics of the amplifiers, and the achievable carrier/noise cancellation in the two loops:

$$S_\phi^{\text{FFA}} = \frac{S_\phi^{\text{main}}}{NS} + \frac{k_\text{B} f^{\text{aux}} T_0 L_{\tau 1}}{P_\text{i}(1-\eta_1^2)(1-\eta_2^2)} + \frac{S_{\phi,\text{1Hz}}^{\text{aux}}}{CS}\frac{\text{1Hz}}{f_\text{m}} \tag{11.6}$$

where S_ϕ^{main} is the phase noise of the main PA in rad^2/Hz, NS is the noise cancellation factor (second interferometer), $k_\text{B} = 1.38\times10^{-23}$ J/K is Boltzmann's constant, f^{aux} is the auxiliary amplifier noise factor, T_0 is the operating temperature (e.g., 293 K), $L_{\tau 1}$ ($\geq$1) is the loss of the delay element τ_1, P_i is the input power to the FFA in W, f_m is the offset (Fourier) frequency and *CS* is the carrier suppression factor (first loop) defined as

$$CS = \frac{\text{Power available at one combiner input} - \text{combiner loss}}{\text{Power input to auxillary amplifier}}$$

$$\approx \frac{P_{\mathrm{i}}}{P_{\mathrm{i}}^{\mathrm{aux}}} \frac{(1-\eta_1^2)(1-\eta_2^2)}{L_{\tau 1}} \qquad \text{(with respect to the } \textit{through} \text{ input)} \qquad (11.7)$$

Here $P_{\mathrm{i}}^{\mathrm{aux}}$ is the power at the input to the auxiliary amplifier in Watts.

$S_{\phi,1\mathrm{Hz}}^{\mathrm{aux}}$ is the auxiliary amplifier phase noise "1 Hz intercept" in rad^2/Hz, from which the flicker noise of the auxiliary amplifier at other offset frequencies is determined, assuming a $1/f$ model. If the noise does not have a $1/f$ characteristic, then (11.6) can be modified accordingly. The white noise generated in the auxiliary amplifier as given by its noise figure is handled by the second term in (11.6).

We can see from (11.6) that if the carrier suppression is perfect (then $CS \rightarrow \infty$), the phase noise is limited only by the NF of the auxiliary amplifier (which can be 1 dB or so) and the balance at the output combiner. For example, if $CS \rightarrow \infty$ in the first loop but only a 20-dB balance is achieved in the second loop, then no more than 20-dB noise reduction is possible. Alternatively, the noise cancellation loop may be perfectly balanced, but poor carrier suppression in the first loop will introduce excess noise from the auxiliary amplifier (primarily from the flicker noise introduced by the carrier oscillator). Note that if the auxiliary amplifier has lower flicker noise than the main amplifier (i.e., $S_{\phi,1\mathrm{Hz}}^{\mathrm{aux}} < S_{\phi,1\mathrm{Hz}}^{\mathrm{main}}$) then the overall phase noise reduction can exceed the carrier suppression in the first loop. Such a scenario may be encountered, for example, if the main power amplifier is of a different technology than the low-power auxiliary amplifier.

The PM and AM noise of the phase and amplitude matching element γ_1 provides another limit to the noise floor. Since γ_1 (as shown in Figure 11.23) is not in series with the main amplifier, its noise contribution cannot be cancelled at the output combiner. Insertion of γ_1 prior to (and in series with) the main amplifier reduces the net amplifier gain by the insertion loss of γ_1. If this gain reduction can be tolerated, then placing γ_1 prior to the main amplifier is recommended, but care must be taken with the bandwidth of any automatic balance control circuit to prevent noise injection.

11.5.2.2 Carrier and Distortion Cancellation

Achieving desired levels of carrier cancellation can be difficult. Figure 11.24 shows the summation of two sinusoids, which are nominally the same frequency and

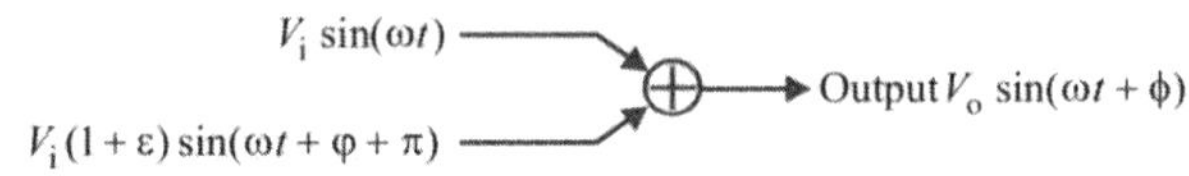

$$V_o = V_i\sqrt{1+(1+\varepsilon)^2 - 2(1+\varepsilon)\cos\varphi}$$

$$\phi = \tan^{-1}\left[\frac{(1+\varepsilon)\sin\varphi}{1-(1+\varepsilon)\cos\varphi}\right]$$

Figure 11.24 Sum of two sinusoids of nominally equal frequency and amplitude and opposite phase.

amplitude but opposite phase. If ε is the amplitude mismatch and φ is the phase error, then the resultant has amplitude and phase as illustrated in Figure 11.24.

For the carrier cancellation loop we can express *CS* in terms of the amplitude and phase mismatch seen by (with respect to the through input) the auxiliary amplifier from the inputs to combiner η_2 (weighted by appropriate transmission factors) as

$$CS = -10\log_{10}\left[1+(1+\varepsilon)^2 - 2(1+\varepsilon)\cos\phi\right] \quad \text{dB} \tag{11.8}$$

Figure 11.25 shows the effect of mismatch on *CS*.

For example, if *CS* = 40 dB is required [determined from known $S^{\text{aux}}_{\phi,1\text{Hz}}$ and required S^{FFA}_{ϕ} from (11.6)] then amplitude match better than 0.01 and phase balance of 0.5° are needed simultaneously.

NS can be expressed in an identical manner to *CS* in (11.8) so the noise cancellation loop demands are equally as stringent. Thus, for a feedforward noise reduction circuit, some form of automatic control is needed to maintain loop balance

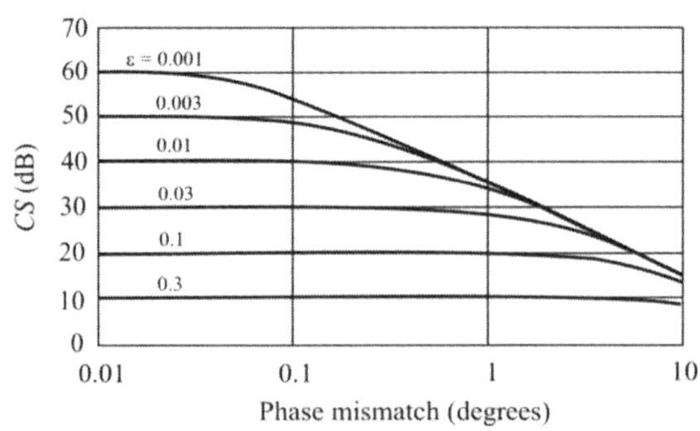

Figure 11.25 Effect of phase and amplitude mismatch on carrier cancellation.

over temperature, frequency range, and time if the noise performance is to be maintained.

11.5.2.3 Automatic Loop Balance

The feedforward circuit as shown in Figure 11.23 has no feedback loop and therefore is inherently stable. However, the addition of automatic loop balance control compromises the guaranteed stability of the system.

Maintaining a carrier null at the input to the auxiliary amplifier is reasonably straightforward, and is usually performed by phase- and amplitude-sensitive detectors driving control signals into voltage-variable phase shifters and attenuators in γ_1. These control loops often need only be narrowband, to account for parameter drift over temperature and time, which generally are slow processes. For systems that require wide operating bandwidth, including most EW systems, delay matching becomes critical, as the control systems usually require single-quadrant operation and AM-PM channel isolation demands are high.

Maintaining a balance at the noise cancellation node is more complicated, since the goal is to cancel all the noise. A solution by Seidel [16] used injection of a pilot tone into the main amplifier's spectrum, then synchronous detection with the observed tones at the inputs of the noise cancellation coupler. Such a system adds significant complexity, requires microwave switching, and places demands on the stability of the output coupler.

11.5.2.4 AM Performance of Feedforward Amplifiers

The feedforward amplifier offers simultaneous PM and AM noise reduction. An equation similar to (11.6) can be written for the AM spectrum:

$$S_{\mathrm{a}}^{\mathrm{FFA}} = \frac{S_{\mathrm{a}}^{\mathrm{main}}}{NS} + \frac{k_{\mathrm{B}} f^{\mathrm{aux}} T_0 L_{\tau 1}}{P_{\mathrm{i}}(1-\eta_1^2)(1-\eta_2^2)} + \frac{S_{\mathrm{a,1Hz}}^{\mathrm{aux}}}{CS} \frac{1\mathrm{Hz}}{f_{\mathrm{m}}} \tag{11.9}$$

We see that the white noise limit is once again imposed by the input losses and auxiliary amplifier NF, while the flicker AM component is determined by the AM performance of both amplifiers and the carrier and noise suppression achieved.

11.5.2.5 Bandwidth of Feedforward Amplifier

As mentioned previously, the operating frequency range of the feedforward amplifier is determined by delay matching in the two cancellation loops. In the microwave region (where wavelengths approach a few cm) delay matching becomes critical and stability of the automatic control loops requires length matching to within a few mm. Further, phase linearity (i.e., delay flatness) of all components is important. For fixed frequency use (rare for EW applications), or operation below microwave frequencies, delay matching is not difficult. Often, adjustment of a phase shifter is all that is required. The noise reduction bandwidth of the feedforward amplifier is inherently wide, to limits imposed by delay matching. In any case, it is apparent that feedforward has the capability to remove device flicker PM and AM noise to the limits imposed by cancellation error.

11.5.2.6 Summary of Feedforward Distortion Reduction

Overall, feedforward offers significant potential for both phase and amplitude noise reduction. The noise floor of the FFA approaches the thermal limit set by the auxiliary amplifier NF, with a net amplifier NF of a few decibels achievable with very low flicker noise. However, subtle phase and amplitude balance is required in two loops to achieve the potential noise performance. This usually requires automatic control, which adds significantly to system complexity and compromises the guaranteed stability of the system. Further, delay and gain matching over significant frequency range is critical in both loops for operation.

11.5.3 Feedback Distortion Reduction

Feedback has been used extensively for reducing the PM and AM noise in oscillators and two-port devices such as amplifiers. In general, PM and AM noise reduction are addressed separately, often with two similar but independent control systems.

11.5.3.1 Feedback Amplifier Phase Noise Reduction

A feedback control system to reduce the phase noise of an amplifier is shown in Figure 11.26 [17]. A phase detector (for example, a double-balanced mixer)

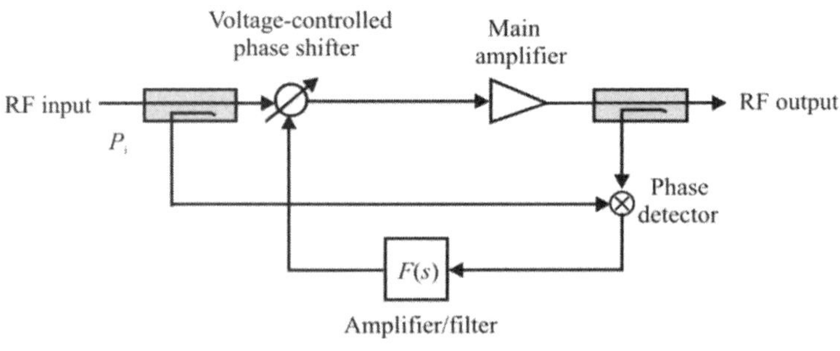

Figure 11.26 Feedback amplifier phase noise reduction.

compares the phase of the signals at the input and output of the amplifier. With inputs in quadrature, the phase detector outputs a voltage spectrum that represents the phase noise added by the amplifier. After baseband amplification and appropriate filtering (to ensure stability) the voltage signal drives a phase shifter to reduce the noise. The noise reduction bandwidth is limited by the baseband amplifier frequency response. Wideband opamps have gain-bandwidth products approaching 1 GHz, but the stability demands of the control system (i.e., adequate phase and gain margins) usually demand that closed loop gain falls to unity at a few MHz due to circuit parasitics. However, since phase noise typically has a $1/f$ signature, the loop gain required at large frequency offsets will not be as great as closer to the carrier. To ease demands on the baseband amplifier, phase detector conversion sensitivity and phase shifter phase/voltage slope should be as high as possible.

The noise floor of this system is generally dominated by the noise performance of the phase detector. Other noise limits include the baseband amplifier voltage noise and available loop gain (the phase shifter is the actuator of the control system, hence its noise contribution is reduced by loop gain). An approximation to the phase noise of the feedback amplifier of Figure 11.26 is given by

$$S_\phi^{\text{FBA}} = \frac{S_\phi^{\text{amp}} + S_\phi^{\text{VCP}}}{\left|1-\gamma(s)\right|^2} + \left|\frac{\gamma(s)}{1-\gamma(s)}\right|^2 S_\phi^{\text{PD}} \tag{11.10}$$

where S_ϕ^{amp} is the power amplifier phase noise, S_ϕ^{VCP} is the phase noise of the voltage-controlled phase shifter, S_ϕ^{PD} is the phase detector noise floor (all in rad^2/Hz) and $\gamma(s)$ is the open-loop gain given by

$$\gamma(s) = K_{\text{VCP}} K_{\text{PD}} F(s) \tag{11.11}$$

Here K_{VCP} is the conversion sensitivity of the phase shifter in rad/V and $F(s)$ is the transfer function of the baseband amplifier/filter. Note that K_{PD} is usually related to power and hence is a function of power amplifier characteristics. The noise performance of the baseband amplifier/filter can be lumped into S_ϕ^{amp} if required.

In general, the phase noise performance of mixers is better than amplifiers at a given operating frequency, enabling 20- to 30-dB noise reduction. For operation over an extended frequency range, normally the case in EW applications, delay matching is important, as the control system gain and stability are functions of the phase shift seen at the phase detector (with a double-balanced mixer, gain varies as

$\sim \cos\Delta\varphi$ [with $\Delta\varphi$ the deviation from quadrature]; if $|\varphi|$ exceeds 90°, feedback becomes positive).

11.5.3.2 Feedback Amplitude Noise Reduction

A circuit similar to Figure 11.26 can be used to reduce the AM noise of the amplifier by replacing the phase shifter with a controllable attenuator and using an amplitude-sensitive detector. In general, the amplitude detector cannot discern between AM noise generated in the device and that present in the input signal. This enables automatic level/gain control (ALC, AGC) circuits to be used that use feedback to reduce level fluctuations.

11.5.3.3 Transistor Active Feedback Noise Reduction

At the device level, noise can be sensed and reduced with negative feedback. Figure 11.27 shows a BJT amplifier in a CE configuration. Current fluctuations in the transistor contribute to phase and amplitude noise. An unbypassed emitter resistor (R_E ~3.3 Ω) reduces noise [18, 19], but further noise improvement is achieved by

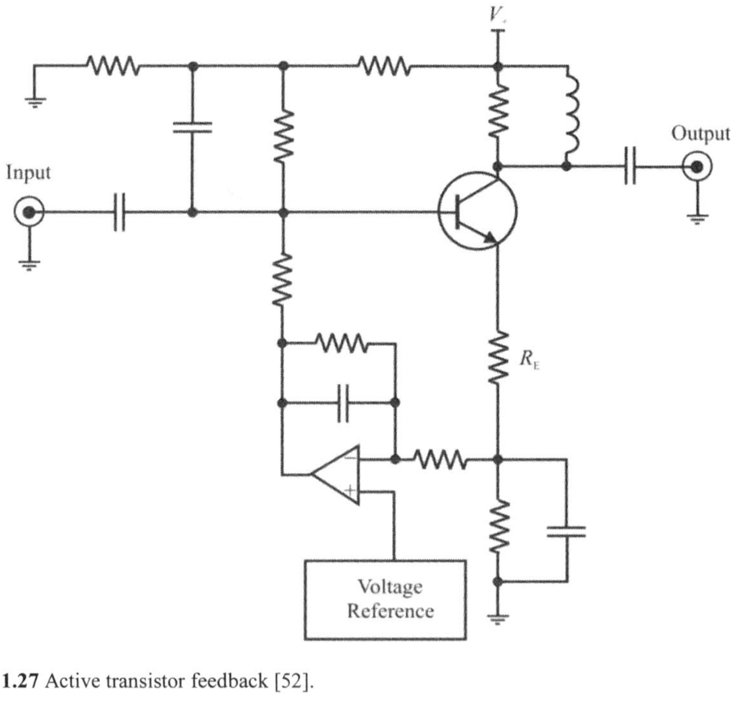

Figure 11.27 Active transistor feedback [52].

sensing the emitter current and feeding back a signal to the base terminal. PM and AM noise reduction of 20-dB has been demonstrated [20].

11.6 Other Amplifier Linearization Techniques

The amplifier schemes seen so far not only provide improved linearity and associated reduction in IMD but also offer the benefit of noise reduction. In this section we offer some other amplifier linearization methods.

11.6.1 Cartesian Feedback

Cartesian feedback in the form shown in Figure 11.28 is a baseband linearization technique [21]. The idea is to I-Q modulate the carrier before power amplification. The distorted signal is then fed back through an I-Q demodulator; the separated I and Q components are fed back to perform the linearization. Normally the limiting factor in this system is the nonlinearities of the downconverting mixers.

Since the I and Q channels are the same, I and Q can be easily matched, unlike polar feedback whose paths are very different, making matching difficult. Perhaps the main disadvantage of this scheme is the narrow bandwidth that's inherent in baseband feedback systems.

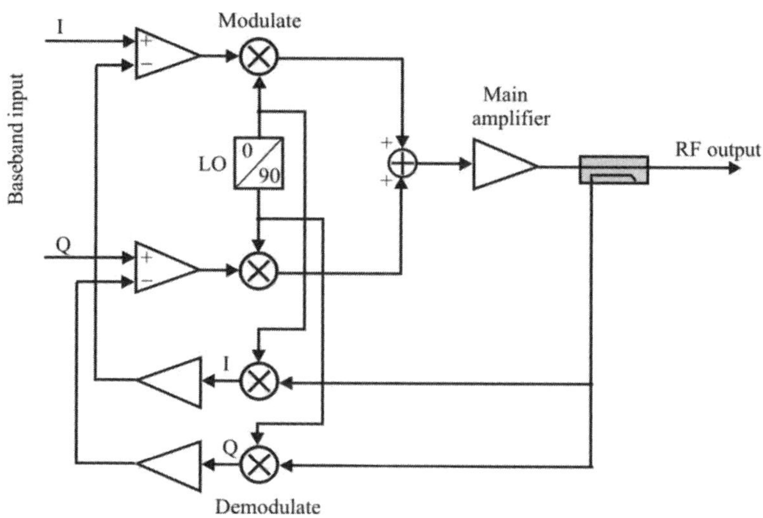

Figure 11.28 Cartesian feedback.

A major design issue in Cartesian feedback systems is referred to as phase alignment. Figure 11.28 shows a typical Cartesian feedback system. The loop drivers feed the baseband inputs of the upconversion mixer, which in turn drives the power amplifier. Some means of coupling the output of the power amplifier to the downconversion mixer is employed, and the output of this mixer is used to close the feedback system.

11.6.1.1 Impact of Phase Misalignment on Stability

Ideally, a Cartesian feedback system functions as two identical, decoupled feedback loops: one for the I component, and one for the Q component. This corresponds to the case of $\phi = 0$ in Figure 11.29. In practice, however, this state of affairs must be actively enforced. Delay through the PA, phase shifts of the RF carrier due to the reactive load of the antenna, and mismatched interconnect lengths between the LO source and the two mixers all manifest as an effective nonzero ϕ. Worse, the exact value of ϕ varies with temperature, process variations, output power, and carrier frequency. A Cartesian feedback system in which ϕ is nonzero is said to have phase misalignment. In this state the two feedback loops are coupled, and the stability of the system is compromised.

The impact of phase misalignment on system stability can be seen mathematically. We start by observing that the demodulated symbol S' is rotated relative to S by an amount equal to the phase misalignment ϕ. To see this, we write Cartesian components of the demodulated symbol

$$I' = (I \sin \omega t + Q \cos \omega t)\sin(\omega t + \phi) \tag{11.12}$$

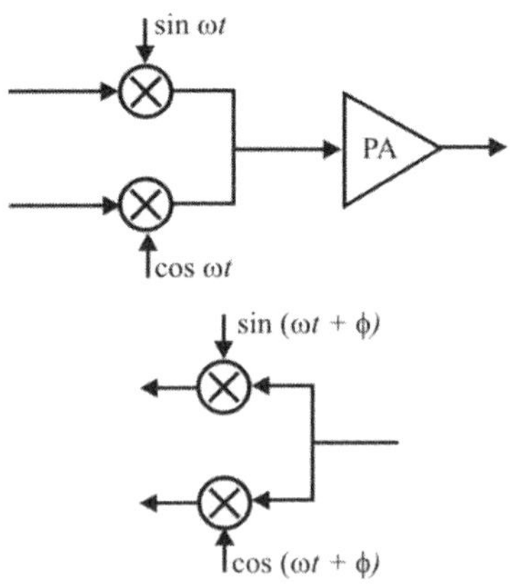

Figure 11.29 Cartesian feedback phase alignment.

$$Q' = (I \sin \omega t + Q \cos \omega t) \cos(\omega t + \phi) \tag{11.13}$$

where ω is the carrier frequency. Using trigonometric identities and assuming frequency components at 2ω are filtered out, we arrive at S'

$$I' = \frac{1}{2}(I \cos \phi + Q \sin \phi) \tag{11.14}$$

$$Q' = \frac{1}{2}(-I \sin \phi + Q \cos \phi) \tag{11.15}$$

We see that for $\phi \neq 0$, an excitation on the I input of the modulator results in a signal on the Q' downconverter output (and similarly for Q and I'). Accordingly, we say that the two loops are coupled.

11.6.1.2 Summary

Cartesian feedback is one of the most popular linearization techniques and is widely used. It manages the linearity of an entire transmit chain, including the upconverter and amplifier stages. This architecture is such that the technique is well suited to integration as an SoC.

The nature of the feedback process means that the amount of correction afforded by the control system decreases with increasing modulation bandwidth. Intermodulation improvement typically on the order of 30-dB is possible over a 25-kHz modulation bandwidth (a representative tactical CNR bandwidth, and therefore a typical EW bandwidth), and more than 10-dB improvement has been achieved for CDMA-type bandwidths in excess of 1 MHz. Furthermore, efficiencies of transmitters employing Cartesian loop feedback can be as high as 70%.

CFB Strengths and Weaknesses

The seminal strengths (S) and weaknesses (W) of Cartesian feedback linearization can be summarized as:

- W: Bandwidth limitation
- W: Stability concerns
- S: Low complexity
- S: Highly resistant to drift and aging
- SS: Robust to poor characterization of PA
- W: Low-loss delay lines and summations critical

- S: SoC possibilities

11.6.2 Constant Envelope Amplifiers

The simplest modulation formats to amplify are FSK and OOK. In both cases, when the amplifier needs to deliver power, it is at a constant level and hence at a fixed operating point on the amplifier power transfer characteristic. This means that the amplifier only needs to be optimized to deliver power efficiently at that output level and for a fixed input level and there can be variations in this output level. The linearity of the transfer characteristic is thus unimportant. As a result, most simple FM transmitters use class C biased amplifier designs with efficiencies of between 50% and 70%. The same can be said for the EW systems that target the associated simple receivers.

The situation is complicated somewhat when the carrier needs to be pulsed on and off, (i.e., with OOK or TDM), and the rise and fall times of the pulse edges need to be carefully controlled to minimize spectral pollution. Rapid changes in RF signals tend to produce wideband products in the spectrum. In this case, more design effort is needed to control (linearize) the power transfer characteristic of the amplifier.

11.6.3 Predistortion

It is desirable to run high-power devices, such as EW RF PAs and traveling-wave tube amplifiers (TWTAs), at high output power to maximize the transmitted power against resource utilization. In addition to backing-off the power level as discussed, another method to compensate for the intolerable IMD performance characteristics under saturation of the amplifier is to use *predistortion*. The concept is illustrated in Figure 11.30. Processing of the amplifier input (in this case) adds some power to the input signal to bring the PA output up to a level that it would be at if the amplifier response were linear as shown.

The general configuration for predistortion methods is illustrated in Figure 11.31. This topology can be implemented with either analog or digital components to perform the predistortion. The linearizer attempts to precondition the signal going to the power amplifier with an inverse characteristic so the result at the output of the amplifier is flat.

While the IMD performance of such a system can be greatly improved with this method, the phase noise is not reduced. In fact, with the series arrangement as shown in Figure 11.31, phase noise may indeed be significantly worse than with the amplifier alone, due to the noise characteristics of the linearizer. AM noise may be improved or degraded, depending on the saturation characteristics of the PA.

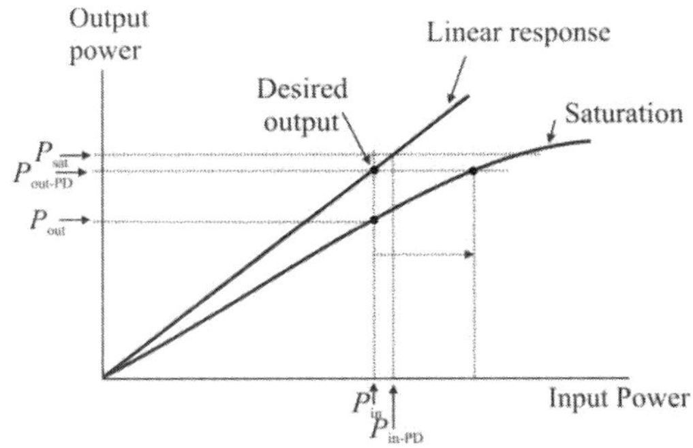

Figure 11.30 Predistortion by adding input power for compensation.

Figure 11.31 The linearizer attempts to precondition the signal going to the power amplifier with an inverse characteristic so the result at the output of the amplifier is flat.

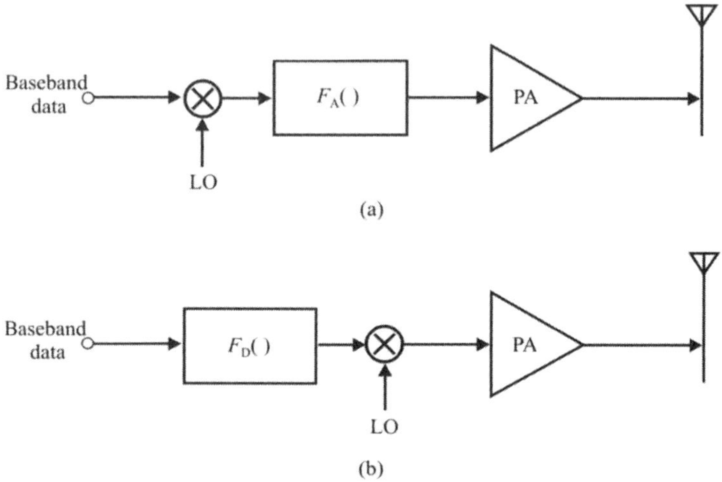

Figure 11.32 (a) predistortion at RF and (b) predistortion at baseband.

Some advanced linearization topologies combine predistortion and feedforward. If the predistortion network is incorporated within a feedforward network, then the noise of the linearizer can be removed.

The predistortion can occur either at baseband or RF as illustrated in Figure 11.32. $F_A()$, shown in Figure 11.32(a), represents predistortion functionality at RF while $F_D()$ in Figure 11.32(b) predistorts the baseband data prior to mixing.

11.6.3.1 Analog Predistortion

Two possible topologies for analog predistortion are illustrated in Figures 11.33 and 11.34. In Figure 11.33, the forward-biased nonlinear characteristics of the diodes are used to approximate an exponential response into the main amplifier. Predistortion in Figure 11.34 adds delay to a replica of the input RF signal as well as filtering of the input signal itself.

11.6.3.2 Adaptive Predistortion

When the predistortion is adjusted dynamically, it is called *adaptive predistortion*. One particular topology to implement adaptive predistortion is shown in Figure 11.35. The system estimator determines an estimate of what the output of the PA is versus its input, generating an error estimate $\varepsilon_{sys}(t)$. This error estimate is used to adjust the predistorter as well as the next error estimate. The PA output is compared to (subtracted from) the input to compute a symbol error $\varepsilon_{sym}(t)$, which is also used

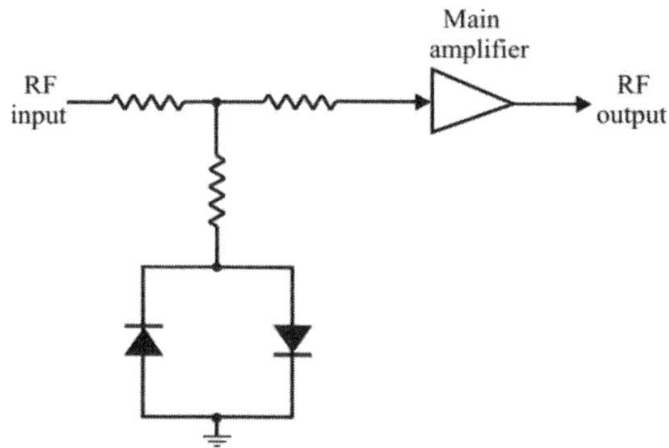

Figure 11.33 Analog predistortion possible realizations.

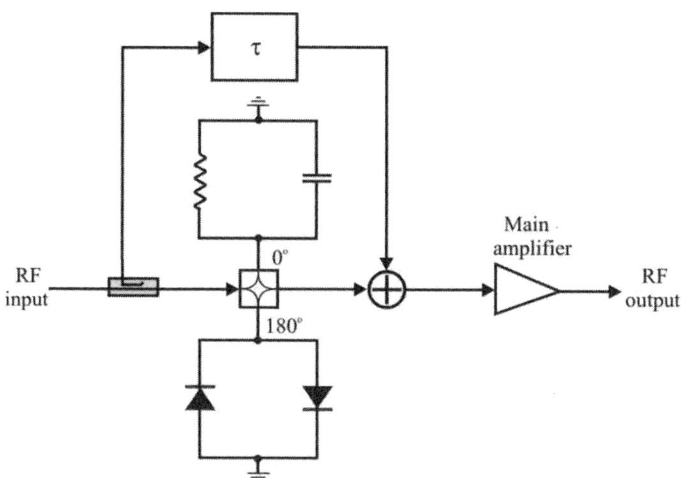

Figure 11.34 Analog predistortion.

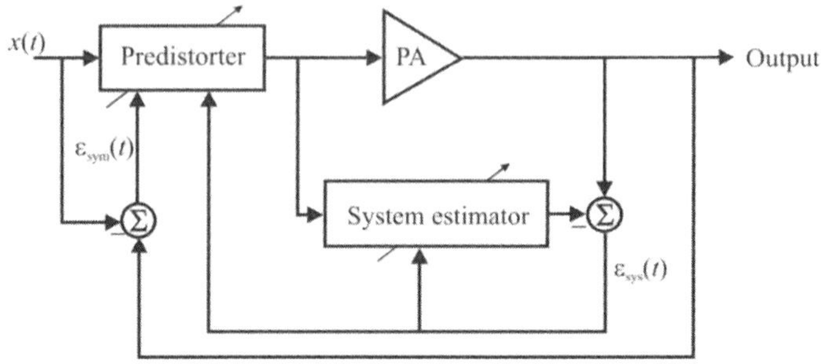

Figure 11.35 Adaptive predistortion.

to adjust the predistorter to produce a better match between the PA output symbol with the input symbol.

11.6.3.3 Digital Predistortion

It is possible (desirable?) to implement the predistortion digitally to the baseband data stream. This is illustrated in Figure 11.36. The gain adjustments for the I and

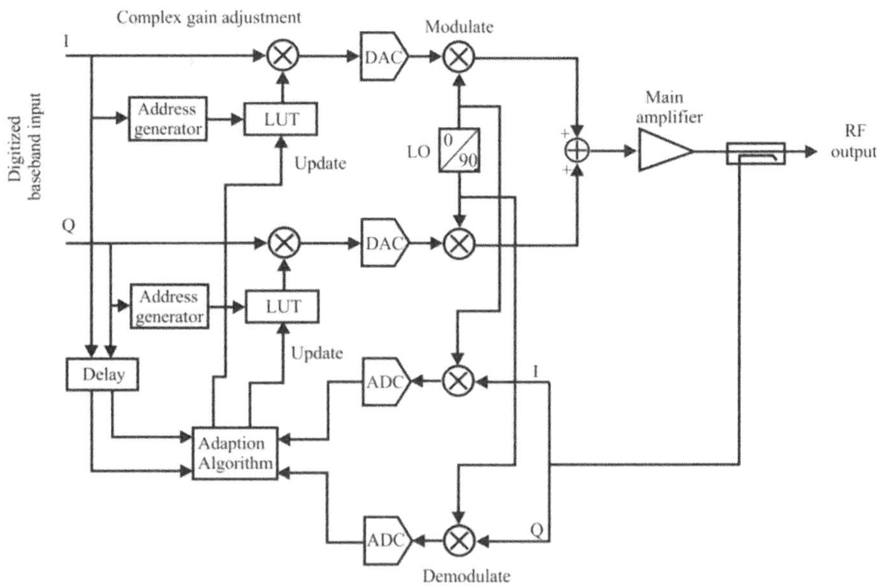

Figure 11.36 Digital predistortion.

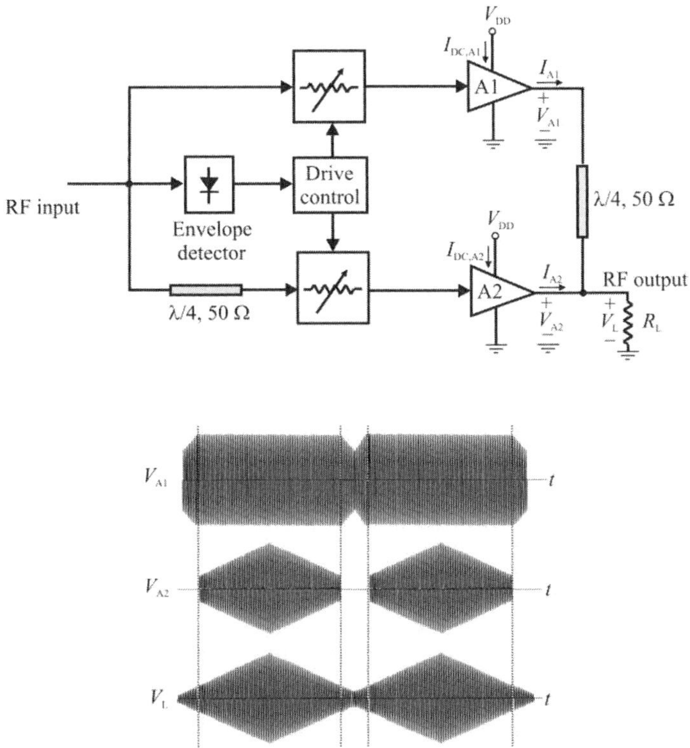

Figure 11.37 Doherty amplifier.

Q baseband signals are stored in look-up tables (LUTs) that are addressed based on the PA output and the adaption algorithm in use.

The obvious advantage of this configuration is the flexibility that ensues with using a general-purpose DSP for the predistortion processing. However, the complexity increases and this approach incurs the power overhead of a DSP chip.

11.6.4 Doherty Amplifier Architecture

One technique for efficiency preservation involves adding a second amplifier to the main amplifier. The main amplifier is operated in class B mode while the auxiliary amplifier is operated in class C. The resulting configuration is called the Doherty architecture, and is illustrated in Figure 11.37. Its operation is summarized in Figure 11.37. As the main amplifier (A1) approaches saturation, the second amplifier (A2) starts to put out power. Ideally the sum of the two power sources exactly compensate

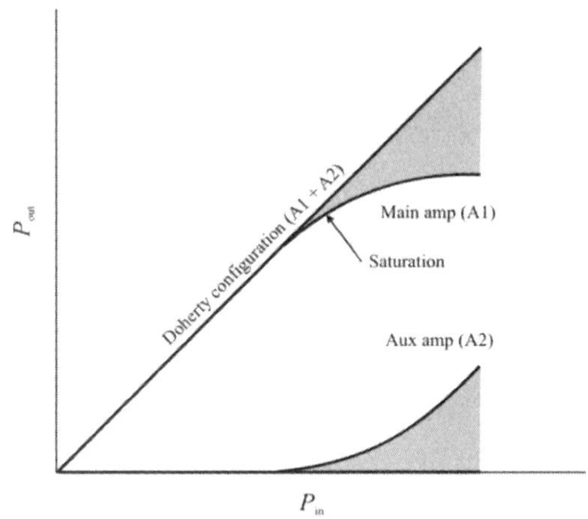

Figure 11.38 Doherty amplifier 2 operation.

one another and the output is the sum of the two, causing the transfer characteristic to remain linear.

Named after William Doherty, a Bell Labs researcher, the Doherty amplifier comprises a class B carrier stage in parallel with a class C peaking stage. In the input, half the input signal drives one device in the class B; half the other. On the output, the signals are summed after the carrier channel is passed through a $\lambda/4$ TL. Somewhat like a class G or H amp, the class B amp sustains the output most of the time, but the class C amp cuts in on high signal peaks. The ideal overall transfer characteristic is shown in Figure 11.38.

The principal advantages of the Doherty amplifier configuration are its efficiency and simplicity. It is most useful for communication signals that have a large *peak to average ratio* (PAR), such as modern digital communication signals to counter the corresponding EW signals. Its utility for signals with low PAR is limited and other simpler amplifier topologies are more appropriate.

Its largest single limitation for EW applications is its relative narrow bandwidth. The $\lambda/4$ lines have rather narrow bandwidths, thus restricting the range over which they operate. This imposes the requirement for having several such $\lambda / 4$ lines and switching them in as appropriate.

The benefit of the Doherty is increased efficiency, relative to a pure class B. The efficiency performance is indicated in Figure 11.39. α is the power division ratio of the two amplifiers at the peak output power (PEP).

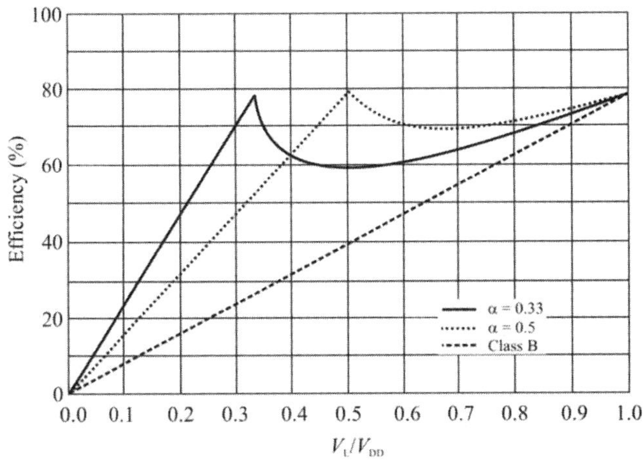

Figure 11.39 Doherty amplifier efficiency.

11.6.5 Envelope Elimination and Restoral

From the above discussions it should be clear that linear operation of an amplifier is an important notion in many circumstances. When there are substantial amplitude variations in a signal to be amplified, such as in AM and many forms of PM, linear amplifiers are required to avoid unacceptable distortions in the amplified signal. For constant-amplitude signals, amplifiers are typically operated in saturation where they are the most efficient. Distortion is not an issue in these circumstances.

One way to efficiently amplify a signal that contains amplitude variations is called *envelope elimination and restoration*. A simplified flow diagram of how the process works is shown in Figure 11.40. A limiter removes the amplitude variations in the signal and the resultant constant modulus signal is amplified through its own amplification chain. Since the signal has constant amplitude, a saturated amplifier can be used in this path. A separate channel is provided that is preceded by an amplitude detector. This slowly varying signal is amplified with a nonsaturated, and thus less efficient, amplifier. These two components are then combined at the output of the amplifier.

11.6.5.1 EER design issues

- Phase matching between the two signal paths is critical.
- Polar feedback is a possibility.
- Restoring the envelope in a power-efficient way is very challenging.

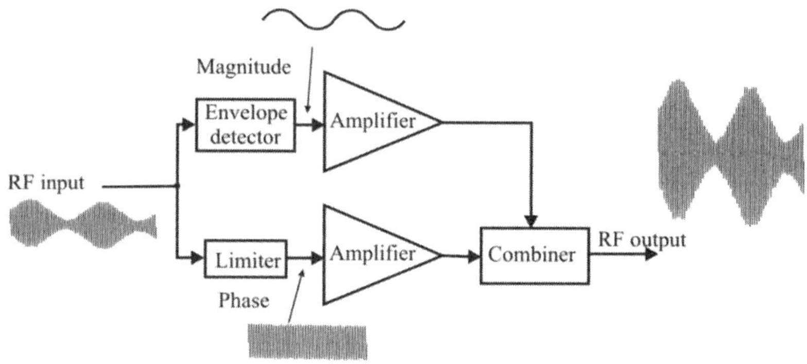

Figure 11.40 Amplifier efficiency preservation by envelope elimination and restoral.

11.6.6 LINC

Linear amplification with nonlinear components (LINC) is different from most other techniques of linearization of power amplifiers, because no feedback from the output of the PA is used [22]. There are no feedforward loops either. The PA can be highly nonlinear. The baseband processing accepts a gain and phase modulated input signal, and generate two wideband constant envelope phase modulated signals. These signals are up-converted through two well matched nonlinear amplifier chains and summed. The complex signals are generated such that all undesired out-of-band components are in exact anti-phase in the two amplifier chains and cancel at the output, while the wanted components are in phase and reinforced (see Figure 11.41).

The generation of two wideband constant envelope phase modulated signals $s_1(t)$ and $s_2(t)$ have to be accurate; DSP technology allows $s_1(t)$ and $s_2(t)$ to be so generated. Thus, the linearity performance of the technique is determined by the gain and phase match between the two amplifiers. The better these amplifiers are matched the better the linearity that ensues and the better noise performance.

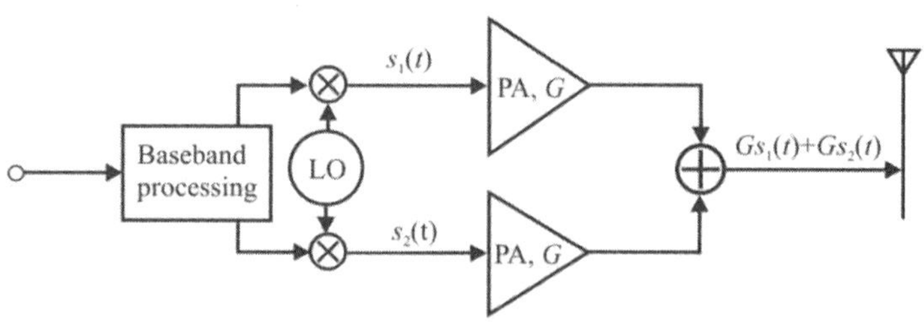

Figure 11.41 Illustration of LINC method to linearize power amplifiers

The input signal $s(t)$ is a complex representation of a bandlimited signal and can be written as

$$s(t) = r(t)e^{j\phi(t)}; \quad 0 < r(t) < r_{max} \tag{11.16}$$

This signal is split into two signals, $s_1(t)$ and $s_2(t)$, with modulated phase and constant amplitudes as described in. Thus

$$s_1(t) = s(t) - e(t); \; s_2(t) = s(t) + e(t); \; |s_1(t)| = |s_2(t)| = r_{max} \tag{11.17}$$

where $e(t)$ is in quadrature to the source signals $s_1(t)$ and $s_2(t)$:

$$e(t) = js(t)\sqrt{\frac{r_{max}^2}{|s(t)|^2} - 1} \tag{11.18}$$

The output is given by:

$$s_{out}(t) = 2Gs(t) \tag{11.19}$$

The quadrature signal $e(t)$ is added to one leg of the forward loop and subtracted from the other leg of the forward loop to give a constant envelope signal as shown in Figure 11.42.

The principal disadvantage with this approach is the generation of two constant envelope signals is complicated and additionally good power combining with low loss and high isolation is very difficult to achieve.

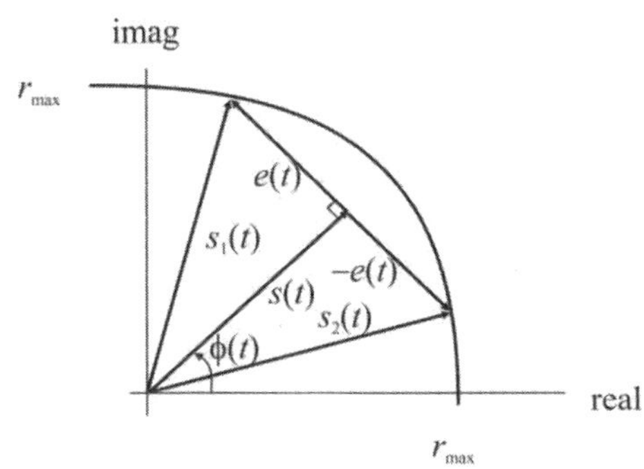

Figure 11.42 Constant envelope signals.

11.7 Load-Pull Techniques and Their Applications in Power Amplifier Design

Matching network design based on two-port parameters as discussed in Chapter 9 assumed that the two-port in question is linear so that parameter superposition principles apply. Such is frequently not the case for power amplifiers so other techniques often must be used. One of the techniques for characterizing generally nonlinear high-power devices is referred to as *load pull*. We discuss such techniques in this section.

11.7.1 Basics of Load-Pull

Essentially a load-pull system consists of a load-tuner, active or passive [23–44], and the controlling mechanism to precisely set the tuner impedance to achieve the desired impedance. Similarly, a source-pull system has a tuner at the input (see Figure 11.43).

Load-pull analysis is used to construct a set of contours (typically on a Smith chart) that determine the maximum power output achievable with a given load impedance. That is, load-pull is a process that enables synthesis of varying impedance environments at the output port of (usually) an active device in applications where transistor device performance needs to be experimentally determined. Load-pull refers specifically to presenting a priori known impedance to the nonlinear device in a precisely controlled fashion in order to determine optimal performance of the device. In the context of PA design, the best loading conditions depend on the distortions and nonlinearity exhibited by the device. These loading conditions are significantly different from the linear case, where the optimum loading conditions are directly identified from the *s*-parameters. Load-pull techniques aid in the identification of optimum loading conditions experimentally, while physically changing load reflection coefficient, Γ_L, as shown in Figure 11.43, for the extraction of design parameters, such as output power, the DC to RF power conversion efficiency, the operating power gain and gain compression, and the

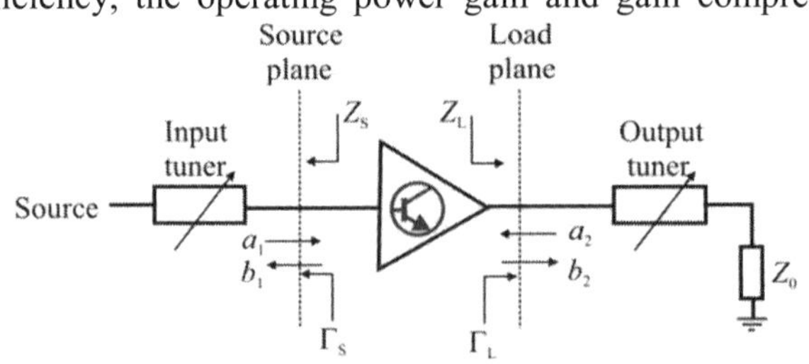

Figure 11.43 Depiction of reflection coefficients at the load and source ports.

power-added efficiency, from transistor devices. The load-pull technique allows for the analysis of active device performance under varying loading conditions and allows for the design of matching circuits. These contours are very useful in assessing the actual impedance a device should see when it is used in an amplifier.

The desired matching impedance, Z_L, the incident and reflected traveling waves, a_2 and b_2, at the output port and reflection coefficient, Γ_L, are related as

$$\Gamma_L = \frac{a_2}{b_2} \tag{11.20}$$

$$\Gamma_L = \frac{Z_L - Z_0}{Z_L + Z_0} \tag{11.21}$$

where Z_0 is the characteristic impedance of the system in which the device is going to be used (typically 50 Ω).

The contours are constructed as follows:

> Initially the closest match at the output is found for the optimum output impedance. This is done by adjusting both the input and the output tuners to get the maximum constant output power. This forms the center of the load-pull loci and is a single impedance (see Figure 11.44).
>
> Next the output impedances are changed and the input tuner is adjusted to provide conjugate matching and thus constant input power. This is repeated as many times as desired and for each constant output power point a set of loci are generated that provide the impedances which provide that power. A few of these loci are shown on Figure 11.44.

Note that each contour represents the maximum output power achievable with a given load impedance. Load-pull contours are useful in determining the actual impedance a device should see when used in an amplifier. They also predict graphically what will happen when the load changes from the optimum load due to any anomalies such as aging and other variations in the impedances at the output (e.g., temperature). They provide a valuable aid in the design and evaluation of power amplifiers.

A precautionary step needs to be taken, however. Load-pull tuners can accurately synthesize a given load impedance; this impedance is only at a known frequency. The harmonic impedances that result from particular settings of the

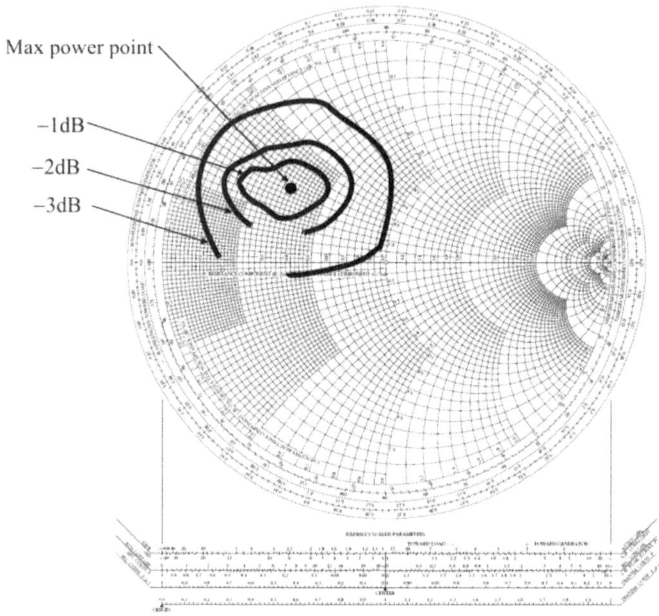

Figure 11.44 Load-pull contours.

tuners are usually not characterized. The efficiency of a device can be strongly dependent on the harmonic impedances, so load-pull characterization needs to be carefully interpreted if contours of constant efficiency are measured instead of constant output power.

11.7.2 Common Load-Pull Techniques

The load-pull techniques can be categorized as either passive or active. In the passive technique, the desired impedance is synthesized by varying the reflection coefficient of the impedance controlling element as depicted in Figure 11.45 [1]. The reflection coefficient is varied by tuning the passive tuner's phase and/or amplitude attached to the output of the device.

The main advantages of the passive technique are:

- Rapid impedance synthesis,
- Relatively higher power-handling capability and the measurements of high-power devices without any non-linear effect,
- Ease of use,

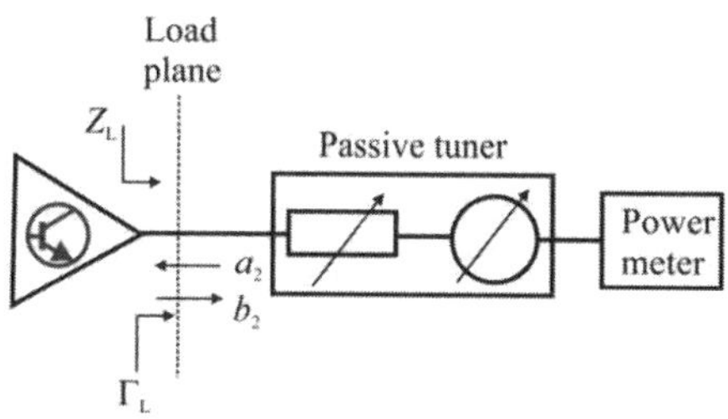

Figure 11.45 Passive load-pull technique.

- Low maintenance cost,
- Relatively low implementation cost,
- Absence of any oscillation.

The main disadvantage of this technique is the limitation of the synthesized impedances in term of the magnitude of the associated reflection coefficient.

A passive load-pull structure can typically synthesize reflection coefficients with a magnitude of up to only 0.85 or so. This limitation becomes critical in the case of a device that requires load impedances near the edge of Smith chart, such as for the characterization of high-power transistor devices.

In all cases the input power is assumed to be sufficient to drive the device to a maximum current or a maximum voltage. The input power is continuously tuned to keep it constant by the passive tuner at the input. The function of the input tuner is to adjust the matching such that the large signal input power is always a constant, even when the output impedance and the output tuner are adjusted.

To overcome this limitation, load-pull systems based on active topologies have been developed. Active load-pull systems consist of either active open-loop or active closed-loop architectures. In both techniques, the reflection coefficient is synthesized at an access plane by injecting a signal. In principle it is done by controlling the complex gain around the active structure.

In the active open-loop technique, shown in Figure 11.46, the reflection coefficient, Γ_L, presented to the active device access, is synthesized by controlling the variable attenuator ATT and phase-shifter in order to fix the magnitude and phase of the traveling wave a_2. Then, the synthesized reflection coefficient depends on ATT, phase and the power delivered by the RF generator. For this reason, in the case of power sweep or during load-pull, the travelling wave, b_2, is not constant so methods must be employed to synthesize a given Γ_L.

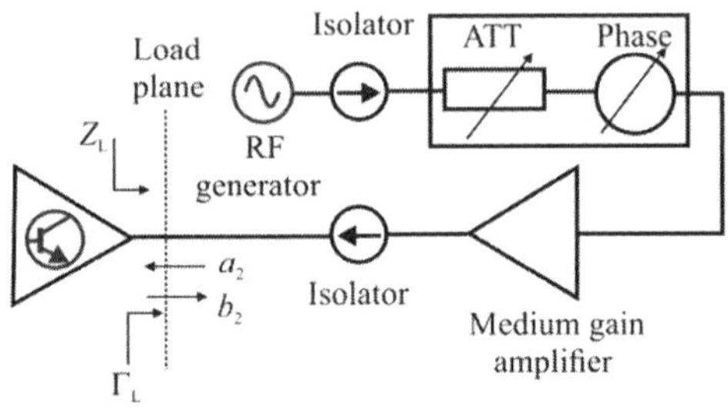

Figure 11.46 Active open-loop load-pull technique.

11.7.3 PA Design

As an example, this section shows the application of harmonic load-pull in the design of an inverse class-F PA. For this purpose, a prematched harmonic load-pull setup was used for characterizing a 10W GaN HEMT from CREE. The transistor was biased below the threshold voltage at $V_{GS} = -3.5$V and $V_{DS} = 34$V and operated at the WiMAX carrier frequency of 2.45 GHz. The harmonic and the fundamental terminations were fixed simultaneously through several iterations. The optimal performance was achieved for an input power (P_{in}) of 24.65 dBm. The results obtained for class F mode optimal PAE are shown in Figure 11.47. We see that the

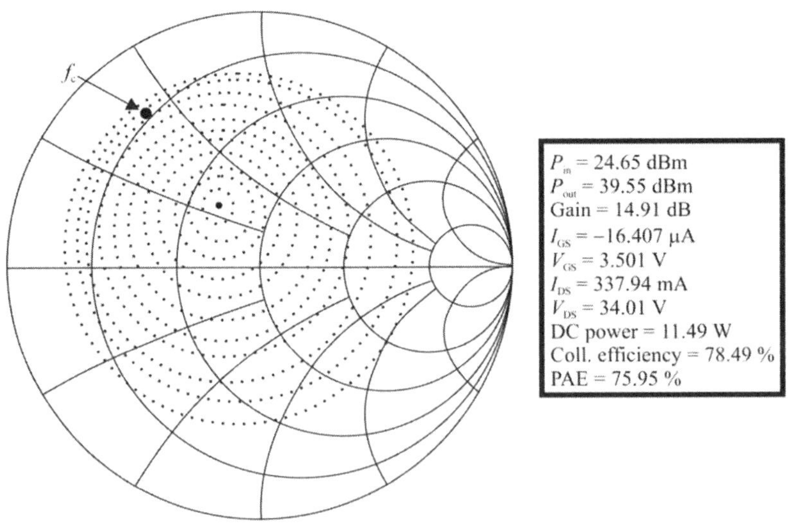

Figure 11.47 Load-pull data. [30].

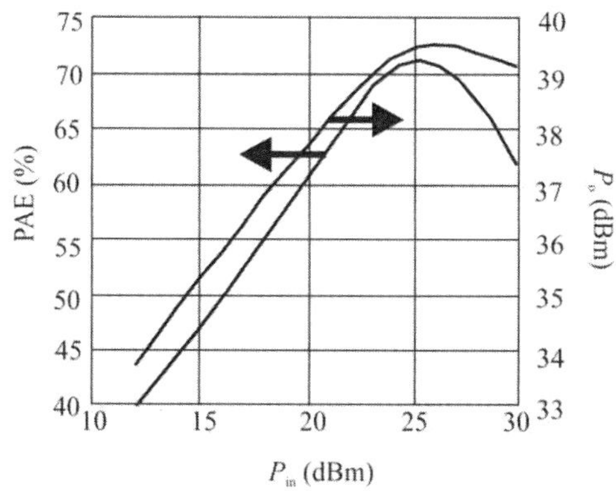

Figure 11.48 P_o and PAE vs P_{in} for example.

measured gain (G) was 11.9 dB and the output power (P_{out}) was 39.55 dBm. At these conditions, the measured drain efficiency was 78.5% and power added efficiency (PAE) was 75.95%. It should be noted that these measured efficiencies were at the transistor package plane without considering the loss of the output matching network.

This class F^{-1} PA was prototyped based on the load-pull data. Measured results for this prototype are given in Figures 11.48 and 11.49 [30]. It is evident from these figures that for a drive power of 24.65 dBm, the PA delivers an output power of 39.4 dBm with a PAE of to 71.2%. These results are very close to the results achieved from the load-pull measurement when losses in the matching networks are considered.

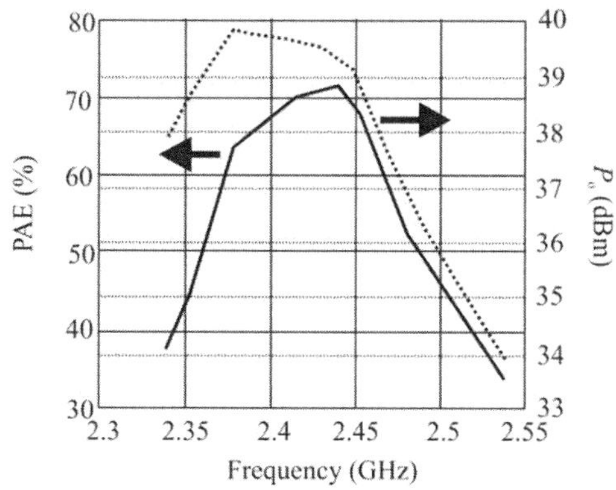

Figure 11.49 P_o and PAE for example.

Table 11.1 Comparison of Load-Pull and Prototype Measurements

Design Parameter	Load Pull Measurements	Prototype Measurements
P_{in} (dBm)	24.65	25
P_{out} (dBm)	39.55	39.08
G (dB)	14.91	14.08
V_{GS} (V)	−3.5	−3.2
I_{DS} (mA)	337.94	428
V_{DS} (V)	34	25
η_D (%)	78.49	75.61
η_{PA} (%)	75.95	72.66
Γ_L @ f_0	$0.745\angle 126.46$	$0.745\angle 126.46$
Γ_L @ $2f_0$	$0.920\angle 122.74$	$0.920\angle 122.74$
Γ_L @ $3f_0$	$0.876\angle 100.01$	$0.876\angle 100.01$
Γ_S @ f_0	$0.883\angle 224.43$	$0.883\angle 224.43$

Table 11.1 gives the comparison between the load-pull characterization data and the measured results of the PA prototype. It is self-explanatory and evident that the measured performance and the load-pull data are in excellent agreement.

11.7.4 Conclusion

In this section we reviewed and discussed common load-pull techniques and systems used in the design of PAs. The on-wafer devices require special care and it has been highlighted here. Finally, an example of a class F^{-1} PA design was presented to show the usefulness of load-pull techniques.

11.8 Concluding Remarks

It should be clear at this point that the primary reason for considering linearity issues in power amplifiers is the amount of energy spillover into adjacent or other close-by channels that wastes power and can, and does, cause unintended interference. Often, closely located friendly communication networks are attempting to use these channels and the jamming hinders this.

There are some methods that can be engaged to minimize this interference, and we discussed some of them in this chapter. There is nothing available, however, that can totally eliminate cochannel interference caused by EW activities. It should be noted that simply communicating in a channel interferes so some degree in the adjacent channels; it's a matter of this degree. Normally EW systems employ considerably more emitted power than communication devices.

11.8.1 Linearization Techniques Summary

A summary of the weaknesses (W) and strengths (S) of the linearity approaches discussed in the chapter is as follows:

- Power backoff: simplicity (S); low power efficiency (W).
- Predistortion: conceptually clear (S); requires good PA model (W).
- Adaptive predistortion: No drift problem (S); introduces complexity (W).
- Feedforward: No stability worries (S); matching and drift concerns (W).
- Feedback: Stability concerns (W)
- EER: Possibly high-efficiency (S); adds another power amp problem (W).
- LINC: Conceptually appealing (S); fundamental implementation issues (W).
- CFB: simplicity (S), robust to poor PA model (SS); stability concerns (W).

References

[1] Kenington, P. B., *High Linearity RF Amplifier Design*, Norwood, MA: Artech House, 2000, p. 34.
[2] Kenington, P. B., *High-Linearity RF Amplifier Design,* Norwood, MA: Artech House, 2000.
[3] Fuller, L., https://people.rit.edu/lffeee/mosfet_l.pdf, accessed March 2009.
[4] Saraswat, http://web.stanford.edu/class/ee316/MOSFET_Handout5.pdf, accessed April 2010.
[5] McNeilage, C., E. N. Ivano, P. R. Stockwell, and J. H. Searls, "Review of Feedback and Feedforward Noise Reduction Techniques," *IEEE Frequency Control Symposium,* 1998.
[6] Black, H. S., U.S. Patent 1,686,792, Oct. 9, 1929.
[7] Maclean, D. J. H., "Broadband Feedback Amplifiers," Research Studies Press, Chichester, 1982.
[8] Black, H. S., "Stabilized Feedback Amplifiers," *Bell System Technical Journal*, Vol. 13, No. 1, 1934, pp. 1–18.
[9] Jurgen, R. K., "Feedforward Correction: A Late Blooming Design," *IEEE Spectrum*, April 1972, pp. 41–43.
[10] Cardinal, J-S., and F. M. Ghannouchi, "A New Adaptive Double Envelop Feedback (ADEF) Linearizer for Mobile Radio Power Amplifiers," *Proceedings IEEE Microwave Theory and Technics Symposium,* 1994, pp. 573–576.
[11] Seidel, H., "A Feedforward Experiment Applied to an L-4 Carrier System Amplifier," *IEEE Transactions on Communication Techniques,* Vol. 19, No. 3, 1971, pp. 320–325.
[12] Ivanov, E. N., M. E. Tobar, and R. A. Woode, "Tunable Microwave Oscillator for Low Phase Noise Applications," *Proceedings IEEE International Frequency Control Symposium,* 1997, pp. 985–993.
[13] Bennett, T. J., and R. F. Clements, "Feedforward: An Alternative Approach to Amplifier Linearization," *Radio and Electrical Engineering,* Vol. 44, No. 5, 1974, pp. 257–262.
[14] Kenington, P. B., M. A. Beach, A. Bateman, and J. P. McGeehan, U.S. Patent 5,334,946, issued 2 Aug. 1994.
[15] Seidel, H., H., H. R. Beurrier, and A. N. Friedman, "Error-Controlled High-Power Linear Amplifiers at VHF," *Bell System Technical Journal*, Vol. 47, No. 5, 1968, pp. 651–722.
[16] Seidel H., 1971), "A Microwave Feed-Forward Experiment," *Bell System Technical Journal*, Vol. 50, No. 9, 1971, pp. 2879–2916.

[17] Driscoll, M. M., and R. W. Weinert (1995), "Spectral Performance of Sapphire Dielectric Resonator Controlled Oscillators Operating in the 80K to 275K Temperature Range," *Proceedings IEEE International Frequency Control Symposium,* 1995, pp. 401–412.

[18] Halford, D., A. E. Wainwright, and J. A. Barnes, "Flicker Noise of Phase in RF Amplifiers and Frequency Multipliers: Characterization, Cause and Cure," *Proceedings IEEE Frequency Control Symposium,* 1968, pp. 340–341.

[19] Andressen, S. G., and J. K. Nesheim, "Phase Noise of Various Frequency Doublers," *IEEE Transaction on Instrumentation Measurements*, Vol. 22, June 1973, pp. 185–188.

[20] Walls, F. L., E. S. Ferre-Pikal, and S. R. Jefferts, "The Origin of 1/f AM and PM Noise in Bipolar Junction Transistor Amplifiers," *Proceeding of the IEEE 1995 International Frequency Control Symposium*, 1995, pp. 294–304.

[21] Dawson, J. L., and T. H. Lee, "Cartesian Feedback for RF Power Amplifier Linearization," *Proceeding 2004 American Control Conference*, Boston, MA, June 30 - July 2, 2004, pp. 361–366.

[22] Patel, J., "Adaptive Digital Predistortion Linearizer for Power Amplifiers in Military UHF Satellite," MSEE Thesis, University of South Florida, Scholar Commons, FL, March 29, 2004.

[23] Verspecht, J., J. Horn, L. Betts, D. Gunyan, R. Pollard, C. Gillease, and D. E. Root, "Extension of X-Parameters to Include Long Term Dynamic Memory Effects," *Proceedings IEEE MTT-S IMS Digitization,* Boston, MA, June 2009, pp. 741–744.

[24] Ghannouchi, F. M., M. S. Hashmi, S. Bensmida, and M. Helaoui, "Loop Enhanced Source- and Load-Pull Technique for High Reflection Factor Synthesis," *IEEE Transactions on Microwave Theory and Techniques,* Vol. 58, No. 11, Nov. 2011, pp. 2952–2959.

[25] Hashmi, M. S., A. L. Clarke, S. P. Woodington, J. Lees, J. Benedikt, and P. J. Tasker, "An Accurate Calibrateable Multi-harmonic Active Load-Pull System based on the Envelope Load-Pull Concept," *IEEE Transactions on Microwave Theory and Techniques,* Vol. 58, No. 3, March 2011, pp. 656–664.

[26] Stancliff, R. B. and D. P. Poulin, "Harmonic Load Pull," *IEEE MTT-S Symposium Digest,* June 1979, pp. 185–187.

[27] Di-Luan, L., and F. M. Ghannouchi, "Multi-Tone Characterization and Design of FET Resistive Mixers Based on Combined Active Source-pull/Load-pull Techniques," *IEEE Transactions on Microwave Theory and Techniques*, Vol. 46, No. 9, Sep. 1998, pp. 1201–12011.

[28] Takayama, T., "A New Load-Pull Characterization Method for Microwave Power Transistors," *IEEE MTT-S Symposium Digest*, June 1976, pp. 218–220.

[29] Bava, G. P., U. Pisani, and V. Pozzolo, "Active Load Technique for Load-Pull Characterization at Microwave Frequencies," *IEEE Electronic Letters,* Vol. 18, No. 4, 18 February 1982, pp. 178–180.

[30] Hashmi, M. S., A. L. Clarke, S. P. Woodington, J. Lees, J. Benedikt, and P. J. Tasker, "Electronic Multi-Harmonic Load-Pull System for Experimentally Driven Power Amplifier Design Optimization," *IEEE MTT-S Symposium Digest*, June 2009, pp. 1549–1552.

[31] Ghannouchi, F. M., F. Beauregard and A. B. Kouki, "Power Added Efficiency and Gain Improvement in MESFET Amplifiers Using an Active Harmonic Loading Technique," *Microwave and Optical Technology Letters,* Vol. 7, No. 13, Sep. 1994, pp. 625–627.

[32] Teppati, V., A. Ferrero, and U. Pisani, "Recent Advances in Realtime Load-Pull Systems," *IEEE Transactions on Instrumentation and Measurements,* Vol. 57, No. 11, Nov. 2011.

[33] Bouysse, P., J. Nebus, J. Coupat, and J. Villotte, "A Novel Accurate Load Pull Setup Allowing the Characterization of Highly Mismatched Power Transistors," *IEEE Transactions on Microwave Theory and Techniques,* Vol. 42, No. 2, Feb. 1994, pp. 327–332.

[34] Ghannouchi, F. M., Z. Guoxiang, and F. Beauregard, "Simultaneous Load-Pull of Intermodulation and Output Power under Two-Tone Excitation for Accurate Sspa's Design," *IEEE Transactions on Microwave Theory and Techniques,* Vol. 42, No. 6, June 1994, pp. 929–934.

[35] Maury Microwave Corporation, "Device Characterization with Harmonic Load and Source Pull," Application Note: 5C-044, December 2000.

[36] Focus Microwave, "Load Pull Measurements on Transistors with Harmonic Impedance Control," Technical Note, August 1999.

[37] M. S. Hashmi, A. L. Clarke, J. Lees, M. Helaoui, P. J. Tasker, and F. M. Ghannouchi, "Agile Harmonic Envelope Load-Pull System Enabling Reliable and Rapid Device Characterization," *IOP Journal of Measurement Science and Technology,* April 2010, Vol. 21, 055109.
[38] Colantonio, P., A. Ferrero, F. Giannini, E. Limiti, and V. Teppati, "Harmonic Load/Source Pull Strategies for High Efficiency PAs Design," *IEEE MTT-S Symposium Digest*, June 2003, pp. 1807–1810.
[39] Spirito, M., M. J. Plek, F. van Rijs, S. J. C. H. Theeuwen, D. Hartskeerl, and L. C. N. de Vreede, "Active Harmonic Load-Pull for On-Wafer Out-of-Band Device Linearity Optimization," *IEEE Transactions on Microwave Theory and Techniques*, Vol. 54, No. 12, Dec. 2006.
[40] Hashmi, M. S., F. M. Ghannouchi, P. J. Tasker, and K. Rawat, "Highly Reflective Load-Pull," *IEEE Microwave Magazine*, June 2011.
[41] Mueller, J. E., and B. Gyselinckx, "Comparison of Active versus -passive On-Wafer Load-Pull Characterization of Microwave MMWave Power Devices," *IEEE MTT-S Symposium Digest,* June 1994, pp.1077–1080.
[42] Focus Microwave, "Comparing Harmonic Load-Pull Techniques with Regards to Power Added Efficiency," Application Note, May 2007.
[43] Simpson, G., "Harmonic Load-Pull with High Gamma Tuners," *Microwave Journal*, Vol. 51, No. 5, May 2011, p. 232.
[44] De Groote, F., J. Verspecht, D. Barataud, and J-P. Teyssier, "An Improved Coupling Method for Time Domain Load-Pull Measurements," *IEEE European Microwave Conference*, Paris, France, Oct. 2005, p. 4.

Chapter 12

Thermal Management of RF Power Amplifiers

12.1 Introduction

Heat has long been an issue for system designers, but the problem has recently become severe [1]. Designers must view thermal management not as an afterthought but as an issue they must deal with from the beginning of the design process. Heat's continuous flow from one material to another creates a temperature gradient across those materials.

There are two primary areas where heat management must be addressed in an EW system: (a) in the power supplies of the modules, and particularly the HPA power supply and (b) the HPA itself, and especially the output stage(s).

Designing a thermal-management system requires the evaluation of the means for moving the heat and the manner and location in which the design rejects the heat early in the design cycle to avoid causing severe problems at the system level. This consideration is important because the thermal operating range—that is, the temperatures the system can tolerate—is more limited, and any approach that may be employed at the system level is likely to be more expensive than one implemented at the chip/component level.

In this chapter we consider the ways that heat is removed from electronic devices in EW systems. This can be accomplished in four ways [2]:

- *Conduction* - The amplifier baseplate is mounted to the heat sink. The heat sink must be designed such that the maximum specified baseplate temperature is not exceeded, normally +71°C.
- *Convection* - The amplifier contains a very large finned heat sink. The surface area is large enough to dissipate the heat in a still air environment.

- *Forced Air* - Cooling air is directed through a finned heat sink in the amplifier.
- *Liquid* - The cooling fluid is pumped through a chill plate in the amplifier. The cooling heat exchanger is external to the amplifier.

All of these approaches require a heat sink thermally attached to the device. Therefore, we examine the characteristics of heat sinks in this chapter, and review their ability to remove heat.

12.1.1 Passive versus Active Cooling

With the exception of fans, which usually work in conjunction with heat sinks, most of today's thermal-management systems are passive types. Conduction-based thermal-management systems, such as thermal-interface materials, improve the flow of heat from one location to another, greatly enhancing the efficiency of the overall thermal-management system. However, convection-based systems have the drawback of allowing heat to flow in an uncontrolled manner from one level to the next. These systems have served the field well but have the drawback of removing not only the heat that is limiting device performance but also any of the heat from the surrounding area, which more than likely is not limiting the device or system performance.

One approach to this problem is to use an active device inside the electronic package for localized thermal management. To reduce the cooling necessary at the system and building levels, however, we must reduce the amount of heat you extract from the die. Cooling the die generally keeps their operating frequency near its peak. However, the temperature within one of the hot spots on the die and not the temperature across the entire die typically limits this peak frequency. Instead of extracting the heat from the die as a whole, we could extract the heat only from the hot spot. In this way, we are dealing with a smaller system-level problem and subsequently have a smaller thermal management problem.

Diamond and graphite foam have emerged as very promising materials for managing heat removal in electronic devices and systems [3]. We briefly discuss these approaches.

This chapter is structured as follows. First, we examine heat flow in electrical circuits and its properties. We then enter into a discussion of heat sinks, particularly for active power devices, where we include a description of a selection procedure. That is followed by a examination of the thermal properties of PCBs. A few of the newer materials used for heat removal are discussed next—in particular graphic foams and diamond composites. Next, we discuss thermoelectric cooling approaches for heat removal. The last topic presented is a discussion of enclosure fans, their properties, and selection procedures.

12.2 Heat Flow

The heat equation is an important partial differential equation describing the distribution of heat, or variation in temperature, in a given region over time. For a function $u(x, y, z, t)$, which is a measurement of the temperature, *T*, of three spatial variables—*x*, *y*, and *z*—and the time variable, *t*, the heat equation is

$$\frac{\partial \vec{u}}{\partial t} = \alpha \nabla^2 \vec{u} \tag{12.1}$$

where α is the *thermal diffusivity* of material. The heat equation describes the physical distribution and time variation of the temperature through a material. Expanding (12.1) we get

$$\frac{\partial \vec{u}}{\partial t} - \alpha \left(\frac{\partial^2 u}{\partial x^2} + \frac{\partial^2 u}{\partial y^2} + \frac{\partial^2 u}{\partial z^2} \right) = 0 \tag{12.2}$$

α is given by the thermal conductivity, denoted by λ, divided by density and specific heat capacity at constant pressure. That is,

$$\alpha = \frac{\lambda}{\rho c_p} \tag{12.3}$$

where

λ is thermal conductivity (W/m/K)
ρ is density (kg/m^3)
c_p is specific heat capacity (J/kg/K)

It indicates the ability of a material to conduct thermal energy relative to its ability to store thermal energy and it has the unit of m^2/s. Taken together, ρc_p can be considered the volumetric heat capacity [J/(m^3/K)].

As we see in the heat equation, (12.1), thermal diffusivity is the ratio of the time derivative of temperature to its curvature, quantifying the rate at which temperature concavity is "smoothed out." In a sense, thermal diffusivity is the measure of thermal inertia. In a substance with high thermal diffusivity, heat moves rapidly through it because the substance conducts heat quickly relative to its volumetric heat capacity or "thermal bulk."

The other important governing equation for heat flow is

$$Q = U + W \tag{12.4}$$

where Q is heat flow, U is the change in the internal energy of the system, and W is work on the system. That is, heat flow is equal to the work on the system plus the change in internal energy of the system. Hence, the change in heat is equivalent to the heat flowing into the system. Combining these two equations, and in the absence of any work, U is proportional to the change in temperature. Thus, we arrive at $Q = \Delta T$, where ΔT is the change in temperature. For passive systems, cooling by the conduction of heat is a linear function of temperature and a constant related to the material properties of the solid. This constant, λ, may be a function of many variables, including temperature, power, and voltage.

The primary systems for passive cooling of electronic systems are thermal interface materials, heat spreaders, and heat sinks. Each performs a different function for removing heat from a system. Heat sinks may be an environment, such as water or air, or an object that absorbs and then dissipates heat while in physical or thermal contact. This dissipation may occur through direct or radiant transfer of heat. Heat-sink performance is a function of material, geometry, and the overall surface-heat-transfer coefficient along with the temperature of the heat sink. Generally, we can improve forced-convection heat-sink thermal performance by increasing the thermal conductivity of the heat-sink materials and increasing the surface area.

Thermal-interface material fills the gaps between thermal-transfer surfaces, such as microprocessors and heat sinks, to increase thermal-transfer efficiency. Air, a poor conductor, normally fills these gaps. The most common thermal-interface material is white paste or thermal grease—typically, silicone oil encapsulating aluminum oxide, zinc oxide, or boron nitride. Heat spreaders are most often simply metal plates having high thermal conductivity. Designers also use carbon-based heat spreaders with anisotropic characteristics. They act as heat exchangers, moving heat between localized heat and a secondary, larger heat exchanger. Heat moves through all these passive components only with a temperature difference from a higher to a lower temperature. The rate of flow is proportional to the difference in temperature. Both active and passive approaches can cause or assist in this flow.

In Figure 12.1 a region is illustrated that has two different temperatures on each side. The distance of heat transfer is defined as Δx, which is perpendicular to area A. The rate of heat transferred through the material is Q, from temperature T_1 to temperature T_2, when $T_1 > T_2$. Thermal conductivity, λ, is a bulk property that describes the ability of a material to transfer heat. It is given by

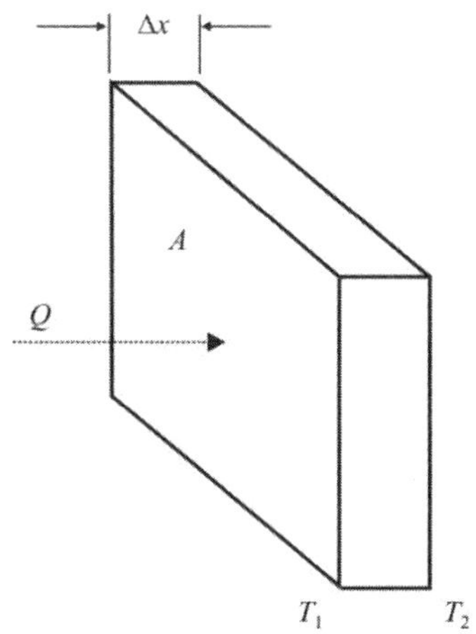

Figure 12.1 Conduction heat transfer process from hot (T_1) to cold (T_2) surfaces.

$$\lambda = \frac{Q\Delta x}{A(T_2 - T_1)} \tag{12.5}$$

Thermal conductivity of materials plays a significant role in the cooling of electronics equipment. From the die where the heat is generated to the cabinet where the electronics are housed, conduction heat transfer and, subsequently, thermal conductivity are the integral components of the overall thermal management process.

The path of heat from the die to the outside environment is a complicated process that must be understood when designing a thermal solution. In the past, many devices were able to operate without requiring an external cooling device like a heat sink. In these devices, the conduction resistance from the die to the board needed to be optimized, as the primary heat transfer path was into the PCB. As power levels increased, heat transfer solely into the board became inadequate. Much of the heat is now dissipated directly into the environment through the top surface of the component. In these new higher-powered devices, low junction-to-case resistance is important, as is the design of the attached heat sink.

To determine the importance of material thermal conductivity in a specific thermal management application (e.g. a heat sink), it is important to separate the overall thermal resistance associated with conduction heat transfer into three parts: interfacial, spreading, and conduction resistances.

An interface material enhances the thermal contact between imperfect mating surfaces. A highly thermally conductive material, with good surface wetting ability, will reduce *interfacial resistance.*

Spreading resistance is used to describe the thermal resistance associated with a small heat source coupled to a larger heat sink. Among other factors, the thermal conductivity of the base of the heat sink directly impacts *spreading resistance.*

Conduction resistance is a measure of the internal thermal resistance in a heat sink as heat travels from the base to the fins, where it dissipates into the environment. In regard to heat sink design, conduction resistance is less important in natural convection and low air flow conditions, becoming more important as flow rates increase.

Common units of thermal conductivity are W/m/K and Btu/hr-ft-°F.

12.3 Heat Removal from Power Amplifiers

A PA can be modeled with the thermal model depicted in Figure 12.2. P_d represents the power dissipated by the device (Watts). T is the temperature of the heat sink (°C) and θ_t represents the thermal resistance of the heat sink (°C/W). C_t is the thermal capacitance of the transistor. We can write the branch equation for the thermal resistance as

$$T_2 - T_1 = P_d\theta_t \tag{12.6}$$

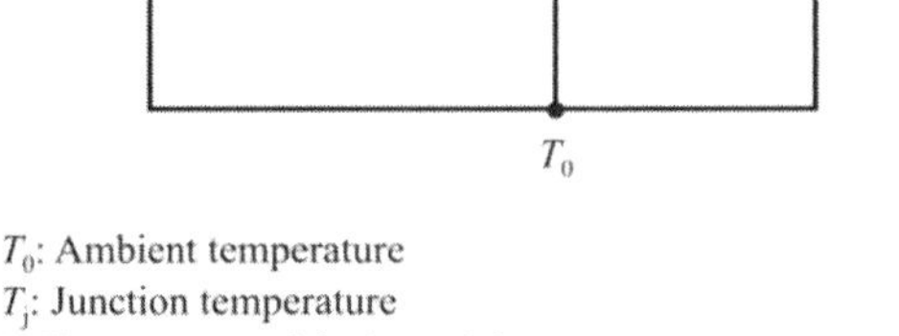

T_0: Ambient temperature
T_j: Junction temperature
T: Temperature of the heat sink
θ_j: Junction resistance between the transistor and heat sink (rather large)
θ_t: Thermal resistance of the heat sink
C_t: Thermal capacitance of the transistor

Figure 12.2 Thermal model.

so that

$$\theta_{\mathrm{t}} = \frac{T_2 - T_1}{P_{\mathrm{d}}} \tag{12.7}$$

The branch equation for the thermal capacitance is

$$P_{\mathrm{d}} = C_{\mathrm{t}} \frac{d}{dt}(T - T_0) \tag{12.8}$$

We know that

$$\sum P_{\mathrm{d}_i} = 0 \tag{12.9}$$

and that

$$\sum T_i = 0 \tag{12.10}$$

Consider the power flow depicted in Figure 12.3.

$$P_{\mathrm{d}} = \frac{T - T_{\mathrm{j}}}{\theta_{\mathrm{j}}}$$

implies that

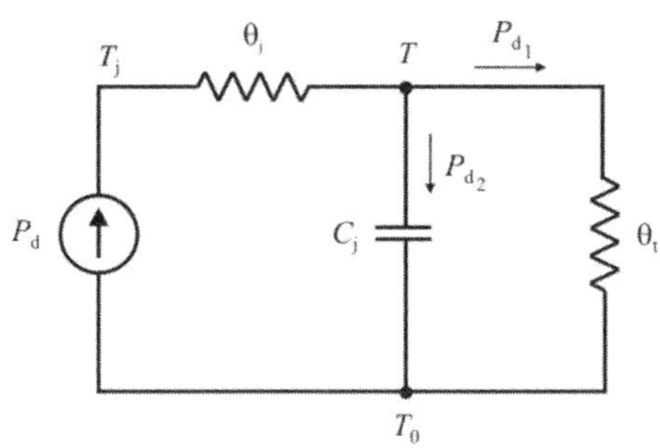

Figure 12.3 Thermal model.

$$P_{d_1} = \frac{T - T_0}{\theta_t}$$

and

$$P_{d_2} = C_t \frac{d}{dt}(T - T_0) = C_t \frac{dT}{dt} = C_t \dot{T}$$

We can also write

$$P_d = \frac{T - T_0}{\theta_t} + C_t \dot{T} \tag{12.11}$$

or

$$T(t) + \theta_t C_t \dot{T}(t) = P_d \theta_t + T_0 \tag{12.12}$$

Assuming a unit step input, the homogeneous solution to (12.12) is

$$T_h(t) = Ae^{-t/\tau} \tag{12.13}$$

with

$$\tau = \theta_t C_t \tag{12.14}$$

and we have

$$T(t) = Ae^{-t/\tau} + B \tag{12.15}$$

To determine A and B in (12.15), we consider the boundary conditions. At $t \to \infty$, $P_{d_2} \to 0$. Therefore

$$\begin{aligned} T_\infty &= P_d \theta_t + T_0 \\ B = T_\infty &= \theta_d R_t + T_0 \end{aligned} \tag{12.16}$$

The initial conditions are

$$T(0) = T_0 = A + B$$

so that

$$A = T_0 - T_\infty = -P_d\theta_t \tag{12.17}$$

Thus

$$T(t) = P_d\theta_t(1 - e^{-t/C_t\theta_t}) + T_0 \tag{12.18}$$

and

$$T_j(t) = T(t) + \theta_j P_d \tag{12.19}$$

Equation (12.19) is sketched in Figure 12.4. The response is a monotonically increasing exponential with an initial value given by T_0 and a final value given by T_∞ (12.16).

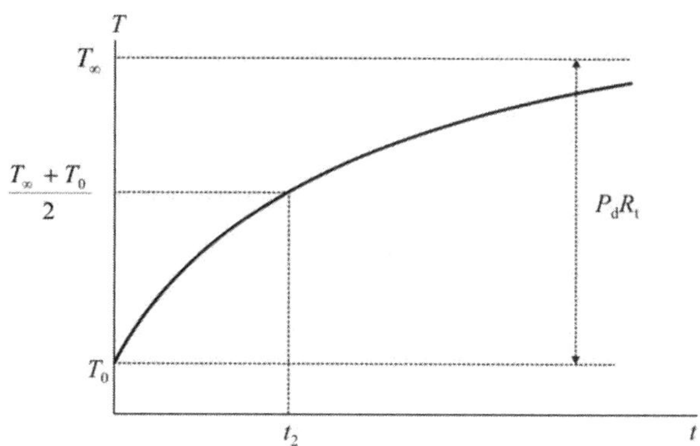

Figure 12.4 Thermal response.

12.4 Thermal Analysis and Design in Electronic Circuits

12.4.1 Introduction

This section briefly discusses the issues involved with heat generation and removal in electronic components. Instructions are given for selecting an appropriate heat sink when required [4].

12.4.2 Heat Removal in Resistors

A typical resistor has a maximum power rating given when the ambient temperature is 25°C. The allowable power dissipated has to be reduced when the ambient temperature is higher. At the maximum operating temperature (typically between 100 and 150°C) the allowable power dissipation becomes zero. For temperatures in between the power is linearly derated in proportion to temperature. This is illustrated in Figure 12.5.

It might be inferred from this discussion that the allowable power dissipation increases when the ambient temperature is less than 25°C. This is not generally true and depends more on allowable internal current densities than absolute temperatures. Do not make assumptions –always check information from the manufacturer about operation the part in this mode. As an example, consider a common ¼ watt carbon film resistor that can dissipate up to ¼ watt when the ambient temperature is 25°C and has a maximum operating temperature of 125°C. For each degree that the ambient temperature is above 25°C the maximum power rating of the resistor decreases by 0.0025 W. Thus, if the maximum ambient temperature is 60°C then the maximum allowable power dissipation of the resistor is decreased to 0.16 W. One thing to keep in mind is that when dissipating maximum

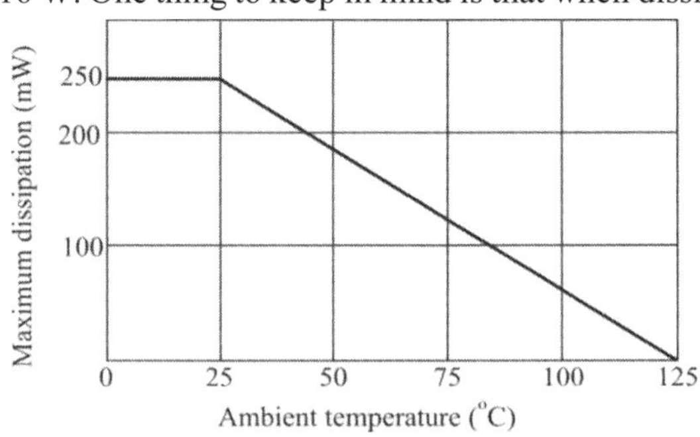

Figure 12.5 Power derating plot for a ¼ W resistor.

allowable power, the part is extremely hot and will quickly burn a finger that touches it. For example, a ¼ watt resistor dissipating its maximum power will have a surface temperature well over 100°C!

It is common practice to design for a power dissipation of no more than about one-half the rated power dissipation at the maximum operating temperature. This practice significantly reduces the maximum operating temperature and extends the useful life of the component considerably. In the previous example, the resistor would not be made to dissipate more than about 0.08 W.

The first thing to do in the design is to determine the actual maximum power the resistor will dissipate and to know the maximum design temperature (typically 60°C but could be over 100°C). Then the appropriate power rating for the resistor is selected so that the maximum power dissipated is about half the rated power dissipation at the maximum operating temperature. Note that high power resistors are intended to be mounted on a suitable heat sink. A 100 W resistor will seriously overheat if not properly mounted. Always study the manufacture's recommended methods for heat removal.

12.4.3 Heat Removal in Solid-State Devices

Managing power dissipation in semiconductor components is the same as for resistors. The most important operating parameter is the junction temperature. If the junction temperature is too high the part will not work properly and may even be destroyed. The typical maximum junction temperature is 125°C although in some special cases the temperature may be as high as 175°C. As with resistors, the useful operational life can be extended considerably if the maximum junction temperature is limited to much less than the rated value. A good value is from 25 to 50°C less than rated.

The case of many semiconductor components is designed so that a suitable heat sink can be mounted. The allowable amount of power that can be dissipated depends on the ability of the heat sink to conduct the heat away. Because the manufacturer cannot know what heat sink you are going to use, the manufacturer rates the power dissipation in terms of an infinite heat sink that holds the case temperature at 25°C. This does not mean that the power dissipation could go to infinity as there is the internal thermal resistance to the case. Power dissipation must still be limited so that the junction temperature is no higher than the maximum rated.

The "Ohm's law" of temperature is: the junction temperature is equal to the power dissipated multiplied by the thermal resistance to ambient plus the ambient temperature. In equation form this is as follows

$$T_{\mathrm{j}} = P_{\mathrm{d}}\theta_{\mathrm{ja}} + T_{\mathrm{a}} \tag{12.20}$$

Temperature is in degrees C, power dissipation is in Watts, and thermal resistance is in Watts per degree C. Note that temperature is analogous to electrical voltage, power is analogous to electrical current, and thermal resistance is analogous to electrical resistance. The symbol, *Q*, is often used in thermal calculations to represent heat flow, i.e. power (not to be confused with the use of *Q* as an indicator of the quality of reactive components, or the quiescent operating point of an amplifier circuit).

Equation (12.20) can also be expressed as

$$\theta_{\text{jamax}} = (T_{\text{jmax}} - T_{\text{amax}}) / P_{\text{dmax}} \tag{12.21}$$

$$P_{\text{dmax}} = (T_{\text{jmax}} - T_{\text{amax}}) / \theta_{\text{ja}} \tag{12.22}$$

$$T_{\text{amax}} = T_{\text{jmax}} - P_{\text{dmax}} \theta_{\text{ja}} \tag{12.23}$$

Equation (12.20) is used for analyzing an existing system. Equation (12.21) is used for designing the heat sink system. Equations (12.22) and (12.23) are useful to establish design limits.

The thermal resistance from the junction to ambient is generally composed of two or three components. In the situation of when no heat sink is used, the total thermal resistance is the sum of the junction to case thermal resistance and the case to ambient thermal resistance as shown below.

$$\theta_{\text{ja}} = \theta_{\text{jc}} + \theta_{\text{ca}} \tag{12.24}$$

The value of a heat sink is that it reduces the thermal resistance from the case to ambient. When a heat sink is used the total thermal resistance is the sum of the internal junction to case thermal resistance plus the thermal resistance from the case to the heat sink plus the thermal resistance of the heat sink to ambient.

$$\theta_{\text{ja}} = \theta_{\text{jc}} + \theta_{\text{cs}} + \theta_{\text{sa}} \tag{12.25}$$

The thermal resistance from the case to the sink is made small—typically less than 1°C/W)—by using what is known as heat sink compound which is thermally conductive and fills the inherent air gaps. Without heat sink compound the thermal resistance between the case and the sink can be several times larger.

Table 12.1 shows some typical values of power dissipation and thermal resistances for common transistor cases. Always check the manufacturer's data for a specific part as these numbers can vary a lot depending on internal construction. For example, integrated circuits tend to have much higher thermal resistances from

Table 12.1 Typical Semiconductor Mounting Case Thermal Data (see Figure 12.6)

	P_{dmax}, (W)			θ (°C/W)			
Case	**Infinite HS***	**Good HS**	**No HS**	θ_{jc}	θ_{ca}	θ_{cs}	θ_{sa}
TO-3	100	60	5	0.5	18	0.5	0.5
TO-220	25	15	2	3	45	1	2
TO-5	4	2.5	1	20	75	5	15
TO-18	2	1.2	0.5	40	160	10	30
TO-92	1	0.6	0.33	60	240	40	60

*HS = heat sink

the junction to the case than plain transistors. The cases listed in Table 12.1 are illustrated in Figure 12.6.

A common mistake is to choose a part with a higher power rating than is needed and then assume that all is well. The power rating of *all* components is based on proper methods being applied to remove the heat. Fail to remove the heat and the component will malfunction or be destroyed. Pay close attention to the manufacture's application information for parts that must dissipate heat. In some cases, the component leads form part of the heat sinking system and must not be too short.

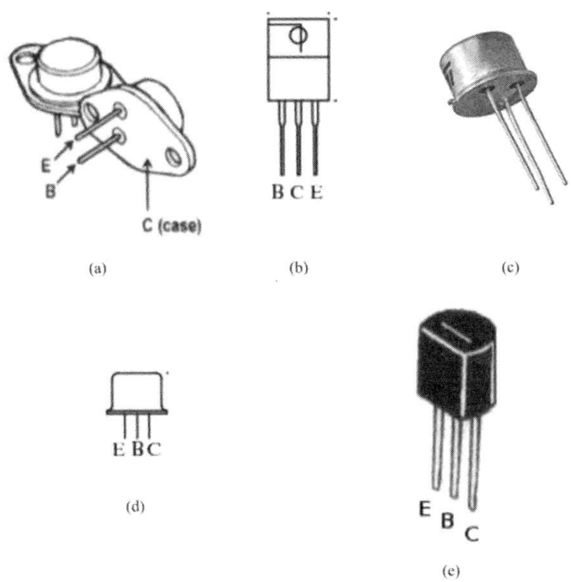

Figure 12.6 Transistor cases (a) TO3, (b) TO220, (c) TO5, (d) TO18, and (e) TO92.

12.5 Heat Sinks

The amount of heat that a power semiconductor or module generates depends upon its mode of operation, bias level, signal amplitude, and the shape of the applied signal. For example, power transistors used as Class A amplifiers for sine or complex waves or pass transistors and modules for regulating dc power supplies generate much more heat than square-wave oscillators, Class AB amplifiers, Class B square-wave amplifiers, and PWM (pulse width modulation) amplifiers. This is perhaps easiest to see when considering a power transistor that operates with square waves, the collector voltage is high with low current, then high current with low voltage during the saturation and cutoff periods. The result is very low average power dissipation. But, in either case, these devices need some type of heat sink to help dissipate the internal power they generate.

A silicon power transistor or high-current diode has an allowable maximum junction temperature of usually 150 °C or a little more. If the heat-generating device does not connect to a greater mass or surface area, the amount of junction current and voltage as given in a typical specification sheet could hardly be approached before the device would exceed the maximum collector junction temperature. The reason is that the device specification sheets list their characteristics with the device mounted on a heat sink of considerable size, usually called an "infinite" heat sink.

The temperature of the air surrounding the heat-generating device must be much less than the maximum allowable junction temperature so the device does not overheat. The heat sink attempts to reduce the junction temperature to that of the surrounding medium. If the heat sink were thermally perfect, the transistor junction temperature could reach the same as the lower ambient. However, in practice this is not possible. The thermal conductivity path from the transistor junction to the ambient air contains physical connections between the junction and case, case and heat sink (through an insulator, if used), and heat sink and ambient. These connections are not thermally perfect and therefore produce a temperature differential. They are thermal resistances, and each thermal resistance has a unique value. The value is expressed as a temperature in degrees centigrade per watt of dissipation (°C/W). Due to these resistances, a temperature differential always appears between the collector junction and ambient. This variable must be kept to a minimum.

The main requirements of heat sinks, formulated possibly not in the order of descending significance, are:

- High thermal conductivity;
- Low cost;
- Low electrical conductivity;

- Thermal expansion coefficient of the heat sink should be equal to that of the material to be cooled;
- Low loss tangent;
- Small weight.

The purpose of a heat sink is to provide a low thermal resistance from the case of a component to the ambient air. Heat is both conducted to the air and radiated. A heat sink is constructed with a high thermal conductivity material such as aluminum and is made so as to have a large surface area for efficient heat transfer to the surrounding air. Heat sinks are often black anodized to increase thermal radiation. In the absence of forced air, convective air currents will conduct thermal energy away from the heat sink. Thus, it is very important for heat sinks to be physically mounted such that convective air currents can exist. A heat sink in a closed box is of little use.

Thermal energy is never destroyed. It always accumulates (i.e. temperature rises) when it cannot dissipate into the surroundings. Fans are often used to improve thermal transfer. Heat sinks are typically rated in terms of temperature rise for a given power and given air flow. This data is generally provided as a plot. It is wrong to ask the question, "What size heat sink will get rid of 5 watts?" The answer could range from small to huge depending on what is actually needed. Figure 12.7 is a typical plot used to present heat sink data. Note that there are two curves and that each curve uses a different set of axes as shown. One curve shows the temperature rise for a given power dissipation in free convection. This is the curve that is most commonly used. When forced air flow is needed, the other curve shows the effective

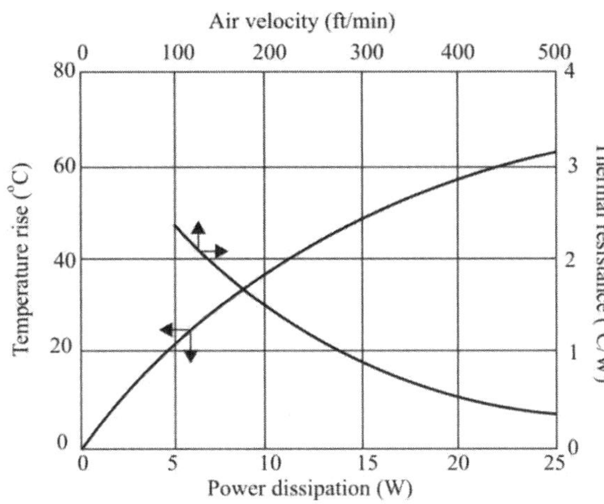

Figure 12.7 Typical heat sink data.

thermal resistance for a given air flow. Use whichever curve is applicable and ignore the other.

The following is the proper method to use in selecting a heat sink:

1. Calculate the maximum power dissipation, P_d, that will occur in the device.
2. Lookup the maximum allowable internal temperature (i.e. junction temperature for solid state devices) in the data book.
3. Lookup the thermal resistance, θ_{jc}, from the internal device to the case.
4. Specify the actual maximum internal temperature to design to, T_{jmax}. This can be equal to that in the previous step but for long device life the maximum temperature should be many tens of degrees lower.
5. Calculate the maximum allowable case temperature by $T_{cmax} = T_{jmax} - P_d\theta_{jc}$.
6. Specify the maximum ambient air temperature that the device must operate in, T_{amax}.
7. If the application is for natural convection then calculate the maximum allowable temperature rise of the heat sink: $T_{rmax} = T_{amax} - T_{cmax}$. If this number is small (less than about 10 degrees) then it may be very difficult to accomplish heat removal. It is impossible if the result is negative. If the application uses forced air then calculate the maximum allowable thermal resistance of the heat sink:

$$\theta_{samax} = \frac{T_{jmax} - T_{amax}}{P_{dmax}} - \theta_{jc} \tag{12.26}$$

8. Choose a heat sink that will physically mount the device and that has a temperature rise or thermal resistance (as shown on a

Figure 12.8 This is a typical heat sink designed for either natural air convection or forced-air cooling and comes in a wide variety of shapes and sizes. Manufacturers usually provide specs for both uses.

plot) for the operating P_d and expected air flow (from convection to forced) less than T_{rmax} or θ_{samax}. When in doubt, choose the larger heat sink.

The effective thermal resistance of the heat sink can be calculated as $\theta_{sa} = T_{rise} / P_d$. T_{rise} is taken from the performance plot. Notice that the effective thermal resistance can vary a lot depending on how the heat sink is used. This is why the manufacturer provides a plot rather than a specific value.

When forced air is a requirement then there needs to be some means for detecting either that the air flow is insufficient or the temperature of the case has exceeded some specification so that that some kind of shut down can take place to prevent failure.

The most commonly used type of commercially available heat sink is the finned extruded aluminum style, as shown in Figure 12.8. It is usually painted flat black with a bare exposed area for mounting the semiconductor device, or it may be completely black anodized. It can be punched to accommodate any pin or mounting arrangement. This basic shape of radiator is most desirable for forced air and natural convection cooling because much surface area is exposed to the surrounding air within a comparatively small volume. Approximately 80 in^2 of radiating surface can be contained in a volume of only 4.5 × 1.5 × 3 in. These same heat sinks can be given a serrated surface to increase the available surface area per unit volume even more. Also, some heat sinks come in "pin-grid arrays" and other geometries for special applications.

Heat sinks are devices that enhance heat dissipation from a hot surface, usually the case of a heat generating component, to a cooler ambient, usually air [5]. For the following discussions, air is assumed to be the cooling fluid. In most situations, heat transfer across the interface between the solid surface and the coolant air is the least efficient within the system, and the solid-air interface represents the greatest barrier for heat dissipation. A heat sink lowers this barrier mainly by increasing the surface area that is in direct contact with the coolant. This allows more heat to be dissipated and/or lowers the device operating temperature. The primary purpose of a heat sink is to maintain the device temperature below the maximum allowable temperature specified by the device manufacturers.

12.5.1 Thermal Circuit

Before discussing the heat sink selection process, it is necessary to define common terms and establish the concept of a thermal circuit. The objective is to provide basic fundamentals of heat transfer for those readers who are not familiar with the subject. Notations and definitions of the terms are as follows:

Q: total power or rate of heat dissipation in Watts, represent the rate of heat dissipated by the electronic component during operation. For the purpose of selecting a heat sink, the maximum operating power dissipation issued.
T_j: maximum junction temperature of the device in °C. Allowable T_j values range from 115°C in typical microelectronics applications to as high as 180°C for some electronic control devices. In special and military applications, 65°C to 80°C are not uncommon.
T_c: case temperature of the device in °C. Since the case temperature of a device depends on the location of measurement, it usually represents the maximum local temperature of the case.
T_s: sink temperature in °C. Again, this represents the maximum temperature of a heat sink at the location closest to the device.
T_a: ambient air temperature in °C.

Using temperatures and the rate of heat dissipation, a quantitative measure of heat transfer efficiency across two locations of a thermal component can be expressed in terms of thermal resistance θ, defined as

$$\theta = \Delta T / Q \tag{12.27}$$

were ΔT is the temperature difference between the two locations. The unit of thermal resistance is in °C/W, indicating the temperature rise per unit rate of heat dissipation. This thermal resistance is analogous to the electrical resistance R_e, given by Ohm's law:

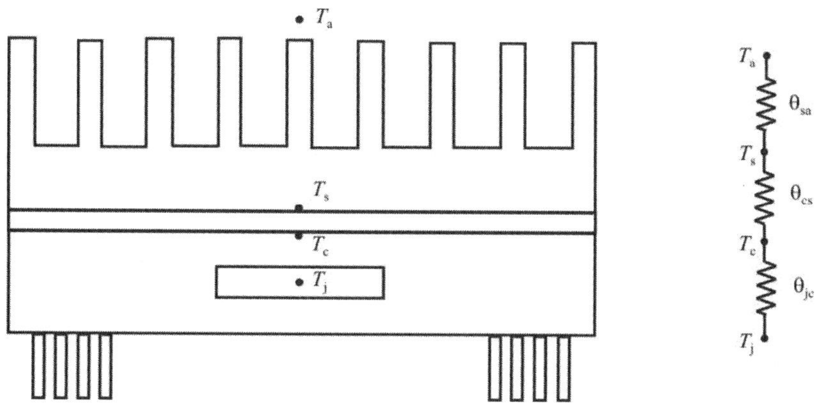

Figure 12.9 Thermal resistance equivalent circuit.

$$R_e = \Delta V / I$$

with ΔV being the voltage difference and I the current.

Consider a simple case where a heat sink is mounted on a device package as shown in Figure 12.9. Using the concept of thermal resistance, a simplified thermal circuit of this system can be drawn, as also shown in the figure. In this simplified model, heat flows serially from the junction to the case then across the interface into the heat sink and is finally dissipated from the heat sink to the air stream.

The thermal resistance between the junction and the case of a device is defined as

$$\theta_{jc} = T_{jc} / Q = (T_j - T_c) / Q \tag{12.28}$$

This resistance is specified by the device manufacturer. Although the θ_{jc} value of a given device depends on how and where the cooling mechanism is employed over the package, it is usually given as a constant value. It is also accepted that θ_{jc} is beyond the user's ability to alter or control.

Similarly, case-to-sink and sink-to-ambient resistance are defined as

$$\theta_{cs} = \Delta T_{cs} / Q = (T_c - T_s) / Q \tag{12.29}$$

$$\theta_{sa} = \Delta T_{sa} / Q = (T_s - T_a) / Q \tag{12.30}$$

respectively. Here, θ_{cs} represents the thermal resistance across the interface between the case and the heat sink and is often called the interface resistance. This value can be improved substantially depending on the quality of mating surface finish and/or the choice of interface material. θ_{sa} is heat sink thermal resistance.

Obviously, the total junction-to-ambient resistance is the sum of all three resistances:

$$\theta_{\text{ja}} = \theta_{\text{jc}} + \theta_{\text{cs}} + \theta_{\text{sa}} = (T_{\text{j}} - T_{\text{a}}) / Q \tag{12.31}$$

12.5.2 Required Heat-Sink Thermal Resistance

To begin the heat sink selection, the first step is to determine the heat sink thermal resistance required to satisfy the thermal criteria of the component. By rearranging (12.31), the heat sink resistance is easily obtained as

$$\theta_{\text{sa}} = (T_{\text{s}} - T_{\text{a}}) / Q - \theta_{\text{jc}} - \theta_{\text{cs}} \tag{12.32}$$

In this expression, T_j, Q and θ_{jc} are provided by the device manufacturer, and T_a and θ_{cs} are the user defined parameters.

The ambient air temperature T_a for cooling electronic equipment depends on the operating environment in which the component is expected to be used. Typically, it ranges from 35 to 45°C, if the external air is used, and from 50 to 60°C, if the component is enclosed or is placed in a wake of other heat generating equipment. Extreme environments such as in uncontrolled aircraft bays on high flying aircraft can extend beyond these ranges.

The interface resistance θ_{cs} depends on the surface finish, flatness, applied mounting pressure, contact area and, of course, the type interface material and its thickness. Precise value of this resistance, even for a given type of material and thickness, is difficult to obtain, since it may vary widely with the mounting pressure and other case dependent parameters. However, more reliable data can be obtained directly from material manufacturers or from heat sink manufacturers. Typical values for common interface materials are tabulated in Table 12.2.

With all the parameters on the right side of (12.32) it becomes the required maximum thermal resistance of a heat sink for the application. In other words, the thermal resistance value of a chosen heat sink for the application has to be equal to or less than θ_{sa} value for the junction temperature to be maintained at or below the specified T_j.

12.5.3 Selecting a Suitable Heat Sink

In this section we go through an example of the process of selecting a heat sink for a specific application [6].

Table 12.2 Thermal Properties of Some Interface Materials

Material	Conductivity (W/in/°C)	Thickness (in)	Resistance (in² °C/W)
There-O-Link Thermal Compound	0.010	0.002	0.19
High performance thermal compound	0.030	0.002	0.07
Kon-Dux	0.030	0.005	0.17
A-Dux	0.008	0.004	0.48
1070 Ther-A-Grip	0.014	0.006	0.43
1050 Ther-A-Grip	0.009	0.005	0.57
1080 Ther-A-Grip	0.010	0.002	0.21
1081 Ther-A-Grip	0.019	0.005	0.26
A-Phi 220 @20psi	0.074	0.020	0.27
1897 in Sil-8	0.010	0.008	0.81
1898 in Sil-8	0.008	0.006	0.78

Source: [5]

The actual temperature difference of a BJT depends upon the amount of power the junction dissipates. The average power dissipation for a BJT is approximately

$$P_D = I_C \times V_{CE} \tag{12.33}$$

where

P_D is the power dissipation in watts of constant current generation
I_C is the collector current
V_{CE} is the collector-to-emitter voltage

Consider the characteristics of a single-ended, class A amplifier stage. Suppose the transistor has a peak collector current of 15A and a power dissipation capability of 90W at a case temperature of 25°C. The power dissipation capability decreases linearly from 25°C to zero watts of dissipation at 100°C case temperature (see Figure 12.10).

When the Class A amplifier stage output power is 90W peak, the quiescent collector current is about 1.5A at 30V collector-to-emitter (biased at midpoint on its load line). Using the equation for power dissipation, the class A stage produces 45 W (P_d = 30V × 1.5A = 45 W). The product of collector current and voltage either side of this quiescent state produces less power dissipation, so 45 W is the maximum power dissipation to consider. This value is used below to calculate the heat sink required.

As the amount of power dissipation in the device increases, the size of the heat sink must increase to allow more surface area to be exposed to the ambient. First, the heat sink material and size are considered. Thermal conductivity of the material

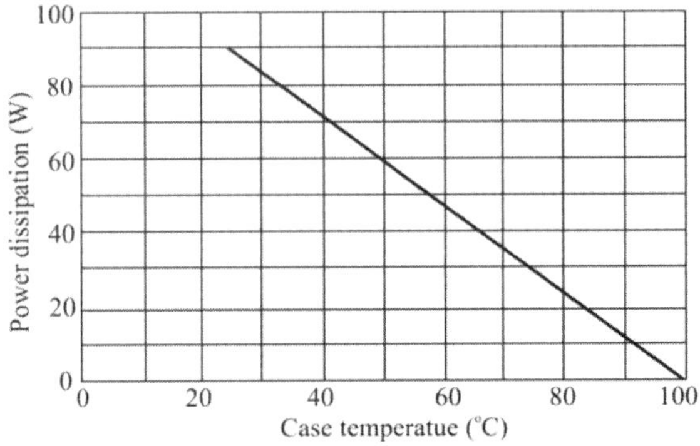

Figure 12.10 Transistor case temperature derating curve. Transistor specifications for power dissipation are typically rated at 25°C and derated to zero at 100°C. The variables for a reliable design must fall under the curve.

should be as high as possible. Copper is about the best for thermal conductivity and aluminum follows a close second. The difference in cost between copper and aluminum exceeds their difference in thermal conductivity, therefore aluminum heat sinks are pretty much the industry standard.

Consider a square of bright aluminum, 1/8 in. thick with a surface area of 5in^2 per side, positioned vertically. From actual measurements, the thermal resistance is about 9°C/W. Refer to Figure 12.11. As the size increases, the thermal resistance reduces until a practical limit is reached of about 140 in^2 of heat sink area. Here the thermal resistance is about 1.4°C/W in free air, and increasing the surface area does

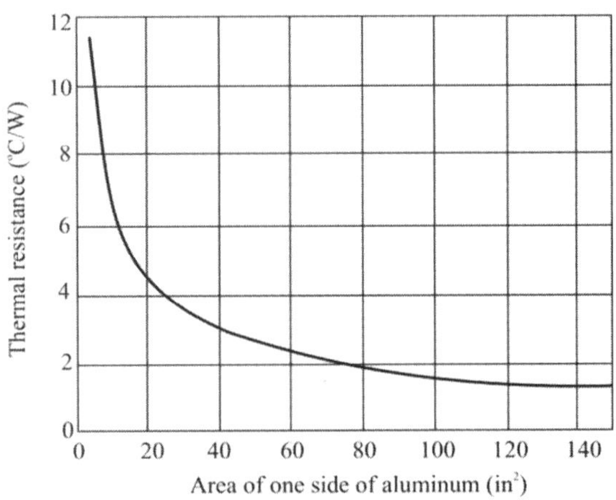

Figure 12.11 Heat transfer of square 1/8" aluminum heat sink.

not significantly decrease the thermal resistance. More than 140 in^2 in this case would be considered an infinite heat sink. Copper positioned vertically, 1/8in thick, by 140 in^2 has a thermal resistance of about 1.0°C/W compared to aluminum at 1.4°C/W.

The thermal resistance drops rapidly from no heat sink to 3.0°C/W at 40 in^2 and levels off to about 1.5°C/W as the heat sink reaches about 100 in^2 (infinite heat sink). Increasing it any further makes little difference.

Now we match the requirements to a proper heat sink. First, according to (12.20), sum the thermal resistances, multiply the sum by the power dissipation of the semiconductor, and add the results to the ambient temperature. If this temperature exceeds the maximum allowable temperature for the device as shown in the derating curve (Figure 12.10), one or more factors in the computation must be changed. For example, the ambient temperature could be lowered or the power dissipation reduced by lowering the input voltage or current. Also, any one of the thermal resistances could be reduced to achieve the same end.

A typical graph supplied with fabricated heat sinks is shown in Figure 12.12, which conveys the thermal resistance characteristics for the heat sink in Figure 12.8. The temperature differential in degrees Celsius from the mounting to ambient air is indicated on the ordinate. The power dissipation capability of the heat sink in watts is plotted along the abscissa. We see that the vertically mounted heat sink has a thermal coefficient of about 1.8°C/W. The transistor can be isolated electrically from the heat sink with an insulator such as mica. The mica insulator, however, has a coefficient of 0.8°C/W to 0.5°C/W, which must be added to the thermal resistance of the heat sink. Silicone grease applied to both sides of the insulator reduces this resistance by half or 0.4 to 0.25°C/W. In any case, the transistor or module should be mounted near the lower edge of a vertically mounted heat sink for best dissipation.

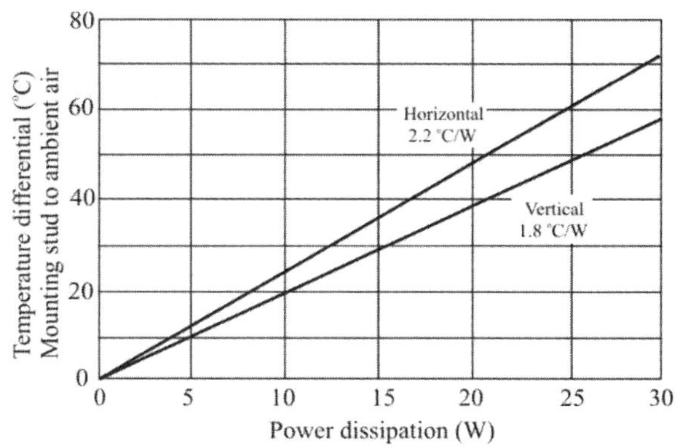

Figure 12.12 Thermal characteristics of finned heat sink shown in Figure 12.8.

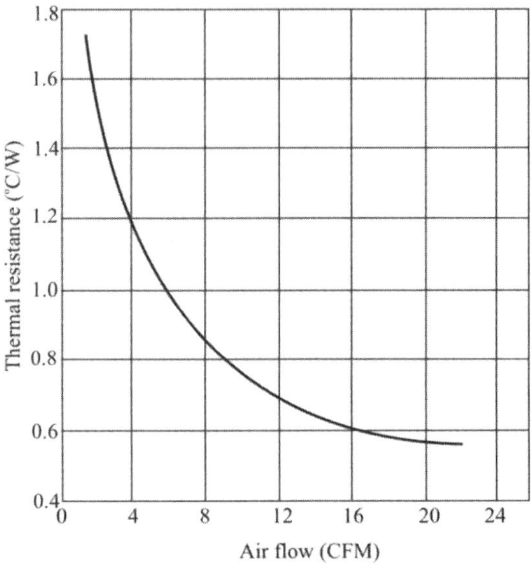

Figure 12.13 Heat sink performance under forced air flow conditions for the heat sink shown in Figure 12.8. A little forced air flow makes a big difference in thermal resistance when space permits a fan or blower.

Another characteristic some manufacturers indicate is the effect of forced airflow for a natural convection heat sink as shown in Figure 12.13. Thermal resistance of the heat sink is shown along the ordinate and the airflow in CFM is indicated along the abscissa. As shown, the thermal resistance of the heat sink can be reduced considerably from 1.7 to about 0.6 °C/W with airflow of only 16 CFM.

A realistic ambient temperature is 50°C. Select a forced airflow of 16 CFM, heat sink θ_{sa} of 0.6°C/W, and an ambient of 50°C. Substitute these values in (12.20) the equation for solving T_j, and the result is 77°C. The derating curve shows an allowable 25 W at 77°C. So forced air is indeed a viable option. However, this means that each transistor or power module requires its own heat sink with airflow of 16 CFM. The forced air cooling made the difference between an impossible and a possible circuit-heat sink combination for a 50°C environment.

Without the forced air, the previous calculations of the single-ended, Class A amplifier dissipated 45W. Select a maximum ambient temperature of 50°C; a transistor thermal resistance θ_{jc} of 0.8°C/W; silicone grease to give a θ_{cs} of 0.1°C/W; and a commercial heat sink as shown in Figure 12.8 with a θ_{sa} of 1.8 °C/W. Now calculate the junction temperature:

$$T_j = 45(0.8 + 0.1 + 1.8) + 50 = 171.5\ ^\circ\text{C}$$

Table 12.3 Range of Volumetric Thermal Resistance

Flow Condition m/s (lfm*)	**Volumetric Resistance cm^3 °C/W**	
Natural Convection	500–800	(30–50)
1.0 (200)	150–250	(10–15)
2.5 (500)	80–150	(5–10)
5.0 (1000)	50–80	(3–5)

*lfm = linear feet per minute

This is not a satisfactory solution: T_j of only 98.6°C is allowed, so we must fine-tune the numbers in order to lower the temperature to a more reasonable level. Although this temperature is under the maximum allowable for the transistor, a look at the derating graph shows that only 2 W of dissipation is allowed at 98 °C.

12.5.4 Heat-Sink Selection

In selecting an appropriate heat sink that meets the required thermal criteria, we need to examine various parameters that affect not only the heat sink performance itself, but also the overall performance of the system. The choice of a particular type of heat sink depends largely on the thermal budget allowed for the heat sink and external conditions surrounding the heat sink. It is to be emphasized that there can never be a single value of thermal resistance assigned to a given heat sink, since the thermal resistance varies with external cooling conditions.

When selecting a heat sink, it is necessary to classify the air flow as natural, low flow mixed, or high flow forced convection. Natural convection occurs when there is no externally induced flow and heat transfer relies solely on the free buoyant flow of air surrounding the heat sink. Forced convection occurs when the flow of air is induced by mechanical means, usually a fan or blower. There is no clear distinction on the flow velocity that separates the mixed and forced flow regimes. It is generally accepted in applications that the effect of buoyant force on the overall heat transfer diminishes to negligible level (under 5%) when the induced air flow velocity excess 1–2 m/s (200 to 400 lfm).

The next step is to determine the required volume of a heat sink. Table 12.3 shows approximate ranges of volumetric thermal resistance of a typical heat sink under different flow conditions.

The volume of a heat sink for a given low condition can be obtained by dividing the volumetric thermal resistance by the required thermal resistance. Table 12.3 should be used only as a guide for estimation purposes in the beginning of the selection process. The actual resistance values may vary outside the above range depending on many additional parameters, such as actual dimensions of the heat

Table 12.4 Fin Spacing (mm/in) Versus Flow and Fin Length

	Fin Length (in)			
Flow Conditions (m/s, lfm)	75, 3.0	150, 12.0	225, 9.0	300, 12.0
Natural convection	12.5, 0.25	7.5, 0.30	10, 0.38	13, 0.50
1.0 (200)	4.0, 0.15	5.0, 0.20	12.0, 0.24	7.0, 0.27
2.5 (500)	2.5, 0.10	3.3, 0.13	4.0, 0.16	5.0, 0.20
5.0 (1000)	2.0, 0.08	2.5, 0.10	3.0, 0.12	3.5, 0.14

sink, type of the heat sink, flow configuration, orientation, surface finish, altitude, etc. The smaller values shown in Table 12.3 correspond to heat sink volume of approximately 100 to 200cm^3 (5 to 10in^3) and the larger ones to roughly 1000cm^3 (60in^3).

The above tabulated ranges assume that the design has been optimized for a given flow condition. Although there are many parameters to be considered in optimizing a heat sink, one of the most critical parameters is the fin density. In a planar fin heat sink, optimum fin spacing is strongly related to two parameters: flow velocity and fin length in the direction of the flow. Table 12.4 may be used as a guide for determining the optimum fin spacing of a planar fin heat sink in typical applications.

The average performance of a typical heat sink is linearly proportional to the width of a heat sink in the direction perpendicular to the flow, and approximately proportional to the square root of the fin length in the direction parallel to the flow. For example, an increase in the width of a heat sink by a factor of two would increase the heat dissipation capability by a factor of two, whereas an increase the heat dissipation capability by a factor of 1.4. Therefore, if the choice is available, it is beneficial to increase the width rather than the length of the heat sink. Also, the effect of radiation heat transfer is very important in natural convection, as it can be responsible of up to 25% of the total heat dissipation. Unless the component is facing a hotter surface nearby, it is imperative to have the heat sink surfaces painted or anodized to enhance radiation.

12.5.5 Heat Sink Types

Heat sinks can be classified in terms of manufacturing methods and their final form shapes. The most common types of air-cooled heat sinks include:

- ***Stampings:*** Copper or aluminum sheet metals are stamped into desired shapes. They are used in traditional air cooling of electronic components and offer a low-cost solution to low density thermal problems. They are suitable for high volume production, because advanced tooling with high speed stamping would lower costs. Additional labor-saving options, such as taps, clips, and interface materials, can be factory applied to help to reduce the board assembly costs.

- ***Extrusion:*** These allow the formation of elaborate two-dimensional shapes capable of dissipating large heat loads. They may be cut, machined, and options added. A cross-cutting will produce omni-directional, rectangular pin fin heat sinks, and incorporating serrated fins improves the performance by approximately 10 to 20%, but with a slower extrusion rate. Extrusion limits, such as the fin height-to-gap fin thickness, usually dictate the flexibility in design options. Typical fin height-to-gap aspect ratio of up to 6 and a minimum fin thickness of 1.3mm, are attainable with a standard extrusion. A 10 to 1 aspect ratio and a fin thickness of 0.8" can be achieved with special die design features. However, as the aspect ratio increases, the extrusion tolerance is compromised.
- ***Bonded/Fabricated Fins:*** Most air-cooled heat sinks are convection limited, and the overall thermal performance of an air-cooled heat sink can often be improved significantly if more surface area can be exposed to the air stream. These high-performance heat sinks utilize thermally conductive aluminum-filled epoxy to bond planar fins onto a grooved extrusion base plate. This process allows for a much greater fin height-to-gap aspect ratio of 20 to 40, greatly increasing the cooling capacity without increasing volume requirements.
- ***Castings:*** Sand, lost core and die casting processes are available with or without vacuum assistance, in aluminum or copper/bronze. This technology is used in high density pin fin heat sinks which provide maximum performance when using impingement cooling.
- ***Folded Fins:*** Corrugated sheet metal in either aluminum or copper increases surface area and, hence, the volumetric performance. The heat sink is then attached to either a base plate or directly to the heating surface via epoxying or brazing. It is not suitable for high profile heat sinks because of the availability and fin efficiency. Hence, it allows high performance heat sinks to be fabricated for applications.

Figure 12.14 shows the typical range of cost functions for different types of heat sinks in terms of required thermal resistance.

The performance of different heat sink types varies dramatically with the airflow through the heat sink. To quantify the effectiveness of different types of heat sinks, the volumetric heat transfer efficiency can be defined as

$$\eta = \frac{Q}{\dot{m} C_{\mathrm{p}} \Delta T_{\mathrm{sa}}} \tag{12.34}$$

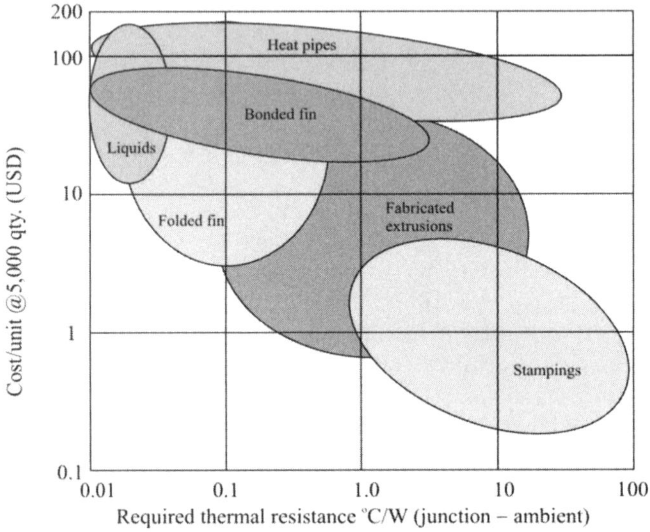

Figure 12.14 Cost versus required thermal resistance.

where, $\dot{m}$ is the mass flowrate through the heat sink, C_p is the heat capacity of the fluid, and ΔT_{sa} is the average temperature difference between the heat sink and the ambient air. The heat transfer efficiencies have been measured for a wide range of heat sink configurations, and their ranges are listed in Table 12.5. The improved thermal performance is generally associated with additional costs in either material or manufacturing, or both.

12.5.6 Thermal Performance Graph

Performance graphs typical of those published by heat sink vendors are shown in Figure 12.7. The graphs are a composite of two separate curves which have been combined into a single figure. It is assumed that the device to be cooled is properly

Table 12.5 Range of Heat Transfer Efficiencies

Heat Sink Type	η Range (%)
Stamping and flat plates	10–18
Impingement flow fan heat sinks	25–32
Fully ducted extrusions	45–58
Ducted pin fin, bonded & folded fin	78–90

mounted, and the heat sink is in its normally used mounting orientation with respect to the direction of air flow. The first plot traveling from the lower left to the upper right is the natural convection curve of heat sink temperature rise, ΔT_{sa}, versus Q. The natural convection curves also assume that the heat sink is painted or anodized black. The curve from the upper left to lower right is the forced convection curve of thermal resistance versus air velocity. In forced convection, ΔT_{sa} is linearly proportional to Q, hence θ_{sa} is independent of Q and becomes a function only of the flow velocity. However, the natural convection phenomenon is non-linear, making it necessary to present ΔT_{sa} as a function of Q.

We can use the performance graphs to identify the heat sink and, for forced convection applications, to determine the minimum flow velocity that satisfies the thermal requirements. If the required thermal resistance in a force convection application is 8°C/W, for example, the above sample thermal resistance versus flow velocity curve indicates that the velocity needs to be at or greater than 2.4 m/s (470 lfm). For natural convection applications, the required thermal resistance θ_{sa} can be multiplied by Q to yield the maximum allowable ΔT_{sa}. The temperature rise of a chosen heat sink must be equal to or less than the maximum allowable ΔT_{sa} at the same Q.

Readers are reminded that the natural convection curves assume an optimal orientation of the heat sink with respect to the gravity. Also, the flow velocity in the forced convection graph represents the approach flow velocity without accounting for the effect of flow bypass. There have been a limited number of investigations [2, 3] on the subject of flow bypass. These studies show that flow bypass may reduce the performance of a heat sink by as much as 50% for the same upstream flow velocity. For further consultation on this subject, readers are referred to the cited references.

When a device is substantially smaller than the base plate of a heat sink, there is an additional thermal resistance, called the spreading resistance that needs to be considered in the selection process. Performance graphs generally assume that the heat is evenly distributed over the entire base area of the heat sink, and therefore, do not account for the additional temperature rise caused by a smaller heat source. This spreading resistance could typically be 5 to 30% of the total heat sink resistance, and can be estimated by using the simple analytical expression developed in [7].

Another design criterion that needs to be considered in the selection of a heat sink is the altitude effect. While the air temperature of an indoor environment is normally controlled and is not affected by the altitude change, the outdoor air pressure does change with the altitude. Since many electronic systems are installed at an elevated altitude, it is necessary to derate the heat sink performance mainly due to the lower air density caused by the lower air pressure at higher altitude. Table 12.6 shows the performance derating factors for typical heat sinks at high altitudes.

Table 12.6 Altitude Derating Factors

Altitude m/ft	Factor
0, sea level	1.00
1000/3000	0.95
1500/5000	0.90
2000/7000	0.86
3000/10000	0.80
3500/12000	0.75

For example, in order to determine the actual thermal performance of a heat sink at altitudes other than the seal level, the thermal resistance values read off from the performance graphs should be divided by the derating factor before the values are compared with the required thermal resistance.

12.6 Parallel Transistors for High Power Applications

With high power applications it may be impossible to find a suitable heat-sink for a particular transistor, then one solution would be to use a different power transistor, or different case (package) type if available. Another alternative is to use two or more transistors connected in parallel, sharing the total power between them [8]. This can be a cheaper option than a single very expensive heat-sink. This concept is illustrated in Figure 12.15.

12.6.1 Thermal Runaway

In many modern circuits power MOSFETs are preferred to BJTs because of the BJT's problem of thermal runaway. This is a process where current flow rises as a natural effect in semiconductors as the temperature of the device increases. This rise

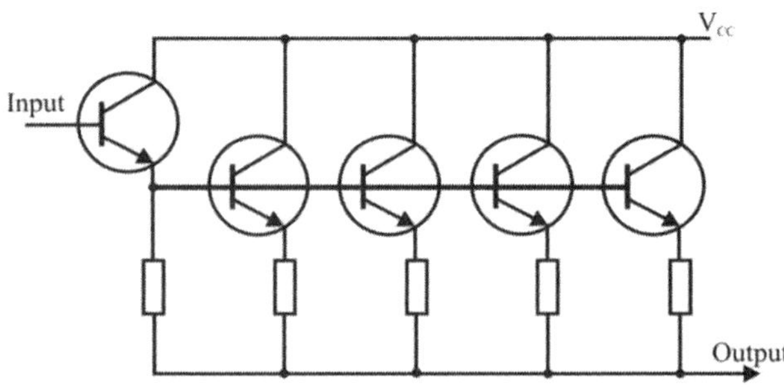

Figure 12.15 Power Transistors Connected in Parallel.

in temperature then leads to a further increase in current flow and a subsequent further rise in temperature, until the rise in temperature and current spirals out of control and the device is destroyed.

When several poorly matched transistors are connected in parallel, the transistor initially passing the most current will get hotter, while the others, passing less current, get cooler. Therefore, the hotter transistor can be in danger of thermal runaway. However, BJTs, carefully matched, may still be preferable to MOSFETs for some high voltage applications.

12.7 Thermal Considerations of PCBs

One of the most important components for thermal management is the printed-circuit board (PCB). In a high-power, high-frequency circuit such as an RF PA, heat can build up around the amplifier's active devices. This heat must be properly channeled away from the active devices and safely dissipated through device packages, circuit ground connections, heat sinks, equipment chassis, and ambient air. The choice of PCB materials is critical for the overall thermal management of a high-power RF/microwave design. In this section we discuss some of the considerations when analyzing the PCB material selection.

A circuit material's capability to control temperature rises as a function of applied and dissipated power is determined in large part to its power-handling capability. As is common with most electronic components, elevated operating temperatures result in shorter operating lifetimes and, often, degraded electrical performance. High temperature can result in damaged and degraded performance, whether it is high environmental ambient temperature, or the temperature of the circuitry and its components is elevated because of high-power operation. Depending upon the amount of power that a circuit must dissipate, maintaining that circuit at a lower temperature usually enables improved reliability and better (or at least more predictable) performance.

A PCB at elevated temperatures will expand and contract with changes in temperature, expanding in three all axes (length, width, and thickness) as the temperature increases. The amount of this expansion for the change in temperature is characterized by a PCB material's *coefficient of thermal expansion* (CTE). Because a PCB is typically formed of one or more dielectric layers laminated with copper (to form transmission lines and a ground plane), the material's linear CTE in the x and y directions is usually designed to match the CTE of copper (about 17 ppm/°C). Thus, the materials expand and contract together with changes in temperature, minimizing stress on the junction of the two materials.

The CTE in the z axis (the thickness) of the dielectric material is usually designed for a low value to minimize dimensional changes with temperature and maintain the integrity of *plated through holes* (PTHs), also known as *vias*. The PTHs provide paths from the top to the bottom of the circuit board as needed for ground connections, as well as for interconnecting multilayer circuit boards.

In addition to these mechanical changes, temperature can also affect the electrical performance of a PCB. The relative dielectric constant of a PCB laminate, for example, varies as a function of temperature, as defined by a parameter known as the *thermal coefficient of dielectric constant*. This parameter indicates changes in the dielectric constant with temperature (typically in ppm/°C). Because the impedance of high-frequency transmission lines is determined not only by the thickness of the substrate material but also by its dielectric constant, changes in *z*-axis CTE and dielectric constant as a function of temperature can significantly impact the impedance of microstrip and stripline transmission lines fabricated on that material.

Microwave circuits rely on tightly matched impedances between components and circuit junctions to minimize reflections. In a PA, impedance-matching circuitry is used to make transitions from the typically low impedances of a power transistor to the typically 50 Ω characteristic impedance of an RF/microwave circuit or system (transmission line to the antenna, for example). Changes in transmission line impedance due to temperature effects can alter the frequency response of a high-frequency amplifier.

Another parameter to consider when analyzing heat removal in PCBs is the glass transition temperature, denoted by T_g. This is the temperature at which some materials can change states. It indicates the temperature at which dramatic changes take place in a material's CTE behavior; see Figure 12.16 for example. Because a material can undergo such a drastic change in CTE, it can become mechanically and electrically unstable when operating above T_g, and in operation should always be maintained below that temperature.

A PCB's *maximum operating temperature* (MOT) is another critical temperature-related parameter. The MOT is a rating that the Underwriters' Laboratory (UL) gives to a unique PCB construction made by a particular PCB fabricator using specific materials. The MOT is the maximum temperature at which the PCB can be operated for an indefinite period of time without significant degradation to critical performance attributes of the circuit.

A PCB material's thermal conductivity is a relative indicator of a laminate's effectiveness in dissipating heat. It is measured in watts of power per meter of material per Kelvin. Analogous to electrical conductivity and the flow of electricity through a material, thermal conductivity is used to predict the rate of energy loss as heat through a given material. The reciprocal of thermal conductivity is thermal resistivity, or the ability of a material to resist the flow of heat. Both the thermal conductivity and the thermal resistivity were discussed previously.

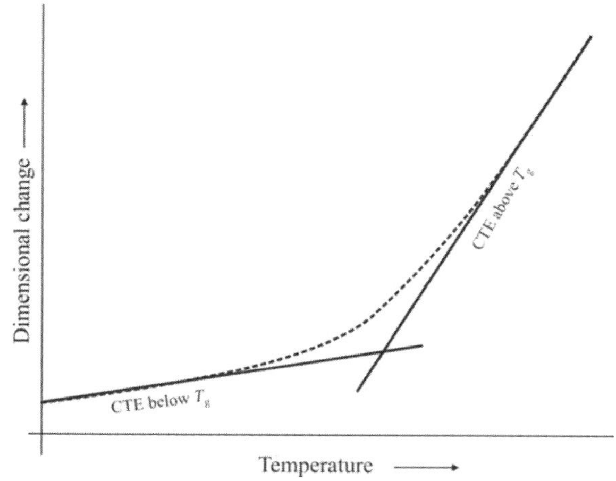

Figure 12.16 A PCB material's CTE behavior can change drastically above the material's glass transition temperature, T_g, and can become mechanically and electrically unstable above that temperature.

The thermal conductivity of a material depends on various properties. For example, glass is a poor thermal conductor, with an extremely low thermal conductivity of 1.1 W/(m-K). Conversely, copper provides very low resistance to the flow of heat, with a thermal conductivity of 401 W/(m-K). Because the thermal conductivity of PCB dielectric materials is low [the thermal conductivity of the very popular high-T_g FR-4 circuit material is typically about 0.24 W/(m-K)], heat can easily build up on the conductive traces of a high-power PCB, which are typically formed of copper. But selecting a PCB material with a higher value of thermal conductivity allows for operation of a circuit at higher power levels.

Table 12.7 offers a comparison of typical PCB laminate materials, including RT/duroid 6035HTC laminate material from Rogers Corp that is a relative newcomer. As the table shows, it has a considerably higher thermal conductivity

Table 12.7 Thermal Conductivities of PCB Laminates

Laminate Material	Thermal Conductivity (W/(m-K)
High Tg FR-4	0.24
Nearly pure PFTE	0.20
Ceramic filled PTFE	0.50
RO43508TM	0.62
RP4360TM	0.80
RT/duroid®6035HTC	1.40

Source: [9]

than FR-4 (about 6×) and even several lowloss, high-frequency laminates as represented by the other entries in Table 12.7. It consists of a ceramic-filled PTFE (polytetrafluoroethylene teflon) composite dielectric available with standard or reverse-treated, electrodeposited (ED) copper foil. It has a relative dielectric constant of 3.50 at 10 GHz in the z-axis which is maintained within a 0.05 tolerance across a board to maintain consist impedance in transmission lines. The CTE in the x- and y-axes is 19 ppm/C, closely matched to that of copper.

Management of the thermal properties of a PCB extends to more than selecting a laminate with high thermal conductivity. The dissipation factor, which is the loss attributed to the dielectric material, also contributes to heat generation. Transmission lines, such as microstrip or stripline circuits used to carry high frequency, high power signals also have losses. Higher insertion losses result in more heat being generated in the transmission lines at higher power levels. Because of the skin effect, the roughness of the copper conductor on a PCB can contribute to increased insertion loss, especially at higher frequencies.

The choice of a PCB material's dielectric constant will determine the size and density of an RF/microwave circuit, since the dimensions of microwave transmission-lines are based on the wavelengths of the signals to be handled. At higher relative dielectric constants, the dimensions of the transmission lines needed to achieve a given impedance grow smaller. In addition to the ground-plane spacing, a PCB's power-handling capability is limited to an extent by the width and insertion loss of its conductors. For a PA, for example, for improved thermal flow choosing a PCB material with a lower relative dielectric constant can lead to wider transmission lines for a given impedance. Using a PCB material with a higher relative dielectric constant may create thermal hotspots in high-power circuits due to finer transmission line dimensions and a more densely spaced circuit. In addition, minimizing the insertion loss of the transmission lines and optimizing the gain of an amplifier can result from selecting a material with low dissipation factor.

Reference [9] documents the results of tests on some PCB laminates to demonstrate their power handling capability. Three materials, a high-T_g FR-4 laminate, a RO4350B laminate and the RT/duroid 6035HTC laminate both of the latter two from Rogers Corp., were tested with the same test circuits, but with each fed with test signals at the same frequency and power levels. The high-T_g FR-4 exhibited the highest rise in temperature above ambient, jumping to +109°C (+229°F) or a rise of +84°C above ambient. The RO4350B laminate exhibited a +56°C rise above ambient, increasing from +25°C to +82°C (+180°F). The RT/duroid 6035HTC rose only +36°C above ambient (from +25°C to +62°C) under the same test conditions.

Further testing on RO4003C laminate from Rogers and RT/duroid 6035HTC with 1-oz ED copper and 2-oz. ED copper at 800 MHz, but otherwise all other test conditions the same produced the results shown in Figure 12.17. A temperature rise

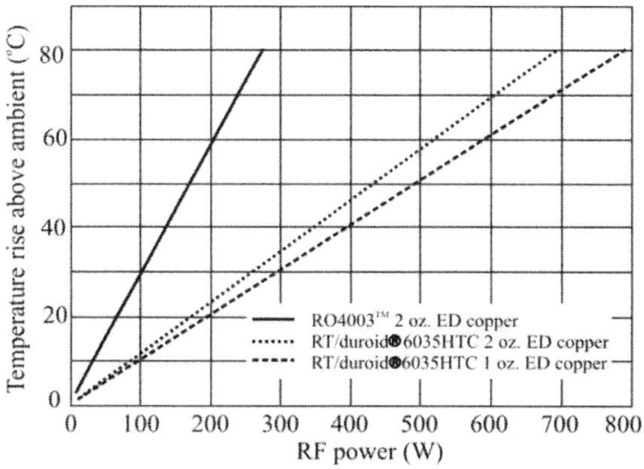

Figure 12.17 PCB power handling. Example of assuming a +105°C MOT and +25°C ambient temperature, 20mil microstrip, RF power versus temperature rise, 800 MHz.

of +80°C above ambient for all three laminates occurred at about 280W for the RO4003C laminate, about 700 W for the RT/duroid 6035HTC with 2-oz. copper, and almost 800W for the RT/duroid 6035HTC with 1-oz. copper.

When tested at 2GHz, shown in Figure 12.18, for the same increase in temperature, the RO4003C's power-handling capability dropped to about 140W, while the RT/duroid 6035HTC with 2-oz. copper handled about 380W and the RT/duroid 6035HTC with 1-oz. copper handled more than 400W. The improved

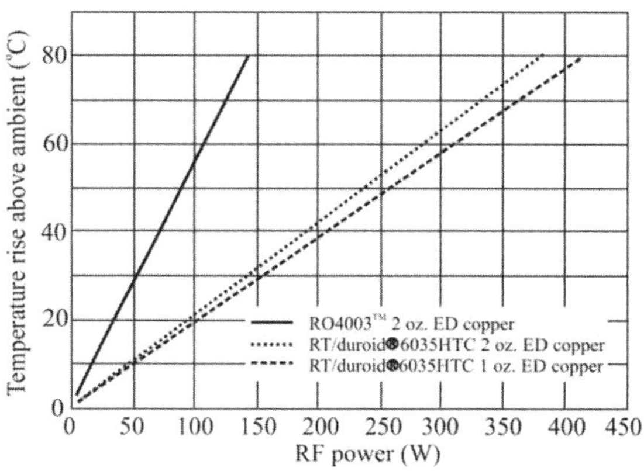

Figure 12.18 PCB power handling. Example of assuming a +105°C MOT and +25°C ambient temperature, 20 mil microstrip, RF power versus temperature rise, 2 GHz.

performance of the RT/duroid 6035HTC with 1-oz. copper over the same dielectric with heavier cladding is because the copper surface is smoother (and hence lower insertion loss) for the RT/duroid 6035HTC with 1-oz. copper.

12.7.1 Summary

As we see from these test results, temperature rises will occur in all PCB materials as a result of high RF power levels at high frequencies. But different materials, and even copper cladding options, have different temperature characteristics. The choice of material should factor in low loss, high thermal conductivity, and stable mechanical characteristics with temperature when considering an acceptable MOT.

12.8 Graphite Foam as Heat Sink Material

The thermal properties of engineered graphite foams have enhanced their consideration as heat sink materials. Materials such as diamond and graphite are good thermal conductors because their crystal structures are nearly perfect with very few crystallographic defects. The graphite foam is derived from a mesophase pitch precursor in which, during the foam's manufacture, mesophase crystals align themselves along the foam's cell walls. When the foam is subsequently graphitized at temperatures 2,800°C, a highly aligned, defect-free graphitic structure better than that found in carbon fiber is obtained. The thermal conductivity of the well-aligned foam ligaments is estimated to be comparable to diamond film (> 1,700 W/m/K). The bulk thermal conductivity of the foam is somewhat less due to the thermal resistance added by more disorder at the cell wall junctions. Current manufacturing methods typically produce material in the 100 to 150 W/m/K range, with bulk conductivities of 187 W/m/K achievable under laboratory conditions.

We present here some of the test results on two typical foam types. Results are compared to the heat-sinking properties of copper and aluminum.

12.8.1 Copper, Aluminum, and Graphite Foam

Solder brazing could be used to attach a foam heat sink to a heated component [10, 11]. The solder method reduced the problematic interfacial resistance when using foams, due to their porous nature. Directly bonding the heat sink to a component has two potential drawbacks. First, the high temperatures common in brazing could damage the electrical component itself. The other issue concerns the complicated replacement or rework of the component. Due to the low tensile strength of foam

Table 12.8 Thermal and Mechanical Properties of Some Heat Sink Materials

Material	Copper (Cu)	Aluminum (Al)	Foam A	Foam B
Thermal Properties				
Bulk thermal conductivity (W/m/K)	391	180	127	175
Specific thermal conductivity [W/(m/K)]/(g/cm^3)	45	63	218	313
z-Plane Thermal Conductivity λ_z (W/m/K)	391	180	125	181
x-y Plane Thermal Conductivity λ_{xy} (W/m/K)	391	180	41	60
Physical Properties				
Density (g/cm^3)	8.9	2.8	0.58	0.56
Tensile strength (MPa)	270	180	1	0.69
Porosity (%)	-	-	73	75

(Table 12.8), there is a greater potential for heat sink damage than with Al or Cu [12].

To avoid these problems, the foam heat sink can be soldered to an aluminum or copper carrier plate. This foam-and-plate assembly can then be mounted to a component in a standard fashion. The carrier plate allows sufficient pressure to be applied to the interface material, ensuring low contact resistance.

12.8.2 Thermal Properties of Graphite Foam

Because of the special structure of graphite foam, there are several prominent thermal properties in the graphite foam. The graphite foam exhibits high effective thermal conductivity [up to 182 W/(m/K)] and low density (0.2–0.6 g/cm^3). The data in Table 12.8 show that the thermal conductivity in the *z* plane is much larger than the one in the *x-y* plane. It implies that the high thermal conductivity of the graphite foam only exists in a certain direction. This is a disadvantage of the graphite foam. Apparently, the heat inside the graphite lattice is transferred down the graphite lattice fast, because of the very stiff nature of the covalent bonds. Moreover, the position and vibration of atoms in the neighboring planes may impede the vibration of atoms in the plane of interest. The crystal perfection controls the thermal performance.

12.8.3 Results

As expected, the traditional copper and aluminum heat sinks performed similarly. Their main difference was the higher thermal conductivity of copper, which reduced spreading resistance.

Table 12.9 Specific Thermal Test Results

	Free Stream Velocity (m/s)		
	1.5	2.5	3.5
	$\Theta_{heater\text{-}air}$		
Copper	1.637	0.986	0.710
Aluminum	1.677	1.016	0.740
Foam A	1.746	1.144	0.937
	$\Psi_{heater\text{-}base}$		
Coper	0.089	0.089	0.099
Aluminum	0.108	0.099	0.099
Foam A	0.158	0.178	0.197
	$\Psi_{base\text{-}air}$		
Copper	1.548	0.897	0.611
Aluminum	1.568	0.917	0.641
Foam A	1.588	0.966	0.740

During slow velocity flow conditions, the lower heat transfer rate means that convection thermal resistance makes up a large portion of the overall θ_{ja} (thermal resistance, junction-to-ambient). As airflow speed increases, the convection resistance decreases, and the internal heat sink conduction resistance is more of a factor in the overall θ_{ja} value. This behavior is seen in Table 12.9 and when comparing the different heat sink materials. As illustrated in Figure 12.19, the graphite heat sink's thermal performance was only 12% lower than aluminum at low flow rates. However, the performance difference increased to 25-30% as the flow rate increased. Due to the lack of a solder joint, the foam heat sink experienced a larger interfacial resistance when compared to the solid heat sinks. This difference can be seen when comparing $\Psi_{heater\text{-}base}$ in Table 12.9. To decouple the effect of

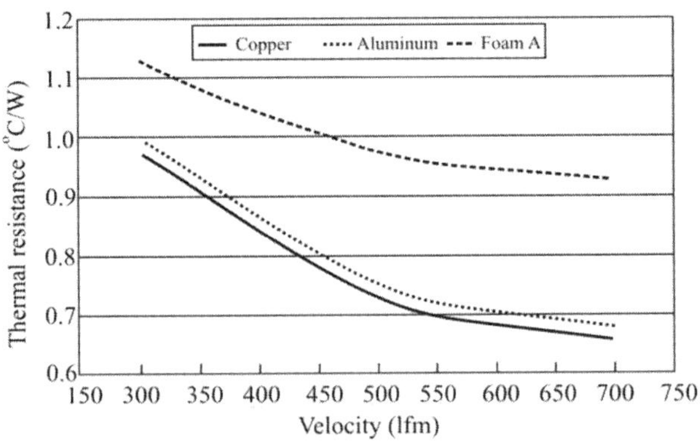

Figure 12.19 Heat sink thermal resistance as a function of free stream velocity and material.

interfacial resistance, $\Psi_{base\text{-}air}$ can be calculated. When ignoring interfacial resistance in this manner foam performs within 1 percent of aluminum at 1.5 m/s, and within 15 percent at 3.5 m/s.

Graphite foam-derived heat sinks show promise in specific applications, but exhibit several drawbacks in mainstream electronics cooling. Due to the frail nature of graphite foam, unique precautions must be taken during the handling and use of these heat sinks. When coupled to a copper base plate, graphite foam can perform with acceptably small thermal spreading resistances. However, the foam's lower thermal conductivity reduces thermal performance at high flow velocities compared to a traditional copper heat sink.

The mechanical attachment needed to ensure acceptable thermal interface performance without soldering or brazing also hinders foam-based heat sinks from being explored in mainstream applications. Despite these challenges, the thermal performance-to-weight ratio of foam is very attractive and well-suited to the aerospace and military industries, where cost and ease of use come second to weight and performance.

12.8.4 Conclusion

Many materials are available to address the thermal dissipation needs of modern components. Traditional materials such as aluminum and copper are the traditional materials for this purpose; however, there are new technologies that offer improvements in cost, weight, or thermal conductivity. The initial question when addressing the material solution is a classic thermodynamics problem: what effect does conductivity have on the overall thermal resistance in my application? Only once this is answered can the benefits of cost, weight, and manufacture be addressed.

12.9 Thermal Conductivity of Diamond Composites

12.9.1 Introduction

Most materials displaying high thermal conductivity, above 100W/(m/K) at room temperature, have diamond-like lattices; among them are diamond, cubic boron nitride, and others, in particular, SiC, BeO, BP, AlN, BeS, GaN, Si, AlP, and GaP [13 – 15].

In applications where fairly high thermal conductivities combined with low electrical conductivities are of prime importance, SiC- and AlN-based ceramics are widely used. The progress in the development of technologies of production of nanosized crystals we are currently witnessing has generated a surge of interest in the investigation and application potential of nanomaterials, including aspects associated with fabrication of high thermal conductivity composites.

Having a high thermal conductivity [up to 2,200 W/(m/K)], diamond occupies a prominent place among the materials that offer promise for developing high-efficiency heat sinks for high-frequency, high power transistors, among other applications that require high levels of heat dissipation.

For composites designed to feature high thermal conductivity levels we ordinarily use high quality diamonds whose single crystals demonstrate a high value of this parameter, and nonferrous metals, such as copper λ = 380 W/(m/K), silver λ = 420 W/(m/K), and aluminum λ = 240 W/(m/K) [16].

An additional requirement imposed on high thermal conductivity composites is that its linear thermal expansion coefficient be equal to that of the material to be cooled—primarily that of semiconductors (Table 12.10).

The problem consists essentially in that in metals heat transport is facilitated by electrons, and that in diamond, by phonons. Therefore, in composites with a metallic binder, energy is transferred between electrons and phonons. It is assumed that the diamond surface is coated by a very thin carbide layer that matches electrons with phonons.

The characteristics of the composite materials developed for use in heat sinks are determined primarily by the quality of the filler/binder boundary. The carbide-forming agents added to diamond and copper or silver are most frequently Ti, Cr, B, and Si. Transition metal carbides (NbC, TiC, WC), nitrides, and borides may be called metalloceramics, because they combine electronic conduction with high hardness. It is in these layers on the diamond surface that electron-phonon interaction arises.

It has been shown [17] that by using an atomic alloy of copper with chromium and Ib diamonds 177–210 μm in size it is possible to increase thermal conductivity

Table 12.10 Thermal Properties of Materials Commonly Used in the Fabrication of Heat Sinks

Material	Thermal Conductivity, λ W/(m/K)	Thermal Expansion Coefficient, $\alpha \times 10^6$, K^{-1}
Synthetic diamond Ib (C)	1,000–2,200	1.1
Silicon (Si)	153	3.8
Silicon Carbide (SiC)	280–400	3.8
Aluminum Nitride (AIN)	285	4.3
Indium Phosphide (InP)	68	4.3
Gallium Arsenide (GaAs)	55	5.9
Copper (Cu)	380	12.5
Silver (Ag)	420	19.8
Aluminum (Al)	240	22

up to 740W/(m·K), to be contrasted with 200W/(m·K) featured by a copper-diamond composite.

This idea is frequently used to improve the thermal conductivity of the diamond-binder boundary. It was proposed [18] to apply titanium coating to particulate diamond from a NaCl melt at 1,000 °C, or NaCl+KCl+$CaCl_2$ at 750 °C, an approach reportedly producing good results. The best agent to mediate diamond sintering at high pressures (6.5 GPa[6]) and temperatures (1,600–1,800°C) and produce composites with very good abrasive resistance was shown to be titanium carbide and TiC0 [19].

Ekimov et al. [20] demonstrated that Ti and Si present in diamond composites prepared at a pressure of 2 GPa, rather than being only a filler, form a binder. In such composites, thermal conductivity depends only weakly on the size of diamond grains and reaches a value as high as 600 W/(m·K) when employed as a SiC-Si matrix.

The data pertaining to the sintering parameters exhibit considerable scatter, both in pressure and in temperature. Published research results dealing with sintering of detonation nanodiamonds (DND) at a high pressure are few in number. It appears obvious, however, that microcrystalline diamonds and detonation nanodiamonds differ markedly in thermodynamic properties. Viewed in the context of possible applications, of most interest would be the possibility of obtaining large (up to 600 μm in size) single crystals by a suitable thermal treatment of detonation nanodiamonds.

12.9.2 Methods Employed to Develop High Thermal-Conductivity Composite Materials

Among the most promising approaches to developing composite materials exhibiting high thermal conductivity are sintering[7] at high-pressures and temperatures, infiltration of a melted metal into a matrix, and spark plasma sintering (SPS), a method that has appeared recently. Consider these methods in more detail.

12.9.2.1 High-Pressure Sintering

The definition "high pressure" assumes, as a rule, pressures above 1–1.5 GPa, i.e., levels in excess of the ultimate strength of hardened steels. Materials employed in high-pressure applications are tungsten carbide-based hard alloys. Equipment

---

[6] The pascal (symbol: Pa) is the SI unit of pressure defined as one newton per square meter. At sea level, 1 atmosphere is about 101.3 Pa. 1 GPa = 10^9Pa.

[7] *Sintering* is the synthetic manufacture of solid products using controlled heating of powdered raw materials. Application of the proper sintering temperature results in the adhesion of the powder grains to each other without melting the material. This proper temperature is typically 2/3 the melting point of the particular material.

enjoying the widest use in synthesis of superhard materials and their compaction are high-pressure chambers of the truncated hemisphere anvil and toroid types, multi-plunger split-sphere, and belt-type equipment. The pressures that can be reached with these chambers range from 4.5 GPa to 12 GPa.

Photographs of a toroid-type chamber can be found in [20], and description of the split-sphere apparatus, for instance, in [21], with a variety of designs illustrated in [22]. The principal shortcoming of the high-pressure approach lies in the size of the chamber and, accordingly, that of the samples, and in heavy labor consumption. Toroid-type chambers are commonly used to fabricate cylindrical samples with a diameter and height of not over 3–5 mm. In most cases, high pressure is employed in fabrication of abrasive composites rather than use of high thermal-conductivity materials.

The high-pressure method is instrumental in producing composites at temperatures above 1,000 °C, without the need for making the operating volume air-tight and providing means for its evacuation. Apart from this, by properly varying the pressure and maintaining a constant temperature, we can control phase transitions in such materials as, for instance, diamond and cubic boron nitride.

12.9.2.2 Infiltration

This process consists essentially in infiltrating a liquid binder into the filler material. High thermal conductivity composite materials are usually prepared with diamond particulates used as a filler and metals (aluminum, copper, silver) for a binder. Infiltration can be driven by squeezing a metal into the filler mechanically (a process called squeeze casting) or through a conveniently applied pressure drop (gas pressure-assisted process). Because these metals do not wet diamond, carbide-forming agents are added to the melt to form a structure of the type filler - barrier layer - binder. The rate of formation of the barrier layer in this case is limited by that of transport of the carbide-forming agent through the melt to the diamond surface. A drawback of this method lies in the slow rate of the process, as well as in the presence of an impurity, the carbide-forming agent, in the binder metal, which degrades its thermal conductivity.

An alternative way of cutting the infiltration time consists in prior formation of a barrier layer on the diamond surface [23], with subsequent infiltration of the melted binder metal. A method was proposed [6] for deposition of a titanium coating on particulate nanodiamonds from NaCl melt at 1,000°C, or from a NaCl + KCl + $CaCl_2$ blend at 750°C. Interaction of SiO_2 with diamond in vacuum at 1,350°C produces a uniform polycrystalline SiC coating 60 nm thick on diamond [24].

Infiltration can be used to advantage in fabricating large blocks of composites in an inert gas environment or under reduced pressure. The infiltration method is most expedient in production of large bulk heat sinks.

12.9.2.3 Spark Plasma Sintering

SPS is a fairly new approach to fabrication of composites from a variety of raw materials. Its main difference from the well-known methods employed in high pressure and temperature sintering lies in the possibility this approach offers of performing experiments in vacuum or an inert atmosphere and under pressure. A pulsed high-power current (500–1,000 kW) is passed through the sample to initiate spark discharge between particles of the powder to be sintered. This gives rise to local heating and efficient sintering of the composite, because what is actually being heated in the process is the sample material itself rather than the surrounding heater, as is the case with standard sintering techniques. The process could be conventionally characterized by two temperatures, more specifically, the average temperature of the material being sintered (it is usually not higher than 2,500 °C) and the local temperature at the discharge spot, which is substantially higher. The sintering process takes up, on the average, a few tens of minutes, whereas local heating lasts not longer than one millisecond.

This technique can be employed to advantage to sinter a broad variety of materials including high strength metals (W, Co), ceramics (TiN, Al_2O_3), ZrO_2- and Si_3N_4-based composites, diamond-metal composites, and so on.

12.9.3 Issues with the Application of Metal/Diamond Composites as Heat Sink Materials

As already mentioned, the linear thermal expansion coefficient of diamond is 1.18×10^{-6} K^{-1}, and that of copper, 12.6×10^{-6} K^{-1}. It is desirable that the composition of the composite intended for use as heat sink material have a thermal expansion coefficient as close as possible to that of the semiconductor of interest. An ideal heat sink material should have a thermal expansion coefficient (CTE) from 3.5×10^{-6} K^{-1} to 6×10^{-6} K^{-1}, an interval typical of the widely used semiconductors, such as Si, InP, GaAs, etc., and a high thermal conductivity.

Weber et al. [25] showed that adding to a diamond/copper composite boron with a concentration of 0.001 to 0.1 atom per copper atom brought the thermal expansion coefficient to the level of $\sim 7 \times 10^{-6}$ K^{-1}, and for chromium added in a similar concentration it was 10×10^{-6} K^{-1}.

The difficulties arising in an attempt to optimize the composition of a composite so as to reach a high thermal conductivity and a CTE consistent with that of semiconductors could be possibly resolved by adding a third or even a fourth component to the composite, provided that this would not result in a noticeable degradation of the thermal conductivity. The high electrical conductivity of such heat sinks, comparable with that of copper, would also pose a serious problem.

CVD-produced diamond plates may provide a serious competition to the heat sink materials based on diamond/copper composites. Industry has already produced by the CVD method samples of heat sinks with a thermal conductivity as high as, and even in excess of 1,000W/m/K, which measure 10 × 10mm and are up to 1mm thick. Their price, however, is presently fairly high.

12.9.4 Summary

In the limited space that we have here we could not cover all the aspects of the problem of fabrication of diamond-based composites. Our purpose was to provide an introduction to the topic, to show that such composites hold promise as materials for forming quality heat sinks.

We considered all of the principal methods presently in use in fabrication of diamond-based composites, namely, infiltration, high pressure and temperature sintering, and spark plasma sintering, an approach which has lately been attracting considerable interest. Their merits and shortcomings were considered.

12.10 Thermoelectric Cooling

12.10.1 Introduction

Active or passive cooling techniques can be used to remove heat. *Thermoelectric coolers* (TECs) are active systems; passive systems include thermal-interface materials, heat spreaders, and heat sinks [26].

A TEC is an active thermal-management device that can provide additional heat pumping and temperature stabilization. Figure 12.20 shows a simple example of the type of heat pumping and temperature control available from a TEC. TECs use the Peltier effect to create a heat flux between the junctions of two types of materials. Peltier coolers, heaters, and thermoelectric heat pumps are solid-state active devices that transfer heat from one side of a device to the other side against the temperature gradient—from cold to hot—as they consume electrical energy. People also refer to devices that operate in this manner as Peltier devices, Peltier diodes, Peltier heat pumps, solid-state refrigerators, or TECs.

The most basic representation of the operational space for a thermoelectric cooling device is a load line (Figure 12.21). The load line represents the ΔT and Q_{pumped} (heat-removed) conditions possible for a TEC's drive current. At the maximum drive current for the module, the maximum power the device can pump,

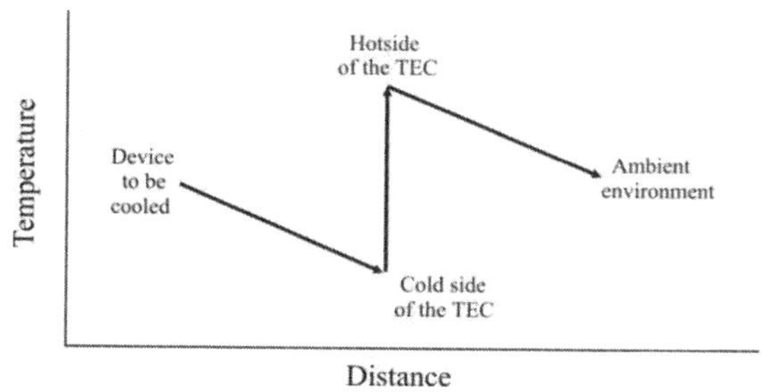

Figure 12.20 TEC temperature control.

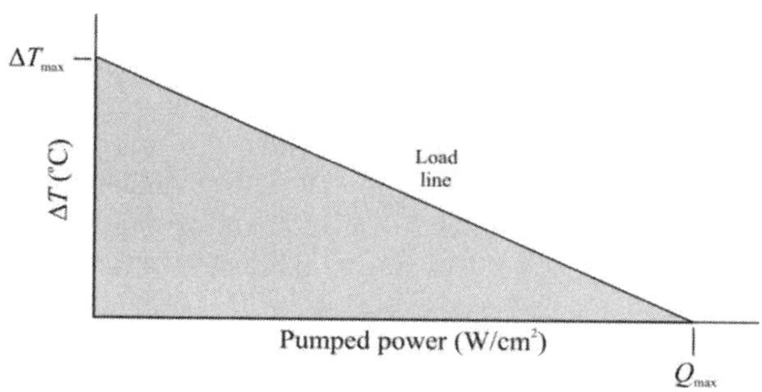

Figure 12.21 TEC Load line. The load line represents the change in temperature and pumped-power conditions possible for a TEC drive current and defines the operational region for the TEC.

Q_{max}, and the maximum temperature difference that the device can sustain between its top and bottom plates, ΔT_{max}, generate the load line. The ΔT_{max} condition occurs when the device reaches a zero-Q condition—that is, when no heat is flowing through it. We can theoretically calculate the value with

$$\Delta T_{\max} = \frac{\zeta^2 T_C^2}{2\lambda\rho} = \frac{\zeta^2 T_C^2}{2K\theta} \qquad (12.35)$$

where ζ is the Seebeck coefficient[8], λ is the thermal conductivity, ρ is the electrical resistivity, T_C is the cold-junction temperature, K is the thermal conductance, and θ is the resistance.

The Q_{max} condition occurs when there is no temperature difference between the top and the bottom of the TEC:

$$Q_{\max} = \frac{A\zeta^2 T_C^2}{2\rho L} = \frac{\zeta^2 T_C^2}{2\theta} \qquad (12.36)$$

where A is the area of the device and L is the length, or thickness, of the thermoelectric material. We can graph the two parameters on a chart as ΔT_{max} at $Q = 0$; Q_{max} at $\Delta T = 0$. The line connecting them is a load line. The resulting load line defines the operational space for TECs and is the best and usual way to illustrate their performance.

12.10.2 TEC Design Issues

One of the drawbacks of TECs is that they consume power while cooling. This power adds to whatever power the cooling zone is pumping out, so the system dissipates more heat at the larger system level with a TEC in operation than without one. If you use TECs to cool devices in the same manner as passive thermal elements—in other words, everywhere—you will end with a larger problem at the system level than you just solved at the chip level. A more cost-effective and efficient approach—and one that is possible only with TECs—would be to cool only what is necessary. In other words, scale the thermal-management system to the size of the heat problem.

The *coefficient of performance* of a TEC is given by

---

[8] The Seebeck coefficient of a material is a measure of the magnitude of an induced thermoelectric voltage in response to a temperature difference across that material, as induced by the Seebeck effect. The SI unit of the Seebeck coefficient is volts per kelvin (V/K), although it is more often given in microvolts per kelvin (μV/K).

$$COP = \frac{Q_{\text{pumped}}}{P_{\text{i}}} \tag{12.37}$$

P_i is the input power.

A TEC pumps a certain amount of heat, Q, and adds a particular amount of heat, $Q \times COP$, to move this heat. This situation results from the inherent inefficiency in all engines. As a result, bulk TECs are often available as systems that include the device itself and a heat-transfer device, such as a fan, heat sink, or heat pipe. The value of the TEC in this case is that it can deliver subambient temperatures and provide active temperature control, but at the cost of increasing the system-level heat-transfer problem.

Because heat in the passive case flows linearly, any material between the TEC and the heat source has a temperature drop across it. This decrease increases the temperature difference that the TEC must pull. The TEC acts as a heat pump moving heat in a manner whose efficiency depends upon the temperature difference it must generate. Minimizing this temperature difference improves the efficiency of the cooler and reduces the additional heat at the system level.

Integrating a TEC close to the heat source is the key to improving the TEC's operational efficiency. Adding a heat-transfer system defeats the purpose of the integration. As such, you must pay careful attention to the characteristics of the heat-transfer problem, the design of the TEC, and the design of its package. When you address these issues—ideally during product design and development—you can achieve significant performance improvements.

12.10.3 Localized Thermal Management

Dense electronic systems generate a lot of heat that can lead to a significant rise in temperatures, causing device and system-level failures. The answer to these problems has always been to use a larger fan or a larger heat sink to move the heat from the electronic package and into the system environment. However, the use of fans and heat sinks merely spreads the heat into other systems, such as enclosed equipment racks, which then require a thermal-management approach of their own. Heat from these racks usually spills into the system room or IT data center. These rooms, with people working in them, are sensitive to high temperatures. Tackling this problem requires the use of expensive air conditioners, which cool everything in the room to the lowest possible temperature.

The most efficient thermal-management system involves embedding thermal-management functions at the source of the heat to remove only the heat that is detrimental to the system's performance and then passing on that reduced heat in a controlled manner to the next level. Figure 12.22 compares the cost of implementing thermal management with the level at which the technique occurs. Implementing

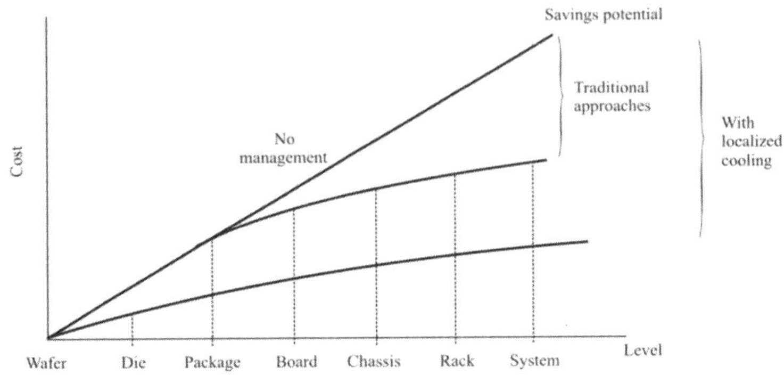

Figure 12.22 Cooling efficiency. Localized cooing at the system level can be more efficient than removing heat from the servers to the server room and then cooling the entire room with air conditioning.

heat sinks, fans, and large-scale cooling creates an energy-savings potential. Introducing localized cooling in the overall thermal-management design translates to a greater cost savings potential at the rack and system levels. Today's electronic systems typically employ only passive elements for thermal management. This approach removes heat uniformly from across the die. The point of heat removal, however, is to reduce the peak temperature on the die to improve performance. The heat associated with the peak temperatures on a die is only a small fraction of the total removed heat. Removing this excess heat leads to problems at the system level.

Solving the thermal-management issues at the die and system levels requires a paradigm shift. Integrating localized thermal management that combines active and passive components within electronic systems allows flattening of the power map on a die and minimizes the waste the system must dispose of. This type of selective removal of heat simplifies issues at the system level and reduces cost.

12.11 Enclosure Fans

We have mentioned at several places in the above discussion that providing forced air across heat sinks can significantly improve on the amount of heat removed. In this section we discuss some of the concepts on how to provide that airflow in electronic enclosures. It can be accomplished using fans, and we now discuss the basic properties of fans [27].

Figure 12.23 Axial flow fan.

12.11.1 Fan/Blower Descriptions

Fans can be thought of as low-pressure air pumps that utilize power from a motor to output a volumetric flow of air at a given pressure. A propeller converts torque from the motor to increase static pressure across the fan rotor and to increase the kinetic energy of the air particles. The motors are typically permanent split capacitor AC induction motors or brushless DC motors.

Cooling fans are available in many configurations, however axial and centrifugal flow (also called blowers) fans are the most common used to cool electronic applications [28].

Electronic cooling fans are used to draw or flow air from the outside of a electronic enclosure into the electronics area, expel heated air from inside of the enclosure, or move air across a heat sink or electronic device to accelerate the removal of heat energy from the device. The use of cooling fans and/or other hardware and technologies to cool electronics is often referred to as *active cooling*.

12.11.2 Types of Fans and Blowers

Air moving devices are generally available as either a fan (as illustrated in Figure 12.23) or a centrifugal blower (as illustrated in Figure 12.24). The main difference between fans and blowers is their flow and pressure characteristics. Fans deliver air in an overall direction that is parallel to the fan blade axis (orthogonal to the plane within which the blade rotates) and can be designed to deliver a high flow rate, but tend to work against low pressure. Blowers tend to deliver air in a direction that is orthogonal to the blower axis at a relatively low flow rate, but against high pressure.

There are several types of fans, some of the most common being propeller, tube axial, and vane axial styles. Propeller fans are the simplest type of fan, consisting of only a motor and propeller. One problem with propeller fans is that tip vortices are produced by the pressure differential across the airfoil section.

Figure 12.24 Typical blower or radial flow fan.

A tube axial fan (the most common type in electronic cooling systems, Figure 12.23) is similar in construction and operation to a propeller fan, but also has a venturi around the propeller to reduce the vortices. The vane axial fan has vanes that trail behind the propeller in the airflow to straighten the swirling flow created as the air is accelerated.

Centrifugal blowers may have a forward curved wheel, a backward curved wheel, or be of the squirrel cage variety (see Figure 12.25). Straight blade blowers are designed to deliver higher pressures for pressurizing and long duct or pipe runs. Backward inclined blower wheels are available with a flat blade design or a more efficient airfoil design blade where each blade has a cross section similar to an airplane wing. These wheel types are commonly used for general ventilation, forced cooling at higher pressures and in dust collection systems where the fan is on the clean air side of the dust collector. Forward curve multi-vane blower wheels are for moving large volumes of air at lower pressures. Typically used for general ventilation or forced cooling at slower speeds.

12.11.3 Basic Aerodynamics

Fans are of such common use that a basic understanding of the aerodynamics is appropriate. The blades of a fan propeller resemble the wing of an airplane and as such follow the same aerodynamic laws. For example, a fan blade produces lift when

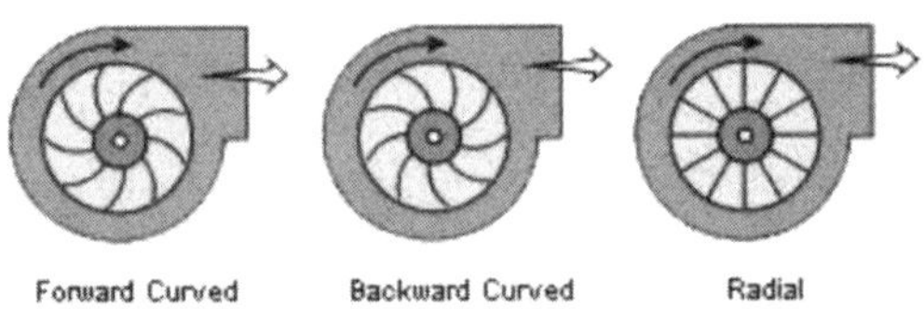

Figure 12.25 Fans: Centrifugal flow fan types (blower). For the applications of these types, see text.

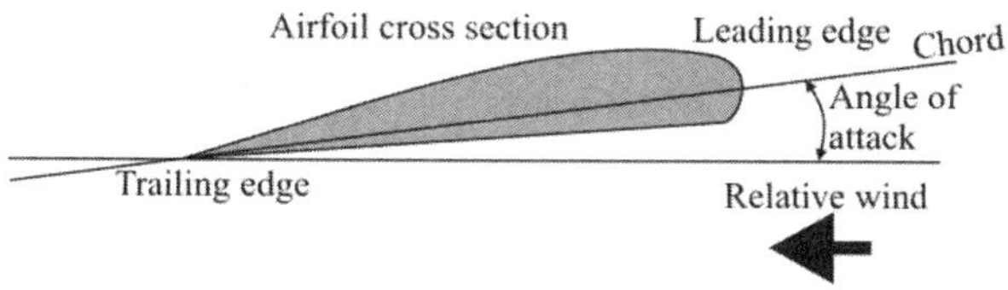

Figure 12.26 Nomenclature for an airfoil. The chord is a line extending from the leading edge to the trailing edge. The angle of attack is the angle formed between the relative wind and the chord.

the chord is elevated from the direction of the relative wind as shown in Figure 12.26.

The elevation angle is referred to as the *angle of attack* (AOA). The greatest airflow delivery from a fan occurs when the AOA is at a minimum, but the pressure differential across the fan is zero. As the AOA is increased, the airflow delivery decreases and the pressure differential increases. The airflow can decrease to nearly zero, but will also deliver the maximum pressure differential in this condition, which is called the *shut-off point*. When an AOA is reached where the air no longer flows smoothly and begins to separate from the blades, an "aerodynamic stall" condition exists.

Since a fan is a constant volume device, it moves the same volumetric flowrate of air irrespective of the air density. However, the mass flowrate does change as the density changes. This becomes important when equipment is expected to operate at altitudes significantly greater than sea level a common application for many EW situations. Therefore, the volumetric flow rate required at high altitude where the air density is low will be greater than that required to achieve the same cooling as at sea level where the density is relatively higher.

12.11.4 Fan Curve

All of the important aerodynamic aspects of a fan are indicated on a fan curve such as is shown in Figure 12.27. In contrast to an airplane wing, there is life after stalling in a fan. A stalled fan continues to deliver air, but at an increased static pressure and a decreased volumetric flow rate, and also at the cost of an increase in noise. If noise is not a consideration, the fan can be utilized in this condition.

At the shut-off point, the fan is in the condition of the maximum potential and least kinetic energy. At free delivery, the fan is in the condition of the maximum kinetic and least potential energy. Although neither of these extreme conditions is likely to occur in practice, they can be useful parameters in comparing fans.

The governing principle is that any given fan can only deliver one flow at one pressure in a particular system. This "operating point" is determined by the intersection of the fan static pressure curve and the system pressure curve. Figure 12.27 illustrates the operating points of both high and low resistance systems, as

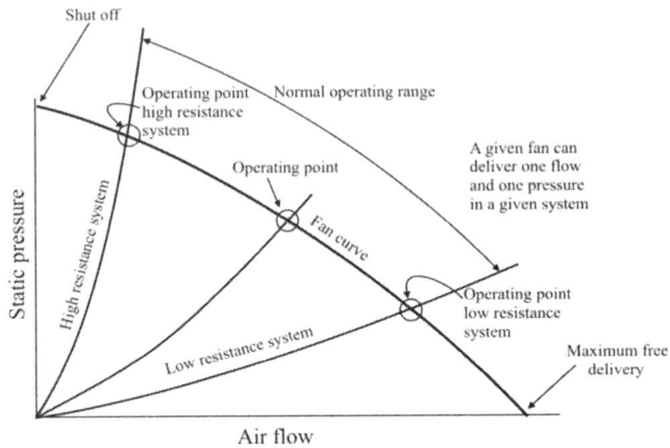

Figure 12.27 Fan curve. Fan/system interaction.

well as one half way in between. It is usually best to use a fan that will give an operating point toward the high flow, low pressure end of the performance curve (the bottom curve) to maintain propeller efficiency and to avoid propeller stall. Each configuration should be analyzed for possible reduction in the overall resistance to airflow. Other considerations, such as available space and power, noise, reliability, and operating environment should also be brought to bear on fan choice as required for the particular application.

12.11.5 Steps to Fan Selection

12.11.5.1 Estimate the Required Airflow

Before selecting a fan, we must obtain as accurate an estimate as possible of the heat to be dissipated, because the overall system air temperature differential above the inlet ambient is directly proportional to the heat dissipated. We can then estimate the amount of required cooling air.

The heat transfer equation is given by

$$Q = \dot{m} C_p \Delta T \tag{12.38}$$

where

Q is the amount of heat transferred to the system, Watts
C_p is the specific heat of the air, $\text{J/(kg} \cdot \text{K)}$

$\dot{m}$ is the mass flowrate of the air, kg/sec
ΔT is the desired air temperature differential (cabinet to ambient outside air), K

The relationship between mass flow rate and volumetric flow rate is

$$\dot{m} = \rho G \tag{12.39}$$

where

G is the volumetric flowrate, m^3/sec
ρ is the air density, kg/m^3

The required volumetric flow rate is calculated from

$$G = \frac{Q}{\rho C_p T} \tag{12.40}$$

This yields a rough estimate of the airflow needed to dissipate a given amount of heat at sea level. As previously noted, it is the mass flowrate of air, not its volumetric flowrate that governs the amount of cooling.

12.11.5.2 Estimate the Actual Airflow

The preceding steps indicated the necessary procedure to estimate the *required* airflow in order to limit the desired overall air temperature rise ΔT. However, we also indicated that the *actual* operating airflow is determined by the intersection of the fan curve and the system resistance curve. There are three options available for estimating this operating point:

- Experimental measurement using a thermal/mechanical mockup of the system;
- Calculation of the operating point using airflow network methods;
- Use *computational fluid dynamics* (CFD) software (available from commercial software companies).

A complete system resistance curve can be developed using the experimental procedure by measuring the total airflow for specific fans or measuring several pressure-airflow data pairs. The latter experimental method will then require the superposition of the selected fan pressure vs. airflow curve and system resistance curve to obtain the operating airflow.

The airflow network procedure provides acceptable results when the geometry is simple and the flow path within the cabinet is known can be estimated.

In many practical applications, however, complex three-dimensional flow paths must be examined that are not initially known. In these situations, CFD software is often used. The fan performance curve can be supplied as an input and the operating point and system resistance are then determined. CFD works by numerically solving the governing equations of flow and heat transfer in three dimensions and takes into account the effects of turbulence and gravity. CFD can be used to study the performance of fans in series and parallel arrangements as well as optimize the location with respect to other objects inside the cabinet. Both of the computational procedures require a static pressure vs. airflow curve for the fan in question.

All of these procedures require a system resistance curve of the type shown in Figure 12.27. System resistance curves are usually expressed as a non-linear expression of pressure vs. airflow:

$$\Delta P = K\rho G^{n} \tag{12.41}$$

where

ΔP is the system pressure loss
K is a load factor specific to the system
ρ is the density of air
G is the airflow rate
n is a constant that varies between 1 and 2 depending on whether the flow is completely laminar ($n = 1$) or completely turbulent ($n = 2$)

If the estimated value of the actual airflow is significantly less than the required value, the enclosure system should be examined for regions where the airflow is better; that is where the airflow resistance could be reduced. Should this review fail to provide an answer, a different fan or perhaps multiple fans should be considered. A different fan can be found simply by reviewing the catalogs of the various fan vendors. Using multiple fans is a little more complex.

12.11.6 Multiple Fans

Combining fans in series or parallel can sometimes achieve the desired airflow. Parallel operation is when two or more fans are blowing together side by side. The performance of two fans in parallel will increase the volume flowrate (double at maximum delivery). Systems with low resistance are best for using parallel fans. A fan curve for multiple, identical fans in parallel may be constructed by scaling the fan curve airflow axis data in direct proportion to the number of fans.

In series operation, the fans are stacked one upon the other, resulting in an increase of static pressure (doubling at shut-off, but less elsewhere). Series fans are best for systems with high resistance. A fan curve for multiple, identical fans in series can be constructed by scaling the fan curve pressure axis data in direct proportion to the number of fans.

With both series and parallel operation certain areas of the combined performance curve will be unstable [particularly with multiple fans (5, 6, 7, and so forth)], and should be avoided. This instability is often unpredictable and is a function of the fan and motor construction and the operating point.

Fan placement in the enclosure is also important to consider. Pressurizing the enclosure (fan blowing in) is the preferred method, since incoming air can be readily filtered. In addition, a pressurized enclosure will deter dust entering through cracks or crevices. The fan is also handling cooler, denser air, and it will therefore have a somewhat higher pressure capability (this may be a very slight advantage for low heat dissipating systems). An important feature of a pressurized system is that the fan life and reliability are increased due to the fan ambient temperature being lower. The biggest disadvantage of pressurization is that heat generated by the fan is dissipated into the enclosure.

12.11.7 Fan Laws

The fan laws permit the determination of the output of a given fan under other conditions of speed or density, or converting the known performance of an air mover of one size to that of another similar unit of a different size.

Geometrically similar fans can be characterized by the following four equations:

Volumetric Flowrate: $G = K_{\mathrm{q}} N D^3$

Mass Flowrate: $\dot{m} = K_{\mathrm{m}} \rho N D^3$

Pressure: $P = K_{\mathrm{p}} \rho N^2 D^2$

Power: $HP = K_{\mathrm{HP}} \rho N^3 D^5$

where:

K is the constant for geometrically and dynamically similar operation
G is the volumetric flowrate
$\dot{m}$ is the mass flowrate
N is the fan speed in RPM
D is the fan diameter

Table 12.11 Fan Law Equations

Constants	Variable	Fan Laws
Diameter (D) Density (ρ)	Speed (N)	$G_2 = G_1(N_2 / N_1)$ $P_2 = P_1(N_2 / N_1)^2$ $HP_2 = HP_1(N_2 / N_1)^3$
Speed (N) Density (ρ)	Diameter (D)	$G_2 = G_1(D_2 / D_1)^3$ $P_2 = P_1(D_2 / D_1)^2$ $HP_2 = HP_1(D_2 / D_1)^5$
Diameter (D) Speed (N) Volumetric Flowrate (G)	Density (ρ)	$P_2 = P_1(\rho_2 / \rho_1)^2$ $HP_2 = HP_1(\rho_2 / \rho_1)^5$
Diameter (D) Mass Flowrate $(\dot{m})$	Density (ρ)	$G_2 = G_1(\rho_2 / \rho_1)$ $P_2 = P_1(\rho_2 / \rho_1)$ $N_2 = N_1(\rho_2 / \rho_1)$ $HP_2 = HP_1(N_2 / N_1)^2$

HP is the power output

From these relationships, it is possible to calculate a fan performance at other conditions. Table 12.11 is a summary of the fan law equations in a form useful for fan analysis.

12.11.8 Cabinet Cooling Hints

In addition to selecting a fan, there usually are some choices in the location of the fan or fans. Figure 12.28 can be used as a guide. All of these common-sense principles apply to both chassis enclosures as well as cabinets and room installations.

12.12 Concluding Remarks

In this chapter the thermal management of electronic circuits, and in particular power amplifiers, was addressed. The basic heat flow properties were covered first with an examination of the physics involved and the appropriate mathematics

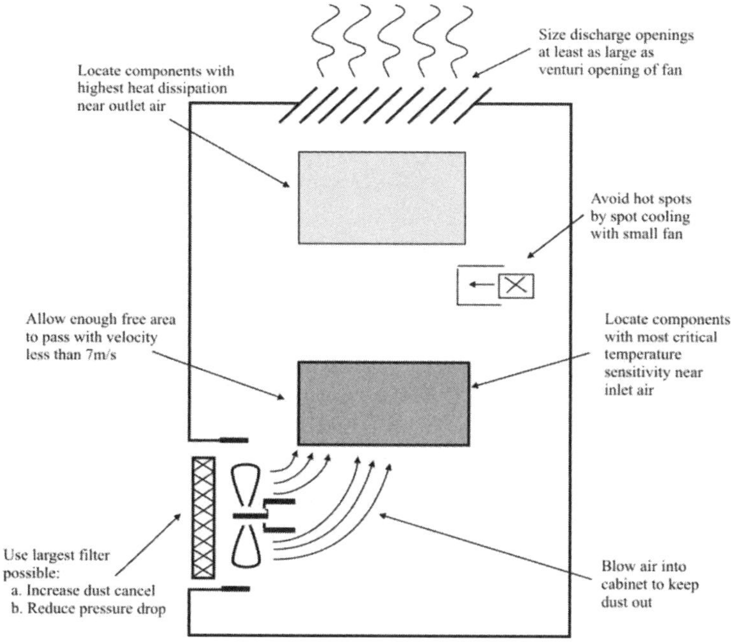

Figure 12.28 Cooling hints.

examined. That was followed by a presentation on heat sinks and how they can be used to cool components. The heat characteristics of PCBs was discussed, along with an examination of some new approaches for heat removal. Last, the properties of fans for heat removal was presented.

Thermal management in EW systems is a critical area of concern and must be considered early in the design process. This is particularly true when power amplification in involved, and active EW systems certainly are involved with that.

References

[1] Active Heat Removal Cools Electronics Hot Spots 76C5EE36-6EB9-4809-98DA-108C609EE4A2 – Mar. 18, 2010, accessed May 2013.

[2] http://www.rf-amplifiers.com/index.php?topic=cooling, accessed Dec. 2009.

[3] Manz, B., "How Far Can We Take GaN Technology?" The Journal of Electronic Defense, Vol. 38, No. 3, Mar. 2015, pp. 41-46.

[4] Kuhn, K. A., http://www.kennethkuhn.com/students/thermal.pdf, April 14, 2003, rev. Apr. 22, 2007, accessed Aug. 2010.

[5] Lee, S., June 1st, 1995. https://www.electronics-cooling.com/1995/06/how-to-select-a-heat-sink/#, accessed Nov. 2009.

[6] Gvorki, J. R., http://www.designworldonline.com/How-to-Select-a-Suitable-Heat-Sink/, November 11, 2009, accessed Mar. 2011.

[7] Song, S., S. Lee, and V. Au, "*Closed Form Equation for Thermal Constriction/Spreading Resistances with Variable Resistance Boundary Condition,*" *Proceedings of the 1994 IEPS Technical Conference*, 1994, pp. 111–121.

[8] http://www.learnabout-electronics.org/Amplifiers/amplifiers51.php, accessed Nov. 2013.

[9] Coonrod, J., *Microwaves and RF,* http://mwrf.com/materials/thermal-design-begins-circuit-board, May 18, 2011, accessed Jan. 2014.

[10] http://www.ecnmag.com/articles/2010/04/comparing-impact-different-heat-sink-materials-cooling performance, accessed May 2013.

[11] http://www.ep.liu.se/ecp/057/vol3/001/ecp57vol3_001.pdf, accessed Jul. 2015.

[12] Klett, J., "High Conductivity Graphite Foams," Oak Ridge National Laboratory, 2003.

[13] www.mdpi.com/1996-1944/2/4/2467/pdf, *Materials 2009*, Vol. 2, pp. 2467–2495, www.mdpi.com/journal/materials, accessed Sep. 2011.

[14] Sergey V., S. V. Kidalov, and F. M. Shakhov Dec. 21, 2009.

[15] Slack, G.A., "Nonmetallic Crystals with High Thermal Conductivity," Vol. 34, *Journal of Physics and Chemical Solids,* 1973, pp. 321–335.

[16] Lienhard, J.H., IV; Lienhard, J.H., V. *A Heat Transfer Textbook.* 3th Ed., Cambridge, MA: Phlogiston Press, 2005, pp. 698–699.

[17] Schubert, T., L. Ciupiński,, W. Zieliński,, A. Michalski, T. Weißgärber, and B. Kieback, "Interfacial Characterization of Cu/Diamond Composites Prepared by Powder Metallurgy for Heat Sink Applications," *Scripta Materialia* , 2008, Vol. 58, pp. 263–266.

[18] Breval, E., J. Cheng, D. K. Agrawal, "Development of Titanium Coatings on Particulate Diamond," *Journal of the American Ceramic Society* 2000, Vol. 83, *Materials* 2009, No. 2, pp. 2106–2108.

[19] Hong, S. M., M. Akaishi, and S. Yamaoka, "High-Pressure Synthesis of Heat-Resistant Diamond Composite Using a Diamond–TiC0.6 Powder Mixture," *Journal American Ceramics Society,* 1999, Vol. 82, pp. 2497–2501.

[20] Ekimov, E.A., N. V. Suetin, A. F. Popovich, and V. G. Ralchenko, "Thermal Conductivity of Diamond Composites Sintered under High Pressures," *Diamond and Related Materials*, 2008, Vol. 17, pp. 838–843.

[21] Abbaschian, R., H., Zhu, C. Clarke, "High Pressure-High Temperature Growth of Diamond Crystals Using Split Sphere Apparatus," *Diamond and Related Materials*, 2005, Vol. 14, 1916–1919.

[22] Abyzov, A.M.; Kidalov, S.V.; Shakhov, F.M. "Composite Material of Diamond-Copper with High Thermal Conductivity,"*Material Science Transactions, (Materialovedenie)*, Vol. 5, 2008, pp. 24–28.

[23] Miyamoto, Y., T. Kashiwagi, K. Hirota, O. Yamaguchi, H. Moriguchi, K. Tsuduki, A. Ikegaya, "Fabrication of New Cemented Carbide Containing Diamond Coated with Nanometer-Sized SiC Particles," *Journal of the American Ceramic Society*, 2003, No. 86, pp. 73–76.

[24] Weber, L., R. Tavangar, "On the Influence of Active Element Content on the Thermal Conductivity and Thermal Expansion of Cu–X (X = Cr, B) Diamond Composites," *Scripta Materialia.* 2007, Vol. 57, pp. 988–991.

[25] Active Heat Removal Cools Electronics Hot Spots, March 18, 2010.

[26] Turner, M., http://www.electronics-cooling.com/1996/05/all-you-need-to-know-about-fans/, accessed July 2015.

[27] http://www.engineersedge.com/industrial-equipment/electronic-cooling-fans.htm, accessed Sep. 2013.

[28] Ellison,. N., Fan Cooled Enclosure Analysis Using a First OrderMethod, *Electronics Cooling*, Vol. 1, No. 2, October 1995, pp. 16-19.

Chapter 13

Power Supplies for RF Power Amplifiers

13.1 Introduction

Power supplies serve extensive EW system applications. Basic distinctions between linear and switching power supplies make a difference, depending on the specific application. Each type has its own advantages and disadvantages [1, 2].

The term *power supply* is a misnomer. Power supplies do not create or supply power. Rather, power supplies are electronic or electrical equipment that change the form of available energy into a form useful to equipment. Some examples of sources of electrical energy are: wall outlets, solar panels, batteries, and wind. A more recent technology is energy harvesting—energy extraction from our environment around us. The process of changing electrical energy form is called power conversion and a better term for power supply is *power converter*.

For example, in the typical home, power is available at electrical wall outlets in the form of 120 volts, 60 Hz alternating current (in the United States). This means that electrical current with RMS amplitude of 120V is constantly reversing in direction 60 times per second. This is the form of electrical power generated by rotating machinery at the utility company.

But this form of electrical energy is not directly useful for electronic equipment, which typically requires a constant voltage level of some amount, perhaps 12 *volts direct current* (VDC). The role of the power supply, in this case, is to change the form of electrical energy available from the primary supply (e.g., wall alternating voltage) to a constant voltage at 12 VDC.

13.1.1 Types of Power Supplies

Power supplies can change voltage up or down, rectify voltage, provide isolation, provide a constant current, provide a constant power, or frequency.

They can also be classified according to switching or linear type. A switching supply uses high-frequency switching and energy storage devices such as inductors and capacitors to chop the incoming energy into small building blocks and then recompose the building blocks into the required form for use. A linear supply uses a series element such as a pass transistor to absorb the unneeded energy that is not used by the load. A switching supply is generally light and efficient, but the input current and output voltage cleanliness is not so great. A linear supply produces very clean output voltage, but it is generally large and heavy.

The architecture shown in Figure 13.1 applies to all power supplies that convert AC to DC. The input transformer normally converts a (relatively) high voltage AC source (e.g., 120V or 240V in the United States, 440V in the military) to a lower AC value. It also provides some isolation of the power supply from line fluctuations and noise. This AC voltage is converted into unipolar waveforms by the rectifier. The output of the rectifier may be smoothed by an LC circuit, removing much of the remaining ripple. If the application does not require regulation of the power supply DC level, then the voltage regulator is not required. Otherwise the resulting DC voltage/current is then fed to a regulator. The regulator is where the two main types of supplies differ. A linear supply uses active devices to regulate the DC voltage level while a switching (nonlinear) supply converts the DC level into a switched waveform (a square wave) to perform the regulation. The switched waveform is reconverted to DC after regulation is applied.

13.1.2 Linear Supplies: Basic Concept

The incoming AC voltage is stepped down to a lower AC voltage. For example, 120 VAC is stepped down to 24 VAC. The 24 VAC is then rectified through a full-wave bridge rectifier, usually with a high-current, low-voltage bridge. A filter capacitor is used to maintain a constant DC level with minimum ripple at the output of this rectifier.

The output voltage is controlled by a power transistor operating in its linear region. It acts as a variable resistor in series with the load. The power transistor receives its control from a circuit that senses output voltage. The control circuit modifies the transistor bias to maintain a constant voltage output, regardless of changes in the load current.

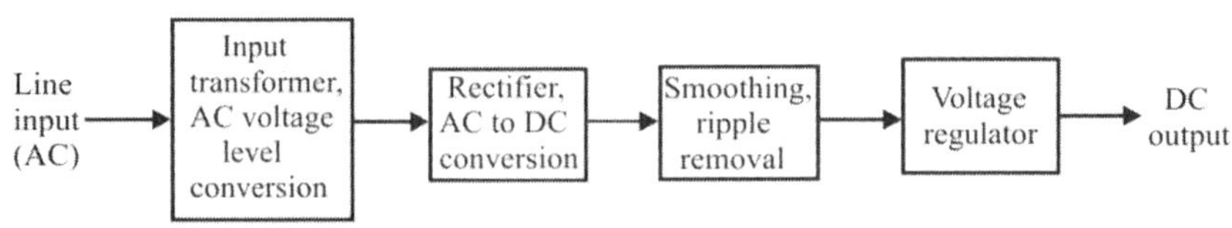

Figure 13.1 Flow diagram of power supply architecture.

13.1.3 Switching Supplies: Basic Concept

The incoming ac voltage is rectified and filtered to produce a high-voltage dc. A low-current, high-voltage bridge rectifier (that may not require a heat sink) can be used, as opposed to the linear bridge.

A power transistor — a MOSFET — is connected in series with the transformer. The MOSFET serves as an on-off switch and switches at a preset frequency. While the MOSFET is switching, the magnetic field in the transformer is building and collapsing, allowing energy to transfer to the secondary side.

The magnetic energy received by the secondary windings of the transformer is then full-wave-rectified and reconstructed into the proper dc level. A sample of parameters (V_{out}, I_{load}, and so forth) can be sent back to the primary side to serve as input to the *pulse-width modulator* (PWM). The PWM circuit modifies the length of time that the MOSFET is switched on in order to maintain output regulation. For example, in a switching power supply producing 12 VDC and powering a 3A load, an increase in the load to 4A causes the output voltage to drop slightly. The feedback circuit detects the voltage drop and passes it to the PWM, which turns the MOSFET on for a longer period (i.e., it increases the duty cycle), causing more magnetic energy to transfer to the secondary side until the output voltage reaches its predetermined value.

13.1.3.1 Switching Frequency

Frequency for a switching power supply usually ranges between 30 kHz and 150 kHz, but it can be much higher. Frequency for linear power supplies is the same as the line frequency (60 Hz in North America). Switching frequency selection depends on the application for which the power supply is designed. Because high-frequency switching occurs at f_0, harmonics are generated at $3f_0$, $5f_0$, $7f_0$, ... The selection of the frequency has to be such that none of these harmonics will interfere with the load. With power supplies for receivers and amplifiers, for example, the switching frequency should be selected so as not to interfere with VHF, UHF, or the IF used.

13.1.4 Advantages and Disadvantages

Linear — one advantage of linear power supplies may be familiarity, because they have been available for many years. They are known to be relatively noise-free and reasonably reliable. They are generally easy to design and fairly inexpensive to manufacture.

Because of the large transformers required, linear power supplies are generally heavy, which may be either an advantage or a disadvantage, depending on the need to balance weight distribution in a given application. As a general rule of thumb, a

16V output linear power supply weighs about one pound per ampere. A possible disadvantage of linear power supplies relates to the power transistor used to regulate the load. Because the power transistor operates in its linear region, and all the output current must pass through it, it requires large heat sinks to dissipate energy loss. (Recall that the power transistor is in series with the load and acts as a variable resistor.) Except in rare instances where heat is wanted to warm interior space, the inefficiency of linear power supplies — 50% — has to be considered a disadvantage.

Switching power supplies — although switching power supplies have been available for a number of years, higher production costs compared to linear power supplies, have limited their use in some applications. Early switching power supplies used discrete components to control pulse width and BJTs instead of MOSFETs as main switch components. As a result, the disadvantages of switching power supplies once included uneven reliability and radiated EMI (electrical noise). Although they were known to be noisy, unreliable, and difficult to mass produce, switching power supplies had the advantage of being lighter and smaller than their linear counterparts. In the last few years, big improvements in PWM and MOSFET design have been made. When all design considerations have been taken into account, switching power supplies are highly reliable and virtually noise-free. Production costs have come down because application-specific components are being designed for use in switching power supplies.

Switching power supplies are about 80%–90% efficient. Higher efficiency usually is an advantage, because heat normally is considered to be wasted energy (at the least) and potentially damaging to nearby electronic components.

13.1.5 Summary

Switching power supplies are gaining in popularity mostly because of their smaller size, lighter weight, and efficiency. Reliability and noise characteristics are becoming less of an issue as customers learn about the latest product developments. When assessing efficiency, size, and cost of shipping, one has to consider the alternative to a linear power supply: the switching power supply.

The structure of this chapter is as follows. We begin with a discussion of perhaps the simplest form of power supply (except for batteries) that is the unregulated power supply. Such power supplies are useful in some simple applications where power is not critical. Then we include a discussion on power regulators and how they are used in power supplies. That is followed with a section on power distribution in EW systems including some material on decoupling and bypassing for noise reduction. We then discuss using ground vias for PCB layout. A side-note is then included on a method for obtaining –5V from a +12V power supply. –5 is a recurring requirement in MOSFET-based systems for gate biasing and +12V is a common power level in such systems. With the circuit shown, a separate –5V supply isn't needed. The next section discusses the fundamental issues in EMI/RFI/EMC considerations. The last section in this chapter provides some tips for implementation of power supplies in-high power EW systems.

13.2 Unregulated Power Supplies

13.2.1 Introduction

The aim of a DC power supply is to provide the required level of DC power to the load using an AC supply at the input. Different applications require different attributes, but more often than not these days DC power supplies provide an accurate output voltage—this is regulated using electronic circuitry so that it provides a constant output voltage over a wide range of output loads.

In most power supplies there are number of different elements (see Figure 13.1). These may not all be present in every design.

- *Input transformer*: The input transformer is used to transform the incoming line voltage down to the required level for the power supply. Typically, the input transformer provides a step-down function. It also isolates the output circuit from the line supply.
- *Rectifier*: The power supply rectifier converts the incoming signal from an AC format into raw DC. Either half-wave or more commonly full-wave rectifiers may be used as they make use of both halves of the incoming AC signal.
- *Smoothing*: The raw DC from the rectifier is far from constant falling to zero when the AC waveform crossed the zero axis, and then rising to its peak. The addition of a reservoir capacitor here fills in the troughs in the waveform, enabling the next stage of the power supply to operate. Large value capacitors are normally used within this stage.
- *Regulator*: This stage of the power supply takes the smoothed voltage and uses a regulator circuit to provide a constant output virtually regardless of the output current and any minor fluctuations in the input level.

13.2.2 Diodes

Linear power supplies utilize diodes to convert bipolar waveforms into unipolar waveforms; that is, from AC waveforms to DC waveforms. We therefore include a brief introduction to the characteristics of diodes.

The diode rectifier circuits can be used for both power rectification and signal detection as well as many other uses. The chief differences are the frequency and the power levels involved. Our interest here is in the power supply application where higher power levels are normally involved.

Diode rectifier circuits obviously rely on diodes for their operation. Diodes are devices that only pass current in one direction. When Ambrose Fleming invented the first diode—a thermionic device / vacuum tube—he called it a valve because of its one-way action. Semiconductor devices are now used, and these provide exactly the same function.

The diode has a characteristic similar to that shown in Figure 13.2. In the forward direction, the so-called forward bias region, a small voltage is required across the diode before it conducts—the turn on voltage. The actual voltage depends on the type of diode rectifier and the material used. For a standard silicon diode rectifier this turn on voltage is around 0.6 volts.

In the reverse direction, the so-called reverse-bias region, little current flows through the diode and the diode rectifier will ultimately break down. The breakdown voltage is normally well in excess of the turn on voltage—the scales in Figure 13.2 have been compressed in the reverse direction to illustrate reverse breakdown.

There are many different types of rectifier diode that can be used—each one with its own properties, advantages, and disadvantages.

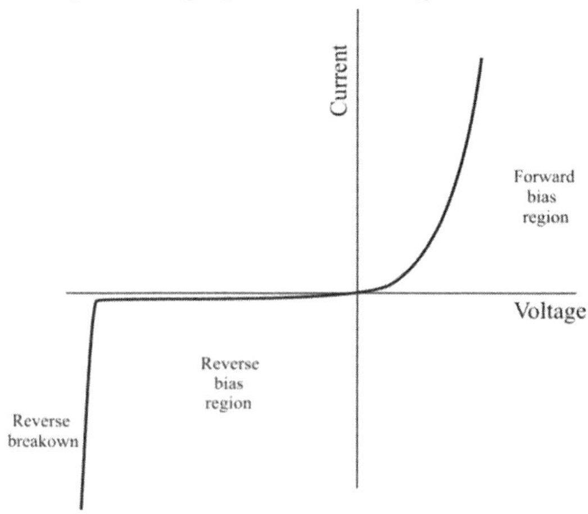

Figure 13.2 Diode rectifier characteristic.

For power rectification applications, power diodes of Schottky diodes are normally used. For signal rectification small point contact diodes, signal diodes, or Schottky diodes may be used. The Schottky diode has the advantage that it only requires a forward voltage of around 0.2 – 0.3V for forward conduction. This is particularly useful when detecting small radio signals, and when used as a power rectifier the power losses are reduced. However, the reverse leakage characteristics are not as good as normal silicon diodes.

13.2.2.1 Diode Rectifier Operation

The action of the diode is to allow current to flow in only one direction. Therefore, an alternating waveform is applied to a diode, then it will only allow conduction over half the waveform. The remaining half is blocked (see Figure 13.3).

13.2.3 Unregulated Power Supplies

The power supply is not a source of energy, but rather is an electronic circuit designed to change energy from the form of an available source to the form required by the load. In effect, the power supply is an interface matching the source to the load. The power supply will always obey the law of conservation of energy.

An unregulated power supply is a simple power supply whose output is a function of the input source voltage, output load, and ambient temperature. There is no circuitry added to regulate the output, compensating for these effects.

There are several different configurations of diode rectifier circuit that can be used. These different configurations each have their own advantages and disadvantages and are therefore applicable to different applications.

13.2.3.1 Types of Unregulated Power Supplies

There are three types of unregulated power supplies:

- Half-wave rectified;
- Full-wave center tap rectified;
- Full-wave bridge rectified.

Half-Wave Rectified Supply

The half-wave rectifier circuit is the most straightforward rectifier circuit that can be used.

Figure 13.3 Diode rectifier operation.

Although the half-wave rectifier circuit may not provide the level of performance needed for many applications, its simplicity lends it to use in many circumstances.

In view of this advantage, the half-wave rectifier circuit can often be seen within large items of electronics equipment.

The half-wave rectifier circuit can be used in a number of different applications. The half-wave rectifier circuit normally utilizes a single diode. This passes one half of the cycle and blocks the other. In this way only half of the cycle is used, but current is only allowed to flow in one direction (see Figure 13.4).

The half-wave rectifier circuit can often be used with a transformer if it is to be used for powering equipment in any way. Normally in this application the input alternating waveform is provided via a transformer. This is used to provide the required input voltage.

Half-Wave Rectifier Diode Requirements

When designing a half wave rectifier circuit, it is necessary to ensure that the diode can provide the required performance. While there are very many parameters that define individual diodes, and these may need to be considered for a given design, some of the major parameters are detailed below:

- *Forward current*: It is necessary that the diode can handle the levels of average current and peak current flowing through it in a half-wave rectifier circuit. The current will peak as a result of the capacitor smoothing circuit. As the current only flows as the capacitor charges up, the current is in short bursts, which are much higher than the average current.

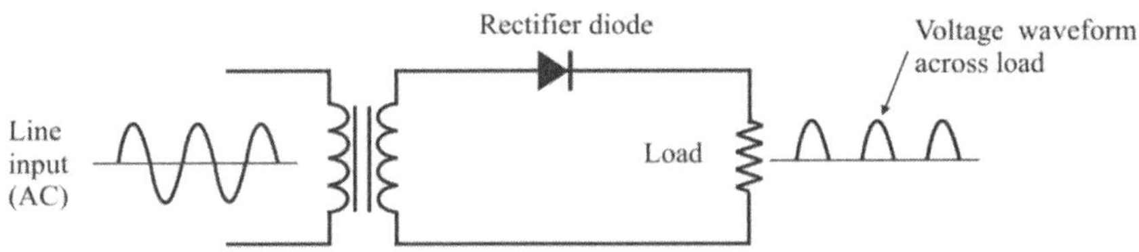

Figure 13.4 Half-wave rectifier.

- *Peak inverse voltage*: The diode must be able to reliably withstand the peak reverse or inverse voltages that appear across it. The peak voltages are not just the output voltage, but higher. The peak inverse voltage rating of the diode should be at least $2\times\sqrt{2}$ times the RMS voltage of the input. This is because the output is normally smoothed by a capacitor, and this will take a value that is the peak of the input waveform. This will be $\sqrt{2}$ times the RMS voltage. With this voltage on the output, the input waveform on the blocked half of the cycle will fall and reach a peak value at the bottom of the crest of $\sqrt{2}$ times the RMS value. The maximum reverse value seen across the rectifier diode is the sum of these two voltages.

There should also be a significant margin, especially when used in a line power supply. This is because voltage spikes can appear on the line.

This supply uses a transformer with a single secondary winding followed by a single, uncontrolled, rectifying diode. The rectifier conducts only during the positive half cycle. This circuit has the minimum parts count. However, the transformer is utilized only during the positive half cycle, resulting in greater RMS heating for a given average current output when compared to the full-wave bridge.

The rectifier diode must be rated for at least twice the maximum output voltage and the circuit experiences one diode drop of voltage and dissipation loss.

- *Half-wave rectifier*: This is the simplest form of rectifier. Often using only a single diode is blocks half the cycle and allows through the other. As such only half of the waveform is used.

While the advantage of this circuit is its simplicity, the drawback is the fact that there is longer between successive peaks of the rectified signal. This makes smoothing less effective and more difficult to achieve high levels of ripple rejection.

Full-Wave Rectifier Supply

The full-wave rectifier circuit is one that is widely used for power supplies and many other areas where a full wave rectification is required.

The full-wave rectifier circuit is used in most rectifier applications because of the advantages it offers. While it is a little more complicated, this normally outweighs the disadvantages. However sometimes it may not be optimum or necessary to use a full wave rectifier circuit.

The full-wave center tap supply, illustrated in Figure 13.5, has a transformer with two equal turn count secondary windings, each one followed by a single

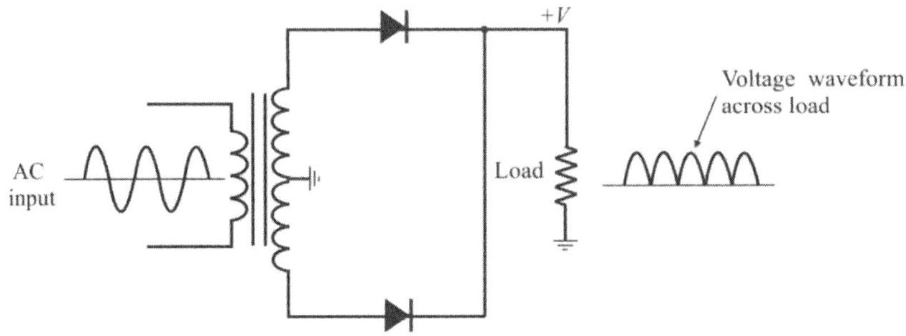

Figure 13.5 Full-wave rectified supply.

uncontrolled, rectifying diode with the other end of the diodes tied together. Each secondary winding of the transformer is utilized only for a half of the cycle, so the transformer must be larger than with the full-wave bridge rectifier. This circuit is suitable for higher power levels than the half-wave rectifier and has the advantage of only one diode voltage drop to the output.

- *Full-wave rectifier*: This form of rectifier circuit uses both halves of the waveform. This makes this form of rectifier more effective, and as there is conduction over both halves of the cycle, smoothing becomes much easier and more effective. The two-diode version of the full-wave rectifier circuit requires a center tap in the transformer. When vacuum tubes/thermionic valves were used, this option was widely used in view of the cost of the valves. However, with semiconductors, a four-diode bridge circuit saves on the cost of the center tapped transformer and is equally effective.

- *Two diode full-wave rectifier*: The two-diode full-wave rectifier circuit is not so widely used with semiconductor diodes as it requires the use of a center tapped transformer. However, this rectifier circuit was widely used in the days of thermionic valves/vacuum tubes. As a rectifier circuit using four valves would be large, the two-diode version was much more preferable. The basic circuit is shown in Figure 13.5.

  The two-diode format for a full-wave rectifier circuit is one of the simplest forms of full wave rectifier circuit in terms of the actual component count.

  The two-diode full-wave rectifier circuit was widely used in the days of thermionic valves or vacuum tubes because it only required the use of two diodes—these could be encapsulated in one device, thereby reducing cost and size.

The basic full-wave rectifier circuit using two diodes and a center-tapped transformer is shown in Figure 13.5. This circuit is very easy to implement, although it does need a center-tapped transformer.

The current flow within the circuit can be seen from Figure 13.5. We can see that the current in each half of the secondary winding is only used for half the cycle. This makes for very inefficient use of the transformer in terms of cost and resources.

- *Transformer output voltage half what it could be*: Utilizing a center tap in the transformer means that only half the full voltage across the two halves of the wind together can be utilized.
- *Increased heating losses*: Because each half of the transformer is used for half of the time, the current through each winding is twice what it would be if a true half-wave rectifier such as a bridge rectifier were used. As heating losses are given by $\text{Loss} = I^2R$, there is four times the heat dissipated for half the time. Over the complete cycle this means that there is twice the heating loss of an equivalent full-wave bridge rectifier circuit.

As a result of these points, to create a full-wave bridge rectifier using the two-diode full-wave rectifier system would require a transformer $\sqrt{2}$ times the size of the one needed for the bridge rectifier. This would cost more as well as being heavier and bulkier. With bridge rectifiers, discussed next, now costing very little, this is the preferred option for most applications.

The full-wave bridge supply has a transformer with a single secondary winding, followed by four rectifying diodes in a bridge configuration, as shown in Figure 13.6. The secondary winding of the transformer is utilized in both half cycles, so the transformer can be smaller than with the full-wave center-tap rectifier. This circuit is suitable for higher power levels than the half-wave rectifier and has the advantage

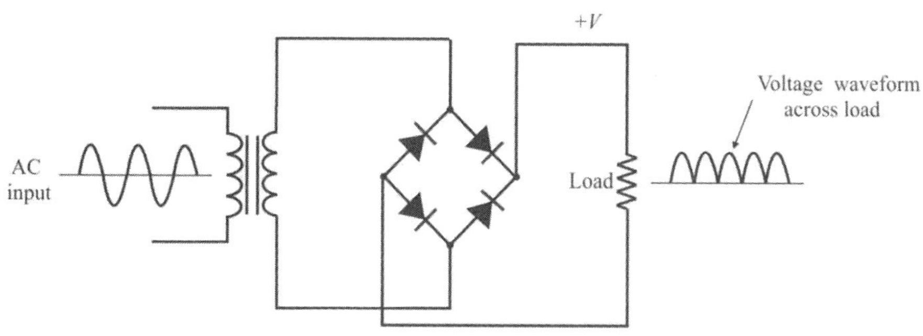

Figure 13.6 Bridge rectifier.

of simplicity. However, it has two diode voltage drops to the output and twice the diode dissipation loss.

The full-wave bridge rectifier circuit configuration is far more widely used these days. It offers a more efficient use of the transformer as well as not requiring a center-tapped transformer. The additional cost is two additional diodes—not an expensive addition these days. Often four diode bridges can be bought as single items, making construction of the overall circuit much simpler.

Split Supply Bridge Rectifier

For many circuits like operational amplifiers, split supplies may be needed. It is possible to create a split supply for these and other applications very easily using a full-wave bridge rectifier, which is illustrated in Figure 13.7. The circuit operates effectively and efficiently because both halves of the input waveform are used in each section of the transformer secondary winding.

13.2.3.2 Comparison of Full- and Half-Wave Rectifiers

The choice of diode rectifier circuit depends on the application. While the full-wave rectifier circuits, and in particular the bridge rectifier circuits are possibly the most widely used, half wave rectifier circuits may offer a better option in some circumstances.

A further advantage when used in a power supply is that the resulting output is much easier to smooth. When using a smoothing capacitor, the time between the peaks is much greater for a half-wave rectifier than for a full-wave rectifier.

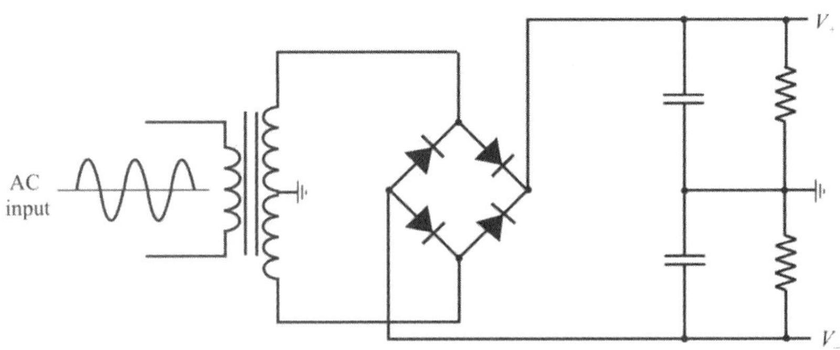

Figure 13.7 Full-wave bridge rectifier providing dual polarity/split supply

Advantages/Disadvantages

- Utilizes both halves of the AC waveform;
- Easier to provide smoothing as a result of ripple frequency;
- More complicated than half-wave rectifier;
- The twice frequency hum on an audio circuit may be more audible.

In view of their advantages, full-wave rectifier circuits are virtually always used in preference to half-wave circuits. The increased efficiency coupled with the better smoothing ability arising from the shorter time between peaks means that their advantages outweigh the disadvantages by a large margin. Only occasionally, often for low requirement supplies would a half-wave rectifier be used in preference to the full-wave rectifier circuit.

13.2.3.3 Capacitor Smoothing Circuits

Rectifiers are normally used in circuits that require a steady voltage to be supplied. To provide a steady DC output. The raw rectified DC requires a smoothing capacitor circuit to enable the rectified DC to be smoothed so that it can be used to power electronics circuits without large levels of voltage variation.

Capacitor Smoothing Basics

The raw DC supplied by a rectifier on its own would consist of a series of half sine waves with the voltage varying between zero and $\sqrt{2}$ times the RMS voltage (ignoring any diode and other losses). A supply of this nature would not be of any use for powering circuits because any analog circuits would have the huge level of ripple superimposed on the output, and any digital circuits would not function because the power would be removed every half cycle.

To smooth the output of the rectifier a reservoir capacitor is used—placed across the output of the reciter and in parallel with the load. This capacitor charges up when the voltage from the rectifier rises above that of the capacitor and then as the rectifier voltage falls, the capacitor provides the required current from its stored charge. See Figure 13.8.

It should be remembered that the only way discharge path for the capacitor, apart from internal leakage, is through the load to the rectifier/smoothing system. The diodes prevent backflow through the transformer, and so forth.

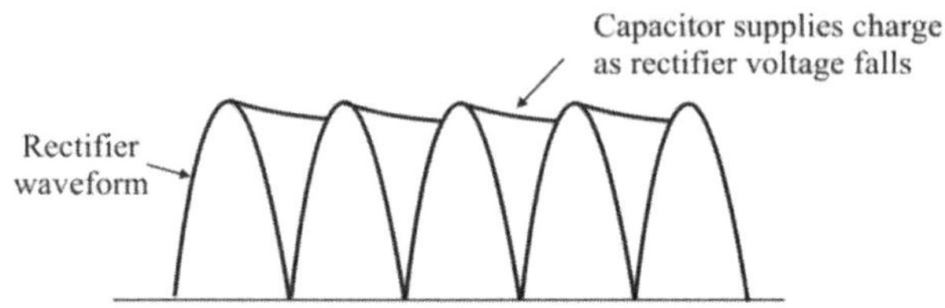

Figure 13.8 Smoothing action of a reservoir capacitor

Smoothing Capacitor Value

The choice of the capacitor value needs to fulfill several requirements. As there will always be some ripple on the output of a rectifier using a smoothing capacitor circuit, it is necessary to be able to estimate the approximate value. Overspecifying a capacitor too much will add extra cost, size, and weight while under-specifying it will lead to poor performance.

Time Constant

In the first case the value must be chosen so that its time constant is very much longer than the time interval between the successive peaks of the rectified waveform:

$$R_L \times C >> 1/f$$

where:

R_L is the overall resistance of the load for the supply
C is the value of capacitor in Farads
f is the ripple frequency—this will be twice the line frequency when a full-wave rectifier is used.

Smoothing Capacitor Ripple Voltage

The ripple for a full-wave rectifier with capacitor smoothing is shown in Figure 13.9. If a half-wave rectifier was used, then half the peaks would be missing and the ripple would be approximately twice the voltage (see Figure 13.10).

For cases where the ripple is small compared to the supply voltage—which is almost always the case—it is possible to calculate the ripple from a knowledge of the circuit conditions. The discharge from the capacitor into the load is in reality exponential. We approximate this nonlinear characteristic as linear as follows:

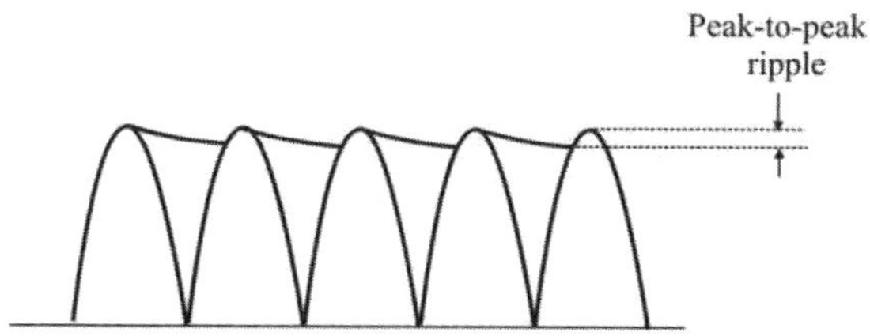

Figure 13.9 Peak to peak ripple for smoothed diode rectifier circuit.

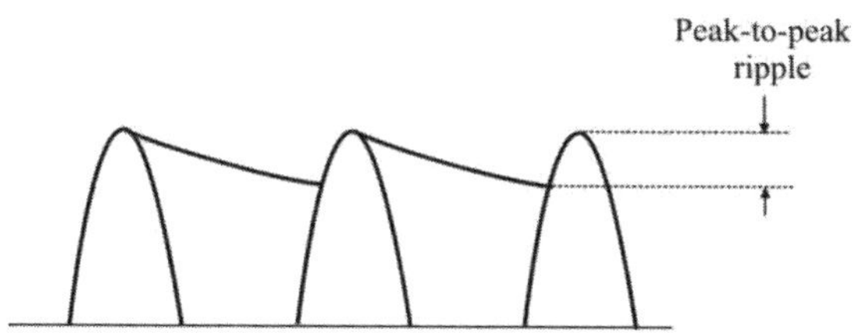

Figure 13.10 Ripple for a half-wave rectifier.

Full-Wave Rectifier

$$V_{\text{ripple}} = I_{\text{L}} / 2fC \tag{13.1}$$

Half-Wave Rectifier

$$V_{\text{ripple}} = I_{\text{L}} / fC \tag{13.2}$$

These equations provide more than sufficient accuracy. Although the capacitor discharge for a purely resistive load is exponential, the inaccuracy introduced by the linear approximation is very small for low values of ripple.

It is also worth remembering that the input to a voltage regulator is not a purely resistive load but a constant current load. Finally, the tolerances of electrolytic capacitors used for rectifier smoothing circuits are large — ±20% at the very best, and this will mask any inaccuracies introduced by the assumptions in (13.1) and (13.2).

Ripple Current

Two of the major specifications of a capacitor are its capacitance and working voltage. However, for applications where large levels of current may flow, as in the case of a rectifier smoothing capacitor in the power supply for a high-power EW output amplifier, a third parameter is of importance— its maximum ripple current.

The ripple current is not just equal to the supply current. There are two scenarios:

- *Capacitor discharge current:* On the discharge cycle, the maximum current supplied by the capacitor occurs as the output from the rectifier circuit falls to zero. At this point all the current from the circuit is supplied by the capacitor. This is equal to the full current of the circuit (see Figure 13.11).

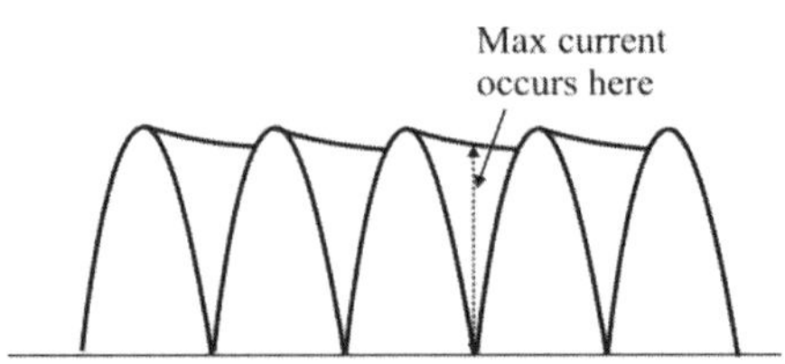

Figure 13.11 Full-wave rectifier max current.

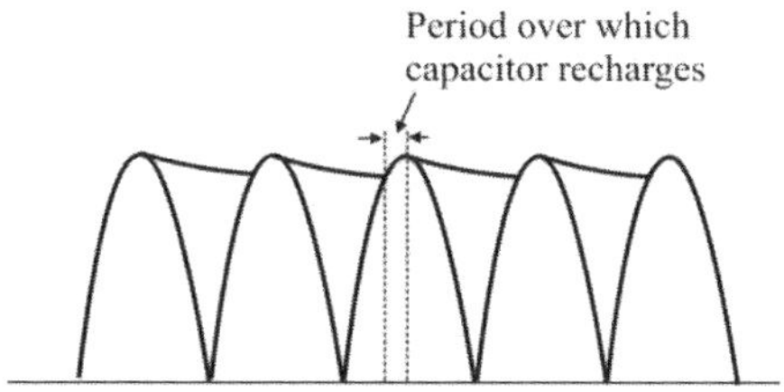

Figure 13.12 Full-wave rectifier capacitor recharge.

- *Capacitor charging current:* On the charge cycle of the smoothing capacitor, the capacitor needs to replace all the lost charge, but it can only achieve this when the voltage from the rectifier exceeds that from the smoothing capacitor. This only occurs over a short period of the cycle. Consequently, the current during this period is much higher. The larger the capacitor, the better it reduces the ripple and the shorter the charge period. See Figure 13.12.

In view of the large currents involved, care must be taken to ensure that the ripple current does not exceed the rated values for the capacitor.

13.2.3.4 Advantages of Unregulated Power Supplies

The advantages of the unregulated supply are low cost and simplicity. There are many applications that do not require precise output voltage. Additionally, we can say that any linear power supply energized off the utility line has an unregulated supply on the front end followed by a regulator.

13.2.3.5 Disadvantages of Unregulated Power Supplies

Disadvantages of an unregulated power supply are as follows:

- If the line voltage increases or decreases, the output voltage will increase or decrease by the same percentage.
- As the load current increases or decreases, the output voltage will decrease or increase respectively due to the finite output impedance of the supply caused by transformer winding resistance, inductance, and trace resistance.

13.3 Power Supply Regulators

Regulators are used in AC to DC converters in order to maintain a constant output voltage or current irrespective to the current or voltage, respectively, at the output.

They are also used to maintain a constant output when the input AC voltage fluctuates. There are two types of regulators: linear and nonlinear (switching). We discuss both in this section.

13.3.1 Linear Power Supply Regulators

Linear power supplies are widely used due to the advantages they offer in terms of overall performance. Linear power supplies are often used in exacting situations where the regulation and removal of noise is paramount.

While linear power supplies may not be as efficient as other types of power supply technology, they offer the best performance and are therefore used in many applications where noise is of great importance. Often audio amplifiers and many other items of electronic equipment use linear power supplies to obtain the best performance.

Linear power supplies gain their name from the fact that they use linear, (i.e., nonswitching) techniques to regulate the voltage output from the power supply. The term linear power supply implies that the power supply is regulated to provide the correct voltage at the output. Sometimes the sensing of the voltage may be accomplished at the output terminals, or on some occasions it may be achieved directly at the load.

For all but the least demanding applications, however, the unregulated design is not able to maintain the output voltage close enough to the prescribed set point as the line voltage and load current changes. Thus, regulation methods have been developed to maintain the output voltage or current at a constant set point.

The first type of regulated design was the linear regulator power supply. The term linear power supply is typically thought of as a type of AC/DC system, providing a regulated output. The linear regulator is the part of the linear regulator power supply that performs the regulation.

13.3.1.1 Linear Regulator Theory

Linear regulators employ a pass element serving as a variable resistor that forms a voltage divider with the load (see Figure 13.13). The pass element functioning as a variable resistor can be semiconductor devices such as a BJT, power MOSFET, IGBT, or an electron tube such as a triode, tetrode, or pentode. Electron tubes would be used in specialized applications where no semiconductor devices are suitable.

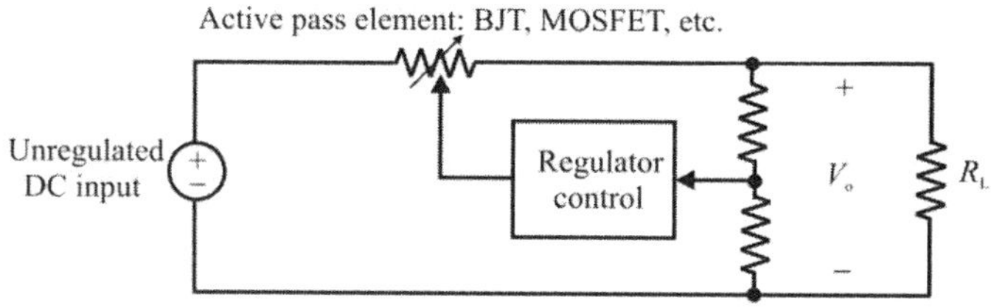

Figure 13.13 Regulated supply.

Power Output Capability

A linear regulator can be designed to regulate power outputs as small as a watt or less. Linear regulators used alone in this fashion are performing DC/DC conversion. Linear power supplies can be designed to provide AC/DC conversion up to tens of kilowatts or even more. In this case the linear regulator is coupled with additional circuitry providing rectification and filtering.

Noise and Ripple

Perhaps the most significant merit of linear power supplies is the cleanliness of the output voltage and the relative lack of electromagnetic emissions. The typical peak to peak output voltage ripple for a linear supply might be 1000× or 60 dB less than the output DC level. So for a 5V output supply the typical peak-to-peak ripple voltage might be 5 mV. A switching supply typically sees about a 100× or 40-dB reduction. It is possible with careful design to achieve a 10,000× or 80-dB reduction in a linear regulated power supply.

Transient Response

The response of linear power supplies to line and load transients is better their switching counterparts because the linear supply does not have a switching frequency to limit the bandwidth. In linear supplies, the regulation bandwidth is typically limited by parasitic device elements.

Weight and Size

The weight and size of the linear power supply is the major disadvantage. AC/DC conversion is done at low frequencies and therefore the transformer must be large to keep the core from saturating. This factor, along with efficiency to be discussed next, is the main reason why linear power supplies have limited use today. For example, a 500W linear power supply might weigh about 50 pounds and occupy a desktop, whereas a 500W switching supply might weight less than 10 pounds and sit on a small part of the desktop.

Efficiency

Linear regulators are typically thought of a being extremely inefficient—but that is not always the case and they can sometimes be more efficient that a switching power supply!

On a first-order basis, the efficiency of a linear regulator is very simple to determine. The efficiency is simply the output voltage divided by the input voltage. If determined effort is made to keep the difference between the pass element input voltage and the output voltage as small as possible, the efficiency can be very good. This type of linear regulator is called a *low-dropout regulator*. For other cases where the operating point is not conducive to maintaining a low dropout, the linear regulator efficiency can suffer greatly.

As an example, if the input voltage in the figure above is 13.6V, and the regulator maintains an output voltage of 12V, for a voltage drop of 1.6V, the efficiency of the linear regulator is 12 V/13.6V = 81.4%. The efficiency in this case is very good by most standards.

As another example, if the input voltage is 5V and the output voltage is 3.3V, for a voltage drop of 1.7V, the efficiency is 3.3V/5V = 66%. The efficiency in this case is not very good by most standards.

As a last example, if the input voltage is 5V and the output voltage is 1.8V, for a pass element voltage drop of 3.2V, the efficiency is 1.8V/5V = 36%. The efficiency in this case is very poor.

13.3.1.2 Types of Linear Regulators

There are two main types of regulators for linear power supplies:

- *Shunt regulator:* The shunt regulator is less widely used as the main element within a voltage regulator. For this form of linear power supply, a variable element is placed across the load. There is a source resistor placed in series with the input, and the shunt regulator is varied to ensure that the voltage across the load remains constant.
- *Series regulator:* This is the most widely used format for a linear power supply. As the name implies a series element is placed in the circuit, and its resistance varied via the control electronics to ensure that the correct output voltage is generated for the current taken.

Both of these types of linear regulator used in power supplies have their uses and can be used in different situations.

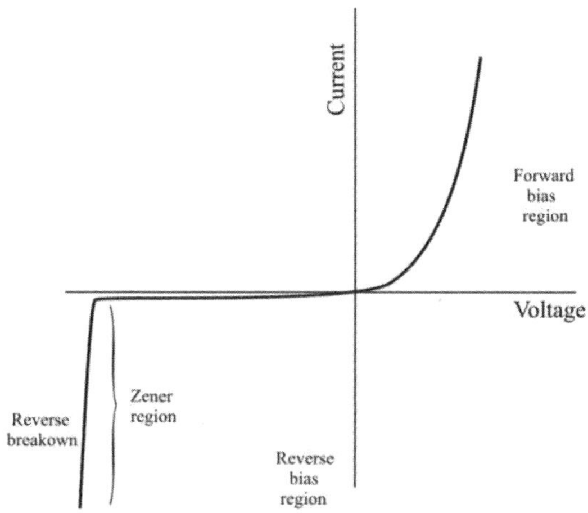

Figure 13.14 Zener diode characteristics.

13.3.1.3 Zener Diodes

A Zener diode is a type of diode that is especially constructed to operate in the Zener range. The Zener range is that portion of the reversed biased region where reverse breakdown occurs (see Figure 13.14). As we see, the voltage is maintained reasonably constant as the current is varied forming a relatively good voltage source.

13.3.1.4 Voltage Regulator

Although the shunt voltage regulator is not widely used to provide the main regulation in many applications, it nevertheless finds uses in many other areas of circuitry. A simple Zener diode circuit provides a prime example of a shunt voltage regulator. As such the shunt voltage regulator is an essential element within linear power supply technology.

Shunt Voltage Regulator Operation

The basic operation of a shunt regulator can be seen from Figure 13.15. Essentially the load is operated with a resistor in series with the voltage source and the shunt regulator then in parallel with the load.

In order to keep the voltage across the load constant, a level of current must be drawn through the series resistor to maintain the required voltage across the load.

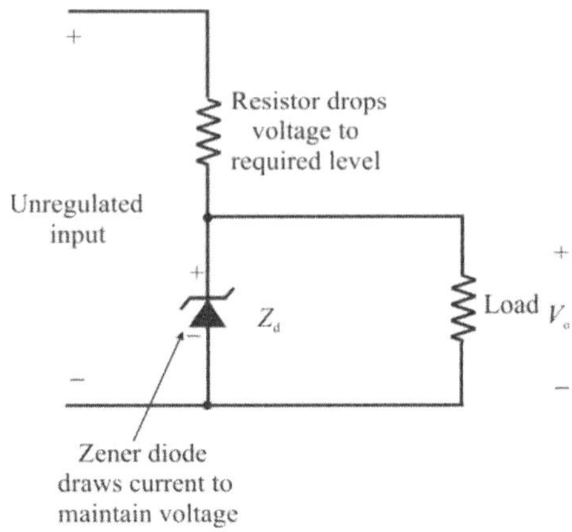

Figure 13.15 Shunt regulator using a Zener diode.

The load will take some and the remaining current is drawn by the shunt voltage regulator.

The circuit is designed so that at maximum load current the shunt regulator draws virtually no current and at minimum load current, the shunt voltage regulator passes the full current.

As a result, shunt regulators are inefficient because maximum current is drawn from the source regardless of the load current (i.e., even when there is no load current).

Zener Diode Shunt Regulator

One of the most common and simple forms of shunt regulator is the simple Zener diode regulator circuit shown below. Its operation is very straightforward. Once over its small minimum current, the Zener diode maintains an almost constant voltage across its terminals. The series resistor drops the voltage from the source to the Zener diode and load. As the Zener diode maintains its voltage, any variations in load current do not affect the voltage across the Zener diode. It takes up the current variations required to ensure the correct drop across the series resistor.

In this shunt voltage regulator circuit, the Zener diode must be capable of dissipating the power from the maximum amount of current it is likely to handle. This is most likely to be a little more than the maximum current supplied to the load as the Zener diode will need to pass all the current when load current is zero. Thus, the total maximum current that will be passed by the diode is the load current plus an allowance for current to maintain the reference voltage when the load is taking its maximum current.

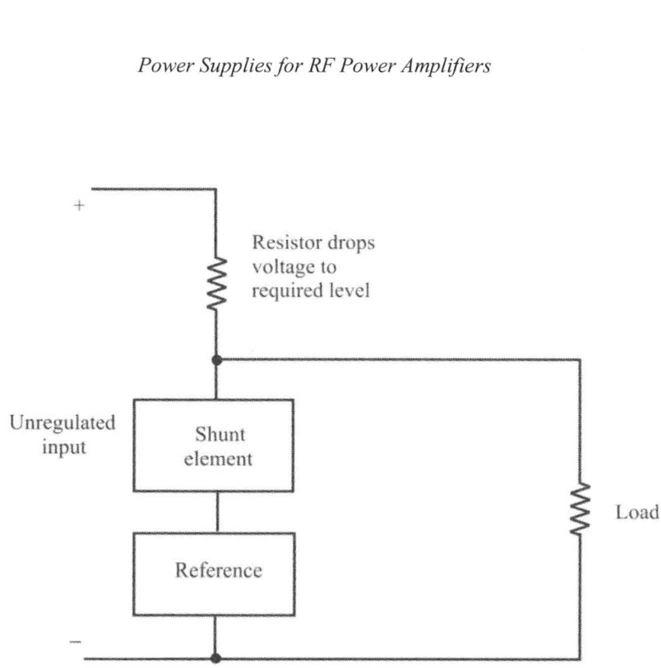

Figure 13.16 Shunt regulator with feedback.

It should also be noted that for the shunt regulator circuit, the series resistance is comprised of the series resistor value, plus any source resistance. In most cases the value of the series resistor will dominate, and the source resistance can be ignored, but this may not always be the case.

Shunt Regulator with Feedback Loop

The shunt voltage regulators discussed above does not have any feedback (i.e. they run in an open-loop manner). As imagined, the performance of this form of shunt regulator is enough for many applications, but much higher levels of performance can be achieved by providing feedback based on the output voltage of the shunt voltage regulator and feeding this back into the system to ensure that the required output voltage is accurately maintained.

Using a shunt voltage regulator with feedback as shown in Figure 13.16, the output voltage is sensed, and the voltage compared to a predetermined reference.

The level of the shunt current is then altered to return the output voltage to the required level.

Full shunt regulators are not widely used because they exhibit low efficiency. Series regulators are widely available and offer higher efficiencies, although they are not as high as switch mode supplies.

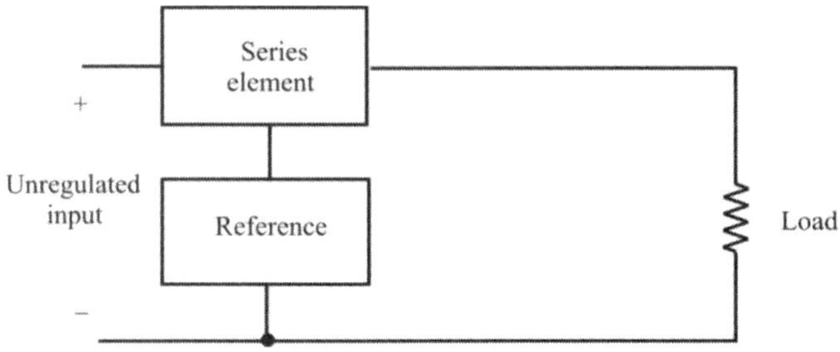

Figure 13.17 Series regulator.

Series Voltage Regulator (Series Pass Regulator)

The series voltage regulator format (or, as it is sometimes called, the series pass regulator) is the most commonly used format for providing the final voltage regulation in a linear voltage regulator circuit (see Figure 13.17). The series voltage regulator or series pass voltage regulator uses a variable element placed in series with the load. By changing the resistance of the series element, the voltage dropped across it can be varied to ensure that the voltage across the load remains constant.

The advantage of the series voltage regulator is that the amount of current drawn is effectively that used by the load, although some will be consumed by any circuitry associated with the regulator. Unlike the shunt regulator, the series regulator does not draw the full current even when the load does not require any current. As a result, the series regulator is considerably more efficient.

Emitter Follower Voltage Regulator

A simple emitter follower series regulator configuration is shown in Figure 13.18. It is one of the simplest implementations of this concept is to use a single-pass transistor in the form of an emitter follower configuration, and a single Zener diode drive by a resistor from the unregulated supply. This provides a simple form of feedback system to ensure the Zener voltage is maintained at the output, albeit with a voltage reduction equal to the base emitter junction voltage—0.6 volts for a silicon transistor.

It is a simple matter to design a series pass voltage regulator circuit like this. Knowing the maximum current required by the load, it is possible to calculate the maximum emitter current. This is achieved by dividing the load current (i.e. transistor emitter current by the β or h_{fe} of the transistor).

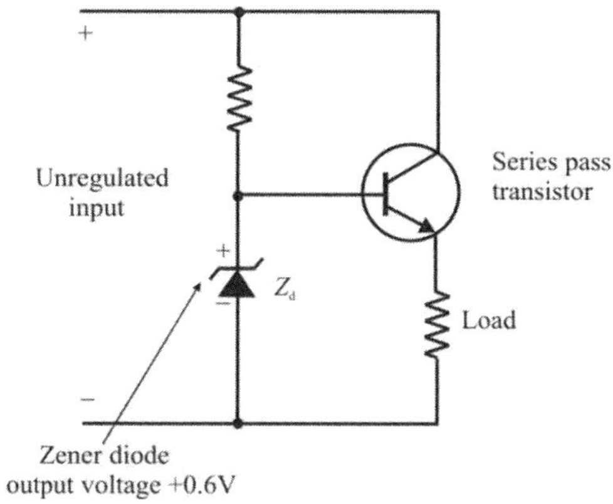

Figure 13.18 Emitter follower regulator.

The Zener diode will generally need a minimum of around 10mA for a small Zener to keep its voltage regulated. The resistor should then be calculated to provide the base drive current and the minimum Zener current from a knowledge of the unregulated voltage, Zener voltage and the current required. [(unregulated voltage–Zener voltage)/current]. A small margin should be added to the current to ensure that there is sufficient room for margin when the load, and hence the transistor base is taking the full current.

The power dissipation capacity for the Zener diode should be calculated for the case when the load current, and hence the base current is zero. In this case the Zener diode will need to take the full current passed by the series resistor.

Series Pass Regulator with Feedback

In order to provide improved levels of performance it is possible to add a more sophisticated feedback network into the regulator circuit.

Using feedback within a voltage regulator enables the output to be sampled and compared with a stable reference voltage. The error is then used to correct the output voltage. In this way, a far higher level of performance can be obtained in terms of the required output voltage as well as ripple and spikes (see Figure 13.19).

It is possible to use a simple two-transistor circuit for a series pass regulator with voltage sensing and feedback. Although it is quite straightforward to use an

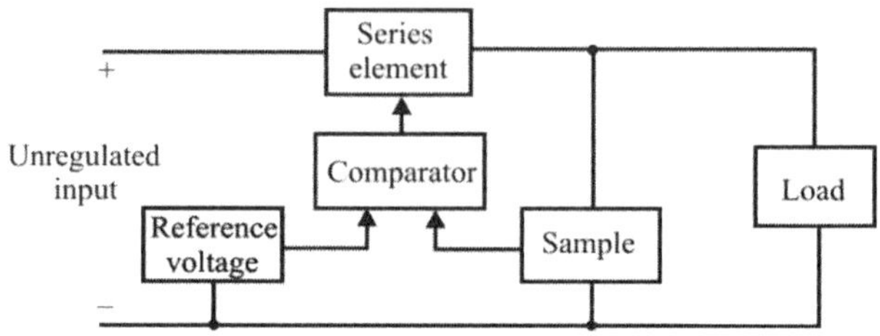

Figure 13.19 Series pass voltage regulator with feedback.

operational amplifier, which will provide higher levels of feedback and hence better regulation, this two-transistor circuit illustrates the principles well.

Two-Transistor Series Pass Regulator

Consider the simple two transistor series pass regulator in Figure 13.20. In this circuit Q_1 forms the series pass transistor. The second transistor, Q_2 acts as theomparator, feeding the error voltage between the reference diode and the sensed output voltage which is a proportion of the output voltage as set by the potentiometer, R_2. The resistor R_1 provides the current for the collector of Q_2 and the reference diode Z_d.

Voltage Reference

Any linear voltage regulator can only be as good as the voltage reference that is used as the basis of the comparison within the system. While a battery could be used in

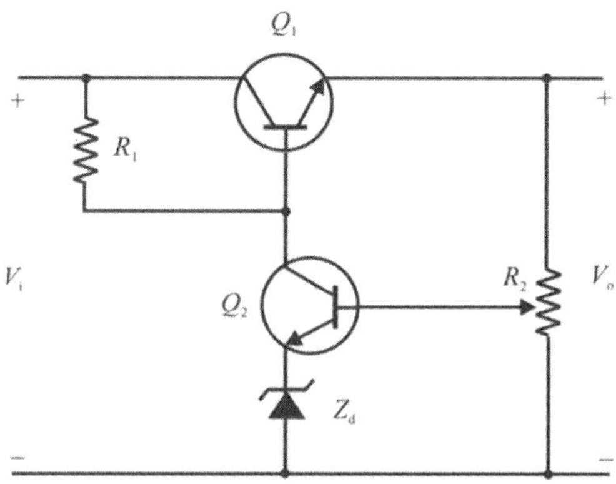

Figure 13.20 Two-transistor series pass regulator.

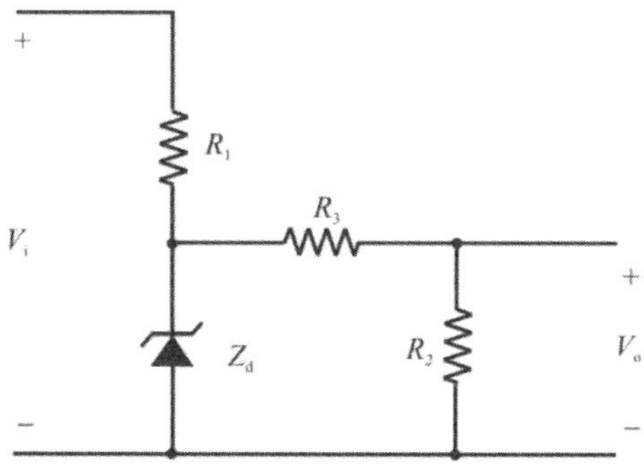

Figure 13.21 Reference source for series pass voltage regulator.

theory, this is not satisfactory for most applications. Instead Zener diode-based references are almost universally used, as illustrated in Figure 13.21.

Integrated circuit regulators and references use sophisticated on-chip combinations of transistors and resistors to obtain temperature-compensated and precise voltage reference sources.

The voltage reference must be driven from the unregulated supply. It cannot be taken from the regulated output as there are start-up issues. At start-up there is no output and therefore the reference output will be zero and this will be maintained until the reference starts up.

Often the output from the reference source is fed via a potential divider. Not only does this reduce the output voltage, which is normally very useful, but it also allows a capacitor to be added to the output to help remove any ripple or noise that may be present. The reduced voltage is also useful because the minimum voltage output is governed by the reference voltage.

Output Sampling

The simple emitter follower series voltage regulator circuit directly compared the output with the voltage reference. In this way the output voltage was equal to that of the reference, neglecting the base emitter voltage drop.

However, it is possible to sample a proportion of the output voltage and compare this to the reference. If this is done, then the output voltage becomes greater than the reference voltage as the negative feedback in the circuit fights to keep the two compared voltages the same (see Figure 13.22).

If for example the reference voltage is 5V and the sampling or potential divider provides 50% of the output voltage, then the output voltage will be maintained at 10V.

The potential division or sampling can be made variable, and in this way, the output voltage can be adjusted to the required value. Normally this method is only

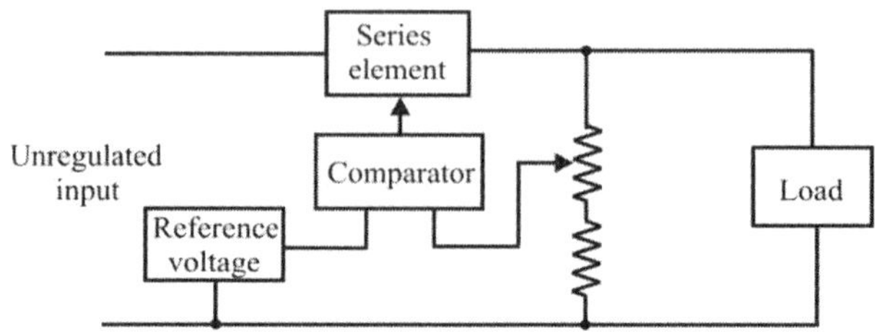

Figure 13.22 Series pass voltage regulator with sampled output.

used for small adjustments as the minimum output level obtained by this method is an output equal to the reference voltage.

We should remember that using a potential divider has the effect of reducing the feedback loop gain. This has the effect of reducing the loop gain and thereby reducing the regulation performance. Normally there is sufficient loop gain for this not to be a major problem except when only a very small proportion of the output is sampled.

Care should also be taken not to increase the voltage of the output to a point where the regulator does not have sufficient drop across it to regulate the output voltage sufficiently.

Low Dropout Series Voltage Regulators

One of the considerations of any regulator is the voltage that must be placed across the series pass element. In addition to the drop across the regulator itself, there must be sufficient voltage to run the drive circuitry. In some circuits, a low dropout regulator is important (i.e. where the level of voltage drop available across the series regulator element is limited). This minimum drop-out voltage can be important and is often a specified parameter in many integrated regulator chips.

While the circuits shown here are simple transistor circuits, the same principles are used in larger circuits and within integrated circuits. The same series pass regulator concepts as well as the reference diode circuits, sampling, and other areas all use the same elements.

13.3.1.5 Summary

We have discussed that linear power supplies excel in producing outputs with very low noise and ripple, low electromagnetic emissions, and have excellent transient response. However, they are by nature large and heavy when compared to switching power supplies. Linear power supplies would be the type of choice when outputs with low ripple and noise content are of paramount concern, and where larger size and substantially greater weight can be tolerated.

13.3.1.6 Linear Power Supply Advantages/Disadvantages

The use of any technology is often a careful balance of several advantages and disadvantages. This is true for linear power supplies, which offer some distinct advantages, but also have their drawbacks.

Linear PSU advantages

- *Low noise:* The use of the linear technology without any switching element means that noise is kept to a minimum and the annoying spikes found in switching power supplies are not found.
- *Established technology:* Linear power supplies have been in widespread use for many years and their technology is well established and understood.

Linear PSU disadvantages

- *Efficiency*: In view of the fact that a linear power supply uses linear technology, it is not particularly efficient. Efficiencies of around 50% are not uncommon, and under some conditions they may offer much lower levels.
- *Size*: The use of linear technology means that the size of a linear power supply tends to be larger than other forms of power supply.
- *Heat dissipation*: The use of a series or parallel (less common) regulating element means that significant amounts of heat are dissipated and this needs to be removed.

Despite the disadvantages, linear power supply technology is still widely used, although it is more widely used where low noise and good regulation are needed. One typical application is for audio amplifiers.

13.3.2 Switching Regulators

The primary advantage of switching-mode power supplies is that, given ideal parts, they can accomplish power conversion and regulation at 100% efficiency. All power loss is due to less than ideal parts and the power loss in the control circuitry. In this section we introduce the fundamentals of how these regulators work. We explore some of the switching-mode power supplies that can be constructed using only a few parts.

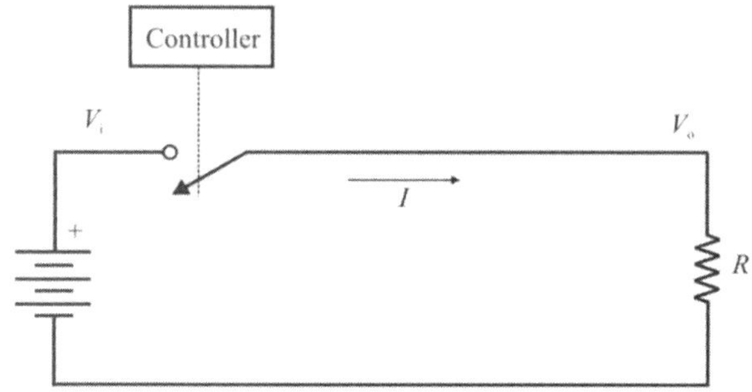

Figure 13.23 Switched circuit.

Consider the circuit shown in Figure 13.23. The DC voltage is switched on and off according to the desired average amount of power to be dissipated by the load R. As we see in Figure 13.24, the longer the switch is in the ON position the higher the *average* voltage level. Suppose that V_i is 12 VDC and the load resistor R is 0.25 Ω. The objective is to open and close the switch so that the average voltage across R is 5 VDC. The waveform of the voltage across R is shown in Figure 13.24(a). If the desired average voltage is 10 VDC, then the switched voltage waveform might look similar to that in Figure 13.24(b) because the switch is ON for a longer time In Figure 13.24, the period, T, of the waveform is $T_{on} + T_{off}$ and the reciprocal of the period is the frequency. For example, a period of 50 μs results in a frequency of 20 kHz. The ratio T_{on}/T is called the *duty cycle* and is denoted by D. The average values of the waveforms over a period are shown by dotted lines.

The average value of a function $v(t)$ is

$$\bar{V}(t) = V_{avg} = \frac{1}{T}\int_0^T v(t)dt \tag{13.3}$$

so in our case $\bar{V} = V_i D$. The mean square value of $V(t)$ is given by

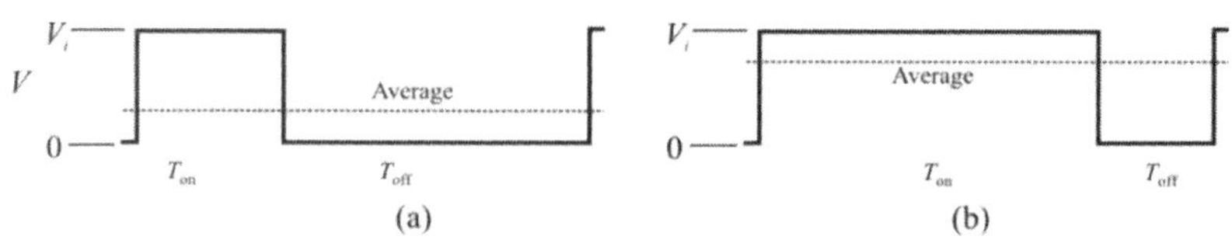

Figure 13.24 Switched waveforms (a) $V_{average} = 5$V and (b) $V_{average} = 10$V.

$$V_{\text{RMS}} = \sqrt{\frac{1}{T}\int_0^T V^2(t)dt} \tag{13.4}$$

In our case $V_{\text{RMS}} = V_{\text{i}}\sqrt{D}$. In our example, V_{i} = 12 VDC and V_{o} = 5 VDC (average) so that

$$D = V_{\text{o}} / V_{\text{i}} = 0.417$$
$$V_{\text{RMS}} = 12 \times \sqrt{0.417} = 7.75$$
$$I_{\text{average}} = 0.417 \times 12 / 0.25 = 20\text{A}$$

and

$$P_{\text{i}} = 12 \times 20 = 240\text{W}$$

This concept forms the basis for all *switched mode power supplies* (SMPS). PWM is used to set the value of the output voltage level. A block diagram for a SMPS is shown in Figure 13.25 although there are many variations on this architecture. Conceptually, the input AC is converted to a filtered but unregulated DC waveform. That is then fed to a high frequency switch (the MOSFET in Figure 13.25) that provides the PWM. This rectangular wave drives a power transformer (isolated configuration), the output of which is rectified and filtered, establishing a DC voltage source. The DC level is compared to a reference and is used to drive the control circuitry that determines the width of the PWM pulse.

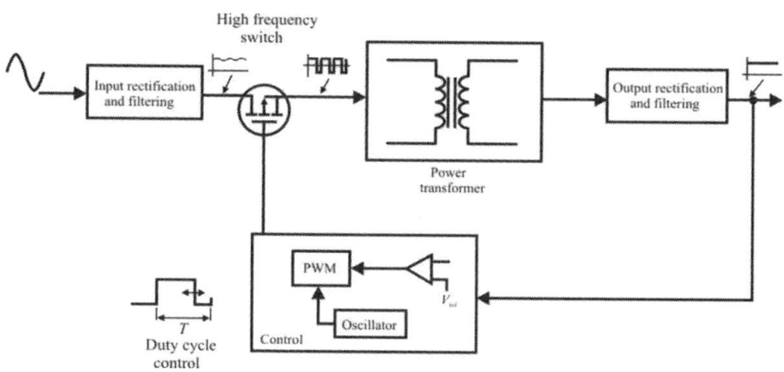

Figure 13.25 Switched-mode power supply.

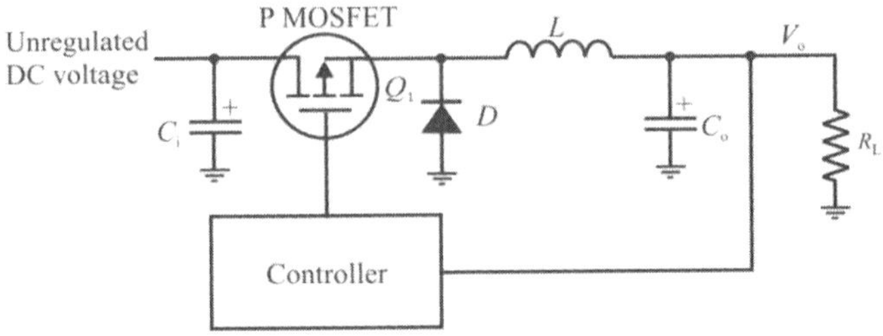

Figure 13.26 Stepdown converter. Typical parameters are V_i = 7 to 24 VDC, V_o = 2 VDC, I_{omax} = 7 A, and f_{sw} = 300 kHz.

Such a configuration can be used to provide either a lower or higher voltage level than the input DC level. When the transformer is not used (nonisolated[9]) and when the level is lower, it is known as a buck converter. Without the transformer (nonisolated) and when the output voltage is higher it is known as a boost converter. We will examine the buck converter in more detail.

Stepdown (buck) switching converters convert a voltage source (typically 8V to 25V) into a lower regulated voltage (typically 0. 5V to 5V). Stepdown converters transfer small packets of energy using a switch, a diode, an inductor and several capacitors. Though substantially larger and noisier (RF noise) than their linear-regulator counterparts, buck converters offer higher efficiency. This is particularly important for battery-powered applications.

A hand-emplaced jamming system is one example of such an application. Such systems have been built and deployed. Another example is an EA system deployed by being ejected out the back of artillery shells in the form of pucks. Several versions of battery-powered ES systems have been built and deployed.

A basic circuit for a stepdown buck converter is shown in Figure 13.26. The MOSFET is controlled by a waveform from the controller that is pulse-width modulated. The width of the pulses that turn the MOSFET on and off are determined by the average output voltage essentially as described above. When the switch is ON, the input voltage causes a current to flow through L and a voltage to be developed across the load [see Figure 13.27(a)]. It also charges C at the same time. D is reversed biased at this point so no current flows through that path. When the MOSFET switch is opened, as in Figure 13.27(b) L attempts to maintain the current (recall that the current through an inductor cannot change instantaneously) so the voltage across L reverses and D becomes forward-biased. A current flow due to the collapsing magnetic field in L, through the load and forward-biased diode. The capacitor smooths the voltage transients due to the field generation and collapsing

[9] Isolated from the line AC source. Such isolation reduces the effects of varying AC line fluctuations.

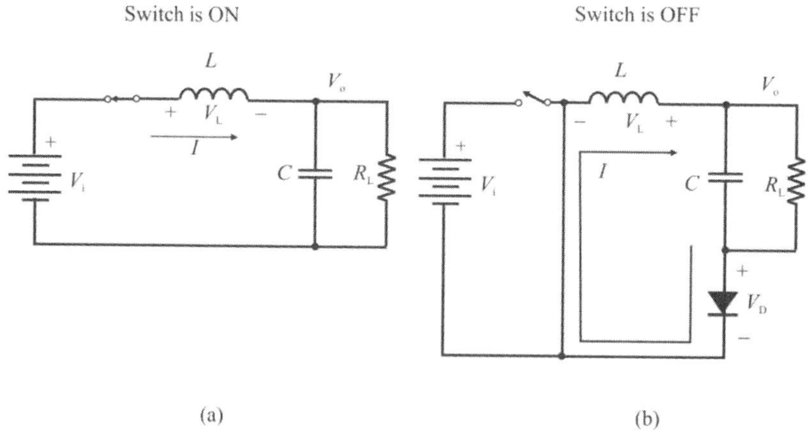

Figure 13.27 Switching action (a) switch is ON and the capacitor is charged and (b) switch is OFF and the capacitor discharges.

in L. This process is repeated for each period of the PWD waveform from the controller.

13.4 Power Supply Components

13.4.1 Power Supply Capacitors and Inductors

In order to function properly, power supplies need components to store and deliver energy. There are two types of energy storage devices used in power supplies: capacitors and inductors.

Unlike resistors, ideal inductors and capacitors only store energy, but never dissipate energy. Therefore, over one complete steady-state switching cycle, the average power dissipated by the device is zero. However, all capacitors and inductors are nonideal, and there may be dissipation effects to consider.

Inductors and capacitors either act as energy reservoirs, or temporarily store buckets of energy for transfer to the output. When acting as a reservoir, the change in inductor current or capacitor voltage change is considered small for analysis purposes. When transferring buckets of energy, the inductor current or capacitor voltage change is large and may be of resonant character.

13.4.1.1 Power Supply Capacitors

A capacitor stores energy in the electric field between two parallel conducting plates. The energy stored is proportional to the square of the voltage across it:

$$E(t) = \frac{1}{2} C V^2(t) \tag{13.5}$$

An important property of capacitors is that they resist the change of voltage across the plates and therefore act as reservoirs of available charge. The current through an ideal capacitor can change instantaneously, but the voltage never can, always maintaining voltage continuity. The current through a capacitor is given by

$$I(t) = C \frac{dV(t)}{dt} = C\dot{V}(t) \tag{13.6}$$

Nonideal power supply capacitors have equivalent series resistance and leakage current. Common types for power supply capacitors are aluminum electrolytic, tantalum, multilayer ceramic, and film. Aluminum and tantalum types are polarity sensitive. They also have an RMS current rating that often determines the amount of capacitance in an application since there must be enough capacitors to handle the RMS current. A little-known fact about multilayer ceramic capacitors is that the capacitance exhibits strong dependence on temperature. For each 10°C or less ambient temperature, the estimated lifetime doubles.

13.4.1.2 Power Supply Inductors

A power inductor is the electrical dual of the power supply capacitor, and stores energy in the magnetic flux established by the geometry of the inductor winding, core material, and the current through the winding. The energy stored in the inductor is proportional to the square of the current through it

$$E(t) = \frac{1}{2} L I^2(t) \tag{13.7}$$

The voltage across an ideal inductor can change instantaneously, but the current never can, always maintaining current continuity. The voltage across an inductor is given by

$$V(t) = L\frac{dI(t)}{dt} = L\dot{I}(t) \tag{13.8}$$

A nonideal inductor has copper and core loss due to the winding resistance and core resistance. These nonidealities need to be considered in any design and taken into account. Common types for power electronics are toroidal shapes employing various magnetic powder formulations and various shapes of different ferrite core materials. These materials are chosen according to the performance and cost requirements and have a current rating, beyond which the permeability reduces greatly, causing the core to saturate.

13.4.1.3 Linear Supply Applications of Capacitors and Inductors

Power supply capacitors enable the smoothing of rectifier outputs through energy storage. A smoothing capacitor bank is often referred to as the bulk capacitance. The energy stored in the bulk capacitance becomes the input to the regulator pass element. Linear power supplies also employ a capacitor at the output of the regulator. The purpose of the output capacitor is to provide control loop stability and holdup energy storage in the event of a momentary loss of input power. Linear power supplies must filter out line frequency noise, and therefore the capacitors and inductors are large. Linear supplies typically do not use power inductors in most applications.

13.4.1.4 Switching Supply Applications of Capacitors and Inductors

Power supply capacitors are also used by switching power supplies as the bulk capacitor and at the output for control stability and holdup. Capacitors at these locations, when also coupled with inductors, can also be configured as low-pass LC filters for ripple voltage reduction on the output, and ripple current reduction on the input, and for averaging the switching output voltage. In switching power supplies, the energy stored in capacitors and inductors is managed by a high frequency power switch. Switching power supplies operate at high-frequency and the capacitors and inductors are therefore small. Capacitors in resonant supplies can also large voltage swings as they resonantly transfer energy packets from the input to the output.

13.4.1.5 Power Supply Conduction Modes

According to (13.7), the energy stored in inductors is proportional to the square of the current through the inductor. During times of light loading and when using passive rectifying diodes rather than synchronous rectification, the current through the inductor can fall to zero. This condition is known as *discontinuous conduction mode* (DCM) operation. When the inductor current never falls to zero, or when the power supply employs synchronous rectification, the condition is said to be in *continuous conduction mode* (CCM).

With synchronous rectification, active switches are used for rectifiers and current flow can actually reverse direction allowing the current to continue to flow. Therefore, unless the synchronous switch is commanded to turn off, the conduction mode will always be continuous.

Advantages of DCM over CCM

- The CCM boost, buck-boost, and flyback topologies have a *right half-plane zero* (RHPZ) in their control to output transfer function. This means that if, for example, the load current increases causing the output voltage to initially dip, the response of these topologies will initially correct in the wrong direction before correcting in the right direction. The right half-plane zero is nearly impossible to compensate for in the compensation loop. As a result, the control loop in these CCM converters is typically made to cross over at a frequency much lower than the RHPZ frequency, resulting in lower transient response bandwidths. The DCM version of the boost, buck-boost, and flyback converters do not have a right half-plane zero and can have higher loop crossover frequency, allowing higher transient response bandwidths.
- In the buck, boost, buck-boost, and all topologies derived from these, the input to output and control to output transfer functions contain single-pole responses while operating in DCM. Converters with only single-pole transfer functions are easier to compensate than converters having a double-pole response.

Disadvantages of DCM over CCM

- While operating in CCM, the DC conversion ratio (the output voltage divided by the input voltage), is independent of the load, to a first-order basis. This characteristic makes DC analysis of converters operating in CCM easier. However, while operating in DCM, the DC conversion ratio is dependent on load, complicating the DC analysis.
- When the inductor current reaches zero while in nonsynchronous operation, the end of the inductor connected to the switch (also called the freewheel end), must immediately transition to the voltage at the other end of the inductor. However, there will always be inductive and capacitive parasitic elements that will cause severe ringing if damping is not implemented. It is possible for this ringing to exceed switch voltage

ratings. It is also possible for this ringing to be a source of radiated and conducted emissions. Lastly, this ringing can produce undesirable noise at the output of the power supply.

- Boost, buck-boost, and derived topologies including the flyback are commonly operated only in DCM to avoid the adverse effects of the RHPZ described earlier. However, to achieve the same power in DCM as in CCM, the peak and RMS currents are substantially higher, resulting in greater losses in the conduction paths and greater ringing because the energy stored in inductances is proportional to the square of the current. Energy stored that is not delivered to the output causes ringing and subsequent losses.
- In DCM, the inductance must be much smaller in value to allow the current to fall to zero before the start of the next cycle. Smaller inductance results in higher RMS and peak inductor currents. Because the RMS and peak currents are greater in DCM than in CCM, the transformers must be sized larger to accommodate greater flux swings and copper and core losses.

13.4.2 Power Semiconductor Switches

All ideal switching supplies continually decompose input power at a given voltage and current into energy packets. The small energy packets are then continually reassembled into output power at a new desired voltage and current. In an ideal supply, the output power is equal to that at the input. In practical applications there will be some energy loss due to waste heat.

The power semiconductor switches perform the decomposition and reassembly process, directing the flow of energy from the input, through the inductor and capacitor energy storage components, to the output.

The power semiconductor switches, depending on the topology and requirements are active transistors, thyristors, or passive diode rectifiers.

13.4.2.1 Ideal Power Semiconductor Switch Properties

The ideal switch has the following properties:

- Infinite breakdown voltage;
- When the switch is off, there is zero current through the switch;
- When switch is on, there is zero voltage across the switch;
- The turn-on and turn-off transition times of ideal switches are zero;
- Since the either the voltage or the current is always zero in an ideal switch, the instantaneous dissipation, which is the product of instantaneous voltage and instantaneous current, is always zero.

13.4.2.2 Nonideal Power Semiconductor Switch Properties

A practical switch will approximate an ideal to the first order. Even so, practical switches will have parasitic elements which cause:

- Finite breakdown voltage;
- Leakage current in the off-state;
- Nonzero voltage across the switch in the on-state;
- Nonzero turn-on and turn-off transition times;
- Since there is some voltage across the switch in the on-state, and the transition times are non-zero, there is non-zero dissipation which must be managed.

13.4.2.3 Power Semiconductor Switch Classification

Power semiconductor devices can be classified in at least two ways:

- By type of charge carrier;
- By active vs. passive function.

Charge Carrier Classification

Power semiconductor devices can have two modes of charge carrier conduction:

- Minority carrier, where charge is carried by electrons in a P-type semiconductor, or by holes in an N-type semiconductor;
- Majority carrier, where charge is carried by electrons in an N-type semiconductor or by holes in a P-type semiconductor.

Understanding what type of charge carrier conduction a device has helps us to understand its most basic properties and suitability for different applications.

Minority Carrier Devices

Minority carrier devices excel for applications involving higher voltage levels. Minority carrier devices used for power semiconductors have a p-n junction where the n region is very lightly doped or intrinsic. Since, under reverse bias conditions, the n region is nearly devoid of charge carriers and therefore has low conductivity, it can sustain a large breakdown voltage.

When the device is forward-biased, minority carriers are injected into this n region, resulting in a great increase in conductivity, so that under forward bias conditions, the on-state voltage is relatively low. This change in the conductivity is referred as *conductivity modulation*. Comparably rated majority carrier devices for high-voltage applications would actually have a higher on-state voltage due to the product of current and channel resistance.

However, minority carrier devices are slow to turn-on and turn-off due to a large stored minority carrier charge that must be inserted or removed prior to the device changing state. This limits the maximum switching frequency to ~100 kHz.

Majority Carrier Devices

Majority carrier devices excel for applications involving lower voltage levels and for higher frequencies. Majority carrier devices have fast switching times because they do not store minority charge. They operate either by electrostatic control of conduction cross-sectional area or by employing a metal semiconductor junction. Neither method involves minority carriers.

Active versus Passive Devices

Power semiconductor switches can further be classified according to whether they are passive responding devices or actively controlling devices.

Passive Devices

Passive devices change conduction state based on the external voltage and current conditions. They do not need to be controlled. For example, when the current through a diode attempts to change direction, the diode turns off. When the diode becomes forward-biased, it turns on. These off and on states do not depend on any control signal to tell the diode to switch.

Active Devices

Active controlling devices are typically three terminal devices and change from a blocking state to a conducting state as commanded by a control input signal. Active devices control the operation of a switching power supply.

13.4.2.4 Minority Carrier Device Family

Minority carrier devices have charge storage times of a few hundred nanoseconds to a few microseconds. Consequently, the switching time is limited to this order of magnitude, and the maximum practical switching frequency of a power supply with minority carrier active switches is limited by the storage time to about 50 to 100 kHz. Some device types have greater minority charge storage and/or more limited methods of minority charge removal, resulting in further frequency limitations. The minority carrier devices in use today are the PIN diode, BJT, IGBT, and the thyristor.

PIN Diode

The power versions of the PN junction diode have broad applicability in switching power supplies. The PIN diode is the same as a PN diode, but with the addition of an intrinsic layer between the P and the N layers. This intrinsic layer provides the breakdown voltage capability of the device since under reverse-bias conditions, it is devoid of charge carriers.

Besides the usual breakdown voltage and rated current, the other main consideration in selecting a diode for rectification use is the reverse recovery time. Line frequency rectifiers can use standard recovery time diodes since those frequencies are 50/60 Hz. Switching power supply applications using minority carrier transistors such as IGBTs or BJTs require diodes with fast recovery times of less than 500 nsec because of the higher frequency of the switching. Switching power supply applications using majority carrier devices such as power MOSFETs require ultra-fast recovery times of less than 100 nsec because the switching frequencies are even higher.

Advantages of the PIN Diode:

- Inexpensive;
- Does not require control.

Disadvantages of the PIN Diode:

- Has a reverse recovery time that contributes to power loss.

Typical Maximum Ratings:

- Maximum breakdown voltage, (V_{RRM}): 4500V;
- Maximum current, (I_F): 600 A;
- Maximum apparent power ($V_{RRM} \times I_F$): 2137 kVA.

Bipolar Junction Transistor

The BJT is an older technology. Switching power supply applications for BJTs include switching voltages over 600V at frequencies less than 20 kHz. Due to conductivity modulation effects, the BJT can exhibit lower on state collector-emitter voltage than the drain-source voltage across a comparable MOSFET. The BJT has three terminals: collector, base, and emitter. The BJT conduction state is controlled by the level of current injection into the base terminal. The governing relationship is:

$$I_C = h_{FE} I_B \tag{13.9}$$

The current gain, h_{FE}, for a power BJT may be in the range of 20–100 for lighter collector currents. However, as maximum rated collector current is approached, h_{FE}, degrades to the range of 5–10, making control of the device more difficult. Additionally, V_{CEsat} increases causing much greater conduction losses.

The BJT also has substantial storage charge that limits its ability to turn off quickly. Typical storage time and collector fall time is in the range of 1–5 μs. This turn-off time limits the maximum practical switching frequency of power supplies using BJTs as the power semiconductor switch.

Advantages of the Power BJT:

- In low power and lower switching frequency applications, the slower switching speed can be used to minimize EMI generation, reducing or eliminating the requirements for input EMI filtering. Power supply cost and size reductions result.
- Lower on-state voltage relative to a comparable power MOSFETs for devices with ratings over 600V.

Disadvantages of the Power BJT:

- Substantial control current required, resulting in efficiency issues;
- Substantial turn-off storage time, resulting in slow switching times and therefore relatively slow maximum switching frequencies.

Power BJT Typical Maximum Ratings

- Maximum collector to emitter breakdown voltage, (V_{CEO}): 1200V;
- Maximum collector current, (I_C): 60A;
- Maximum apparent power ($V_{CEO} \times I_C$): 213.8 kVA.

Insulated Gate Bipolar Transistor

The *insulated gate bipolar transistor* (IGBT) is a recent technology. Switching power supply applications are for switching voltages over 600 V at frequencies less than 100 kHz and for higher power levels than either the BJT or power MOSFET can support. The IGBT is a three-terminal hybrid of the BJT and power MOSFET. Internally, the drain of an N-channel power MOSFET sinks current from the base of a PNP BJT (see Figure 13.28). Therefore, the IGBT has a high impedance input similar to a MOSFET. It also has a collector-emitter output that has a large breakdown voltage and a low on-state voltage as in a BJT. IGBT technology has been greatly improved, resulting in shorter turn-off delay times compared to the BJT.

The IGBT is a device that combines the MOSFET gate driving characteristics with the high current and low saturation voltage of a BJT. It acts as a high-frequency, high-current switch that is used in AC/DC inverter, motor control, and in switching mode power supply applications.

It has lower V_{CE} (saturation) voltage that allows it to operate with a higher current density than with bipolar transistor. It can be modeled as a PNP transistor driven by a power MOSFET. The normal gate driving voltage used is in the region of 15V where saturation voltage is obtained so that conduction loss is kept to a minimum.

The safe operating area (SOA) is critical information that shows the maximum operating current and voltage of the device. It shows the forward-bias SOA and reverse-bias SOA when the gate emitter junction is forward-bias and reversed-bias respectively.

The IGBT technology is improving with better switching speed, lower conduction voltage drop, higher current carrying capability and higher reliability.

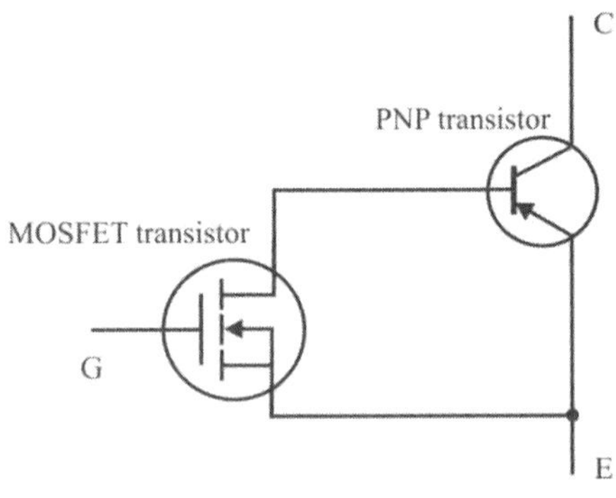

Figure 13.28 IGBT model.

The availability of these devices with current up to 1 kA, voltage ratings up to 2 kV, switching speeds of 200 nsec and on-state voltage down to 2.0V and below have made these devices a popular choice among designers in the world of power electronics.

The advancement of this technology has enabled the inverter system solutions to be implemented for many consumer electronics appliances such as air conditioning controls and refrigerator controls.

Advantages of the IGBT

- Voltage-controlled, high-input impedance device, easier than current control of BJT;
- For devices with higher voltage ratings, the on-state voltage is much lower than that of a comparable MOSFET;
- Power handling capability is ten times better than power MOSFETs or BJTs;
- Shorter delay times relative to the BJT, about 300 nsec to 1500 nsec.

Disadvantages of the IGBT:

- Have a current tail turn-off characteristic that results in longer delay times relative to the power MOSFET.

IGBT Typical Maximum Ratings

- Maximum collector to emitter breakdown voltage, (V_{CEO}): 4000V;
- Maximum collector current, (I_C): 400A;
- Maximum apparent power ($V_{CEO} \times I_C$): 340 kVA.

Thyristors

The most prolific member of the thyristor family is the *silicon-controlled rectifier* (SCR).

Silicon-Controlled Rectifier

The SCR is the oldest semiconductor technology used in power switching applications. The circuit symbol for the SCR is shown in Figure 13.29. SCR applications primarily involve either phase control of AC line power or electronic crowbars where it is desired to rapidly dump energy stored in a large capacitor.

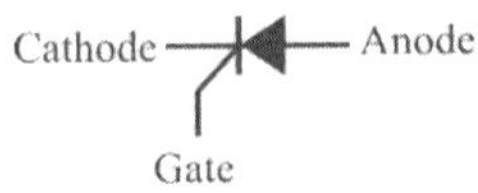

Figure 13.29 SCR.

Because an SCR has a large turn-off time, it is not suitable for switching applications greater than about 400 Hz.

The SCR has three terminals: anode, gate, and cathode. Conduction from the anode to the cathode begins when current is injected into the gate. A conducting SCR looks like a diode, and once conduction begins, the SCR latches on so that removal of the gate drive does not turn the SCR off. The SCR can only be turned off when the anode to cathode current falls below the holding current specification or reverses. Before the SCR can turn off, the minority charge carriers must be removed by the reverse current or recombination effects.

Advantages of the SCR

- Operating characteristics lends the SCR to phase control of AC powers;
- Lowest cost per kVA of all power semiconductor switches;
- Large power handling capability.

Disadvantages of the SCR

- Substantial minority carrier charge storage time, limiting upper switching frequency to 400 Hz or less.

SCR Typical Maximum Ratings

- Maximum anode to cathode breakdown voltage, (V_{AK}): 3800V;
- Maximum anode current, (I_A): 3360A;
- Maximum apparent power ($V_{AK} \times I_A$): 6726 kVA.

13.4.2.4 Majority Carrier Device Family

Majority carrier devices do not store significant minority carriers and therefore have turn-on and turn-off times an order of magnitude faster than minority carrier devices. Switching times of majority carrier devices are less than 200 nsec and frequently very much less. Consequently, the maximum practical switching frequency of a power supply with majority carrier active switches can reach 1 MHz and beyond.

Power MOSFET

We will discuss the N-channel enhancement mode power MOSFET. There is also a P-channel enhancement mode power MOSFET for some power supply applications.

The power MOSFET has three terminals: drain, gate, and source. The gate to source voltage controls the conduction state of the power MOSFET. There is an insulating layer between the gate and an electrostatically controlled conduction channel between the drain and the source. Application of a gate to source voltage greater than the device threshold voltage will cause the power MOSFET to turn on by modulating the geometry of the electrostatic conduction channel.

Power MOSFETs also contain a slow reverse recovery time antiparallel body diode from the drain to the source.

Parameters of the Power MOSFET

- R_{DS}, the on-state drain to source resistance, which causes conduction losses and non-zero voltage drop across the on-state switch;
- C_{DG}, the Miller, drain-to-gate, or reverse-transfer capacitance, which slows the turn-on and turn-off transition time of the transistor switch as well as potentially being a source of instability and dV/dt turn-on problems;
- C_{GS}, the gate to source capacitance, which contributes a delay time to the turn-on and turn-off of the transistor switch.

Advantages of the Power MOSFET

- Voltage controlled, high-input impedance device, easier than current control of BJT;
- Fast switching speed because they are free from minority carrier stored charge.

Disadvantages of the Power MOSFET

- Can be susceptible to thermal runaway in power supply applications requiring constant current;
- Fast transient switching induces greater radiated and conducted emissions;

- Devices with breakdown voltages of 600V or higher typically have greater on-state voltages than comparable minority carrier devices.

Power MOSFET Typical Maximum Ratings

- Maximum drain to source breakdown voltage, (V_{DS}): 4000V;
- Maximum drain current, (I_D): 600A;
- Maximum apparent power, ($V_{DS}{\times}I_D$): 56 kVA.

SiC Power MOSFET

SiC is a semiconductor material for use with power MOSFETs. These devices have an R_{DSon} about one third that of comparable silicon power MOSFETs, allowing for on-state voltages comparable to IGBTs. This feature opens the door to high power applications for power MOSFET's with breakdown voltages over 600V. Since the silicon carbide power MOSFET is a majority carrier device, there is no associated storage time to cause the well-know current tail characteristic of IGBTs. Consequently, SiC power MOSFETs enable high voltage switching at frequencies greater than 50 kHz.

Additionally, the total gate charge on a SiC power MOSFET is actually about three times less than for a comparable silicon power MOSFET, resulting in yet further gains in the upper frequency limit and/or reduction of switching loss.

Furthermore, the SiC material is relatively insensitive to operating temperature, allowing an R_{DSon} that is stable over the operating temperature. The maximum junction temperature is typically 200 °C.

Advantages of the SiC Power MOSFET

- Low on-state drain to source voltage due to low R_{DSon};
- Low total gate charge;
- R_{DSon} changes little as temperature is increased;
- Higher maximum junction temperature.

Disadvantages of the SiC Power MOSFET

- Gate drive requires greater voltage for full enhancement and slightly negative values for reliable cutoff;

- Thoughtful consideration of parasitic layout parameters required due to very fast switch transitions;
- Likely to generate greater emissions due to rapid switch transitions.
- Cost.

GaN MOSFETs

GaN is another semiconductor material for use with power MOSFETs with breakdown ratings of 200V or less. These devices have a total gate charge approximately one fifth that of comparable silicon MOSFETs and an R_{DSon} about half or less. This feature enables switching applications well above 2 MHz in frequency, greatly reducing component size for nonisolated topologies and facilitating large stepdown ratios in buck converters [3].

Advantages of the GaN Power MOSFET

- Very low total gate charge;
- Low R_{DSon} resulting in lower conduction losses;
- R_{DSon} changes less than silicon power MOSFETs with temperature;
- Device is fully enhanced with a gate to source voltage of 5V.

Disadvantages of the GaN Power MOSFET

- Thoughtful consideration of parasitic layout parameters required due to very fast switch transitions;
- Likely to generate greater emissions due to rapid switch transitions.

GaN Power MOSFET Typical Maximum Ratings

- Maximum drain to source breakdown voltage, (V_{DS}): 1200V;
- Maximum drain current, (I_D): 33A;
- Maximum apparent power, ($V_{DS}\times I_D$): 6.6 kVA.

Comparing IGBT and MOSFET Structures

The structures of both IGBTs and MOSFETs are very similar. When it comes to electron current flow, an important difference is the addition of a P-substrate layer beneath the N-substrate layer in the IGBT. In this extra layer, holes are injected into the highly resistive N-layer, creating a carrier overflow. This increase in conductivity within the N-layer helps reduce the total on-state voltage of the IGBT. Unfortunately, it also blocks reverse current flow. As a result, an additional diode (often referred to as a freewheeling diode) gets placed parallel with the IGBT to conduct the current in an opposite direction.

The absence of minority carrier transports allows MOSFETs to switch at higher frequencies. There are, however, two limitations: the transit time of electrons across the drift region and the time required to charge/discharge the input gate and Miller capacitances.

Switching Power

A reduction in on-state voltage can cost the IGBT to experience slower switching speed at turn-off. The reason is that while electron flow can be abruptly halted simply by reducing the gate-emitter voltage below the gate threshold voltage (as is the case with the MOSFET), there's still the matter of the holes that are left in the drift and body regions (there's no terminal connection to remove them). The only way to get them out of there is by sweep-out, which is dependent on voltage across the device and internal recombination. As a result, the device displays a tail current at turn-off until the recombination is complete. This has always been a big drawback for the IGBT.

Conclusion

MOSFETs and IGBTs are fast replacing a large majority of older solid-state and mechanical devices. It's a movement that doesn't look like it's going to slow down any time soon either, especially with the development of SiC material quality. SiC power devices are showing developers advantages like less loss, smaller size, and improved efficiency. Innovations like this will continue to push the limits of MOSFETs and IGBTs into higher-voltage and higher-power applications. As a result, trade-offs and overlaps are likely to continue in many applications. With that being the case, careful analysis of the device itself is perhaps the most logical solution when faced with the task of selecting a transistor for SMPS applications.

Silicon Schottky Diodes

The power versions of the silicon Schottky diode also have broad applicability in switching power supplies. Being majority carrier devices, silicon Schottky diodes do not have minority stored charge and therefore have zero reverse recovery time. However, silicon Schottky diodes have roughly 10× as much junction capacitance, which can have similar effects as reverse recovery time as well as ringing with parasitic inductance in the circuit.

Silicon Schottky diodes have substantially less forward voltage drop than their non-Schottky counterparts, which can be used advantageously to reduce conduction losses or to tweak output voltage for certain topologies.

An undesirable characteristic of silicon Schottky diodes is much greater reverse leakage current, which can be problematic at elevated temperatures, causing substantial dissipation on the device and loss of efficiency.

Advantages of Silicon Schottky Diodes

- Zero reverse recovery time;
- Substantially less forward voltage drop compared to conventional diodes.

Disadvantages of Silicon Schottky Diodes

- Ten times greater junction capacitance than conventional diodes;
- Substantially greater leakage current compared to conventional diodes.

Silicon Schottky Diode Typical Maximum Ratings

- Maximum collector to emitter breakdown voltage, (V_{RRM}): 250V;
- Maximum collector current, (I_F): 140A;
- Maximum apparent power ($V_{RRM} \times I_F$): 24 kVA.

Silicon Carbide Schottky Diodes

SiC Schottky diodes, being majority carrier devices, do not store minority carrier charge and therefore do not have a reverse recovery time. Additionally, the SiC semiconductor has excellent stability characteristics over the operating temperature range as well as increased maximum junction temperature. Unlike the silicon Schottky diode, the forward voltage drop and junction capacitance of a SiC Schottky diode are comparable to the non-Schottky diodes. Like the silicon Schottky diodes, the SiC Schottky diodes have substantially greater leakage current compared to conventional diodes.

Advantages of Silicon Carbide Schottky Diodes

- Zero reverse recovery time;
- Relatively insensitive to operating temperature change;
- Greater maximum junction temperature.

Disadvantages of Silicon Carbide Schottky Diodes

- Substantially greater leakage current compared to conventional diodes.

Silicon Carbide Schottky Diode Typical Maximum Ratings

- Maximum breakdown voltage, (V_{RRM}): 1700V;
- Maximum forward current, (I_F): 25A;
- Maximum apparent power ($V_{RRM} \times I_F$): 42.5 kVA.

Gallium Arsenide Schottky Diode

GaAs Schottky diodes have the lowest junction capacitance of all power semiconductor diodes. Consequently, they are able to operate at switching frequencies substantially higher than other types of power semiconductors. Switching at 5 MHz and beyond is possible.

Advantages of GaAs Schottky Diodes

- Lowest junction capacitance;
- Temperature stable.

Disadvantages of GaAs Schottky Diodes

- Generally higher leakage currents;
- Generally lower breakdown voltages.

GaAs Schottky Diode Typical Maximum Ratings Available

- Only one breakdown rating available: 300V;
- Maximum forward current, (I_F): 29A;
- Maximum apparent power ($V_{RRM} \times I_F$): 13.7 kVA.

13.4.3 Power Supply Isolation

In the context of power supply design, isolation is the electrical and sometimes magnetic separation between two circuits in close proximity to each other. Isolation is usually required when one topology is incompatible with another, causing interference and improper operation one way or the other.

13.4.3.1 Main Purposes of Power Supply Isolation

There are many purposes for isolation. Some of the most common are delineated here.

Safety

Electronic power supplies commonly process dangerous voltages with low impedances. Isolation is often required to prevent operator contact with dangerous voltages. Transformers used within power supplies function as a safety isolation transformer.

Voltage Level Shifting

Power supply designs must often accommodate loads at different potential voltages than the source. Isolation is often required to achieve the potential level shifting. Within the power supplies themselves, voltage level shifting devices may be required to couple energy or signals to circuits at two or more different voltages. Such an example is a gate drive transformer used for high-side switch control.

Enable Step-Up Conversion Using Buck-Derived Topologies

The buck converter has a conversion ratio that is a linear function of duty. This characteristic is very desirable. However, the buck converter is a step-down converter only. To achieve a conversion ratio greater than one without a transformer, we must use a converter topology such as a boost, buck-boost, noninverting buck-boost, SEPIC, and so forth. However, these topologies all have nonlinear conversion ratios. To achieve conversion ratios greater than one with a linear conversion ratio, a transformer isolated derivative of the buck converter can be used. This type of converter can even be used to achieve a buck-boost type characteristic where the input voltage can range greater or less than the output voltage.

Nonbuck derived topologies with nonlinear conversion characteristics can also use transformers for isolation and assisting with step-up and step-down conversion. The most common example is the flyback topology (the transformer in the flyback more correctly functions as a multiple winding inductor).

Obtaining Multiple Output Converters

The isolating characteristics of a transformer also allow the design of power supplies with multiple outputs by adding windings to the transformer along with rectifier and filter components. A very common example of this type of power supply is the desktop computer supply with +12V, +5V, and +3.3V outputs.

Ground Loop Prevention

Isolation between circuits can also be used for the purpose of preventing ground loops. Ground loops occur when two or more circuits share a common return path. When ground loops occur, there is the high likelihood that the signal voltage developed by current from one of the circuits on the return path will disrupt the operation of the other circuit.

Providing Galvanic Isolation

Electrical isolation inherently provides galvanic isolation since galvanic corrosion depends on the conduction of electrical charge.

13.4.3.2 Power Supply Isolation Methods

There are at least three methods for achieving isolation in power supplies: physical separation, transformers, and optical couplers, although isolation can be achieved in many ways. Each of these methods has its advantages and disadvantages and is usually optimal in different situations.

Physical Separation

The most obvious form of power supply isolation is simple physical separation—used when there is no electrical, magnetic, or thermal interaction desired. The separation may employ different types of dielectric mediums between the conducting surfaces.

Transformers

Transformers provide electrical isolation, but allow power to be magnetically coupled through from primary to secondary. They are used when it is desired to transfer power across the power supply isolation boundary and can be designed for safety isolation, voltage level translation, step-up and step-down functions.

Optocouplers

Optocouplers are used to transfer signals across different voltage levels, and without introducing significant parasitics across the boundary. Two types are:

- *IC Package*

The optocoupler in an IC package consists of a light-emitting device and a light receiving device inside the same IC package. Both are semiconductor devices. The efficiency of transfer is described by the *current transfer ratio* (CTR). The CTR ages with input drive level and with operating temperature making careful design with these devices important. In power supplies, they are commonly used to transfer feedback from the secondary to the primary across the power supply isolation boundary. Typical isolation voltages for this type of device are on the order of 3 kV.

- *Optical Fibers*

  High-voltage power supplies can make use of optocouplers with a data transmission fiber between the transmitter and the receiver. This form of optocoupler can achieve tens and even hundreds of kilovolts of isolation while processing control and data signals. Optocouplers use light rather than electronics to provide the feedback.

Power Supply Isolation Parasitics

Isolation is never perfect. It is a poor practice to assume that if conductors do not touch, then they are isolated. The parasitic parameters described in this section are responsible for undesirable isolation boundary performance.

Imperfect Insulators

All insulators have some degree of conductivity, resulting in leakage current. This condition worsens as voltage levels increase because the electric field increases as the voltage level increases. The dielectric properties of insulators varies, depending on the insulating material under consideration.

Surface Tracking

Surface tracking is caused by conduction contaminants on the insulator surfaces between conductors. This type of parasitic is commonly responsible for circuits with extremely high lumped resistance values not working as planned. The greater the voltage potential between the conductors, the greater is the tracking current. If the combination of contaminant surface density is great enough or the voltage potential is great enough, an arc will occur leaving a carbon trail that is difficult to remove. Once the carbon trail is formed, the voltage holdoff capability at that location is compromised.

Dielectric Breakdown

Dielectric breakdown occurs when the electric field strength across a dielectric medium is great enough to cause electrons to be stripped from their normal orbiting locations around the atoms. If the dielectric is not renewable, the reliability of the part is compromised. If the dielectric breakdown occurs along an insulation surface between two conductors, a carbon path is left which will result in surface tracking at that location.

Stray Capacitance

Every insulator material used in isolation applications exhibits capacitance that is a function of the area of and distance between the effective conducting surfaces, as well as the dielectric constant of the material. At frequencies greater than DC, there will be some current flowing through the effective capacitor. This effect causes two circuits to be DC isolated but not AC isolated. If the circuits are physically close enough to each other, and the frequency components are high enough, one circuit will be able to disrupt the operation of the other circuit. This is known as *crosstalk*. Stray capacitance between primary and secondary windings is a common source of objectionable common mode currents and can cause power supplies to fail emissions testing.

Leakage Inductance

Transformers will always have a parasitic called leakage inductance. It is caused by flux from one winding not linking with the other winding(s). Leakage is typically seen as an undesirable parasitic that should be minimized. In resonant supplies it is possible to use it advantageously.

Mutual Inductance

Mutual inductance causes undesired linking between two different circuits. This is usually a problem when there are large magnitude currents with rapid switching transients in close proximity to sensitive nodes of other circuits.

Transformers

Power supplies contain two main circuits: a primary side and a secondary side. The primary side connects to the power source, and the secondary side connects to the load. The transformer is the interface between the two main circuits.

Transformers convert the voltage available on the primary side to the required voltage level on the secondary side. Energy transfers from the primary side to the secondary by the continuous building up and collapsing of a magnetic field. Alternating current passing through the primary winding generates this field. The transfer of energy, from the primary to the secondary, takes place during the build-up and collapse phase of the magnetic field. This electromagnetic energy gets picked up by the secondary winding to generate the required voltage on the secondary side.

The voltage generated on the secondary side is generally proportional to the ratio of number of wire turns between the primary and the secondary windings. A transformer is normally made of a primary winding of copper wire, which is isolated from a secondary winding, and a core, which is made from a ferrous material such as iron or ferrite. Design and construction of a transformer requires consideration of such things as input and output current, voltage, core cross-sectional area and materials, insulation materials and methods, physical size and style, and temperature rise caused by core and wire losses. A transformer that has not been designed correctly may have less efficiency and may be electrically unsafe.

The basic technology behind switching transformers is as follows: as the rate of change of the magnetic field increases in the transformer (i.e., increase in switch frequency), the transformer can be made smaller with smaller cores and wires to produce the same output power. Lighter core materials such as ferrite can be used instead of laminated iron. The resulting transformer assembly becomes much smaller and lighter than its linear counterpart.

13.5 Power and Ground Distribution

A well-performing RF layout requires careful planning of board part placement, power supply routing, supply bypassing, and grounding. These techniques, along with correct placement of filter components, can address PLL and other spurs that generate EMI. We discuss some of these layout issues in this section while EMI factors and issues are presented in Chapter 14. Power distribution is frequently not considered until late in the design process and that is often too late to take into consideration the many issues that power involves.

13.5.1 Fundamentals of Supply Routing and Bypassing

When designing RF circuits, the implementation and layout of the power supply section is often treated as an afterthought when compared with the high-frequency signal path. Without careful consideration, it is easy for the supply voltages around the circuit to become corrupted and noisy, adversely affecting the system performance of the RF circuitry. Proper planning of the PCB layer stack-up, V_{CC}

routing via a star topology, and proper decoupling of the V_{CC} pins will help in achieving the best RF performance possible.

Starting with a sensible PCB layer definition will ease the rest of the layout process. Given a four-layer board, as is commonly used in WLAN routing, a typical stack-up would be to use the top layer for component placement and RF routing, a ground plane on the second layer, power routing on the third layer and whatever signal routing remains on the fourth layer.

Placing an uninterrupted ground plane on the second layer, is imperative for establishing well-controlled impedances for the RF signal paths. It also allows for ground returns to be as short as possible and isolates the first and third layers so that coupling is minimized. While it is possible to effectively use other stack-up approaches (required if a different number of layers is used), the aforementioned approach has been proven to be successful. While it may be tempting to use a large power plane to simplify the routing of the V_{CC} signals, this approach will most definitely lead to degraded system performance. By tying all of the supply voltages together at a large plane, it is impossible to prevent noise transfer from one pin to another.

Instead, using a star topology reduces coupling between the various supply pins in a system. An example of distributing V_{CC} through a star topology is shown in Figure 13.30. A main V_{CC} node is established, from which individual traces branch out to feed each of the RF IC supply pins. Using independent traces for each supply pin presents spatial separation between the pins, thus minimizing the amount of coupling. Each line will also have a finite amount of parasitic inductance associated with it, and that inductance works in our favor to help filter high-frequency noise from the lines.

When using a star topology for V_{CC} routing, it is necessary to properly decouple the supply lines. Decoupling is complicated by the fact that capacitors have parasitic inductance. In practice, a capacitor is represented as a series RLC circuit as shown in Figure 13.31. The capacitance will dominate at low frequencies, but after the *self-resonant frequency* (SRF) at

$$f_{\mathrm{SRF}} = \frac{1}{2\pi\sqrt{LC}} \tag{13.10}$$

the impedance of the capacitor will begin to look inductive. Thus, a capacitor is only useful for decoupling purposes over a frequency range that is near or below its SRF, where the capacitor presents a low impedance at the frequency of interest.

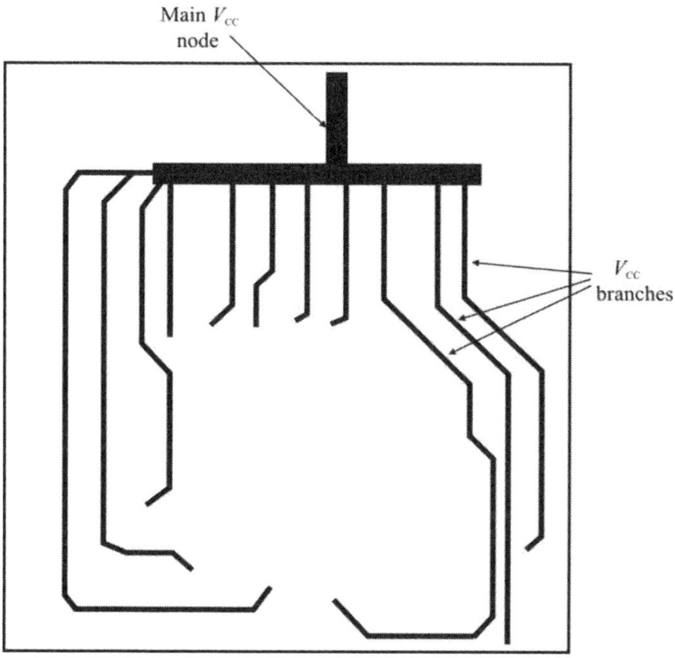

Figure 13.30 V_{CC} star topology.

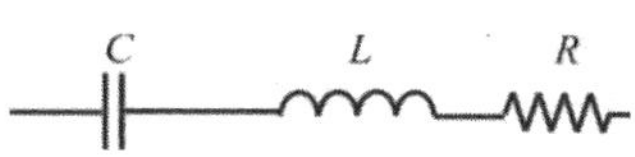

Figure 13.31 Equivalent circuit of a capacitor.

Figure 13.32 shows typical s_{11} performance for various capacitor values. From these plots, the SRFs can be seen by the dip in the graph. We can also see that the higher capacitances provide better decoupling (apparent lower impedance) at lower frequencies than the lower-valued capacitors.

In a good board design, drop as many vias to ground as possible in the RF circuitry section—especially for the exposed grounding paddle of popular IC packages.

At the main node of the V_{CC} star, it is desirable to place a large-valued capacitor such as 2.2 μF. This capacitor has a low SRF but is effective at removing low-frequency noise and creating a stable dc voltage. At each supply pin of the IC, a lower-valued capacitor such as 10-nF should be used to remove any higher-frequency noise that may couple onto the V_{CC} line.

If the part of the circuit that the supply pin is powering is sensitive to noise (i.e., a VCO supply) it may be necessary to place two capacitors close to the IC. For example, using a 100-pF cap in parallel with a 10-nF capacitor will provide a wider frequency range of decoupling and will make the supply less susceptible to noise. Each supply pin should be carefully examined to determine how much decoupling is necessary and at what frequencies the particular circuitry is most vulnerable to noise.

Combining good power supply decoupling techniques with a well-thought-out PCB layer stack-up and careful V_{CC} routing (implementing a star topology (see Figure 13.30) will provide a solid foundation for any RF system design. While there are other factors that can degrade system performance, having a supply as noise free as possible is essential in achieving optimal performance.

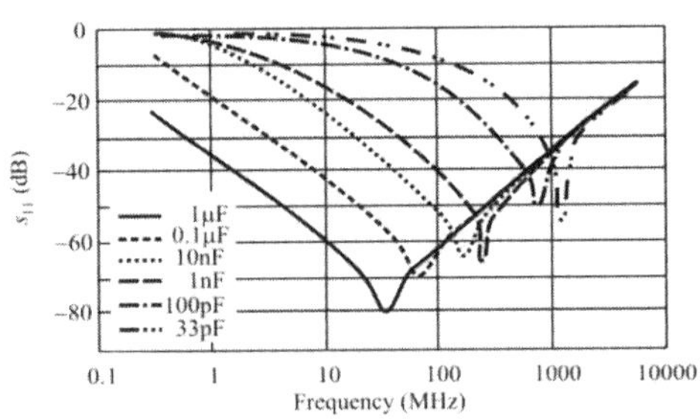

Figure 13.32 Capacitor impedance variation with frequency.

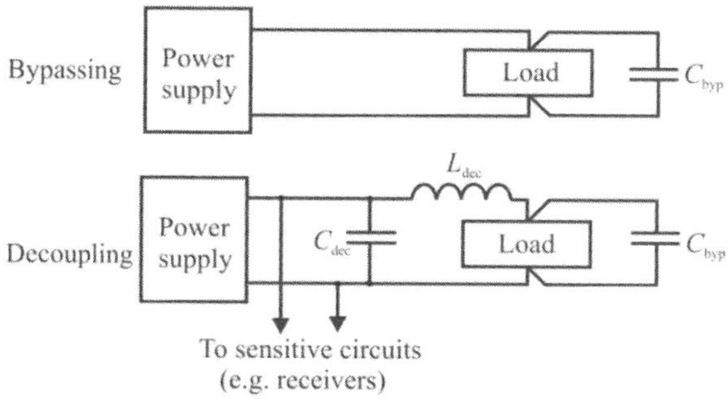

Figure 13.33 Bypassing and decoupling.

13.6 Decoupling and Bypassing for Power Supply Noise Reduction

13.6.1 Introduction

Many of the problems that appear in EW systems, in particular interference and EMC problems, emanate from the nonideal power supply. Noise and spurious oscillations can occur in real power supplies. Rules of thumb are available that can usually be applied successfully to simple problems; however, with a little understanding and forethought even the more obscure problems can be overcome.

13.6.2 Bypassing and Decoupling

Many believe bypassing and decoupling are synonymous. They are not; they are distinct concepts, and each is a solution to a different problem (see Figure 13.33).

Bypassing is the reduction of current flow (usually higher in frequency) in a high impedance path by shunting that path with a bypass device, usually a capacitor (denoted here by C_{byp}). Bypassing is used to reduce the noise current on power supply lines.

Decoupling is the isolation of two circuits on a common line. The decoupling network is usually a low-pass filter and the isolation is rarely equal in both directions. Decoupling is used to prevent transmission of noise from one circuit to another. In Figure 13.33 a bypass capacitor, C_{byp}, is shown along with the decoupling circuit composed of L_{dec} and C_{dec}. This is because in practice bypassing is always used when decoupling.

Most circuits in EW systems require bypassing, not decoupling. Decoupling is frequently used in sensitive receiver front ends, however, since EW receivers are typically extremely sensitive by design. Understanding both concepts is important; we begin with bypassing.

13.6.3 Bypassing

Due to the finite bandwidth of all voltage regulators, their output impedance increases with frequency, which can be modeled as an inductor in series with the output. Typical values lie between 1 μH and 2 μH for a linear three-terminal regulator. The output impedance of switching regulators varies widely, depending on the implementation, and should be measured for each case. The interconnecting leads add about 20 nH per inch. When an active load is connected, the time-varying current demand creates a noise voltage across these inductors. This noise voltage can be reduced in two ways: reduce the rate of change of the current passing through the inductor since

$$v_{\mathrm{L}}(t) = L\frac{di}{dt} \tag{13.11}$$

or reduce the inductance (or both). Bypassing reduces the rate of change of the current through the inductor.

In bypassing, a high-frequency, low-impedance path (a capacitor) is provided for the varying currents from the load that shares as little inductance as possible with the power supply leads. The key to successful bypassing is to properly determine the flow of current from a load and to supply a return path that is not common with any other part of the circuit (see Figure 13.34). The bypass path must have a significantly lower impedance at the frequency of interest than the power supply leads. It is always better to use many small capacitors connected in parallel than one large one. This is because the equivalent series inductance does not vary significantly with capacitance. The parallel bypass paths achieved with the small capacitors results in a much lower total inductance.

Determining the correct return current path is important; it is often not obvious at first glance as demonstrated in Figure 13.35. In this case the standard bypassing is applied, which causes a current to flow in the ground and supply leads, which generates a noise voltage. In Figure 13.36, this situation is remedied.

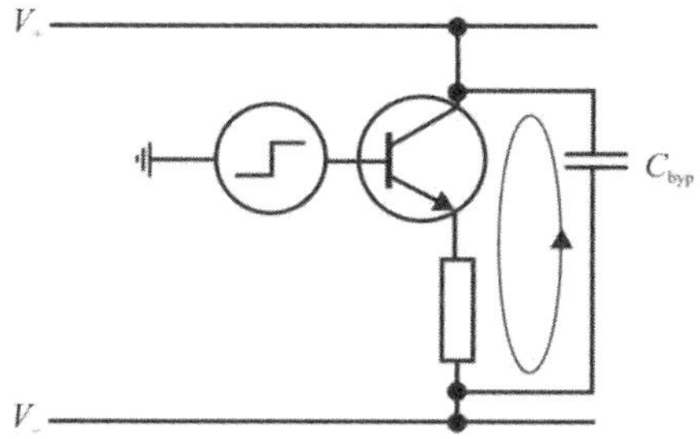

Figure 13.34 Proper bypassing isolates the noise current paths.

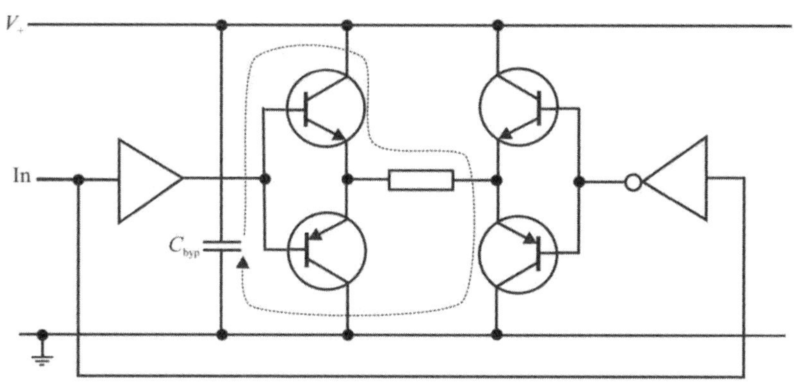

Figure 13.35 Improper bypassing. Bypass current runs through shared supply lines, in this case both the V_+ lines as well as ground.

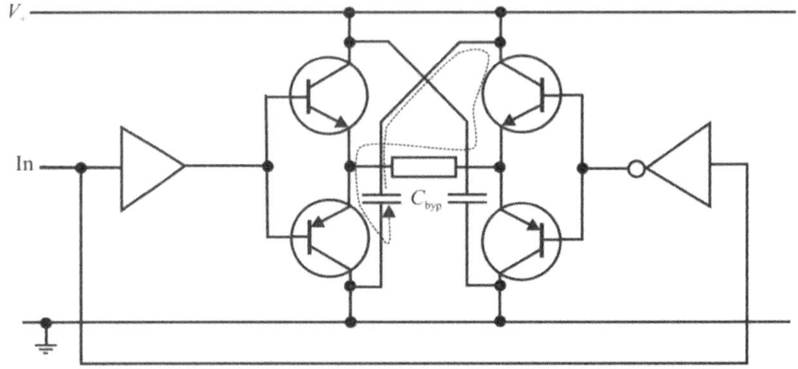

Figure 13.36 Proper bypassing. Here the bypass current avoids both the power supply lines.

13.6.4 Reducing Inductance

As mentioned, one way to reduce the noise voltage developed in the power supply inductance is to reduce that inductance since $v_L = X_L i_L = j\omega L i_L$. To reduce the inductance of a linear regulator, its bandwidth can be increased, or its open-loop output impedance can be decreased. Of course, either of these options requires a freshly designed regulator. There are also two methods for decreasing the inductance of the power supply bus. One is to decrease its self-inductance, and the other is to increase the mutual coupling to its return path (see Figure 13.37). A wire's self-inductance can be reduced by decreasing its length, increasing its radius (a small effect), or running multiple isolated wires.

The coupling shown in Figure 13.37 between the forward and return paths causes the voltage generated in the self-inductance of one path to be canceled by the voltage induced from the coupling from the other path. Of course, with perfect coupling comes perfect cancelation and zero effective inductance. It is best to place the forward and return paths as close as possible. Closely spaced supply planes, as opposed to supply traces, along with a ground plane, causes decreases in the mutual

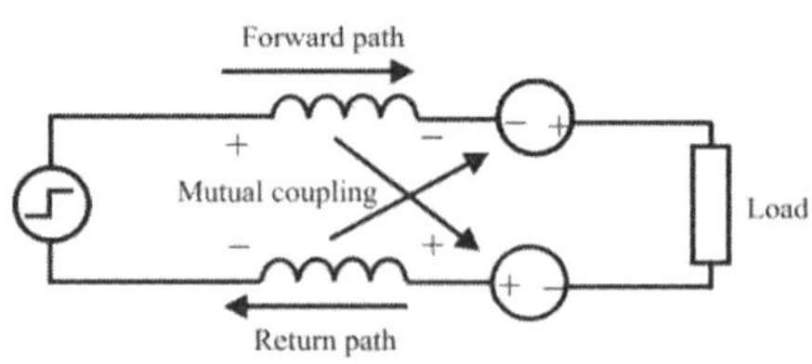

Figure 13.37 Using mutual coupling to reduce inductance.

coupling between paths by decreasing the distance between the planes. If traces must be used their width should be as wide as possible.

When using two-layer boards, it is difficult to lay out power supply planes. Another technique is to use a power grid combined with a ground plane. Many isolated traces to the load from the power supply are thus created, which reduces the self-inductance of the supply. This technique can be very successfully applied on digital boards where some cross coupling does not significantly disrupt signals. On analog boards, even this is difficult to do, especially when isolation between circuits is required (as for example, in the aforementioned RF front ends). In this case, the preferred method is to structure the supply C_{dec} like a tree as illustrated in Figure 13.30. This minimizes the length, and hence the inductance, of the supply C_{dec}. Avoid long serial, or daisy-chained power supply traces.

13.6.5 Decoupling

When it is required to isolate one circuit from the noise of another, the amount of shared supply trace between them should be reduced. If that is not adequate, decoupling should be used. Decoupling decreases noise transmission in two ways (see Figure 13.38). First, since decoupling always consists of a high impedance element in series with the supply line, it assists the bypassing and encourages the noise current to flow through the low impedance bypass element rather than the supply. Second, it acts as a-low pass filter so that the high-frequency content of any current that does pass through the series element will be attenuated, so that the regulator will be able to keep the supply voltage stable.

The series element may either be a resistor (when DC voltage drop is not an issue), or an inductor, or both. Sometimes even a parallel resonant tank circuit may be used if the noise is concentrated at one frequency.

Two parameters are dominant when choosing component values for a decoupling network. First, obviously, is isolation versus frequency. Second, and frequently disregarded, is the impedance seen by the load. This second point is important because a large output impedance will cause the load noise current to be translated into a noise voltage larger than otherwise. Typically, a large series inductor is chosen to give good isolation. This, then, requires a large bypass capacitor in order to keep the output impedance at a reasonable level. Therefore, the

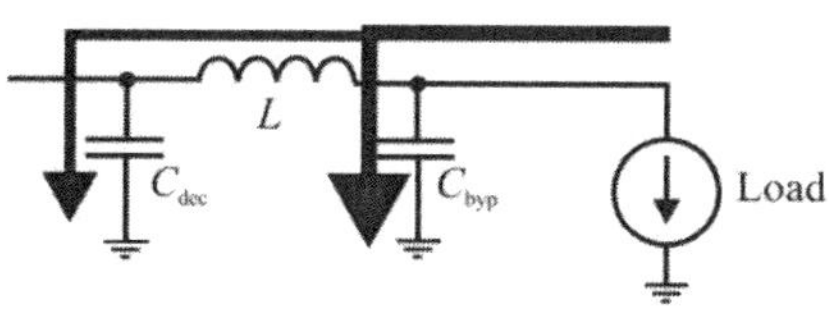

Figure 13.38 Current paths in a typical decoupling network.

decoupling network becomes large and expensive. For both economy and size, the smallest inductance that gives the required isolation should be used. It is sometimes possible to get good isolation, low output impedance, and small component values by using a multisection filter rather than a single series inductor. This allows the cutoff frequency of the filter to increase while keeping the isolation high at the frequency of interest. This eases the requirements on the bypass capacitor.

With the circuit in Figure 13.38, care must be taken to assure that there is negligible noise on the power supply side. If not, C_{dec} will couple the noise to ground, likely creating more problems.

13.6.6 Damping

When we added this bypass capacitor to the stray inductance, the power supply became an underdamped resonant circuit. This leads to ringing and noise peaking on the supplies, which likely will end up on the output. Often, this problem is severely aggravated using LC decoupling networks. For example, consider the decoupling network shown in Figure 13.39. When a 20-mA step is applied, the resulting ringing has a peak amplitude of over 500 mV and a time constant for the decay of over 40 μsec. Care must be taken in designing bypassing and decoupling networks to minimize transient response amplitude and provide damping. Transient response can be approximated with

$$V_{peak} = I_{peak}\sqrt{\frac{L}{C}} \quad \text{for } Q >> 1 \tag{13.12}$$

where for parallel RLC

$$Q = \sqrt{\frac{R^2C}{L}} \text{ and } \tau = 2RC \tag{13.13}$$

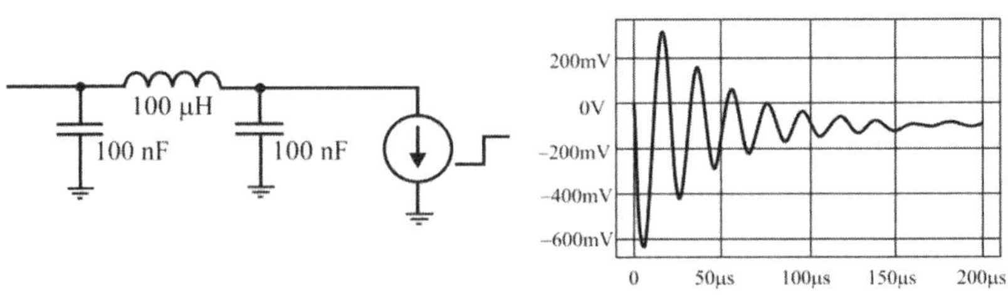

Figure 13.39 A decoupling network and its measured response to a 20-mA current step.

and for series RLC

$$Q = \sqrt{\frac{L}{R^2 C}} \text{ and } \tau = 2\frac{L}{R} \tag{13.14}$$

There are several techniques for providing damping, four of which are shown in Figure 13.40. Method A provides the best isolation and the lowest output impedance at high frequencies. The downside of method A, however, is that it suffers from high DC series resistance, which reduces output voltage and supply regulation. Methods B and D are usually unacceptable for other reasons. When low DC output resistance is required, method C is the preferred damping method. The problem with C is its high-frequency output impedance. By adding series resistance to the bypass capacitor, the amplitude and high=frequency components of the transient response to the load are increased. The amplitude of the critically damped response is twice that of the undamped. When the resistive parasitics in the components are not enough to provide proper damping, consideration should be given to whether the advantages of damping outweigh the disadvantages.

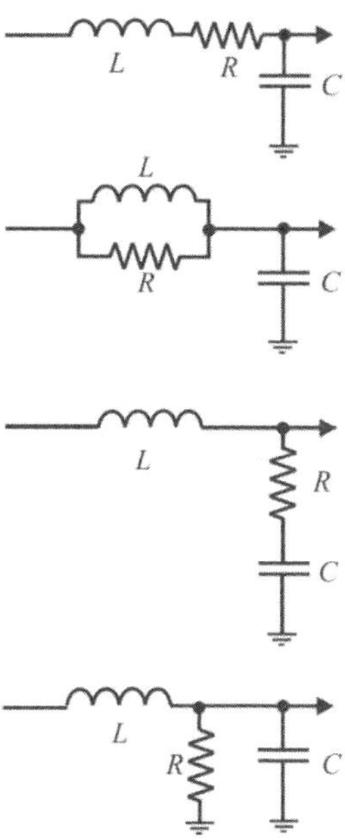

A. Reduced output voltage
Decreased supply regulation

$$R_{\text{crit}} = 2\sqrt{\frac{L}{C}}$$

B. Poor decoupling

$$R_{\text{crit}} = \frac{1}{2}\sqrt{\frac{L}{C}}$$

C. Decreased bypass effectiveness

$$R_{\text{crit}} = 2\sqrt{\frac{L}{C}}$$

D. High power dissipation

$$R_{\text{crit}} = \frac{1}{2}\sqrt{\frac{L}{C}}$$

Figure 13.40 Several methods for damping decoupling networks along with the value of the resistor needed for critical damping.

It is not necessary to provide damping for every bypass capacitor; rather a single large capacitor can be added to each board and damping provided for that capacitor. This is illustrated in Figure 13.41. The single large capacitor must be at least 16 times larger than the sum of all the high-speed bypass capacitance connected to the supply to dominate in the damping resistor calculation. If the capacitor is large and the inductor is small, then often the *equivalent series resistance* (ESR) of the capacitor may be large enough to damp the network.

It is necessary that both the series RCL network consisting of L, R_{damp}, and C_{damp}, and the parallel RCL network consisting of L, R_{damp}, and C_{byp} be properly damped. This requires that both

$$R_{\text{damp}} > 2\sqrt{\frac{L}{C_{\text{damp}}}} \tag{13.15}$$

and

$$R_{\text{damp}} < \frac{1}{2}\sqrt{\frac{L}{C_{\text{byp}}}} \tag{13.16}$$

The requirement that $C_{\text{damp}} > 16C_{\text{byp}}$ assures that a value of R_{damp} exists that satisfies both (13.15) and (13.16). Choosing a larger ratio between C_{damp} and C_{byp} results in a range of resistor values satisfying both inequalities. This is valuable if for no other reason that the value of R_{damp} can only be specified to within some finite tolerance. If C_{damp} is chosen large enough, its ESR will likely satisfy both (13.15) and (13.16), eliminating the need to add an additional discrete resistor to implement R_{damp}.

When damping decoupling networks like the one in Figure 13.39, it is usually a good idea to damp both sides of the network. Damping is typically provided on the power supply side by the large electrolytic capacitors at the output of the regulators.

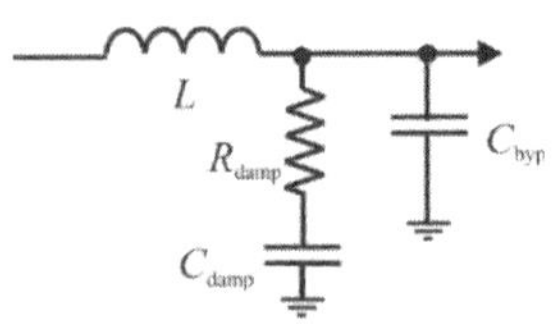

Figure 13.41 The preferred approach for providing damping to a decoupling network.

13.6.7 Noise Reduction Design Procedure

The following design procedure can be used for power supply noise reduction networks.

1. Identify all high derivative (*di*/*dt*) current sources and sinks, as well as their return paths. Multiple paths may exist, all of which must be identified and bypassed. For example, an op-amp sources current from the positive supply and sinks current to the negative supply. A useful rule of thumb limit for the maximum *di*/*dt* for a total board is 10 mA/μsec.
2. Place the forward path as close as possible (e. g., on opposite sides of the board) to the return path and maximize the trace width to minimize inductance by mutual coupling. If possible, lay out power supply busses in a grid or a plane. Avoid long serial supply traces.
3. Bypass all high derivative (*di*/*dt*) current sources and sinks with capacitors that work well at the frequencies of interest. Ceramic capacitors are good for this application because they are inexpensive, small, and work well at high frequencies. Avoid capacitors with formed leads (see Figure 13.42) because of their large *equivalent series inductance* (ESL).
4. If needed, choose the decoupling elements based on required isolation and frequency response requirements. If using a simple series inductor, its value should be as small as possible. Avoid high-Q inductors; in this application low-Q is desirable.
5. Choose a damping resistor and capacitor that satisfy (13.15) and (13.16). Use (13.16) to choose the resistor and (13.15) to choose the capacitor. Be sure to include the capacitor ESR as part of the total damping resistance. When using (13.15) to compute the resistance needed for critical damping include the effective output inductance of the regulator, the parasitic inductance of the wiring, and the effective series inductance of the capacitor along with the decoupling inductance. Choose a capacitor large enough so that the ESR provides adequate damping. Tantalum electrolytic capacitors usually work well in this role. If the ESR of the capacitor is

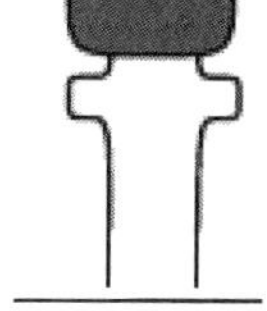

Figure 13.42 Capacitor with formed leads.

not large enough to critically damp the network, then additional resistance should be added to the large capacitor.

13.6.8 Operational Amplifiers

Due to their large open-loop gain, op amps can be quite sensitive to noise on the power supplies. Figure 13.43 shows the block diagram of a typical (μA741) op amp.

Notice that the integrator has its noninverting input referred to the negative supply. This means that any high-frequency noise that exists on the negative supply rail gets coupled to the output through the integrator. In other words, if the negative supply voltage changes abruptly, the integrator will force the output to follow the change. When the entire amplifier is embedded in a global feedback loop (which they frequently are), the error signal at the input will tend to restore the output, but the recovery will be limited by the settling time of the amplifier. As a result, an amplifier of this type may have outstanding low-frequency power supply rejection, but the negative supply rejection is fundamentally limited at high frequencies, and essentially nonexistent at frequencies above the closed-loop bandwidth.

Due to parasitics in the amplifier, the power supply rejection of the positive supply also deteriorates with frequency, but this is less severe than what occurs with the negative supply. This causes an asymmetrical pulse response. Typically, a positive-going pulse edge will have a response determined mainly by the amplifier whereas a negative-going pulse edge will have a response due to both the amplifier and the power supply. To solve this problem, both bypassing and decoupling can be used. With an op amp, it is better to use small resistors (10–50 Ω) rather than inductors as decoupling elements to avoid resonances. The low-frequency power supply rejection will take care of the low-frequency voltage drops due to the resistor.

In Figure 13.43, a secondary feedback loop exists that is not widely recognized in which signal is fed back from the collector of the output PNP transistor through

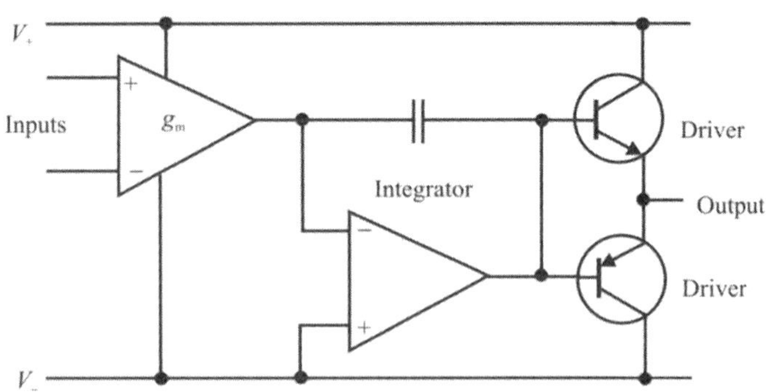

Figure 13.43 Block diagram of a μA741 operational amplifier.

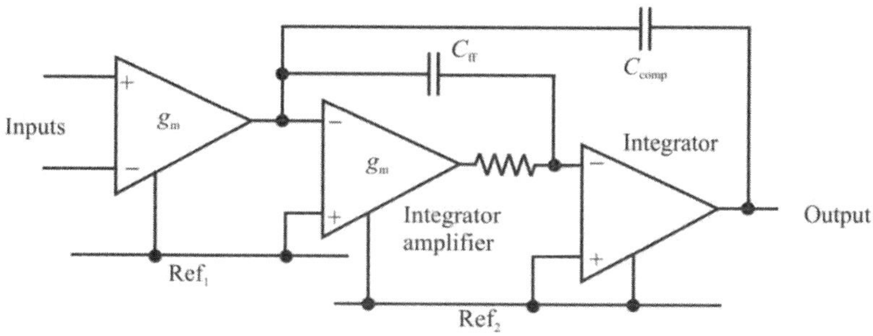

Figure 13.44 A feed-forward amplifier.

the negative supply rail and into the noninverting input of the integrator. The dynamics of this path are dependent on both the load and the negative supply impedance. When the supply looks inductive, it tends to destabilize the amplifier. A capacitive load will aggravate the situation. However, bypassing the negative supply will help alleviate this problem. Note that neither a high-speed op amp nor fast signals are required for this problem to appear.

13.6.9 Feed Forward

Many high-speed op amps use a technique called feed forward as a bandwidth enhancement technique. In a feed-forward op amp, the slow pnp level shift stage is bypassed with a capacitor to provide an alternate high-speed path. A block diagram of a feed-forward amplifier is shown in Figure 13.44.

This means that any signal appearing between the two supplies is effectively inserted in the integrator loop. This problem can usually be minimized by tying the two supplies together with a bypass capacitor, as in Figure 13.45.

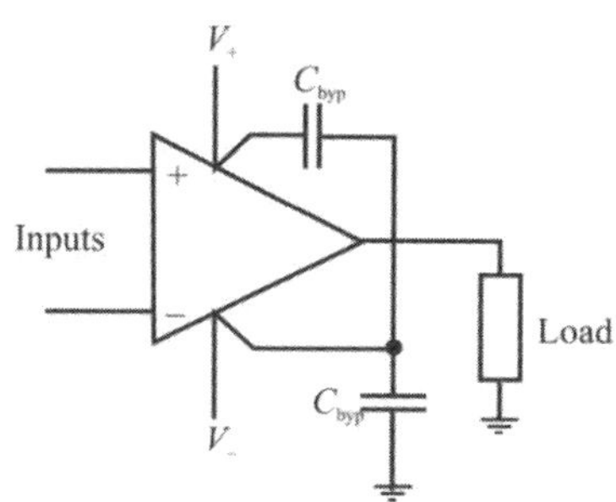

Figure 13.45 Bypassing a feed-forward amplifier.

13.7 RF Grounding Using Ground Vias

Grounding and routing are critical steps in RF board layout and fabrication. These steps will directly impact board parasitic parameters.

There are no unique solutions to ground distribution in RF board design; several approaches may achieve satisfactory system performance. Split ground planes or split traces can be used to separate analog and digital signals, or isolate high-current or high heat-generating sections.

However, a single solid ground plane in a four-layer stack-up board works well. The rule of thumb is to avoid cross interference by using a ground plane to shield the RF section from other circuitry in the board. As described earlier, layer 2 is usually designated as the ground plane while layer 1 is used for components and RF routing.

After grounding arrangements are settled, it is important to route all signal ground returns to the solid ground plane in the shortest path possible. Dropping vias from the top layer ground to the ground plane is a common solution for this task, but vias are quite inductive. The physical model of a via is shown in Figure 13.46. An accurate electrical model is given in Figure 13.47 where L_{via} is via inductance and C_{via} is parasitic capacitance of the PCB pad of a via.

In the grounding technique discussed, the parasitic capacitance can be neglected. A 1.6 mm deep via with a diameter of 0.2 mm offers about 0.75 nH of inductance. The equivalent reactance in the 2.5 GHz and 5.0 GHz WLAN bands is about 12 Ω and 24 Ω, respectively. Therefore, a single via to ground does not provide real grounding for RF signals.

In a good board design, drop as many vias to ground as possible in the RF circuitry section—especially for the exposed grounding paddle of popular IC packages. Otherwise, as an example, undesirable emitter degeneration will occur in receive front-end or power amplifier circuitry. The emitter degeneration leads to diminished gain and degraded noise figure performance. It should be noted that a poorly soldered ground paddle will cause similar problematic effects. In addition, heat dissipation for power amplifiers requires many vias to a solid ground plane. Filtering noise from other stages and constraining locally generated noise to avoid cross-interference between stages through V_{CC} lines are a few of the benefits of using

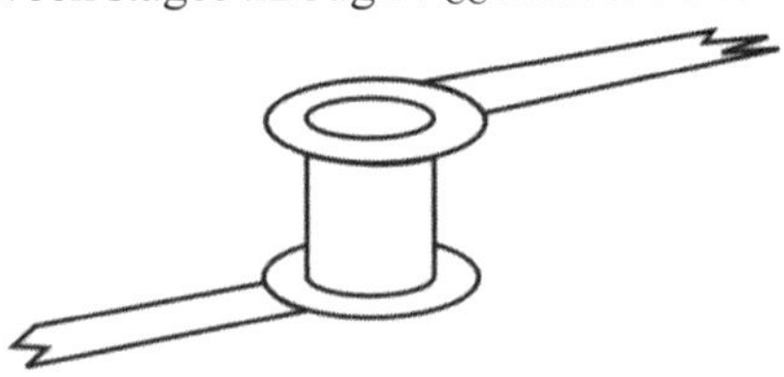

Figure 13.46 Physical model of a via.

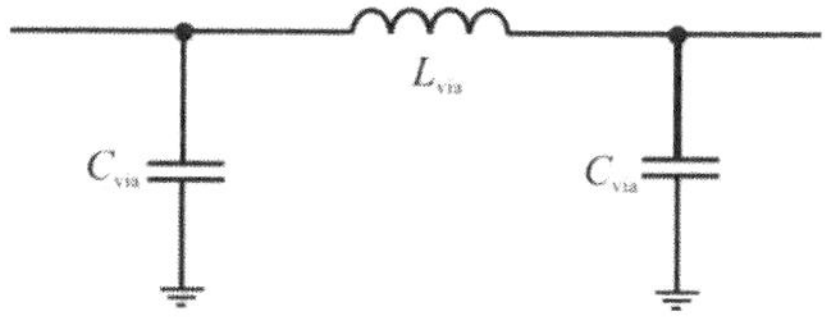

Figure 13.47 Electrical model of a via.

V_{CC} decoupling. However, if the decoupling capacitors share the same ground vias, these vias at the joint end will carry all RF interference from both supplies due to the via inductance to ground. This not only makes decoupling capacitors lose their function, but also provides another path for noise coupling between stages in the system.

As will be discussed later, PLL implementation presents a challenge in system design. Satisfactory spur-level performance may not be achieved without good grounding separation. In current IC designs, all PLLs and VCOs are integrated into the chip; most PLLs use digital current- charge-pump outputs to control the VCO through a loop filter. Usually, a second- or third-order RC loop filter is required to filter the charge pump's digital pulse current to an analog control voltage.

The two capacitors nearest the charge pump output must be grounded directly to the charge pump circuitry ground. This isolates the ground return pulse current path from the VCO ground, minimizing the comparison frequency spurs on the LO.

The third capacitor (for a third-order filter) should directly connect to the VCO ground to prevent control voltages from floating with the digital current. Straying from these principals increases the risk of high comparison spurs.

13.7.1 Managing PLL Spurious with Proper Supply Bypassing and Grounding

Meeting transmit spectral mask requirements can be a challenging component of the design process. Linearity and power consumption must be balanced with enough margin to fall within the appropriate regulatory specifications while maintaining adequate transmit output power. In-band ACPR is considered largely a function of a device's linearity, which can be adapted (within reason) to a particular application. The arduous task of optimizing ACPR in transmitters is often accomplished empirically through bias adjustments, in both IC and PA, coupled with fine-tuning of PA input, output, and interstage matching networks.

However, not all apparent ACPR issues are due to device linearity. In a prime example, a WLAN transmitter can exhibit less than desirable adjacent-channel performance even after extensive tuning and optimization of the power amplifier and PA driver (two of the main contributors to ACPR). Spurs on the LO from the transmitter's PLL can also cause poor ACPR performance. The LO spurs will mix with the modulated baseband signal and the product will be amplified along with the desired channel. This mixing action is only an issue when the PLL spurs are above a certain threshold. When they are below this threshold the ACPR will be dominated by PA nonlinearities.

When the transmit output power and spectral mask performance is linearity limited, we are in a position to trade-off current for linearity and output power, which is the desired scenario. If LO spurs dominate the ACPR performance, then we are spur limited and are required to bias the PA higher to keep its ACPR contribution down for a given P_{out}—which costs more current and offers less flexibility in the design.

This leads to the question of how to limit PLL spurs to an amplitude that does not influence the transmit spectrum. A few techniques can be used once the offending spur has been identified. The first and most tempting solution may be to narrow the PLL's loop-filter bandwidth in an attempt to attenuate the spur. This can work in select cases; however, an example will expose the possible folly in this line of reasoning.

Take the hypothetical situation in Figure 13.48. Suppose a fractional-*N* synthesizer with a 20-MHz comparison frequency is used. If the loop filter is second order, with a cut-off frequency of 200 kHz, a roll-off of roughly 40-dB/decade is nominal, yielding 80-dB of attenuation at 20 MHz.

If the reference spur is measured at –40 dBc, a level that is likely to cause undesired modulation, the mechanism causing the spur probably occurs beyond the influence of the loop filter. (If it were generated prior to the filter, it would have been extremely strong to begin with). Narrowing the filter bandwidth is not likely to improve this spur but will increase PLL lock-time—an undesirable effect in most applications.

The most effective way to combat PLL spurs is by using appropriate grounding, power supply routing, and decoupling techniques. The items discussed at the

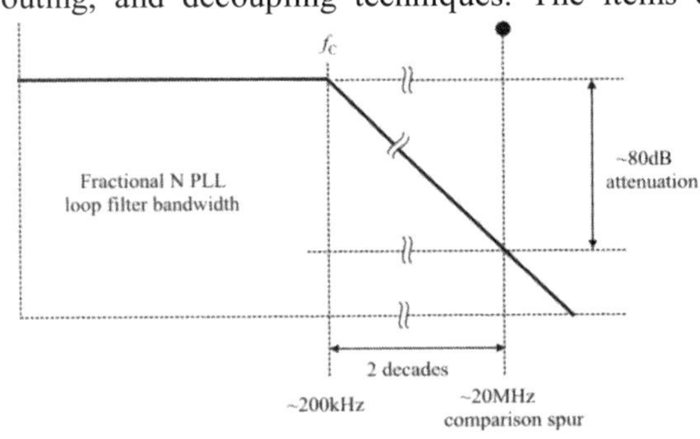

Figure 13.48 Narrowing the PLL's loop filter bandwidth does not always attenuate PLL spurs, and it has the undesirable side effect of increasing PLL lock time.

beginning of this section are a good starting point to mitigate PLL spur issues. A star-topology is imperative due to the relatively large current changes that occur in the charge-pump. The noise generated by current pulses can couple to the power supply of the VCO if isolation is not adequate, and will effectively modulate the VCO at the comparison frequency. This is known as VCO pushing. Isolation can be improved through physical separation of the power.

The two capacitors nearest the charge pump output must be grounded directly to the supply lines, decoupling at each V_{CC} pin, judicious placement of ground vias, and the introduction of series ferrite elements (consider this as a last resort). While not all of these measures may be necessary in every design, each can be used as part of a larger spur-mitigation strategy.

Figure 13.49 demonstrates the effect of an inadequately decoupled VCO power supply. The supply ripple shown directly relates to charge-pump activity that is corrupting the supply line. The corruption can be reduced significantly by increasing the local bypass capacitance.

13.7.2 Conclusion

An understanding of potential issues coupled with careful planning and layout techniques are the cornerstones of a successful radio design. We have explored these issues and presented solutions and strategies to avoid them. In summary, start with a PCB stack-up emphasizing short ground return paths. Route V_{CC} lines in a star pattern, decoupling them globally and locally with appropriate capacitor values. Ground RF sections liberally, keep return inductances to a minimum, and thoughtfully arrange ground pad locations. Using a combination of these techniques can help keep supply noise and VCO pushing to a minimum and increase the likelihood of a successful design.

13.8 Negative –5V Supply for Biasing GaAs FET's

Figure 13.50 illustrates how a –5V bias voltage can be obtained from a positive +12V power supply. The inverter is based on a NE555 circuit followed by a push-pull amplifier and a voltage doubler detector. The oscillation frequency is approximately 32 kHz, which must be well DC filtered at the output so this signal

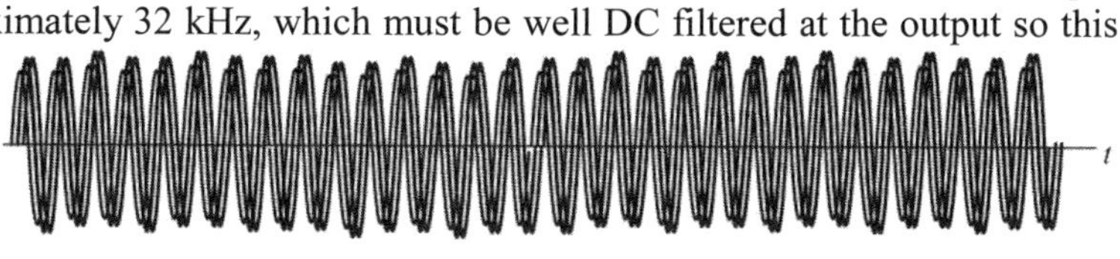

Figure 13.49 An inadequately decoupled VCO power supply yields the noise characteristics shown here.

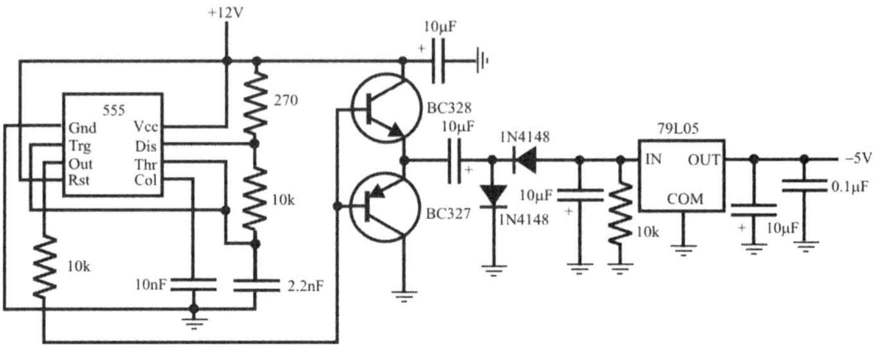

Figure 13.50 Minus 5V power supply from a +12V power supply.

doesn't pass through the bias and get to the RF circuits.

The current output of this configuration is not huge—on the order of 100 mA maximum provided by the negative voltage regulator 79L05. However, as it is primarily intended to bias the gate circuitry in n-channel MOS devices where there is no current absorbed by the FET, not much current is required. Many amplifier stages can be biased with this one circuit.

13.9 Power Supply Implementation

13.9.1 Introduction

There are several issues that must be considered when implementing power supplies. We cover some of these issues in this section.

- *Assuming that it is a perfect black box.* Supplies are less than perfect devices. Not only do converter characteristics and performance affect the load operation, but the load characteristics also affect converter behavior. The same type of interaction also exists at the input of the converter with the source. Finally, the load characteristics can be indirectly reflected to the source through the power supply and vice versa.

- *Assuming that it is a simple aspect of product design.* Converters are often not simple. The design of converters will usually go beyond the skill set of most nonpower electronics engineers. Even the application of off-the-shelf conversion devices can be a challenge as the documentation that accompanies them is usually limited, with little disclosure about the internal design, which may significantly impact compatibility with loads and sources.

- *Allowing it to be an afterthought in the design process.* Since the power supply is not a perfect black box, it cannot be relegated as an afterthought in the design process. Product design must take place keeping converter capabilities in mind. Otherwise it is possible to design a great product, but have no power supply for it.

- *Failure to allocate adequate physical space for it.* Conversion devices need space. While technology for improving efficiency is continually improving allowing for smaller supplies, it is a mistake to not plan for space for converters.

- *Failure to adequately specify the important performance requirements.* There are numerous specifications which a power supply must satisfy. If they are not identified, poor performance is the likely result.

13.9.2 Power Supply is Frequently the Weak Link

Conversion devices are usually the weak link in most electronics products and usually will be the first components to fail. The stress levels on the internal conversion components are usually much higher than in other electronics components. Forms of stress include voltage, current, power, and temperature. All of these forms of stress have major impact on component reliability and lifetime. Therefore, reliability of the product is usually dominated by the power conversion choice. For this reason and others, the converter must not be relegated as an afterthought in the design process.

13.9.3 Correct Perspective for Power Supply Implementation

The power converter must be seen as a critical component in the system. Without a thoughtful implementation, the system will perform poorly at best. Remember that there are usually significant interdependencies between the power supply and the load and the source. Treating it as a black box or as an afterthought will usually lead to late stage and costly surprises. Finding the cheapest implementation usually leads to poor value and prototyping troubles. Apply extra effort to seek the best value which is a compromise of low cost, outstanding performance, small size, and reliability. These factors are typically weighted according to their relative importance in your EW application.

13.10 Concluding Remarks

Power supplies are often the nemesis in EW system design and must be dealt with carefully. EW systems are typically sensitive receiving systems and, in particular, EMI can cause significant problems.

EMC must be addressed in detail all the way through the design process if there is to be any hope at all in the end that EW systems, and in particular, ES systems are to perform as required.

References

[1] Cleary, J. F. (ed.), *GE Transistor Manual*, Syracuse, NY: General Electric, 1964, Ch. 10.
[2] Millman, J., and C. C. Halkias, *Electronic Devices and Circuits*, New York: McGraw-Hill, 1967, Ch. 20.
[3] Manz, B., "How Far Can We Take GaN Technology?" *The Journal of Electronic Defense*, Vol. 38, No. 3, Mar. 2015, pp. 41–46.

Chapter 14

EMI/EMC

14.1 Introduction

We introduce the notions of electromagnetic interference (EMI) and electromagnetic compatibility (EMC) in this chapter. All electronic units emit a degree of electromagnetic fields produced by currents running through the interconnecting wires and/or PCB traces. These fields impinge on other components and wires perhaps in the same chassis or elsewhere. The nature of electric fields is such that they induce currents in these components that may be undesirable, and the result is noise-like interference. When the signals are undesirable, performance can degrade. Therefore, it is normally the case that electronic designs must keep such interfering signals below some threshold.

The above describes radiated coupling. There are other ways that interfering signals are transferred within electronic devices. We will discuss those in this chapter as well.

The EMC performance of electronic equipment today is of great importance simply due to amount of such equipment being used, and as a result it is necessary to include EMC constraints in designs. In order for the unit to pass its EMC testing and be placed on the market, it is necessary for it to conform to the directives and regulations in force. For a unit to be successful, it is necessary for it to be designed to provide a high level of electromagnetic compatibility, EMC performance, and reduction of EMI.

This chapter is organized as follows. We begin with a description of EMI and EMC including a discussion of coupling methods. That is followed by a consideration of the major international EMC standards with emphasis on those in North America and Europe. Next, we address design techniques for EMC compliance. Design techniques for PCBs is considered next, along with specific techniques to incorporate EMC. After that, a discussion of modalities for noise coupling is introduced. Since switched-mode power supplies are becoming common in many types of wireless devices, we include a section of the peculiarities of these power supplies for EMC considerations. Next, we discuss EMC issues for cabling. The last topic covered in this chapter is EMC testing in its various forms.

14.2 Electromagnetic Interference

14.2.1 EMI

All electronic devices radiate some EM waves, which may cause problems for other nearby RF/wireless systems. This is called EMI. This EMI level radiated by a device should be within limits so that nearby systems function properly (see Figure 14.10).

Switching power supply topologies and other circuits in EW systems produce EMI [1, 2]. In the case of power supplies, EMI are signals at frequencies that are multiples of the basic switching frequency. These signals can extend in the spectrum for considerable distances from the switching frequency if they are not properly filtered. Any oscillator in the system can be a likely source for EMI.

14.2.2 EMC

All electronic devices generate electric noise that passes through cables/wires and can cause problems for another device connected with the same electric system. One major source of these signals is the power system to which the device is attached.

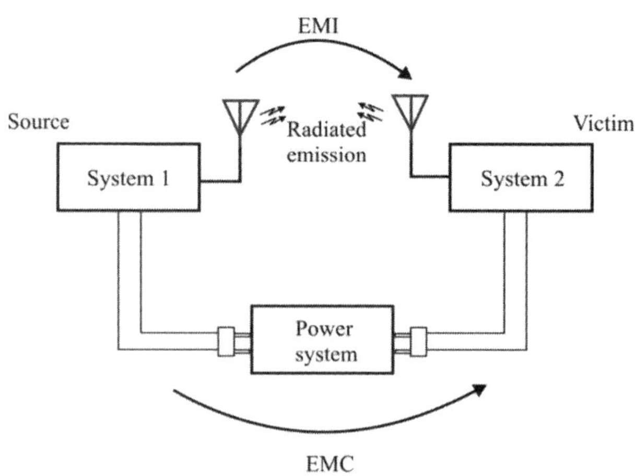

Figure 14.1 EMI/EMC.

This conducted emissions (transients/ripples) should be within limits for proper functioning of the other device/system (see Figure 14.1).

EMC is of increasing importance as the number of wirelessly connected devices increases. Defining what EMC is and understanding the concepts enable electromagnetic compatibility to be achieved from the outset.

EMC is the concept of enabling different electronics devices to operate without mutual interference—EMI—when they are operated in close proximity to each other.

All electronics circuits have the possibility of radiating or picking up unwanted electrical interference that can compromise the operation of one or other of the circuits.

14.2.3 Contribution of Power Supplies to EMI

Switching power supplies have almost completely replaced linear power supplies because of their great advantage in efficiency. However, this comes at the cost of considerably higher EMI. Switching power supplies generate more EMI because they switch large currents at high frequencies—anywhere from 50 kHz to 1 MHz. At these high frequencies optimal power efficiencies and smaller components (primarily magnetics) can be achieved.

The theoretically optimal switching waveform in these power supplies approximates a square wave. However, the sharp edges of these waveforms produce considerable noise at the fundamental switching frequency and its harmonic frequencies. In addition, since the waveform is not sinusoidal, there is an additional edge frequency (the sinusoidal equivalent of the sharp rising and falling edges of the square wave), which generates a pseudofrequency much higher than the fundamental switching frequency of the power supply. Both of the frequencies also have the normal associated harmonics. Therefore, there is a trade-off between power efficiency and EMI.

While the switching MOSFETs, transformers, and output rectifiers are typically the greatest sources of radiated EMI, at high-frequencies even PCB traces act inductively and radiate noise. It is critical for effective EMI suppression to ensure that all these elements have been considered when selecting a power supply.

14.2.4 Introduction to EMI/RFI/EMC

In the context of RF power supplies, noise (EMI), commonly called *radio frequency interference* (RFI), appears as electrical currents. These currents degrade the performance of susceptible devices. EMC is the ability of the source and the receptor to operate properly in a given electromagnetic environment.

There are two modes of EMI — conducted emissions and radiated emissions. Conducted emissions are caused when high-frequency currents flow back through

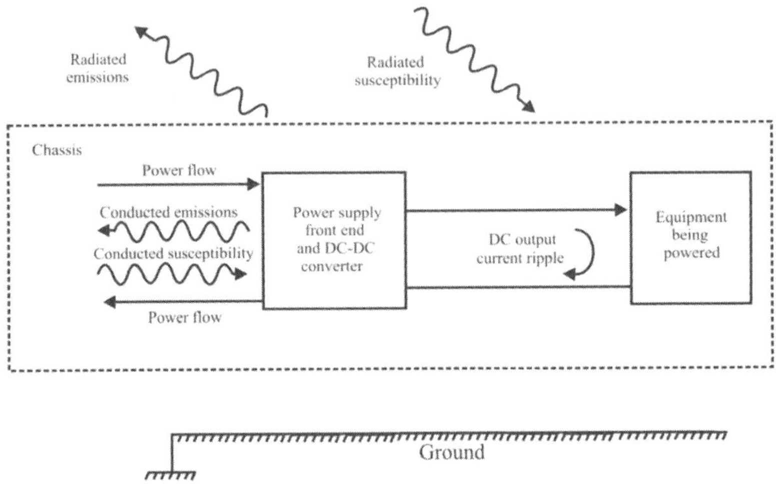

Figure 14.2 In a typical electronic package, current flow from the input of the power supply to the voltage source is referred to as conducted emissions.

the input lines polluting the power source lines (power grid, see Figure 14.2). These currents can be either *common-mode* or *differential-mode* currents. Radiated emissions are electromagnetic fields that are emitted from a unit into the environment.

With radiated emissions, these conductors can be the line and neutral power input lines (the radiated emissions being the radiated result of the high-frequency currents in these lines), or virtually any other set of conductors within power supplies. This second point is what makes tracking down and eliminating sources of radiated EMI such an extreme challenge. Radiated EMI is electromagnetic radiation leaving or entering the product enclosure. Waves leaving are *radiated emissions*. Unwanted waves entering are the source of *radiated susceptibility*.

Noise currents can flow both to and from the power supply's input along the wiring that connects the converter to the energizing source (usually the AC power grid, AC generator, or a battery). Currents exiting the supply are called *conducted emissions*. Radiation from this path can cause interference with other devices, which is why power cables are often routed through ferrite chokes.

Currents entering the supply are called *conducted susceptibilities* (Figure 14.2). Susceptibilities aren't necessarily stopped by the converter, and they can make their way into the device being powered. Specifications for conducted susceptibility are most often required for military equipment, but not so often for industrial or consumer products.

There are two main elements to EMC:

- *Emissions:* The EMI emissions refer to the generation of unwanted electromagnetic energy. These need to be reduced below certain acceptable limits to ensure they do not cause any disruption to other equipment.
- *Susceptibility & immunity:* The susceptibility of an item of electronics to EMI is the way it reacts to unwanted electromagnetic energy. The aim of the design of the circuit is to ensure a sufficiently high level of immunity to these unwanted signals.

14.2.5 EMI: The Fundamentals

There are many forms of EMI that can affect circuits and prevent them from working in the way that was intended. This EMI can arise in a number of ways. EMI can arise from many sources, being either man-made or natural. It can also have a variety of characteristics depending on its source and the nature of the mechanism giving rise to the interference.

By the very name of interference given to it, EMI is an unwanted signal at the signal receiver, and in general methods are sought to reduce the level of the interference.

14.2.5.1 Types of EMI

Different types of EMI can be categorized in a number of ways. There are many ways in which EMI can be carried from one item of equipment to another. Understanding these methods is a key to mitigating the effects.

One way of categorizing the type of EMI is by the way it was created:

- *Man-made EMI:* This type of EMI generally arises from other electronics circuits, although some EMI can arise from switching of large currents, for example. Currents through an inductor cannot change instantaneously, and when attempted radiated interference can arise.
- *Naturally occurring EMI:* This type of EMI can arise from many sources—cosmic noise as well as lightning and other atmospheric types of noise all contribute.

Another method of categorizing the type of EMI is by its duration:

- *Continuous interference:* This type of EMI generally arises from a source such as a circuit that is emitting a continuous signal. However, background noise, which is continuous, may be created in a number of ways, either man-made or naturally occurring.

- *Impulse noise:* Again, this type of EMI may be man-made or naturally occurring. Lightning, ESD, and switching systems all contribute to impulse noise, which is a form of EMI. In general, impulse interference is characterized as short-lived but very broadband.

It is also possible to categorize the different types of EMI by their bandwidth:

- *Narrowband:* Typically, this form of EMI is likely to be a single carrier source — possibly generated by an oscillator of some form. Another form of narrowband EMI is the spurious signals caused by intermodulation and other forms of distortion in a transmitter such as a mobile phone or WiFi router. These spurious signals will appear at different points in the spectrum and may cause interference to another user of the radio spectrum.
- *Broadband:* There are many forms of broadband noise that can be experienced. It can arise from a great variety of sources. Man-made broadband interference can arise from sources such as arc welders where a spark is continuously generated. Naturally occurring broadband noise can be experienced from the sun — it can cause sun-outs for satellite television systems when the sun appears behind the satellite and noise can mask the wanted satellite signal. The frequency spectrum of the sun is very broadband. Fortunately, these episodes are rare and only last for a few minutes.

Apart from understanding the form of the interference, it is also necessary to know how the interference is traveling from the transmitting device to the receiving device. Unfortunately, this is often not easy to discover as many of the signal paths are difficult to define. However good initial design with EMC considerations included alleviates many potential problems.

EW, of course, is also a form of EMI. This type of signal is generated precisely for the purpose of causing EMI in a target receiver.

14.2.6 EMI Coupling Mechanisms

There are many ways in which the electromagnetic interference can be coupled from the source to the receiver. Understanding which coupling method brings the interference to the receiver is key to being able to address EMI problems.

- *Radiated:* This type of EMI coupling is normally experienced when the source and victim are separated by a large distance—typically more than a wavelength. The source radiates a signal that may be wanted or unwanted, and the victim receives it in a way that disrupts its performance.

- *Conducted:* Conducted emissions occur when there is a conduction route along which the signals can traverse. This may be along power cables or other interconnection cabling. The conduction may be in one of two modes:
  - *Common mode:* This type of EMI coupling occurs when the noise appears with the same phase on the two conductors (e.g., out and return for signals), or $+V_e$ and $-V_e$ for power cables.
  - *Differential mode:* This occurs when the noise is out of phase on the two conductors; typically 180° out of phase.

The filtering techniques required will vary according to the type of EMI coupling present. Common mode lines are filtered together. For differential mode they may be filtered together or separately.

- *Coupled:* What is normally termed coupled EMI can be one of two forms: capacitive coupling and magnetic induction (see Figure 14.3).
  - *Capacitive coupling:* This occurs when a changing voltage from the source capacitively transfers a charge to the victim circuitry.
  - *Magnetic coupling:* This type of EMI coupling exists when a varying magnetic field exists between the source and victim—typically two conductors may run close together (less than λ apart). This induces a current in the victim circuitry, thereby transferring the signal from source to victim.

By determining the form of coupling that exists and the way in which it is reaching the victim, it may prove to be that the most effective method of reducing the EMI is by putting measures in place to reduce the coupling and reduce the level of interference to an acceptable level.

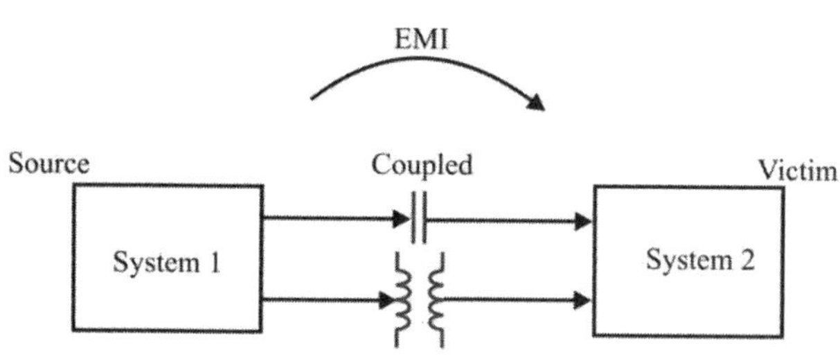

Figure 14.3 EMI due to coupling.

14.3 EMC Standards

With the growing awareness and need to maintain high standards of EMC, many standards have been introduced to help manufacturers meet the levels they need to maintain full electromagnetic compatibility.

Many years ago, the levels of EMC were low and interference often occurred. As a result, it became necessary to introduce EMC standards to ensure the required levels of compatibility were attained to minimize this interference.

EMC is now an integral part of any electronics design process. With standards now implemented and enforced throughout the world, any new product needs to have been tested to ensure it meets the relevant EMC standards. While this presents an additional challenge in the electronics design process, it is essential that good EMC practices are employed and that the EMC performance of the product is sufficient to ensure it operates correctly under all reasonable scenarios.

14.3.1 Evolution of EMC Standards

The basic awareness of the possibilities of interference between various forms of electronics equipment had been around for many years. However, the relatively limited use of electronics by today's standards meant that little was required and undertaken in terms of legislation regarding standards governing the requirements for EMC.

Some of the first EMC standards and legislation were introduced in 1979. The Federal Communications Commission (FCC) in the United States imposed legal limits on the electromagnetic emissions from all digital equipment. These limits were imposed as a result of the growing availability of digital systems including small calculators and forms of digital equipment that were interfering with wired and radio communications and broadcast systems. This legislation included the definition of a number of test methods that were developed to support this EMC legislation.

A further step forward was taken in the 1980s by the European Community. They introduced a new approach to standardizing EMC requirements to enable trade of electronics equipment to be undertaken more freely among the European countries and beyond. One of the major elements of this was the EMC Directive 89/336/EC.

The EMC Directive from the EC was ground-breaking in terms of EMC standards and legislation as it was the first time that limits had been placed on the immunity of the equipment to interference as well as its emissions. As such the EMC Directive recognized both elements of EMC — operating equipment harmoniously is not just a matter of reducing unwanted emissions as intended emissions can also cause interference.

14.3.2 Common EMC Standards

There are several common EMC standards that are widely used. Some of these standards also include other elements apart from just the EMC performance. The most used standards are:

- CISPR 11, which addresses ISM equipment in EN 55011;
- CISPR 16, which addresses methods of measurement;
- CISPR 22, which addresses ITE equipment in EN 55022;
- MIL STD 461F, which addresses EMC test requirements for military equipment;
- FCC Part 15, which addresses ITE equipment.

There are many other standards and specifications that address specific equipment types (e.g., aircraft and automotive) that we won't discuss here.

14.3.2.1 CISPR Applicability

The CISPR standards and specifications are issued by the International Special Committee on Radio Interference, CISPR, which is the organization that sets EMC standards in Europe.

CISPR divides electric equipment into two classes:

- Class B is equipment that is intended for sale/distribution in residential environments;
- Class A is equipment intended for sale/distribution into commercial or industrial environments.

Class A limits are more lenient than class B. Also, the class A limits the radiated emissions as a distance of 10 m while the class B limit the radiated emissions at 3 m.

CISPR 11: EN 55011 EMC Standard

CISPR 11 is a widely used international standard for electromagnetic compatibility within Europe for electromagnetic emissions or disturbances from industrial, scientific and medical (ISM) Equipment. CISPR 11 is maintained by CISPR defined above.

CISPR 11 applies to a very wide variety of equipment including everything from WiFi systems, and microwaves through to arc welders, all of which fall into the industrial, scientific, and medical category that can use the ISM license-free bands like 2.4 GHz.

CISPR 16: EMC Standard for Test Equipment and Methods

CISPR 16 is a series of 14 publications specifying equipment and methods for measuring disturbances and immunity to them at frequencies above 9 kHz.

CISPR 16 specifies all that is needed to undertake the EMC tests in the CISPR series. It defines the characteristics and performance of equipment for the measurement of radio disturbance voltages and currents in the frequency range of 9 kHz to 1 GHz.

The specification includes details of the ancillary apparatus needed for artificial lines networks, current and voltage probes, and coupling units for current injection on cables.

CISPR 22: EN 55022 EMC Standard

CISPR 22 is a widely used standard for electromagnetic compatibility within Europe for information technology equipment (ITE). CISPR 22 is maintained by CISPR. the International Special Committee on Radio Interference.

CISPR 22 is the EU CISPR standard for Information Technology Equipment-Radio Disturbance Characteristics-Limits and Methods of Measurement. CISPR 22 has been adopted for use by most members of the European community.

Summary of CISPR 22 limits

Figures 14.4 and 14.5 show the conducted and radiated EMI limits, respectively, within the CISPR 22 standard.

14.3.2.2 FCC 47 CFR Part 15: FCC Part 15

Part 15 of the FCC Title 47 is often just called FCC part 15 and it relates to EMC. The FCC Part 15 rules and regulations have been designed to align with the European CISPR regulations. There are some differences, but in general the two sets of standards align on the major features.

FCC Part 15 Basics

The scope of FCC Part 15 is split into three sections:

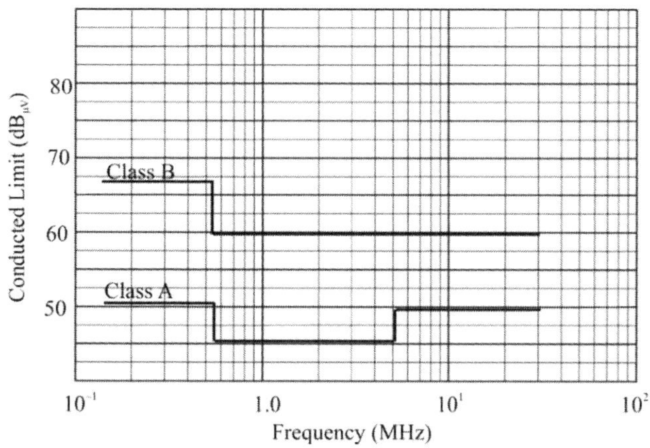

Figure 14.4 CISPR conducted EMI limit.

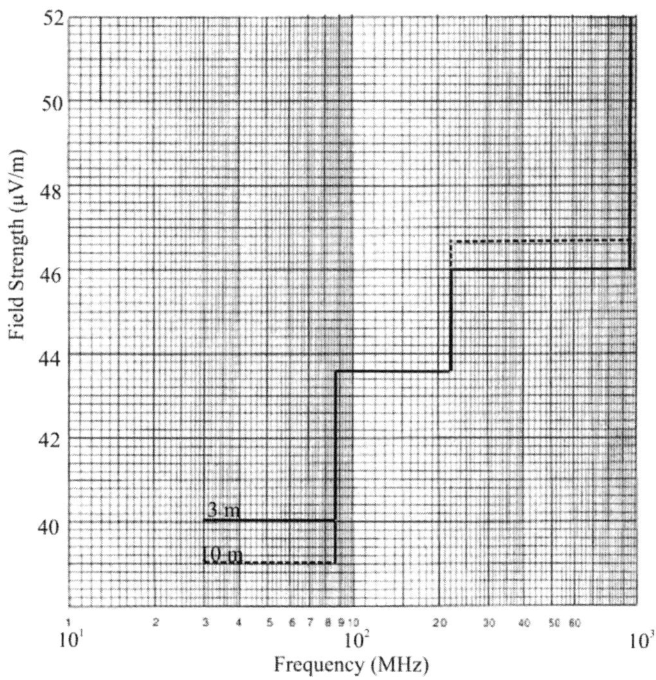

Figure 14.5 CISPR class B 3 m and class A 10 m radiated limit.

- *FCC Part 15A:* This section sets out the regulations and technical specifications for an intentional, unintentional, or incidental radiator that may be operated without an individual license.
- *FCC Part 15B:* FCC Part 15B covers the operation of an intentional or unintentional radiator that must be licensed.
- *FCC Part15 C:* Unless specifically exempted, the operation or marketing of an intentional or unintentional radiator covered by this section is prohibited.

In addition, like the CISPR specifications, there are two classes of device for FCC Part 15:

- *Class A digital device:* Within FCC Part 15, a Class A digital device is one that is marketed for use in commercial, industrial, or business environments.
- *Class B digital device:* Within FCC Part 15, a Class B digital device is one that is marketed for use in a residential or domestic environment.

14.3.2.3 CISPR and FCC Part 15 Comparison

With CISPR 22 widely used within the European community, it is often necessary to relate it to other standards, the main one being the Federal Communications Commission (FCC) in the United States. Here the FCC Part 15 standard is used.

To ensure that either standard can be used to certify digital electronic equipment, FCC Part 15 and CISPR 22 have been made to be relatively harmonious, although there are a few differences.

CISPR 22 differentiates between Class A and Class B equipment and it gives figures for conducted and radiated emissions for each class. In addition, CISPR 22 requires certification over the frequency range of 0.15 MHz to 30 MHz for conducted emissions. On these points there are differences between CISPR 22 and the FCC Part 15.

Also, CISPR 22 has no specified limits for frequencies above 1.0 GHz, and CISPR limits are provided in $dB_{\mu V}$, while the FCC limits are specified in μV, which means that a simple conversion is required for direct comparisons.

In terms of similarities the conducted and radiated emission limits specified in CISPR 22 and FCC Part 15 are close to each other (i.e., within a few decibels of each other over the pertinent frequencies).

In this way it is possible to meet both standards easily for the certification process.

The FCC and CISPR standards have grown very close to being harmonized. Both are broken into Class A and Class B limits. Class B is the more stringent of the two, requiring a device to meet a lower threshold of EMI. CISPR11 applies to ISM (Industrial, Scientific and Medical) devices and CISPR22 applies to ITE (information technology equipment).

A third-party specification usually enforces compliance with standards for conducted and radiated emissions. There are three applicable specifications for conducted emissions:

- FCC part 15, Level A or level B (United States);
- EN 55022, Level A or Level B (Europe);
- MIL-STD-461 for military applications (as specified in CE 102).

Figure 14.6 illustrates the standards for FCC part 15 and EN55022 while the military standard 5 limits are shown in Figure 14.7 [3]. There are significant differences at low frequencies. FCC part 15 starts at 450 kHz. EN 55022 starts at 150 kHz. MIL-STD-461 starts at 10 kHz for CE 102.

14.3.2.4 EMI Compliance

A switching power supply is typically designed to meet emission standards as set forth by the FCC, CISPR, or military standard. As such, an original equipment manufacturer (OEM) user would be well advised to obtain a sample of a candidate power supply and perform a simple, low-cost conducted EMI test to get a preliminary reading on the manufacturer's compliance.

However, it is important to note that these standards are applied by self-declaration. So, in order to show compliance, many power supply manufacturers test only a single unit to demonstrate emission levels beneath the maximum limits.

What this practice does not recognize is that there will inherently be some variation in the EMI levels from unit to unit since component tolerances and manufacturing variations will introduce deviations in the EMI levels during production.

As such, a power supply design which has been qualified by the customer can fail these EMI standards in production, and in some instances by considerable margins. This marginal practice can wreak havoc with a customer's program and cause system EMI failures before and after product release, costing time and money.

In addition, conformance to these standards by the power supply does not guarantee that when it is used in a larger system, the system will pass these EMI tests. The system itself can add noise to the EMI of the power supply. Furthermore, an active load can cause system interactions, which can amplify the EMI further at resonant frequencies. Therefore, it is vital that designers plan for these interactions, and proper EMI performance, as early in the design stage as possible.

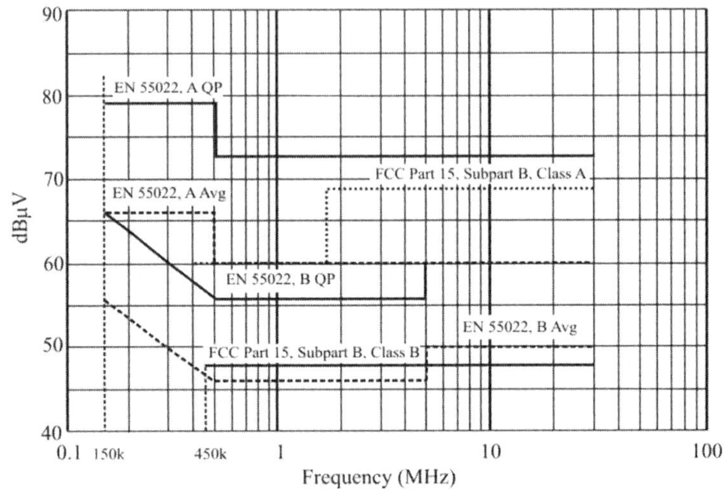

Figure 14.6 The chart shows relevant conducted emission limits using the LISN method.

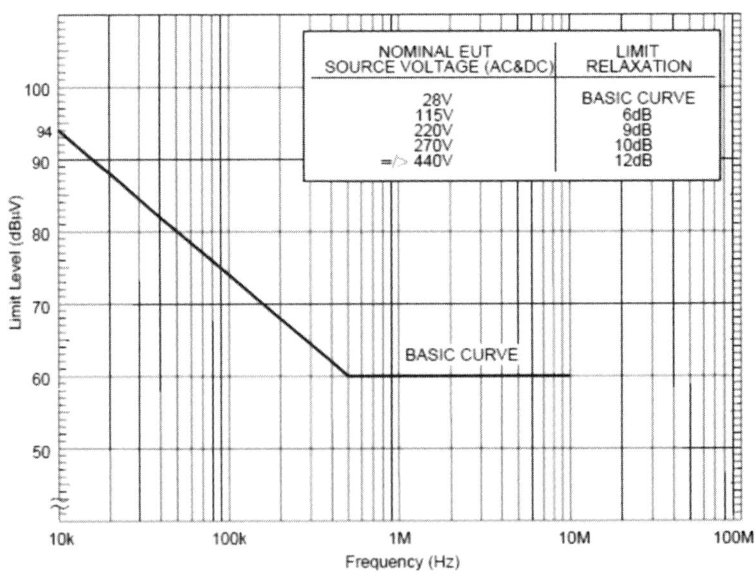

Figure 14.7 CE102 limits (EUT power leads, AC and DC) for all applications. (Source: [3].)

14.3.2.5 EN55032

The EN55022 EMC standard for ITE discussed above was withdrawn on March 5, 2017 and replaced with EN55032. EN55032 covers what is referred to as multimedia equipment which includes what EN55022 called ITE.

Multimedia equipment includes:

- Information technology equipment, audio equipment, video equipment, broadcast receiver equipment, entertainment lighting control equipment, or combinations of these.
- Examples: PC, CD/DVD player, TV/radio receiver, audio amplifier, DMX light controller, and so forth.
- Equipment within the scope of EN55013 or EN55022 or EN55103-1 is within the scope of EN55032.

EN55032 implements a port concept in that the testing procedures depend on the type of ports involved with the multimedia EUT (see Figure 14.8).

Unlike some UL or ANSI standards, the European standards do not grandfather the newest standards. For the European standards this means that when the standard is withdrawn, products must adhere to the latest standards. It will usually have to be re-tested to the latest version of the standard as the withdrawn version is null and void. If the old standard is said to grandfather subsequent releases, then the product does not have to be retested to stay listed and thus it can still be sold in the regions for which that standard applies.

The new EN55032 is unique in a way: although the withdrawn EN55022 does not grandfather the new EN55032, for power supplies, there is no difference in the test procedure or the limits to which the product must adhere for Europe. There is no need to re-test the units to EN55032, but the new standard must feature on the Declaration of Conformity (DoC) of the manufacturer/supplier of complete power supplies.

One of the biggest changes brought in by EN55032 is the frequency coverage for test. It extends the range up to 18 GHz.

14.4 EMC Design Techniques

Good EMC design techniques are not difficult to implement if they are introduced at the earliest stages of design. EMC design from the earliest stages of the product design follows some straightforward and common-sense design approaches.

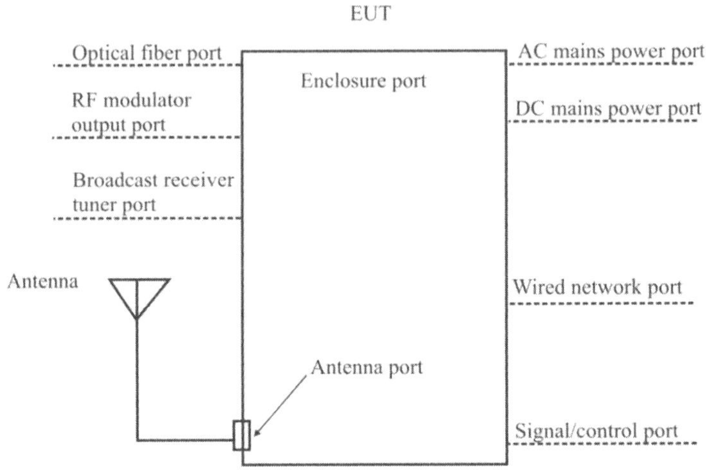

Figure 14.8 Port concept from EN55032.

14.4.1 Basic EMC Design

Any electronic circuit that has signals that change in level will radiate some power as any interconnections and wires will act as radiating antennas, however short they may be. Radiation occurs whenever electrons are accelerated — that is, change their velocity vector. Thus, radiation occurs for alternating currents or on PCB traces that bend around a corner, as examples. Similarly, circuits will tend to pick up radiated signals from other transmitters whether these sources are transmitting intentionally.

EMC design also needs to consider any capacitive and inductive coupling along with unwanted emissions that may be conducted along common lines that go between two or more items of equipment. This includes Earth lines as well.

These EMI problems can affect adjacent pieces of electronics equipment operating alongside one another. With the vast and rapid growth in the usage of electronic equipment, this problem of EMC has become a particularly important topic. As a result, it is necessary to implement EMC concepts from the onset of a new design and implement the various EMC techniques into the whole concept of the product. Only by taking account of the design for EMC aspects at the concept stages of a development can any precautions be implemented correctly and effectively.

14.4.2 Designing for EMC Compliance

When designing an electronic PCB it is necessary to consider a number of issues to ensure that its EMC performance requirements can be met. Trying to fix the EMC performance once the circuit has been designed and built will be far more difficult and costly. Accordingly, there are a number of areas that can be addressed during the design stage to ensure that the EMC performance is optimized. These include:

- Circuit design for minimum radiation;
- EMC filters;
- Circuit partitioning;
- Grounding;
- Screened enclosure;
- Screened lines and cables.

By adopting these precautions, the EMC performance of the circuit can be greatly enhanced. However, it will still need to undergo EMC testing to ensure that it meets the required performance.

14.4.2.1 EMC Circuit Design for Minimum Radiation

One of the chief areas to keep in mind for EMC/EMI performance is the RF radiated emissions arising from connecting cables and the susceptibility to receiving interference. They form the major coupling path for interference in any system. Often these cables need to carry high-frequency signals, possibly data, and this can present some challenges in terms of their EMC/EMI performance.

Since radiation occurs whenever electrons change their velocity vector, any cable will receive and radiate signals, especially when it approaches a quarter wavelength, or odd multiple thereof, because it forms a resonant circuit. However, even when the cable does approach these lengths, EMC can still be a problem.

One solution is to filter the cables entering and leaving the unit. While this does reduce the level of EMI, it may also degrade the performance of the circuit. If high-speed data needs to be carried, then any sharp edges will be removed by the filters (and therefore limiting the bandwidth), and in the worst case, the signal may be attenuated to such a degree that the system does not work. Thus, a careful balance may be required for the filter between the equipment performance and the EMC requirements.

In these circumstances the signals can be carried in a differential format. The signal cables can then be constructed as a twisted pair and could even be shielded. In this way the high-frequency signal can be carried, but its susceptibility to radiation and reception is reduced, because anything received will appear on both lines and canceled out. Additionally, radiation does not occur for the same reason. We will say more about cabling later.

14.4.2.2 EMC Design: Filters

The possibility of introducing EMC filters has already been mentioned. It can form a useful tool for the system designer to use in many instances. EMC filters are particularly useful for lines that only carry low-frequency signals. Power input cables, or other lines that carry status voltages are particularly good candidates for filtering. Here EMC filters can remove any high-frequency components, leaving the low-frequency elements on the line that will not radiate significantly.

EMC filters should be tightly bonded to the chassis at the entry point to the unit. In this way no signals can enter the unit and radiate into it prior to being removed by the filter.

We say more on filter design later.

14.4.2.3 EMC Design: Circuit Partitioning

Circuit partitioning is the process of segregating the parts of a circuit that are sensitive to EMI from those that are less so. It must be accomplished at the very earliest stages of the design in view of the fact that it governs the whole topology of the circuit and the mechanical construction.

The first stage of the partitioning process is to segregate the circuit into EMC critical and noncritical areas. The EMC critical areas are those that contain sources of radiation or may be susceptible to radiation. These areas may include circuits containing high frequency circuitry, low-level analog circuits, and high-speed logic including microprocessor circuits. In EW systems that contain sensitive receiving components, there is a high probability that such areas are susceptible to stray signals.

The noncritical EMC areas are those that contain areas that are unlikely to radiate signals or be susceptible to radiation. Circuits including linear power supplies (not switch-mode power supplies), slow speed circuits, and the like. Once this has been completed, the layout for the design can begin. The critical or sensitive regions can be screened or filters added as necessary at the interfaces to prevent EMI being radiated, or to protect these circuits from the effects of external EMI.

By isolating the EMC critical areas, it is possible to add the relevant measures both at the initial stages of the design, or possibly later. Having an interface provides the possibility for optimizing the overall EMC performance. This may result in the addition of further filtering, screening, and so forth, or it may even enable cost reductions to be made if some of the measures are not required.

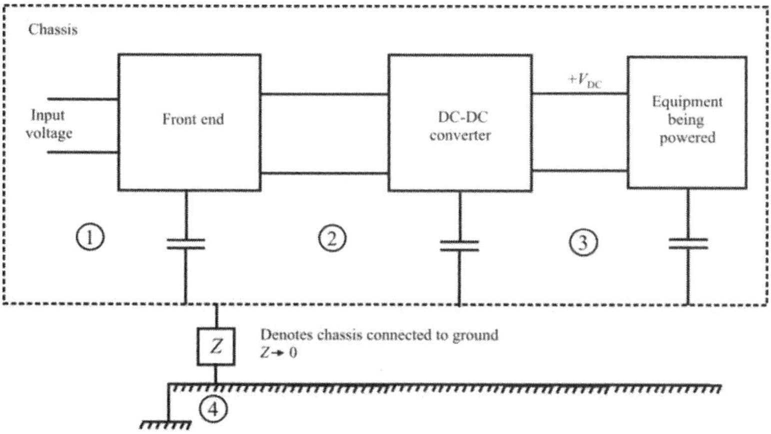

Figure 14.9 Good grounding minimizes all noise sources by presenting noise currents a low-impedance path to Earth. The four places in a power system that are of concern are the front-end input and output (points 1 and 2), which present different potentials, and the DC-DC converter input and output (points 2 and 3).

14.4.2.4 Grounding

Proper grounding reduces the effects of all noise sources, both emissions and susceptibilities, by providing a low-impedance path to ground for EMI currents. There are four significant ground points in a power system (see Figure 14.9). Whether the input is AC or DC, points 1 through 4 are not guaranteed to be at the same potential. The front-end input and output (1 and 2) are at different potentials also, due to interconnect resistance and the inductive elements used in filtering. The converter's input and output (2 and 3) can be at different potentials, due to the transformer floating, without a common input-output ground.

Usually, only the chassis (4) is connected to ground (Earth). Obviously, the lower the value of Z for the EMI frequencies of interest, the more effective the grounding. (Z is typically a complex value that varies with frequency.) It might be useful to look at a block diagram and decide which paths the high-frequency noise currents are likely to take based on how these points are selected and how their connections are implemented.

The ideal situation is for Z to be zero, with all chassis current passing to Earth. A non-ideal situation would be noise currents travelling from the front end to the equipment being powered. During EMI qualification testing, it's important that all noise currents go to Earth, and not to the measurement equipment, as that would affect both common- and differential-mode measurements.

The grounding scheme within a unit is of particular importance for its EMC performance. Poor grounding can lead to current loops that can in turn lead to signals being radiated or picked up within the unit and hence poor EMC performance results.

To help ensure that the Earth or grounding system works satisfactorily, it is worth bearing in mind its function, which is to be a path that enables a current to return to its source. It should obviously have a low impedance, and it should also be direct. Any loops or deviations may cause spurious effects that can give rise to EMC problems.

Designing grounding systems is not trivial. It is more challenging than it appears, but it is essential for a good EMC performance. Lengths must be kept to a minimum because above frequencies of more than a few kilohertz the impedance is dominated by inductance, and lengths of a few centimeters make a significant difference, even at low frequencies.

To overcome these effects, thick wires should be used if possible, and on printed circuit boards ground planes must be used. Critical tracks must be run above the ground plane, and they should be routed so that they do not encounter any breaks in the ground plane. Sometimes it is necessary to have a slot or break in a ground plane, and if this occurs a critical track must be routed over the plane, even if it makes it slightly longer.

These and other approaches can be adopted to ensure that the grounding system is able to reduce the EMC problems to a minimum. Considerable thought should be given to the grounding, as it may not be easy to change at a later time.

14.4.2.5 Screened Enclosure

Although screened enclosures are not preferred from a cost viewpoint, placing the unit in a conductive enclosure that is grounded will significantly improve EMC performance. All filtering can then be undertaken at the interface and the conductive wall will provide a barrier to radiation, thereby improving both the emissions and susceptibility elements of the EMC performance.

Where cost and possibly aesthetics are important, the inside of cabinets can be sprayed with conducive paint; however, the level of resulting screening provided will not be nearly as good as if a fully conductive metal case is used. Where high levels of EMC performance are required, care should be taken to choose a case where the continuity of the screen is not breached. The case should ideally be made with as few elements as possible. At each joint there will be the possibility of radiation passing through. Where joints occur, they should be as tight as possible and they should have good continuity between them.

Some metal cases using a prefabricated style of construction with anodized aluminum panels do not offer good EMC performance, although they are aesthetically more pleasing than some RF tight cases. A balance has to be made depending on the performance required and the EMC goals.

14.4.2.6 Screened Lines and Cables

When lines and cables need to pass into or out of a unit, the cables can be screened to prevent any radiation of the signals being carried or the pickup of external signals. However, when screened cables are needed for EMC requirements, the screen must be bonded to the equipment signal ground as soon as it enters the unit, otherwise unwanted signals may be radiated or picked up and this would compromise the EMC performance.

14.5 EMC Filter Design

In order for an item of electronics equipment to be EMC compliant, it is necessary to incorporate various elements into the design. By designing the circuit to meet the EMC requirements it is possible to significantly reduce the levels of unwanted signals entering and leaving the unit. One of the major ways in which this can be done is to use an EMC filter or a series of filters.

There are many ways in which EMC filters can be incorporated into a unit from a mechanical viewpoint. They may exist as stand-alone EMC filters to be fixed near to the extremities of the unit. They may be mounted on the edge of a PCB. However, one popular method of incorporating an EMC filter into a unit is to incorporate the filter into the connector itself. This has many advantages in terms of convenience and performance.

14.5.1 EMC Filter Methodology

Although circuits may be well screened to prevent any signal radiated or being picked up by the circuit itself, there are always interconnections to and from the circuit. These wires themselves can conduct unwanted signals into and out of the unit. If the unit is to be able to meet its EMC requirements, it is necessary to reduce the levels of unwanted signals that can enter or leave the unit via its interconnections.

In order to enable unwanted signals to be removed, EMC filters need to be placed in the various lines. When the interfering signals have a frequency above that of the signals traveling along the wire, by having a low-pass filter as the EMC filter, only the low frequency signals are allowed to pass, and the high frequency interference signals are removed.

These EMC filters can be in one of a variety of formats. Often, they may be as simple as a resistor or a ferrite placed around a wire. For more exacting requirements, these EMC filters may need to be made up with a number of stages.

The EMC filters may be categorized into two main types. One is where the unwanted energy is absorbed by the EMC filter. The other type of filter rejects the unwanted signal and, in this case, it is reflected back along the line. For EMC filtering the absorptive type is preferred.

14.5.2 EMC Filter Applications

EMC filters are almost always low-pass filters, although on occasion bandpass filters may be used. The reason for this is that typically interfering signals (i.e., ones that are easier to pick up or radiate), tend to be at higher frequencies. These can be filtered by allowing the low frequencies through and rejecting the higher frequencies.

The cut off point for the low-pass filter used as an EMC filter has to be chosen so that it rejects the unwanted frequencies but does not have any undue effect on the wanted signal. Unfortunately, this choice is not always easy and some degradation of the wanted signal may occur.

The EMC filter placement is of importance. EMC filtering can be placed at any level of assembly between segregated areas of circuitry. EMC filters may be placed between segregated areas of a PCB. They may be placed between different boards within a module or subassembly, and an EMC filter may be placed between different modules or subassemblies. A particularly important place for EMC filters is between the equipment and its external environment which makes it particularly effective as it prevents unwanted signals from even entering the equipment. Once unwanted signals enter, they are more difficult to contain.

14.5.3 EMC Filter Design

The EMC filter design is critical to EMC performance. The EMC filter must be capable of providing the required level of attenuation of the unwanted signals while allowing the wanted signals through. In addition to this the EMC filter design must match both the source and load impedances.

Typically for a high-impedance circuit, a capacitor shunted between the line and ground provides better results, while for low-impedance circuits an inductor placed in series with the line provides the best results. Often a single component with a reactance with little effect at frequencies appropriate to the wanted signals, but a much larger effect at the higher frequencies of the unwanted signal, can provide attenuation of up to 30-dB or 40-dB in some cases. To improve the performance of one of these basic filters, further components can be added to make multi-component EMC filters. One precaution: ensure that inductors face a low-impedance sink or source and capacitors face a high impedance.

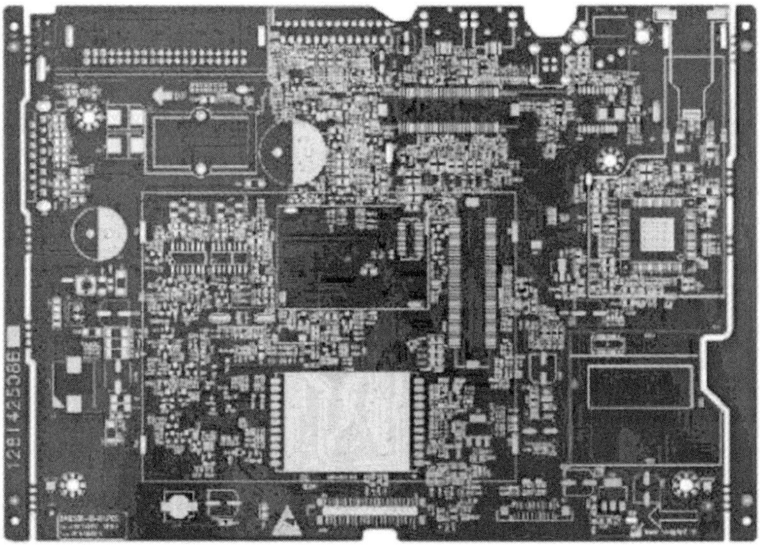

Figure 14.10 Typical multilayer PCB design.

14.6 PCB Design for EMC

The design of the printed circuit board for any equipment has a major impact on its EMC performance and the amount of electromagnetic interference created. One of the key areas of designing a circuit with good EMC performance is that of the PCB design.

A typical multilayered PCB is illustrated in Figure 14.10.

14.6.1 PCB Layout

The PCB design for EMC can enable a circuit board to have good EMC performance, and to help there are a few basic guidelines that can be followed to provide good EMC performance.

PCB traces in a power supply and elsewhere create parasitic inductance loops, which lead to conducted and radiated noise. Also, if multiple loops are in close proximity to each other, they can have a coupling effect that increases EMI levels. Careful attention must be given to the size, placement, and alignment of these loops.

Reducing these loop areas minimizes their capacity to radiate and conduct noise (see Figure 14.11). Also, separating noisy loops from each other will minimize their coupling effect, further reducing their contribution to EMI.

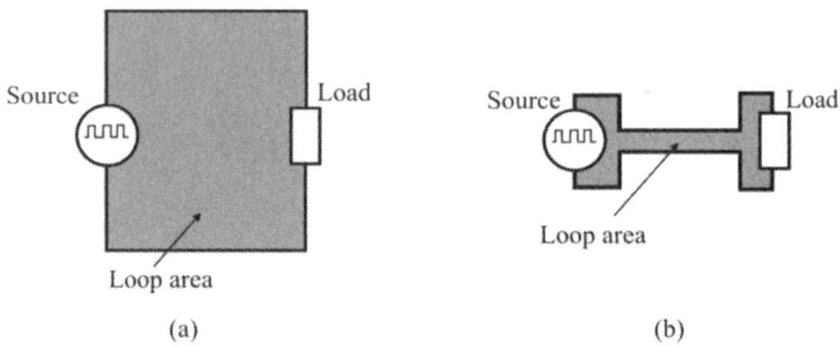

Figure 14.11 Current loops: (a) large current loop and (b) small current loop.

Separating high-current traces from low-current traces increases system stability and reduces EMI. A technique called star-point grounding runs all returns back to a single point. Doing this, or approximating it as closely as the design will allow, reduces the summation of high currents in any single trace, yielding lower EMI.

Also, a linear flow of components on the primary side, from the input line to the transformer, helps reduce coupling the switching waveforms back to the line voltage, ultimately reducing EMI.

Something else to consider is appropriately sized traces. In general, at least 2-ounce copper traces should be used, and the shorter and wider, the better. The size of the trace will be limited by the size constraints of the PCB as well as clearance requirements. However, within those confines, the traces should be made as wide as possible.

Although it is possible to utilize multiple layers to reduce the size of the PCB, when designing a PCB for good EMC performance, this not always the optimal route to take. The PCB design for EMC performance may require coupling to be reduced. This usually means keeping signals apart or the distance between some components to be increased. Although small PCBs with good EMC performance can be designed, care must be taken from the outset and EMC design principles employed from the start.

4.6.1.1 Basic Principles for PCB design for EMC

A four-layer board is often regarded as a good balance between board layout and EMC performance. That said, many boards with more layers can achieve good EMC performance, but require very careful design to achieve it.

Ground Planes Improve EMC Performance

Using one layer within a multilayer board as a ground plane is one technique that is particularly useful. Signal return paths are one of the most difficult issues to resolve in printed circuit boards. It can be difficult to route a ground return satisfactorily from each integrated circuit across other signal layers, and so forth.

The only satisfactory solution is to use a ground plane that provides a low inductance and low resistance common ground, which is a method of providing a short lead length to ground. By having one of the layers in the PCB reserved for a ground plane, it is easy to provide a good path to ground for any signals anywhere on the board. This is done by using vias, which were discussed in Section 13.7.

For some sensitive areas it may be necessary to isolate the ground from the rest of the ground plane to prevent ground currents flowing across that section of circuit. For example, a sensitive section of circuit may need to have its ground isolated and have a single connection to Earth, especially if a higher power section close by may cause Earth currents to flow across the more sensitive section.

Gridding to Create Ground Planes

In some PCBs that may have a limited number of layers, for example one where only two layers are available, a technique referred to as *gridding* may be used to provide good EMC performance. This technique is a close approximation to having a ground plane in a two-layer board; gridding the ground reduces EMI radiation from the signal traces.

Gridding functions by creating a network of orthogonal connections between traces carrying ground. Although the ground plane is not completely contiguous, it sufficiently emulates the ground plane that is used to provide EMC improvements of a four- or more-layer board by providing a ground return path under each of the signal traces and lowers the impedance between the main ICs and the voltage regulation area.

Gridding is achieved by a process of expanding any ground traces and using ground-fill patterns. The aim is to create a network of connections to ground across the PCB. The gridding is achieved by expanding the ground lines to fill up as much of the empty PCB space as possible, even though the ground is not contiguous. Then, all the remaining empty space is filled with ground. In this way as much of the available PCB space is filled with the ground grid as possible while still allowing connections to be made on the layer.

PCB Zoning

Creating different zones on a PCB is another useful design technique to improve EMC and noise in general. PCB zoning is essentially a process of planning where the general location of components for different areas of the circuit is defined before any traces are laid out.

Not only does PCB zoning place like functions on a board in the same general area, as opposed to mixing them together, but it also takes into account the speed of signals in a given area and looks at the optimum location, although EMC is normally not the only consideration involved. Thought is given to the length of lines that may radiate or pick up more noise. For example, one common idea is to place high-speed logic, including microcontrollers, close to the power supply. In this way decoupling of the lines is made easier and the lengths of lines or traces that might radiate or pick up noise are reduced. Functions on the PCB that are not as critical and have slower waveforms are located further away. Typically, analog sections of the board are located even further away as they normally carry lower frequency signals. Planning the areas of the board in this way can have a major impact on the EMC performance of the PCB and it must be done early in the design process.

14.6.1.2 Other Considerations for PCB EMC Design

There are a few other common points to improve the PCB EMC performance:

- *Oscillators:* Particular care must be taken when locating and designing the layout for oscillators. Oscillators are an obvious source of possible radiation. Any oscillator tank loops must be located away from analog circuits, low-speed signals, and connectors. This applies both to the board and to the space inside the box containing the board.
- *System cable assemblies:* Another key point is to design the overall system so that cable assemblies do not pass close to an oscillator or an area that includes high-speed logic, including a microcomputer, after final assembly. Cable assemblies can pick up and carry noise around the overall unit, which can degrade the EMC performance.
- *Keep high speed/noisy lines away from PCB edge:* High-speed lines should be run away from the outside edge of the board. Keeping nonnoisy traces away from areas on the board where they could pick up noise, such as connectors, oscillator circuits, relays, and relay drivers, also helps reduce the problem.
- *Filtering:* In some instances, filtering may be required on certain lines. Ferrite beads can often provide an easy method of limiting high-frequency signals, and good decoupling on the board, especially for the power supply lines, is necessary. Ferrite beads on supply lines prevents high-frequency signals from feeding back into the power supply which, subsequently, get distributed throughout the unit.

- *Filtered connectors:* On some PCBs it may be necessary to use filtered connectors to remove noise. When this is done, the earthing of the connector is important. It should be possible to earth this firmly to the PCB and the chassis.

Many of the EMC issues can be eliminated by good PCB layout and design. In fact, PCB design for EMC performance is always good practice and can prevent many time-consuming investigations and costly rework. If rework is required late in the design cycle it is considerably more costly than if it is built into the design at the outset. PCB design for EMC is therefore one of the keys to a successful design.

14.7 EMI Common Mode and Normal Mode Noise

We discuss some key methods and techniques for resolving and improving common mode and normal mode EMI noise for EMC conformance and conformance test in this section.

14.7.1 EMI Noise Modes

When examining EMC and EMI issues, the mode, or way in which the noise is being carried must be considered (see Figure 14.12).

- *Normal-mode EMI noise:* Normal-mode EMI noise (also referred to as differential mode) is the way we tend to imagine noise being carried on any cable. The noise currents flow out along one conductor and back along the other. This makes a compete circuit for the noise currents. Typically, screening will provide a great deal of immunity against this form of noise as the inner conductor or conductors are screened from pickup or radiation of the noise.

- *Common-mode EMI noise:* Common-mode noise can be a little more difficult to imagine. The currents may flow in the same direction on the lines of a cable—they may even flow along the screening as well, making the screening ineffective. The currents that flow are caused by a current that has leaked out via the floating electrostatic capacitance of the load retained against the Earth and has then returned to the noise source via the Earth. The current may be caused by a direct connection between the load and noise source without going through the Earth.

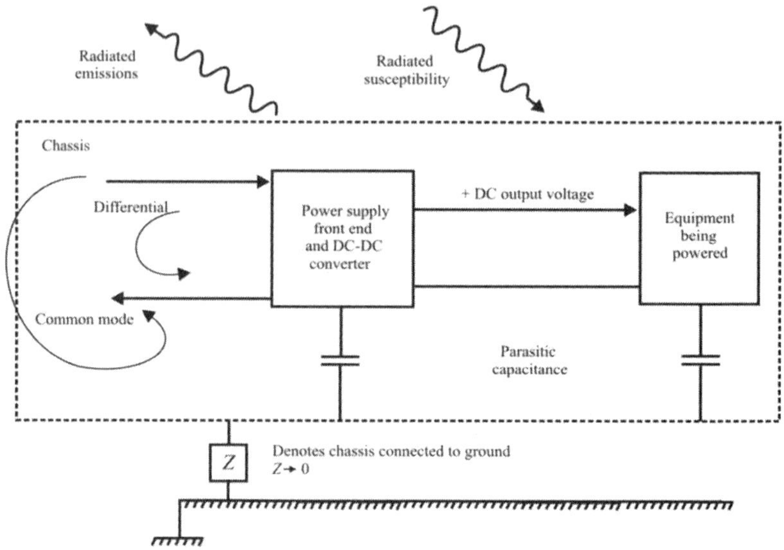

Figure 14.12 In the same basic package, conducted emissions form at the AC or DC input and generate both differential-mode and common-mode noise.

Common mode (CM) emissions occur along two lines of a closed loop simultaneously, in the same direction and in phase (see Figure 14.13). The signal on each line finds a return path through a common ground. Common-mode noise can be mitigated by the addition of a common-mode choke, which has multiple windings on a single core. The addition of capacitors from each line to ground (known as Y-capacitors) will also aid in shunting the unwanted emissions to ground. A quick method of determining if the emission is common-mode is to measure the active and

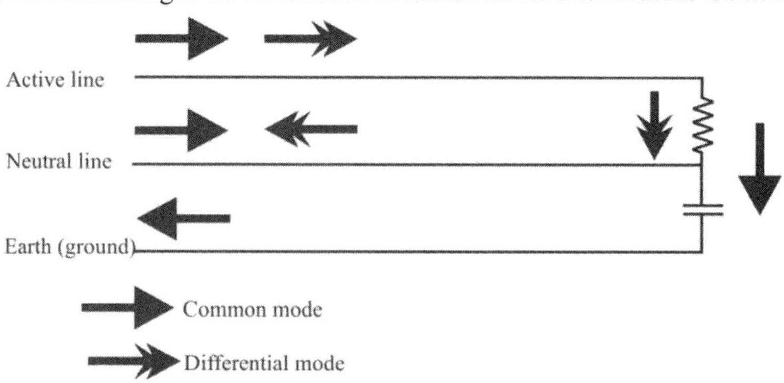

Figure 14.13 Common and differential modes.

neutral lines using a current probe, with both conductors passing through the current probe in the same direction. This will allow the common mode emissions to be measured while differential-mode currents cancel.

Like common-mode emissions, differential-mode (DM) emissions occur along two lines of a closed loop. However, the current flow is in opposite directions. A series EMI inductor may be enough to suppress the unwanted interference signals. Care must be taken to select a component that has a core material that can support the current of the entire line under full load without saturating. The addition of a shunt capacitor across the lines (known as an X-capacitor) may also be required. A quick method of determining if the emission is differential mode is to measure the active and neutral lines using a current probe, with the neutral line conductor passing through the current probe in the opposite direction to the active line conductor. This will allow the differential-mode emissions to be measured while the common-mode currents cancel.

Differential mode currents flow in opposite directions and are out of phase with each other (see Figure 14.12).

14.7.2 EMI Filtering for Conducted Emissions

When selecting an EMI inductor/choke there are three things to consider:

- *Frequency:* the frequency range in which emissions are a problem.
- *Impedance:* usually specified as a minimum value at a given frequency.
- *Current:* the inductor windings and core material should be physically capable of supporting the current.

Conducted disturbances like these will seldom be purely common-mode or differential-mode but probably will be a combination of both. Often several different techniques used in conjunction with each other will be required in order to achieve the desired effect with respect to suppression of emissions. Simple filter circuits can be made from resistors, capacitors, and inductors. Line filters are available that utilize combinations of components in one or more stages. These line filters are designed for all types of applications and may include decoupling capacitors. AC line filtering (either single-phase or three-phase), Ethernet, and even USB filters are available. In the case of high-speed data or signal lines, additional integrated circuitry may be required to filter the EMI noise while maintaining the signal integrity. In many cases the use of such filters can be costly; however, investigations with such filters can help designers work toward a cheaper onboard solution. The diagram shown in Figure 14.14 shows a typical layout of a commercially available line filter circuit.

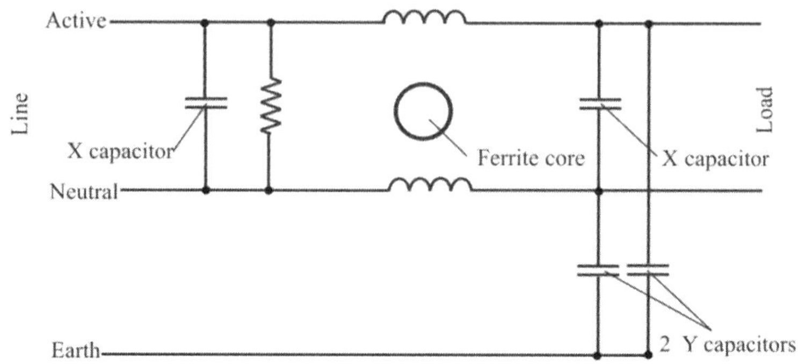

Figure 14.14 Line filter.

It is good engineering practice to consider EMI and conducted emissions along with radiated emissions early in the design stage. Precompliance testing at the product development stage can highlight these types of issues at an early stage. This, in turn, can help mitigate future issues. This results in an overall better product while reducing the overall product development cost.

14.8 Some Specific Concerns with Power Supplies

In this section we will examine some EMI concerns in power supply design.

14.8.1 Ripple Current

A not-necessarily obvious source of conducted EMI is *ripple* in a power supply converter's output. The ripple generally appears at harmonics of the AC line frequency or switching-converter rate.

14.8.2 Leakage Current

A *leakage current* is any current that flows through a path not intended to carry that current. In power supplies, *Earth leakage current* is the current that flows through the ground connection to Earth ground. This leakage current typically travels through a capacitive bridge between line voltage and Earth ground. Agency regulations limit the amount of Earth leakage current that is acceptable in a power supply.

14.8.3 EMI Solution Methods

In this section we identify critical areas to consider when specifying or selecting a power supply.

14.8.3.1 Transformer Design and Construction

The first element to consider for minimizing EMI in a switching power supply design is the transformer. The transformer is typically the noisiest component because it is a large inductive element subjected to high-frequency currents. It is also the component around which the rest of the supply is designed.

Every facet of the transformer's design and construction can affect EMI, and each aspect can be critical. Approaches that can solve EMI issues in one design will often compound them in others. In this way, the design of a transformer for low EMI is often viewed as an art.

We will discuss, in brief, some of the transformer design and construction techniques that should be employed to reduce EMI.

• *Ferrite core selection.* There are peripheral reasons to choose any of the various core types, but typically an RM type core is better at containing electromagnetic fields than an E or EER type core. This is due to the additional core material on all sides of the center leg acting as an electromagnetic shield. Another important consideration is how the core is gapped. By gapping only the center leg of the core, radiated EMI will be further contained. A gapped outer leg will leak radiated EMI through the gap, possibly requiring additional shielding to contain the stray EMI. E and RM cores are illustrated in Figure 14.15.

• The switching drain of the MOSFET will normally be connected directly to the primary winding of the transformer. One design technique that can help reduce EMI in a multilayer winding is making this drain connection to the innermost winding of

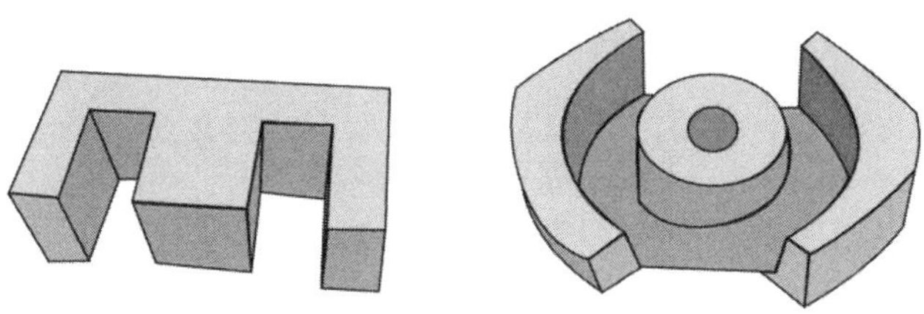

(a) (b)

Figure 14.15 Two of many possible transformer cores: (a) E core and (b) RM core.

the transformer. By doing this, the EMI caused by these high-current, sharp-edged waveforms are contained and canceled by the other windings and shields in the transformer.

• *Winding direction and orientation* play crucial roles in reducing EMI as well. Winding a section from left to right, or vice versa, or starting the winding clockwise or counterclockwise around the core all have an effect on EMI, and so must be considered in the design.

• *Shielding* is commonly used to help further suppress EMI. The use of a Faraday shield[10] between the primary and secondary winding will typically reduce EMI. This shield can be tied to the primary, secondary, or both to achieve optimal suppression. Applying additional tape insulation between primary and secondary windings can also help reduce EMI. Also, a shield can be applied directly to the outside of the core. This shield can be tied to either the output common or the input return for optimal suppression. It is important to note, however, that each shielding layer on a transformer will increase the capacitive coupling, directly increasing the leakage current of the unit.

All of these techniques can be quite effective at reducing EMI, but they can also have undesirable effects. Designing a transformer is a balancing act between EMI suppression and transformer efficiency. Typically, many iterations of transformer construction are required before settling on a final design that performs as desired.

14.8.3.2 Components

Capacitors

As mentioned above, an effective technique for reducing EMI is the use of Y-type and X-type capacitors (see Figure 14.14). Y-capacitors are typically used between the AC line and ground, AC neutral and ground, and between the primary and secondary returns. X-capacitors are placed between the AC line and neutral, and the primary and secondary lines. However, while using X- and Y- capacitors is quite effective at reducing EMI at all frequencies, these components contribute directly to increasing the total leakage current. Therefore, these capacitors can only be used up to the point where the leakage current requirements are not exceeded. Beyond these limits, other solution techniques must be considered.

---

[10] A *Faraday shield* is a conductive material used to contain or control an electric field. This shield can block or filter the transmission of electromagnetic fields through a particular region of space.

Figure 14.16 A common-mode choke is a noise filter designed to reduce common-mode EMI. It is basically composed of two identical differential filters, one on the input line and the other on the input neutral, coupled together by a single core.

Inductors

The common-mode choke is another effective solution for lowering conducted EMI. One is pictured in Figure 14.16. It works similarly to a low-pass filter. These components can be placed before or after the rectified voltage depending on where they achieve optimal suppression. In some designs up to three of these will be used, usually two on the input, and one on the output.

An inductor can be also be used in combination with two capacitors in a π configuration (see Figure 14.17) to create a low-frequency filter (150–700 kHz). In this configuration, the rectified current is passed through a small electrolytic capacitor, then the inductor, and then a large electrolytic capacitor. This can greatly reduce low-frequency conducted EMI.

If the power supply uses *power factor correction* (PFC)[11], a differential mode choke is often used in a π-filter configuration with metal-film capacitors to reduce the differential-mode noise.

These components can be used without contributing to the leakage current.

[11] *Power factor correction* is used to synchronize the current and voltage waveforms to maximize the power delivery of the line source, and to minimize the distortion and harmonics on the AC line. It is the process of adding additional electrical components to the AC input to attempt to bring the *power factor* (PF) as close to 1 (where the current and voltage are in phase) as possible.

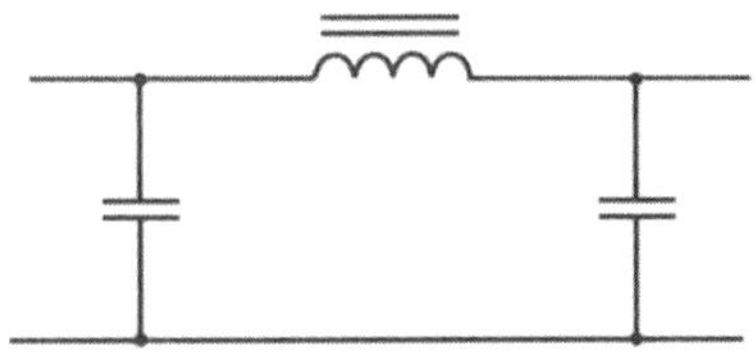

Figure 14.17 A π filter combines both capacitive and inductive filters in a π configuration, which can effectively remove the high-frequency ripple currents that cause EMI.

Switching, Capacitance, and Inductance

Electrical devices switching on and off create almost all electrical noise that is not of thermal origin. (A steady DC current doesn't generate noise except thermal noise in resistors.) Even a passive device such as a rectifier can generate noise. Rectifiers don't turn on and off (that is, conduct and stop conducting) instantaneously. Common rectifiers have a response time of about 1 μs, generating harmonics not only at multiples of the AC frequency being rectified, but also in the megahertz region. These harmonics can find their way to both the input and output of the power supply.

The active devices in a switching power supply (commonly MOSFETs) switch more rapidly (~20 ns), generating higher frequencies, up to 30 MHz, which is the highest frequency specified in commercial standards for measuring conducted interference. (This doesn't mean higher frequencies that could cause interference should be ignored.)

If the front end includes PFC circuitry, there will be additional EMI from the PFC's MOSFETs and diodes.

Multiple conduction paths created by parasitic capacitances within the power supply and the system itself permit high-frequency EMI currents to enter almost any part of the system. These stray capacitances appear not only between parallel wires, but also between conductive surfaces. The parasitic inductance of wiring and board traces contributes to voltage kicks that can propagate through the parasitic capacitances.

Summary

We have reviewed in this section many points of power supply design critical to effective EMI suppression. Among these are:

- Transformer core selection;
- Transformer winding and shielding;

- Component layout and loop size reduction;
- Use of X and Y capacitors;
- Use of inductors and common-mode chokes.

For optimal EMI control at the OEM system level, it is crucial to ensure that all of these points have been considered when selecting an AC/DC switching power supply.

14.9 Cabling

Interconnect cabling is one of the major considerations when discussing EMI/EMC. Such cabling interconnects major electronic modules within these systems. In this section we will present the major types of cabling used in EW systems and consider their predominate characteristics. These types include:

- Twisted pair, both unshielded and shielded;
- Coaxial cable;
- Fiber optics.

Cable is the medium through which information usually moves from one network device to another in a wired network (not the case for WiFi). There are several types of cable that are commonly used with LANs. In some cases, a network will utilize only one type of cable, other networks will use a variety of cable types. The type of cable chosen for a network is related to the network's topology, protocol, and size.

Ethernet is the most prolific standard for cable networks. The IEEE-802.3bz Ethernet standards committee developed 2.5GBaseT and 5GBaseT 100M specifications that were approved in 2016. It is mostly targeting new wiring cabinets that support many WiFi IEEE802.11ac aggregation connections. Table 14.1 contains the current listing of Ethernet standards.

14.9.1 Twisted-Pair

One of the earliest guided transmission media is twisted-pair cables. A twisted-pair cable is comprised of two separate insulated copper wires, which are twisted together and run in parallel. The copper wires for RF signals are typically 1mm in diameter. One of the wires is used to transmit data and the other is the ground reference.

Table 14.1 Ethernet Characteristics

Name	Speed (Mbps)	Lanes per Direction	Bits per Hz	Spectral Bandwidth	Cable Required	Cable Rating (MHz)
10BaseT	10	1	1	10	Cat 3	16
100BaseTX	100	1	3.2	31.25	Cat 5	100
1000BaseT	1,000	4	4	62.5	Cat 5e	100
2.5GBaseT	2,500	4	6.25	100	At 5e	100
5GBaseT	5,000	4	6.25	200	Cat 6	250
10GBaseT	10,000	4	6.25	400	Cat 6A	500

14.9.1.1 Reason for Twisting

All transmissions are prone to noise, interferences, and crosstalk. When the wires are twisted, some part of the noise signals is in the direction of data signals while the other parts are in the opposite directions. Thus, the external waves cancel out due to the different twists. The receiver calculates the difference in the voltages of the two wires for retrieving data. Thus, a much better immunity against noise is achieved.

There are two types of twisted-pair cables: unshielded twisted pair (UTP) and shielded twisted pair (STP). The advantages and disadvantages of shielded and unshielded twisted-pair cable can be debated. Each has their good and bad points. Advocates of STP cable, which includes screened twisted-pair and foil twisted-pair cables, claim that it is superior to UTP cable. Perhaps the greatest advantage of UTP cable, however, is cost.

STP and UTP cable differ in design and manufacture. But their purpose is the same—to provide reliable connectivity of electronic equipment. In theory, both types of cable should do this equally well. The true test comes when each of these cable types performs that task in its respective end-to-end system environment.

14.9.1.2 Unshielded Twisted-Pair

Twisted-pair cabling comes in two varieties: shielded and unshielded. Unshielded twisted-pair (UTP) is the most popular and is generally the best option for low-cost, low performance networks (see Figure 14.18). However, it has been used for networks up to 10 Gbps.

The quality of UTP may vary from telephone-grade wire to extremely high-speed cable. The cable has four pairs of wires inside the jacket. Each pair is twisted with a different number of twists per inch to help eliminate interference from adjacent pairs and other electrical devices. The tighter the twisting, the higher the

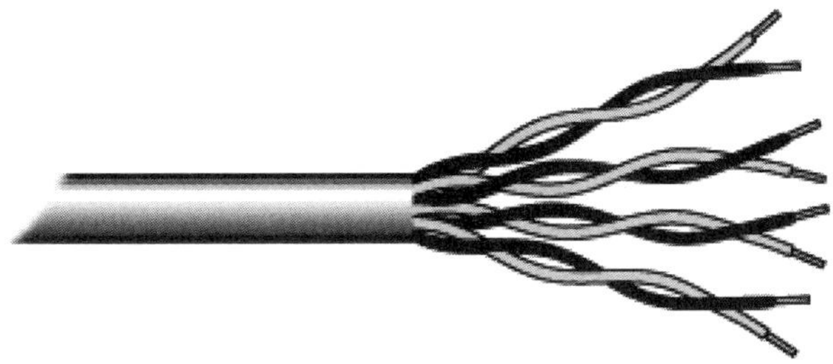

Figure 14.18 Unshielded twisted-pair cable.

supported transmission rate and the greater the cost per foot. The Electronic Industry Association/Telecommunication Industry Association (EIA/TIA) has established standards of UTP and rated ten categories of wire as shown in Table 14.2.

CAT-6 250 MHz UTP cable only supports 10-Gbps short-length applications from 35-55m depending on the crosstalk environment. CAT-6a 500 MHz UTP and also shielded twisted pair have become the most popular usage type in North America as they are certified for 10 Gbps to 100m. Although 10-Gbps Ethernet ports shipments have taken over as the volume leader, newer applications will expand and extend CAT-6a product life cycles.

14.9.1.3 Unshielded Twisted-Pair Connector

The standard connector for unshielded twisted pair cabling is an RJ-45 connector. This is a plastic connector that looks like a large telephone-style connector (see Figure 14.19). A slot allows the RJ-45 to be inserted only one way. RJ stands for

Table 14.2 Categories of Unshielded Twisted-Pair

Category	Speed	Max Length	Use
1	1 Mbps	-	Voice only (telephone wire)
2	4 Mbps	-	Local talk and telephone (rarely used)
3	16 Mbps	100m	10BaseT Ethernet
4	20 Mbps	100m	Token ring (rarely used)
5	100 Mbps (2 pair)	100m	10BaseT Ethernet
5	1,000 Mbps (4 pair)	100m	Gigabit Ethernet
5e	1,000 Mbps	100m	Gigabit Ethernet
6	10Gbps	55m	Gigabit Ethernet
6a	10Gbps	55m	Gigabit Ethernet
7	10Gbps	100m	Gigabit Ethernet

Figure 14.19 RJ-45 connector.

Registered Jack, implying that the connector follows a standard borrowed from the telephone industry. This standard designates which wire goes with each pin inside the connector.

14.9.1.4 Unshielded Twisted-Pair Cable

UTP cable does not rely on physical shielding to block interference. It relies instead on balancing and filtering techniques using media filters, baluns, or both. Noise is induced equally on two conductors and is canceled out at the receiver. With properly designed, manufactured, and installed UTP cable, the network is easier to maintain than one in an STP cable plant, with its shielding continuity and grounding issues.

UTP cable has evolved over the years, and different types are available for different needs. Basic telephone cable, also known as direct-inside wire, is still available. Improvements over the years—variations in the twists or in individual wire sheaths or overall cable jackets—have led to the development of Category 3 cable that is compliant with the EIA/ TIA-568 standard for transmission rates up to 16 MHz. Category 4 UTP cable is specified for signal bandwidths to 20 MHz, and Category 5 UTP for specifications to 100 MHz. CAT-5e 100 MHz UTP cabling usage is in significant decline as it does not support newer emerging applications and signaling technologies. It has been used for primarily 100 Mbps through 1 Gbps Ethernet 1-100m LAN applications. It is used mostly for minor changes in the installed base and not for new datacenter implementations as 10-Gbps Ethernet ports have been quickly ramped up as the primary port type usage. The older CAT-5 UTP cable is only good for 100 Mbps up to 100m and is in the maintenance-to-end-of-life phase of its product life cycle.

Because UTP cable is lightweight, thin, and flexible, as well as versatile, reliable, and inexpensive, millions of nodes have been, and continue to be, wired with this cabling medium. This is especially true for high-data-rate applications. For best performance, this UTP cable should be used as part of a well-engineered structured cabling system.

14.9.1.5 Shielded Twisted-Pair Cable

Although UTP cable is the least expensive cable, it may be susceptible to radio and electrical frequency interference (it should not be too close to electric motors, fluorescent lights, and so forth). If the cable must be placed in environments with significant potential interference, or if it must be placed in extremely sensitive environments that may be susceptible to the electrical current in the UTP, shielded twisted-pair (STP) may be the better solution. Shielded cables can also help to extend the maximum distance of the cables. Table 14.3 lists the current categories of STP cable (see Figure 14.20).

Shielded twisted-pair cable is available in three different configurations:

1. Each pair of wires is individually shielded with foil;
2. There is a foil or braid shield inside the jacket covering all wires (as a group);
3. There is a shield around each individual pair, as well as around the entire group of wires (referred to as double-shield twisted-pair).

STP cable encases the signal-carrying wires in a conducting shield as a means of reducing the potential for EMI. How effective the shielding performs depends on the material used for the shield—its thickness and frequency, the type of EM noise

Table 14.3 Categories of Shielded Twisted-Pair

Category	Speed	Bandwidth	Datarate	Use
5e	1000BaseT/2.5GBaseT	250MHz	1 Gbps	Ethernet
6	5GBaseT/10GBaseT	500MHz	10 Gbps	Ethernet, 55m max distance
7	5GBaseT/10GBaseT	600MHz	10 Gbps	Ethernet
7a	5GBaseT/10GBaseT	1GHz	10 Gbps	Ethernet
8.1	25GBaseT/40GBaseT	2GHz	40 Gbps	Ethernet
8.2	25GBaseT/40GBaseT	2GHz	40 Gbps	Ethernet

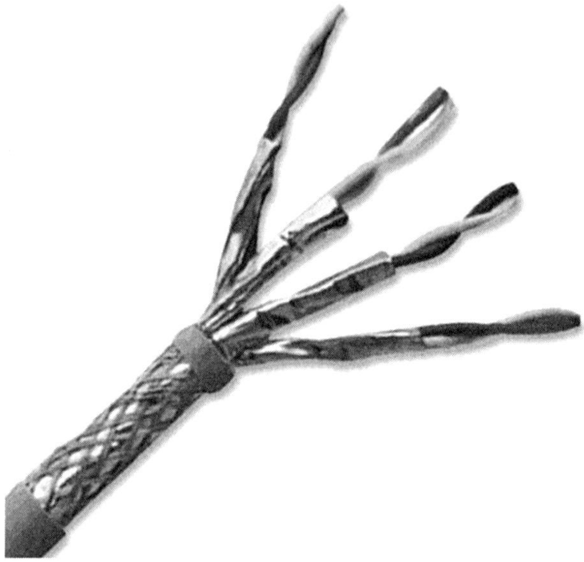

Figure 14.20 Shielded twisted-pair cable.

field present, the distance from the noise source to the shield, any shield discontinuity, the grounding practices employed, and crosstalk.

Some STP cables, for example, use a thick braided shield that makes a cable heavier, thicker, and more difficult to install than its UTP counterpart. Other STP cables use only a thin outer foil shield. These cables, known as screened twisted-pair cables or foil twisted-pair cables, are thinner and less expensive than braided STP cable; however, they are not any easier to install. Unless the minimum bend radius and maximum pulling tension are rigidly observed when these cables are installed, the shield can be torn.

CAT-7 600 MHz individually shielded pair cabling with overall cable shield was initiated by European market leaders and standards bodies for 10 Gbps 100 m Ethernet LAN applications with better crosstalk control performance than CAT-6a. However North American standards groups TIA and EIA-568 have not recognized CAT-7 or CAT-7a standards. CAT-7a 1000 MHz also with individually shielded pairs and overall cable shield is supported by the ISO cabling standard organization. It is the most popular structured cabling type in the western countries of Europe. This cable type appears to support the developing IEEE-802.3bq 25GBaseT and 40GBaseT up to 50 m. Some segments of the North American market have installed CAT-7a cabling for better future-proofing of new applications.

Cat-8 Twisted-Pair Cabling

CAT-8 1.6 GHz and 2GHz STP 1-30 m cabling standards and specifications have been harmonized with national ANSI/TIA-568-C-2.1 and international ISO/IEC 11801-99-1 standards bodies. From the start, there have been two types with CAT-8.1 1.6 GHz backward compatible with CAT-6a connectors targeting 25GBaseT applications. The CAT-8.2 2 GHz specification is not recognized the TIA. CAT-8 cabling types are in production primarily targeting 40GBaseT applications.

The shorter reach will limit applications to be in just telecom rooms and some smaller datacenters, but not the traditional LAN applications. CAT-8 cabling is larger in diameter size, thus taking more room and limiting airflow. It also weighs more, sometimes being too heavy for older datacenter floors and is harder to route in the infrastructure and racks.

There is significant research and development work going on to support nascent Ethernet 100GBaseT roadmap, technologies, and potential products. But the twisted pair cabling reach may be only 1-10-m and would likely not compete well versus datacenter point-to-point twin axial copper cabling in racks or between racks and is already being used for 100G applications in production.

14.9.1.6 Other Considerations

If STP cable is combined with improperly shielded connectors, connecting hardware or outlets, or if the foil shield itself is damaged, overall signal quality will be degraded. This, in turn, can result in degraded emission and immunity performance. Therefore, for a shielded cabling system to totally reduce interference, every component within that system must be fully and seamlessly shielded, as well as properly installed and maintained.

An STP cabling system also requires good grounding and grounding practices because of the presence of the shield. An improperly grounded system can be a primary source of emissions and interference, and can actually exacerbate the EMI problem. Whether this ground is at one end or both ends of the cable run depends on the frequency at which a given application is running. For high-frequency signals, an STP cabling system must be grounded, at minimum, at both ends of the cable run, and it must be continuous. A shield grounded at only one end is not effective against magnetic-field interference.

The length of the ground conductor itself can also cause problems. If it is too long, it no longer acts as a ground. Therefore, because specific grounding requirements depend on the application, a general grounding policy that ensures the best results for an STP cabling system is not possible.

UTP cabling systems do not have this problem. While an STP cabling system is dependent on such factors as physical continuity of the cable shield or installation with adequately shielded and grounded components, a UTP cabling system inherently has fewer potential trouble spots and is easier to install.

With more and more electronic equipment in the typical workspace, EMC becomes increasingly important. Excess radiation from one piece of equipment can adversely affect performance of other devices. In some countries, including the United States and Germany, EMC regulations have existed for years. However, the implementation of the European EMC Directive in 1989 has refocused attention on this problem. This directive, also known as 89/336/European Economic Community, states that all electronic equipment (that is, electronic systems) must comply with the directive, and that systems using this equipment must pass the essential requirements of the directive before they can be sold anywhere in the EEC.

How well do UTP- and STP-based systems stand up to rigorous EMC testing? Contrary to popular assumption, not all STP-based systems can automatically pass EMC tests, while a well-designed UTP cabling system can.

14.9.2 Coaxial Cable

Coaxial cabling has a single copper conductor at its center (see Figure 14.21). A plastic layer provides insulation between the center conductor and a braided metal shield (see Figure 14.21). The braided metal shield helps block any outside interference from fluorescent lights, motors, and other computers from getting to the signal line in the center.

Although coaxial cabling is difficult to install, it is highly resistant to signal interference. In addition, it can support greater cable lengths between network devices than twisted pair cable. The two types of coaxial cabling are thick coaxial and thin coaxial.

Thin coaxial cable is also referred to as thinnet. 10Base2 refers to the specifications for thin coaxial cable carrying Ethernet signals. The 2 refers to the approximate maximum segment length being 200 meters. In actual fact the maximum segment length is 185 meters. Thin coaxial cable has been popular in low-cost/low-performance networks.

Thick coaxial cable is also referred to as thicknet. 10Base5 refers to the specifications for thick coaxial cable carrying Ethernet signals. The 5 refers to the maximum segment length being 500 meters. Thick coaxial cable has an extra

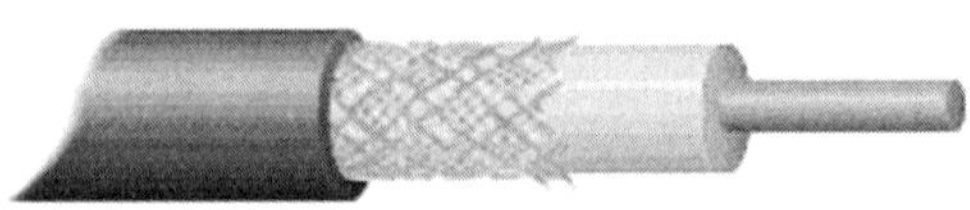

Figure 14.21 Coaxial cable.

protective plastic cover that helps keep moisture away from the center conductor. This makes thick coaxial a great choice when running longer lengths. One disadvantage of thick coaxial is that it does not bend easily and is difficult to install.

14.9.2.1 Coaxial Cable Connectors

The most common type of connector used with coaxial cables is the Bayone-Neill-Concelman (BNC) connector (see Figure 14.22). Different types of adapters are available for BNC connectors, including T-connectors, barrel connectors, and terminators. Connectors on the cable are the weakest points in any network. To help avoid problems, always use the BNC connectors that crimp, rather than screw, onto the cable.

14.9.3 Fiber-Optic Cable

Fiber-optic cabling consists of a center glass core surrounded by several layers of protective materials (see Figure 14.23). It transmits light rather than EM signals eliminating the problem of electrical interference. EMI is not an issue with fiber optic cables, therefore nor is EMC. This makes it ideal for environments that contain large amounts of electrical interference.

Fiber optic cable has the ability to transmit signals over much longer distances than coaxial and twisted pair, although relays may be required. It also has the capability to carry information at vastly greater speeds. This capacity broadens communication possibilities to include services such as video conferencing and interactive services. The cost of fiber optic cabling is comparable to copper cabling; however, it is more difficult to install and modify. 10BaseF refers to the specifications for fiber optic cable carrying Ethernet signals.

The center core of fiber cables is made from glass or plastic fibers (see Figure 14.23). A plastic coating then cushions the fiber center, and Kevlar fibers help to strengthen the cables and prevent breakage. The outer insulating jacket is made of Teflon or PVC.

There are two common types of fiber cables—single-mode and multimode. Multimode cable has a larger diameter; however, both cables provide high

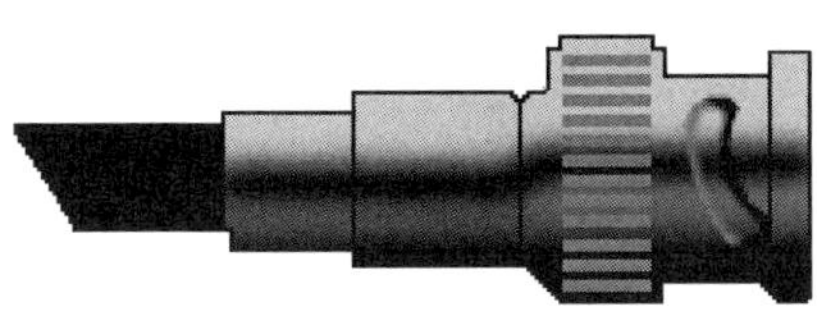

Figure 14.22 BNC connector.

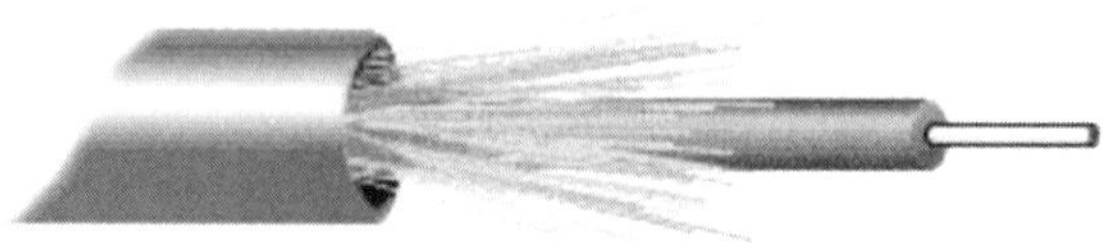

Figure 14.23 Fiber-optic cable.

bandwidth at high speeds. Single mode can provide more distance, but it is more expensive.

14.10 EMI/EMC Testing

14.10.1 Introduction

As we have indicated, EMC is the ability of an electronic device to exist in an EM environment without causing interference to or being interfered with by other electronic devices within that environment. EMC testing is typically broken down into two categories:

- *Emissions:* EM disturbances emanating from an electronic device that may cause interference/malfunction to another electronic device in the same environment. Also known as EMI.

- *Immunity/Susceptibility:* Immunity is the ability of an electronic device to function normally in an EM environment without experiencing interference/malfunction due to the emissions emanating from another electronic device. Susceptibility is basically the opposite of immunity, in that the less a device is immune to electromagnetic interference, the more susceptible it is. Typically, immunity testing is *not* required for consumer/commercial type products intended for sale/distribution in Australia, New Zealand, and North America.

The taxonomy of EMC testing is shown in Figure 14.24.

Note that here we show the testing being performed in an anechoic chamber.[12] While that is quite normal, open-air testing is also used. In open-air testing, control of spurious emissions is much more difficult, however. Measurement of the spectrum to detect such signals is required in order to take them into consideration.

[12] The anechoic chamber is an electromagnetically closed region that has RF absorbing material on all walls, floor, and ceiling. The metal enclosure is electromagnetically sealed that precludes external signals from entering (or exiting). The RF absorption on the walls, floor, and ceiling prevents signals from reflecting.

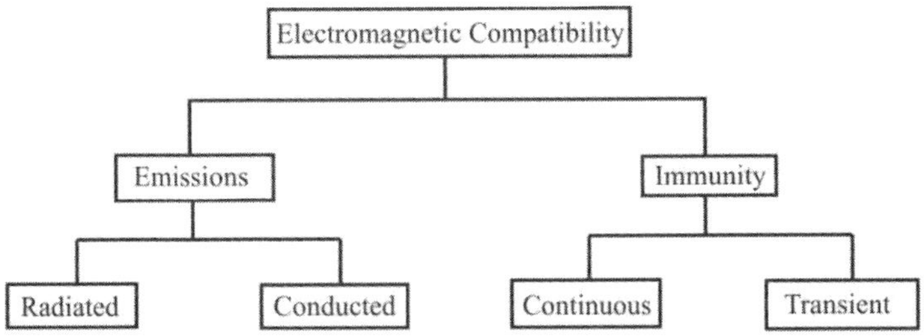

Figure 14.24 Electromagnetic compatibility testing taxonomy.

14.10.2 EM Compatibility Emissions

EMC emissions are further subdivided into two categories:

- Radiated emissions;
- Conducted emissions.

An EM field consists of two components:

- Electric field (E-Field): Usually measured in volts per meter (V/m);
- Magnetic Field (H-Field): Usually measured in amps per meter (A/m).

These two components of an EM field are in themselves two separate fields but not totally separate phenomena. E-fields and H-fields move orthogonally to each other (see Figure 14.25) [4].

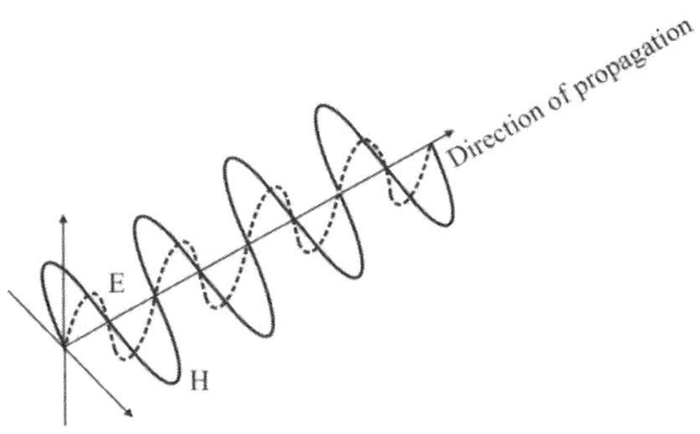

Figure 14.25 E and H propagation. (Source: [4]. ©Artech 2010. Reprinted with permission.)

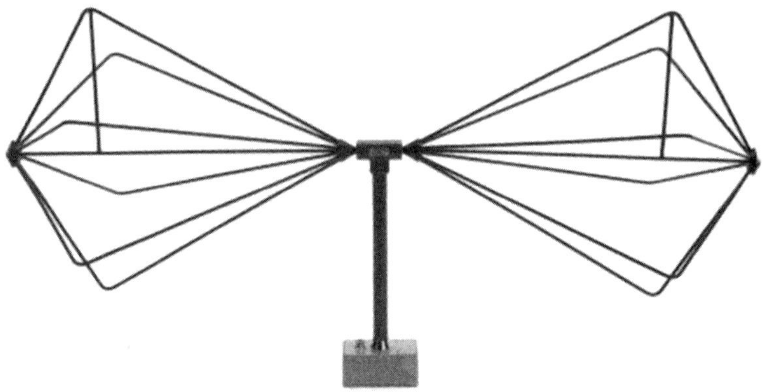

Figure 14.26 EMC test antenna.

Many of the EMC tests require an antenna either for broadcast or receiving the emitted radiation. Figure 14.26 shows a typical antenna used for this purpose. Of course, the antenna must be carefully calibrated so that the antenna characteristics can be removed from consideration when analysing the test results.

The taxonomy of EMC emissions is shown in Figure 14.27. The radiated waves that are examined are the two field components shown in Figure 14.25. Analysis of the conducted emissions is examined depending on whether the signals are continuous or transient.

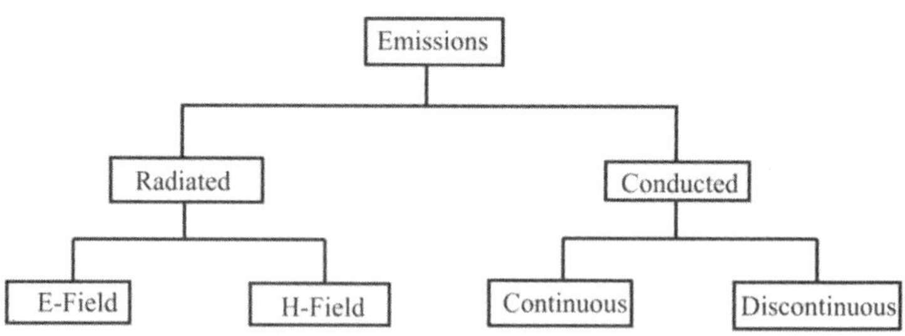

Figure 14.27 Emissions testing.

14.10.3 Radiated Emissions: E-Field

Radiated emissions are EMI that originate from signals generated internally by an electronic or electrical device. Radiated emissions can present challenging compliance issues.

During EMC testing, radiated emissions measurements are made using a spectrum analyser and/or an EMI receiver and a suitable measuring antenna as shown in Figure 14.28. The EUT is placed, under power, in an anechoic chamber.

14.10.4 Radiated Emissions: H-Field

The magnetic component of the EM wave is determined by using a spectrum analyzer and/or an EMI receiver and a suitable measuring antenna. Typical magnetic field antennas include loop antennas as well as specific antennas as per CISPR 15 such as the Van Veen loop. The Van Veen loop antenna is essentially three loop antennae constructed together that measures the magnetic field emissions of a EUT in three axes (x, y and z) (see Figure 14.29).

14.10.5 Conducted Emissions: Both Continuous and Discontinuous

Conducted emissions are EMI, or disturbances, that originate from signals generated internally to the EUT as well. These emissions are then propagated along interconnected cables such as wired ports, telecommunication ports, or power conductors. These emissions can be either continuous (continuously emits at given frequencies) or discontinuous in nature (nonconstant, occurring sporadically). During EMC testing, conducted emissions measurements are made with an EMI receiver via an *impedance stabilization network* (ISN) located within the test chamber (see Figure 14.30).

14.10.5.1 Discontinuous Disturbances/Interference

Discontinuous disturbances are typically caused by thermostatically controlled appliances such as cooker hot plates, refrigerators, or automatically controlled heaters, as well as automatic program-controlled machines such as washing machines. These may include the switching on and off of pumps, motors, and other electrically controlled appliances. All of these represent an event-driven or nonconstant timed switching event of energy.

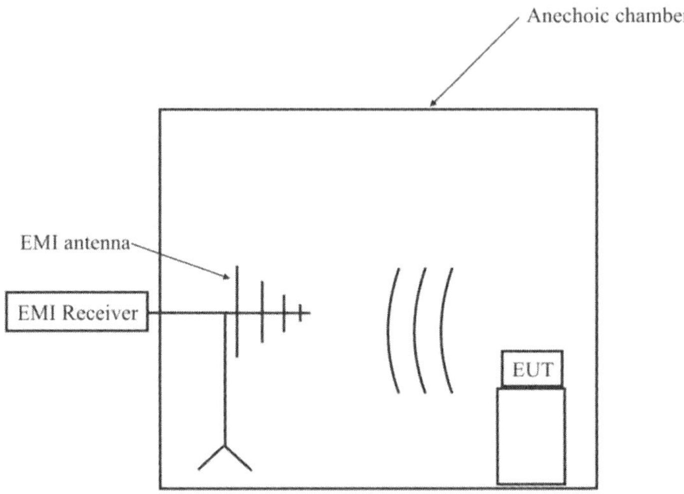

Figure 14.28 Radiated emissions E-field.

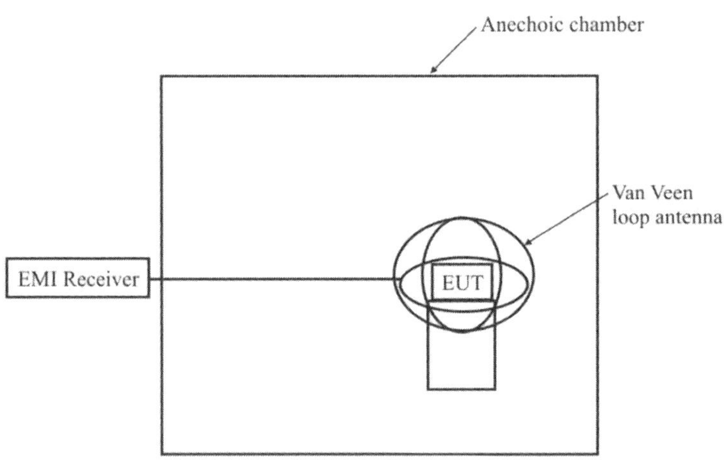

Figure 14.29 Radiated emissions H-field.

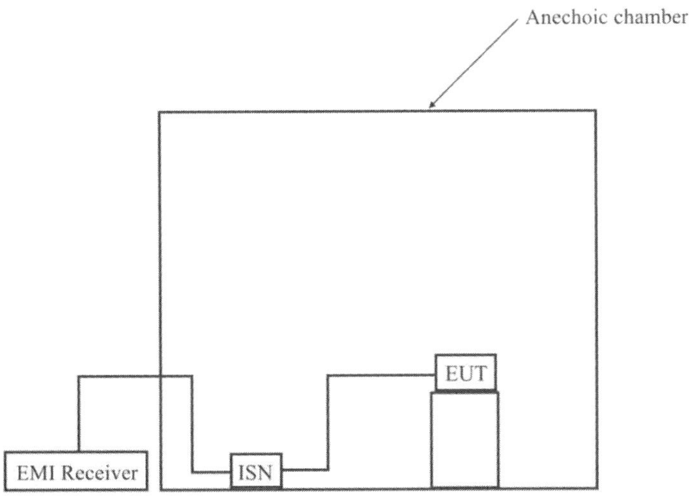

Figure 14.30 Continuous and discontinuous conducted emissions.

14.10.5.2 Continuous Disturbances/Interference

When an electronic device has a source of interference that continuously emits at given frequencies, the interference is called continuous disturbances. Continuous disturbances can be caused by a number of things but typical sources include:

- Switch-mode power supplies;
- Local oscillators;
- Network data traffic (such as Ethernet traffic);
- Electrical motors.

In other words, any constant timed switching event of energy.

Both kinds of conducted emissions/disturbances/interference can be mitigated using similar techniques (primarily filtering); however, it is best to identify whether the emissions are common mode (CM) or differential mode (DM) in nature, as described in Section 14.7, so that the appropriate filter can be employed (see Figure 14.13).

14.10.6 Electromagnetic Compatibility Immunity

EMC immunity testing is either continuous or transient in nature (see Figure 14.31). The EUT is examined for its susceptibility to disruption from EMI from other

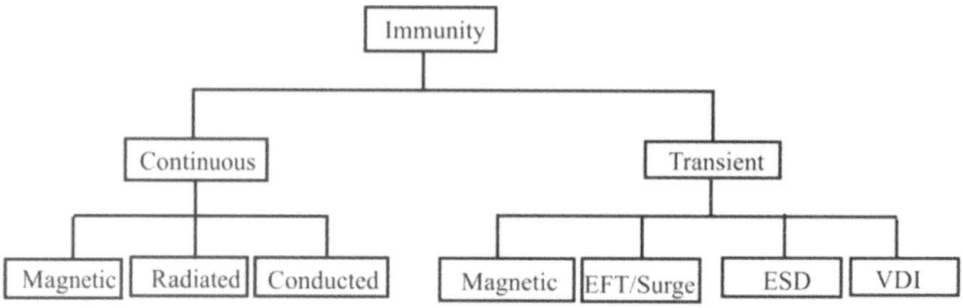

Figure 14.31 Immunity testing taxonomy.

electronic devices in the close environment. Continuous testing is applied to a product to simulate RF proximity that may occur in the real world. Transient phenomena are typically short events that involve bursts of energy.

EMC immunity testing requirements are often split based on how EMI might couple into a device:

- Immunity, enclosure port;
- Immunity, signal ports and telecommunications ports;
- Immunity, input DC power port;
- Immunity, input AC power port.

The test levels, types of interfering signals, and so forth are dependent on the type of device being tested and the standard being applied.

14.10.6.1 Continuous Immunity Testing

Radiated Immunity

RF signal generators, amplifiers, and antennas are used to produce an EM field at varying frequencies (see Figure 14.32). The enclosure port of the EUT and associated cables are exposed to the EM field via a radiating antenna. The radiated test signal is at a specific amplitude and modulation applied for a specific time period. Most standards requiring immunity testing require this testing to be performed.

Conducted Immunity

During conducted immunity testing, an EM field is generated by an RF signal generator and amplifier (see Figure 14.33). This is coupled to the EUT's signal, data, or power port via an injection device commonly a coupling/decoupling network (CDN) is used as an injection device. This kind of conducted immunity testing is continuous in nature and is called radio-frequency continuous conducted in many

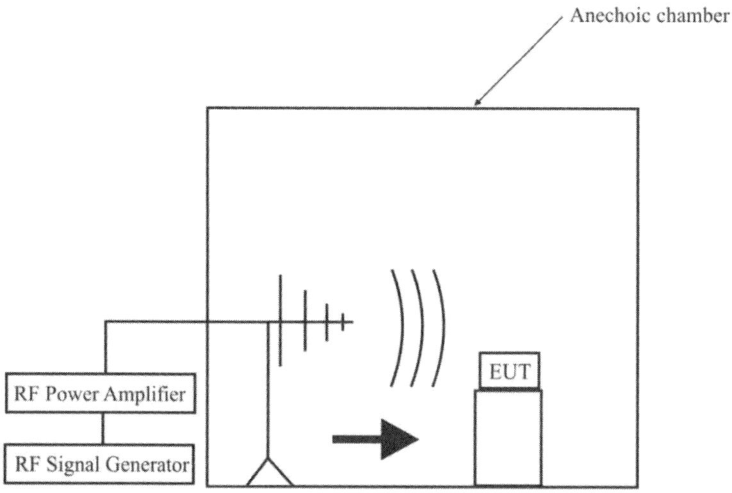

Figure 14.32 Radiated immunity.

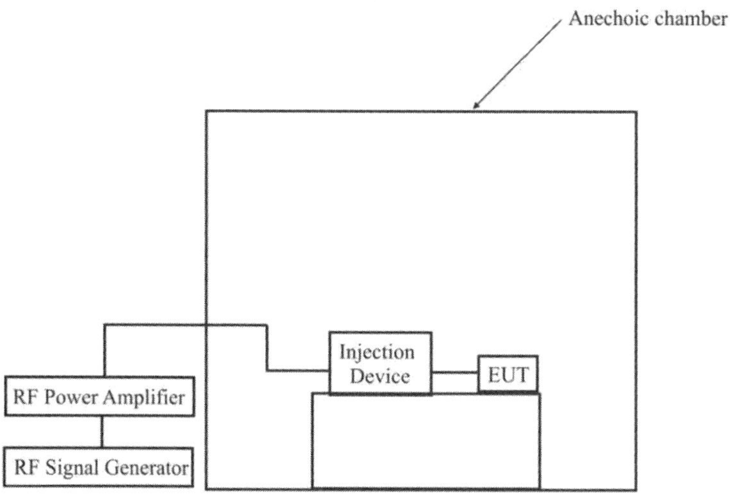

Figure 14.33 Conducted immunity.

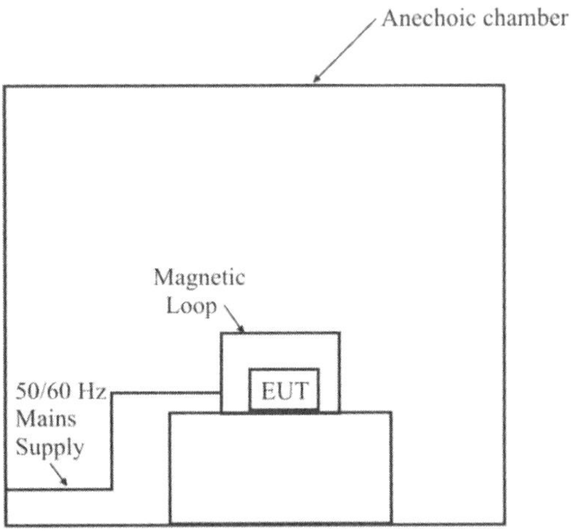

Figure 14.34 Magnetic field immunity testing.

standards. Typically conducted immunity testing is applicable to AC and DC ports and signal cables longer than 3m in length.

Power-Frequency Magnetic Field Immunity

A fluctuating magnetic field is produced by a magnetic coil which oscillates at the line power frequency (50/60 Hz). The EUT is placed inside this fluctuating magnetic field and exposed for enough time as to evaluate the performance of the product (see Figure 14.34). Magnetic field immunity testing is usually only required for magnetically susceptible devices such as tape recorders, computer disk memories, and the like.

14.10.6.2 Transient Immunity Testing

Transient phenomena are short bursts of energy that a EUT will be exposed to for a very short amount of time. Like continuous immunity, transient immunity is applied to a product's enclosure port, signal/data ports, and power ports where applicable (see Figure 14.35).

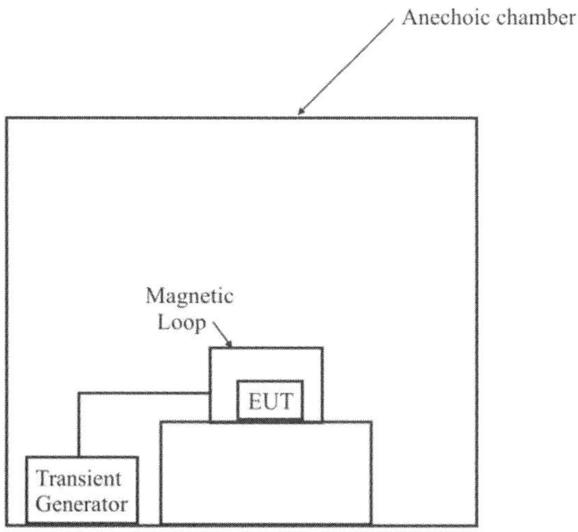

Figure 14.35 Transient immunity testing.

Electro-static Discharge

Electro-static discharge (ESD) pulses are applied directly to the enclosure of a device and indirectly to vertical/horizontal coupling planes in close proximity to the EUT at test levels specific to the standard being applied.

Electrical Fast Transient/Burst

Electrical fast transients (EFTs) are a series of short pulses that are high in amplitude and repetition frequency with a short rise time. Fast transient phenomena are most often caused by high-speed switching events such as interruption of inductive loads and relay contact bounce, and so forth. Typically, fast transient testing is applicable to AC and DC ports and signal cables longer than 3m.

Surges

Surges are a type of transient phenomena produced by high-powered switching events such as magnetic/inductive coupling and lightning. Surge testing on the line port of an EUT is applied at several phase angles of the line supply. Typically surge testing is applicable to AC ports (and sometimes also DC ports) and in some EMC product standards signal cables longer than 30m in length or if the cable is run outside of a building.

Voltage Dips, Short Interruptions, VDI, and Voltage Variations

Faults in the power network are simulated with voltage dips and short interruption (VDI) testing. These faults may be caused by power outages (blackout/brownout events) or by sudden large changes of loads. Voltage variations are typically caused by continuously varying loads connected to the power network. A voltage dip or interruption is a two-dimensional phenomenon that is characterized by the residual voltage (line voltage after the specified dip) and duration (how long the dip in nominal voltage is applied to the product). This test is only applicable to AC input ports of products.

Pulsed Magnetic Field

Like the power frequency magnetic field immunity testing, the EUT is placed inside a magnetic loop. Unlike power frequency magnetic field testing, rather than exposing the EUT to a continuously fluctuating magnetic field (oscillating at 50/60 Hz) the EUT is exposed to a magnetic field pulse provided by a transient generator. The magnetic pulse is high in amplitude, but has a short rise time, after which the performance of the product is evaluated to ensure normal operation.

14.11 Concluding Remarks

The pertinent aspects of EMI/EMC concepts as they apply to EW systems were presented in this chapter. EMC means that a product does not cause excessive interference to other devices in its environment and that other equipment does not cause excessive performance problems for the product. This includes equipment for the military as well as the commercial/industrial settings.

EW systems, in particular, must have exceptional EMC performance due to the nature of their function. Usually, EW systems contain very sensitive receiving equipment that, if excessive interference were present due to EMI/EMC problems, would preclude them from functioning.

We concluded the chapter with a discussion of relevant EMC tests that apply to EW systems All electronic or electrical products sold virtually anywhere in the world must be tested for EMC compliance before they can be marketed or sold. Proof of acceptable performance levels that must be achieved by international agreement is required.

References

[1] http://www.astron1.com/Press%20Releases/ELPAC_EMI_White_Paper.pdf, accessed Jun. 2015.

[2] http://electronicdesign. com/energy/ understanding-emi-noise-power-system-design, accessed Aug. 2012.
[3] Mil Std 461F, 12-2007.
[4] Poisel, R. A., *Introduction to Communication Electronic Warfare Systems*, 2nd Ed, Norwood, MA: Artech House, p. 54, 2010.

Acronyms

2DEG 2-Dimensional Electron Gas

AC Alternating Current
ACPR Adjacent Channel Power Ratio
ADC Analog to Digital Converter
AGC Automatic Gain Control
AlGaN Aluminum Gallium Arsenide
AM Amplitude Modulation
ASK Amplitude Shift Key
AWGN Additive White Gaussian Noise

BASK Binary Amplitude Shift Key
BeO Beryllium Oxide
BER Bit Error Rate
BFSK Binary Frequency Shift Key
BJT Bipolar Junction Transistor

C/I Carrier to Intermodulation Ratio
C2 Command and Control
CA Course Acquisition
CB Citizens Band (Frequency)
CB Common Base (BJT Configuration)
CC Common Collector
CCM Continuous Conduction Mode
CD Common Drain

CDMA Code Division Multiple Access
CE Common Emitter
CFD Computational Fluid Dynamics
CG Common Gate
CMOS Complementary Metal Oxide Semiconductor
COP Coefficient of Performance
CPW Coplanar Waveguide
CS Common Source
CTE Coefficient of Thermal Expansion
CTR Current Transfer Ratio
CW Continuous Wave

DA Distributed Amplifier
DAC Digital to Analog Converter

DC Direct Current

DCM Discontinuous Conduction Mode

DDS Direct Digital Synthesis
DE Directed Energy
DSB Double Sideband
DSB-SC Double Sideband Suppressed Carrier
DSSS Direct Sequence Spread Spectrum

EA Electronic Attack
ED Electrodeposited
EER Envelope Elimination and Restoration
EHF Extremely High Frequency
EM Electromagnetic
EMC Electromagnetic Compatibility
EMCOM Emissions Control
EMI Electromagnetic Interference
EMS Electromagnetic Spectrum
EO Electrooptical
EOB Electronic Order of Battle
EP Electronic Protect
ES Electronic Support
ESL Equivalent Series Inductance
ESR Equivalent Series Resistance
EVM Error Vector Magnitude
EW Electronic Warfare
EWCC EW Command Center

FB Ferrite Bead
FBA Feedback Amplifier
FET Field Effect Transistor
FFA Feed Forward Amplifier
FHSS Frequency Hopped Spread Spectrum
FM Frequency Modulation
FSK Frequency Shift Key

GaAs Gallium Arsenide
GaN Gallium Nitride

GHz	Gigahertz
GPS	Global Positioning System
HBT	Heterojunction Bipolar Junction
HCA	Harmonic Controlled Amplifier
HEMT	High Electron Mobility Transistor
HEMT	High Electron Mobility Transistor
HF	High Frequency
HPA	High Power Amplifier
Hz	Hertz
IC	Integrated Circuit
IF	Intermediate Frequency
IGBT	Insulated Gate Bipolar Transistor
IIP_2	Input 2nd Order Intercept Point
IIP_3	Input 3rd Order Intercept Point
IMD	Intermodulation Distortion
IMP	Intermodulation Product
InGaAs	Indium Gallium Arsenide
InP	Indium Phosphate
IO	Information Operations
IP_2	2nd Order Intercept Point
IP_3	3rd Order Intercept Point
IR	Infrared
ISR	Intelligence, Surveillance, and Reconnaissance
IW	Information Warfare
JFET	Junction FET
JFET	Junction Field Effect Transistor
JP	Joint Publication
JSR	Jam to Signal Ratio
KIL	Kirchoff Current Law
KVL	Kirchoff Voltage Law
LDMOS	Laterally Diffused MOSFET
LF	Low Frequency
LINC	Linear Amplification by Nonlinear Components
LNA	Low Noise Amplifier
LO	Local Oscillator
LO	Local Oscillator
LOS	Line of Sight
LSB	Least Significant Bit
LSB	Lower Sideband
LUT	Look Up Table
MASK	M-ary Amplitude Shift Key
MBE	Molecular Beam Epitaxy
MEMS	Microelectromechanical Systems
MESFET	Metal Semiconductor Field Effect Transistor
MF	Medium Frequency
MHz	Megahertz
MIM	Metal Insulator Metal
MIS	Metal Insulator Semiconductor
MMIC	Monolithic Microwave Integrated Circuit
MOCVD	Metal-Organic Chemical Vapor Deposition
MODFET	Modulation-Doped FET
MOSFET	Metal Oxide Semiconductor FET
MOT	Maximum Operating Temperature
MSB	Most Significant Bit
MSK	Minimum Shift Key
NATO	North Atlantic Treaty Organization
NBFM	Narrowband FM
NPR	Noise Power Ratio
NVIS	Near Vertical Incidence Skywave
OCTC	Open Circuit Time Constant
OEM	Original Equipment Manufacturer
OIP_2	Output 2nd Order Intercept Point
OIP_3	Output 3rd Order Intercept Point
OOK	On Off Key
PA	Power Amplifier
PAE	Power Added Efficiency
PAE	Power Added Efficiency
PAM	Pulse Amplitude Modulation
PCB	Printed Circuit Board
PDM	Pulse Density Modulation
PF	Power Factor
pHEMT	Pseudomorphic HEMT
PLL	Phase Locked Loop
PM	Phase Modulation
psd	Power Spectral Density
PSK	Phase Shift Key
PSTN	Public Switched Telephone Network
PTFE	Polytetraflouroethylene Teflon
PTH	Plated Through Holes
PTT	Push to Talk
PWM	Pulse Width Modulation
QAM	Quadrature Amplitude
QPSK	Quaternary Phase Shift Key
RAM	Random Access Memory
RF	Radio Frequency

RFC	RF Choke
RFI	Radio Frequency Interference
RHPZ	Right Half Plane Zero
RMS	Root Mean Square
ROM	Read Only Memory
SAW	Surface Acoustic Wave
SCR	Silicon Controlled Rectifier
SCTC	Short Circuit Time Constant
SHF	Super High Frequency
Si	Silicon
SiC	Silicon Carbide
SiGe	Silicon Germanium
SMPS	Switched Mode Power Supply
SNR	Signal to Noise Ratio
SoC	System on a Chip
SOI	Signal of Interest
SRF	Self Resonant Frequency
SSB	Single Sideband
TEC	Thermoelectric Cooling
TTP	Tactics, Techniques, and Procedures
TWT	Traveling Wave Tube
TXCO	Temperature Compensated Crystal Oscillator
UHF	Ultra High Frequency
USB	Upper Sideband
VAC	Volts Alternating Current
VCO	Voltage Controlled Oscillator
VDC	Volts Direct Current
VDMOSFET	Vertically Diffused Metal Oxide Semiconductor FET
VFO	Variable Frequency Oscillator
VLF	Very Low Frequency
WLAN	Wideband Local Area Network
ZCS	Zero Current Switching
ZVS	Zero Voltage Switching

About the Author

Richard A. Poisel received a B.S. in electrical engineering from the Milwaukee School of Engineering in 1969 and an M.S. in the same discipline from Purdue University in 1971. He spent three years in the military service from 1971 to 1973. After his service he attended the University of Wisconsin, where he received a Ph.D. in electrical and computer engineering in 1977. From 1977 to 2004, he was with the same government organization, which has had several different names and is currently known as the U.S. Army Research, Development, and Engineering Command, Intelligence and Information Warfare Laboratory. During the 1993–1994 academic year, Dr. Poisel attended the MIT Sloan School of Management as a Sloan Fellow, receiving an M.B.A. Initially a research engineer, Dr. Poisel eventually rose to the role of the director of the laboratory on an acting basis from 1997 to 1999. He was appointed chief scientist in 1999 and was relocated to the Army's Intelligence Center at Ft. Huachuca, Arizona, where he served as a technical advisor to the command group. Retiring from government service in 2004, he served as a Senior Engineering Fellow at Raytheon Missile Systems in Tucson, Arizona, from 2004 to 2011. He is currently a consultant on engineering for electronic warfare applications.

Dr. Poisel holds several patents in the communication Electronic Warfare subject area and is the author of the following, all published by Artech House: *Introduction to Communications Electronic Warfare Systems*, 2nd ed., 2008; *Modern Communications Jamming Principles and Techniques*, 2nd ed., 2011; *Target Acquisition in Communications Electronic Warfare Systems*, 2004; *Foundations of Communications Electronic Warfare*, 2008; and *Electronic Warfare Target Location Methods*, 2005, *Information Warfare and Electronic Warfare Systems*, 2013, *Antenna Systems and Electronic Warfare Applications*, 2012, and *Electronic Warfare Receivers and Receiving Systems*, 2014.

Index

The Artech House Electronic Warfare Library

Dr. Joseph R. Guerci, Series Editor

Activity-Based Intelligence: Principles and Applications, Patrick Biltgen and Stephen Ryan

Advances in Statistical Multisource-Multitarget Information Fusion, Ronald P. S. Mahler

Antenna Systems and Electronic Warfare Applications, Richard A. Poisel

Electronic Intelligence: The Analysis of Radar Signals, Second Edition, Richard G. Wiley

Electronic Warfare for the Digitized Battlefield, Michael R. Frater and Michael Ryan

Electronic Warfare in the Information Age, D. Curtis Schleher

Electronic Warfare Receivers and Receiving Systems, Richard A. Poisel

Electronic Warfare Signal Processing, James Genova

Electronic Warfare Target Location Methods, Richard A. Poisel

EW 101: A First Course in Electronic Warfare, David L. Adamy

EW 103: Tactical Battlefield Communications Electronic Warfare, David L. Adamy

EW 104: EW Against a New Generation of Threats, David L. Adamy

Foundations of Communications Electronic Warfare, Richard A. Poisel

High-Level Data Fusion, Subrata Das

Human-Centered Information Fusion, David L. Hall and John M. Jordan

Information Warfare and Organizational Decision-Making, Alexander Kott, editor

Information Warfare and Electronic Warfare Systems, Richard A. Poisel

Information Warfare Principles and Operations, Edward Waltz

Introduction to Communication Electronic Warfare Systems, Richard A. Poisel

Introduction to Electronic Defense Systems, 3rd Edition, Filippo Neri

Introduction to Modern EW Systems, 2nd Edition, Andrea De Martino

Knowledge Management in the Intelligence Enterprise, Edward Waltz

Mathematical Techniques in Multisensor Data Fusion, Second Edition, David L. Hall and Sonya A. H. McMullen

Military Communications in the Future Battlefield, Marko Suojanen

Modern Communications Jamming Principles and Techniques, Richard A. Poisel

Practical ESM Analysis, Sue Robertson

Principles of Data Fusion Automation, Richard T. Antony

RF Electronics for Electronic Warfare, Richard A. Poisel

Stratagem: Deception and Surprise in War, Barton Whaley

Statistical Multisource-Multitarget Information Fusion, Ronald P. S. Mahler

Tactical Communications for the Digitized Battlefield, Michael Ryan and Michael R. Frater

Target Acquisition in Communication Electronic Warfare Systems, Richard A. Poisel

For further information on these and other Artech House titles, including previously considered out-of-print books now available through our In-Print-Forever® (IPF®) program, contact:

Artech House
685 Canton Street
Norwood, MA 02062
Phone: 781-769-9750
Fax: 781-769-6334
e-mail: artech@artechhouse.com

Artech House
16 Sussex Street
London SW1V 4RW UK
Phone: +44 (0)20-7596-8750
Fax: +44 (0)20-7630-0166
e-mail: artech-uk@artechhouse.com

Find us on the World Wide Web at: www.artechhouse.com